Pigments in Plants

Pigments in Plants

Edited by
Franz-Christian Czygan

With Contributions by
W. Barz · B. H. Davies · C. H. Eugster · H. Falk
H. Grisebach · A. Hager · R. B. Kufner · E. Leistner
B. Liedvogel · H. Mohr · W. Rau · H. Reznik
W. Rüdiger · H. Schneider · P. Sitte · W. Steglich
T. Swain · A. Weber · M. Wettern

139 Figures and 29 Tables

Akademie-Verlag · Berlin

1981

Lizenzausgabe
des Gustav Fischer Verlages Stuttgart · New York
Die 1. Auflage erschien im Gustav Fischer Verlag, Stuttgart 1975
Vertrieb für alle Staaten mit Ausnahme der sozialistischen Länder:
Gustav Fischer Verlag Stuttgart · New York
Vertrieb für die sozialistischen Länder:
Akademie-Verlag · Berlin

Erschienen im Akademie-Verlag, DDR-1080 Berlin, Leipziger Straße 3-4

Lizenznummer: 202 · 100/530/80
Gesamtherstellung: VEB Druckhaus Köthen, DDR-4370 Köthen
Bestellnummer: 762 807 3 (6575) · LSV 1314, 1354
Printed in the German Democratic Republic
DDR 52,- M

Vorwort zur 2. Auflage

Die 1. Auflage des Buches «Farbstoffe in Pflanzen», das seinerzeit die Referate der Sektion «Pflanzenfarbstoffe» der Tagung der Deutschen Botanischen Gesellschaft in Würzburg 1974 zusammenfaßte, fand bei Biologen des In- und Auslandes in kurzer Zeit großes Interesse. So wurde ich 1976 vom Gustav Fischer Verlag gebeten, eine neue Auflage vorzubereiten.

Bewußt habe ich das bewährte Konzept der 1. Auflage übernommen. Wieder werden keine erschöpfenden Übersichten geboten, sondern Forschungsschwerpunkte auf den Gebieten der Physiologie, Biochemie und Chemie pflanzlicher Farbstoffe betont. Allerdings wurde die Neuausgabe nicht nur in ihren einzelnen Abschnitten aktualisiert und erweitert. Auch die Zahl der Referate ist größer geworden. Damit habe ich mir wichtig erscheinende Akzente gesetzt und so sicherlich eine subjektive Auswahl getroffen. Eines der Vorrechte des Herausgebers! Zusätzlich werden Fragen nach Farbstoffen als taxonomisch-systematische Marker gestellt, Fragen zur Chromoplastenbiologie werden hier erstmals in dieser Intensität diskutiert und an einem Beispiel werden Vorkommen sehr unterschiedlicher Farbstoffstrukturen in einer Objektgruppe, den höheren Pilzen, erörtert. Zusammen mit Überlegungen zum Problem «Farbstoffe und Licht» und zu dem nach wie vor ungelösten Rätsel des Abbaus des für die Menschen wichtigsten Farbstoffs, des Chlorophylls, ist der im Vorwort zur 1. Auflage gebundene Blumenstrauß noch bunter geworden.

Ich präsentiere dieses Bouquet in Englisch, zugegeben mehr *nolens* als *volens*. Ich akzeptiere das Englische als *lingua franca* und verkenne nicht deren verbindende Wirkung im Wissenschaftsbereich – und sei es auch nur in Form eines «*Scientific Pidgin English*». Dennoch bleibt auch hier mit dem Verzicht auf die deutsche Sprache eine Möglichkeit, ihr den ihr gebührenden Platz in der wissenschaftlichen Literatur einzuräumen, ungenutzt.

Die Neuauflage der «Farbstoffe in Pflanzen» erscheint in einem Jahr, das für die Behandlung dieser biologischen Fragen ein Jubiläum bedeutet: Vor 75 Jahren publizierte *M. Twsett* seine berühmte Abhandlung über die Auftrennung von Blattfarbstoffen mit Hilfe der Chromatographie (Trav. de la Soc. des Nat. de Varsovie, XIV année, 1903; vgl. auch: Physikalisch-chemische Studien über das Chlorophyll. Die Adsorptionen. Ber. Dtsch. Bot. Ges. **24**, 316, 1906). Erst diese adsorptions-chromatographischen Verfahren ermöglichten die vielen, heute bereits nicht mehr überschaubaren Versuche und Daten zur Chemie und Biologie pflanzlicher Farbstoffe. An den Geburtstag einer, wie ich meine, wichtigen, aber wenig bekannten Publikation sollte nach 70 Jahren ebenfalls erinnert werden, an *E. Stahl*: «Zur Biologie des Chlorophylls. – Laubfarbe und Himmelslicht – Vergilbung und Etiolement» aus dem Gustav Fischer Verlag in Jena (1909). Hier wird in hervorragender Weise nicht nur der damalige Stand des Wissens um das Chlorophyll und seine Funktion referiert, sondern kritisch erörtert, in Frage gestellt und durch neue Theorien erklärt. Die Arbeiten von *M. Twsett* und *E. Stahl* sollten in unserer schnellebigen, viel zu wenig geschichtsbewußten Zeit nicht vergessen werden.

Vergessen möchte ich auch nicht meinen Dank an die Referenten dieser 2. Auflage für ihre Geduld bei der Zusammenarbeit. Meine Entschuldigung gilt vor allen den Kollegen, die sich an die frühzeitigen Ablieferungstermine der Manuskripte hielten und dann besonders lange auf Druckfahnen und Abschlußkorrekturen warten mußten. Ich danke meinen Mitarbeitern für ihre Hilfe verschiedenster Art: besonders Frau Almuth Krüger für das Korrekturlesen und Zeichnen vieler Formeln und meiner Sekretärin Frau Inge Kropp für viele Schreibarbeiten. Ganz besonders fühle ich mich dem Verlag zu Dank verbunden Seine Bereitwilligkeit, auf meine vielen Vorschläge und Wünsche einzugehen, hat die Herausgabe dieses Buches zu einer sehr angenehmen Tätigkeit werden lassen.

Würzburg, zum Jahreswechsel 1978/79

Franz-C. Czygan

Vorwort zur 1. Auflage

Im Rahmen der Tagung der Deutschen Botanischen Gesellschaft in Würzburg vom 23. bis 29. September 1974 beschäftigte sich eine Sektion mit «Farbstoffen in Pflanzen». Es war den Veranstaltern gelungen, führende Wissenschaftler, die sich mit den verschiedensten Farbstoffgruppen befassen, zu Übersichtsreferaten einzuladen. Einem Farbbouquet gleich wurde in den Vorträgen ein Überblick und ein Einblick in die verschiedensten Aspekte der Farbstoff-Forschung gegeben. Der Bogen – ein Poet würde hier sicherlich von einem Regenbogen sprechen – spannte sich von den Flavonoiden (Grisebach, Barz), die heute besonders in ihren Biflavonylen und Flavolignanen auch medizinisch-therapeutische Bedeutung gefunden haben, und den N-haltigen und dem Aminosäuremetabolismus nahe stehenden Betalainen (Reznik) zu den auch physiologisch so überaus wichtigen Porphyrinderivaten (Chlorophylle: Schneider; Phycobiliproteide: Rüdiger) und den verschiedensten Aspekten der Biochemie, Biosynthese und Physiologie der Carotinoide (Davies, Hager, Rau). Eingeschlossen in diesen Kreis sind die weiteren Terpenoid-Farbstoffe, so die erst vor wenigen Jahren entdeckten Diterpen-Pigmente (Eugster), und die schon seit alters her und heute wieder hochaktuellen Vertreter chinoider Farbstoffe (Leistner). Daß diese Abgrenzungen nicht scharf sind und häufig enge Beziehungen zwischen diesen Gruppen bestehen (z.B. Chlorophylle–Diterpene; Terpene–Chinone u.a.m.), versteht sich von selbst und wird in den Referaten deutlich gemacht. Die Tatsache, daß es von der biochemischen Seite her keine zwingenden Gründe geben kann, die farblosen Verbindungen und Vorstufen gleicher Biogenese abzusondern, begründet ihre in vielen Referaten ebenfalls eingehende Erörterung.

Die einzelnen Referenten legten die Schwerpunkte bei «ihren» Pigmenten durchaus unterschiedlich. Dies war zum einen durch die Arbeitsrichtung des Vortragenden bedingt, zum anderen durch die Aktualität der zur Zeit in einzelnen Bereichen besonders untersuchten Teilgebiete. Aber gerade diese Vielfalt und unterschiedliche Betonung der einzelnen Aspekte innerhalb einer Farbstoffgruppe machte den Versuch verständlich, diese Vorträge der Botaniker-Tagung neben ihrer Publikation in den Berichten der Deutschen Botanischen Gesellschaft gesammelt als Buch herauszugeben.

Ermöglicht wurde die Herausgabe nur durch die redaktionelle Hilfe des Vorstandes der Deutschen Botanischen Gesellschaft. Besonders Herrn Professor Dr. H. Lorenzen gebührt mein Dank für vertrauensvolle Zusammenarbeit. Dankbar sei auch der finanziellen Unterstützung durch die Firmen Bionorica KG (Nürnberg), G. Heinemann (Schwäbisch Gmünd), E. Heinse KG (Würzburg), Kneipp-Heilmittel-Werk (Würzburg), W. Maurer GmbH (Saarbrücken) und Sebamat-Chemie GmbH (Bad Salzig) gedacht.

Würzburg, im Januar 1975

Franz-C. Czygan

Contents

Colors and Pigments

Introduction to the First Edition

Franz-Christian Czygan

Institut für Botanik und Pharmazeutische Biologie der Universität Würzburg

The blossoming grain
thickens in late autumn
to evening red
(Painting Collection from the Mustard
Seed Garden, Volume 3, 1701)

Man has been involved with colors and the substances which impart them, thd pigments, since ancient time; he has lived with them, utilized them, and profiteg from them. He has always attempted to alter and beautify his environment usind color. What is more, our forefathers and we ourselves have frequently identified our environment in terms of the concept of «color». A world without color woul not be a world worth living in «for we have life in the colored reflection of brilliance» (Goethe).

With full awareness of the psychic effects, man utilizes color and pigments today more than ever. However, simultaneously with the development to man, Homo sapiens attained an «active» relationship to color. As a seeing animal man has employed pigments of natural origin, i.e., from the realm of plants, animals, and minerals not only to alter his surroundings, but also to alter himself in his environment. Reference is made in this regard to the classical pigments in antiquity: Tyrian purple from the glands of the rock shell and murex, indigo from *Indigofera* species saffron from the stigma of the crocus, or madder dye from the alizarin. These are all pigments whose great value often made them worth their weight in gold or precious stones. Two examples should illustrate that biogenic pigments can also be of great economic importance today. β-carotene, now more economical to produce synthetically, plays an essential role in food chemistry as a yellow-coloring pigment and as pro-vitamin A. The other example is the famous cosmetic of the orient, the henna pigment from *Lawsonia* species, without which our modern hair lotions would not be possible.

Without question we find the psychic qualities of color and pigments to be the most important aspect of their effect. Daily life is not the only place where the psychology of color plays an important role. It has found its greatest power of expression in painting and poetry. The sensation of color is a sensation sui generis which is also closely linked to experiences of the natural world which always appears to us in colored form. The psychic effects of color were investigated as early as the time of Goethe in his treatise on color. In the following period the experience of color was often held to be purely a pleasant or unpleasant sensation, however. Stefanescu-Goanga (1912) was the first to depart from this subjectification of the sensation of color «in order to, based on the fact of the existence of individual consciousness, obtain analyses of the colors as pure qualities of sensation». The sensational response to color, which at the moment constitutes an actual area of research in psychology and even includes color tests for personality determination, is closely linked to the mode of appearance in which colors confront us in our environment, e.g., in colored organisms. Thus we differentiate colors which are based on pigments or dyes from those which are dependent on definite structures. Structural colors, created by the interference of light at thin layers, give us the varied colored impressions of animal scales and feathers.

Is there a reason for publishing a review of plant dyes or pigments? Are there, in addition to the quality which influences man psychically, other common elements of the biogenic pigments which differ substantially in their chemical and biochemical structure, particularly the pigments originating from plants? This question can certainly only be answered inadequately on the basis of our current knowledge. The most important common aspect of most biogenic pigments is unquestionably the effect which is due to their optical properties. Whereas the effect in the animal kingdom is that of warning, impressing, or attracting; in the realm of blossoms and fruits it is closely linked with propagation of the species. The properties common to all pigments already stop at this point, however. Although some groups of pigments definitely have important, or even life-important, physiologic functions in the organisms which produce them (e.g., chlorophyll), final proof of a metabolic role for other pigments is still lacking as in the case of the carotenoids which are discussed as light-blocking substances or photoacceptors. The extent to which these pigments with secondary metabolic properties involve *biochemical fossils* is still an open question. However, β-carotene, which is possibly «useless» for plants, actually has an important physiologic function as an effector in the metabolism of many animals and man.

Despite few common aspects and many differences in the biochemistry, biogenesis, and possible functions of these groups of compounds, they continually stimulate attempts to approach the problem of biogenic pigments from the biologic-biochemical standpoint. With full awareness of the fact that many aspects and many pigment groups have received too little or even no treatment, I submit this review with the same «courage to exclude» that every editor must have today when publishing a review. If, in addition to presenting the current state of pigment research, these papers also point out the vare spots in our knowledge of the biology and, above all, the possible function and significance of these classes of compounds in the organisms themselves, then the purpose of this volume will be fulfilled. That in addition the occupation with plant pigments from their biochemistry to their physiology and psychology gives satisfaction to the esthetes among their researchers, and which scientist today is not an esthete, may only be a personal wish of the editor.

Farben und Farbstoffe

Einführung zur 1. Auflage

Franz-Christian Czygan

Institut für Botanik und Pharmazeutische Biologie der Universität Würzburg

Die blühenden Ähren
verdichten sich im Spätherbst
zu abendlichem Rot
(Aus: Malereiensammlung aus dem
Senfkorngarten, 3. Band, 1701)

Seit uralten Zeiten hat sich der Mensch mit Farben und ihren Trägern, den Farbstoffen, beschäftigt, mit ihnen gelebt, sie gebraucht und genutzt. Schon immer hat er versucht, seine Umwelt durch Farben zu verändern, zu verschönern. Darüber hinaus haben unsere Vorväter und wir häufig unsere Umwelt mit dem Begriff «Farbe» identifiziert. Eine Welt ohne Farbe wäre keine belebenswerte Welt, «denn am farbigen Abglanz haben wir das Leben» (Goethe).

Ihrer psychischen Wirkung voll bewußt bedient man sich der Farbe und der Farbstoffe heute mehr denn je. Aber gleichzeitig mit der Menschwerdung, mit der Ausprägung des Sich-seiner-bewußt-seins hat der Homo sapiens ein «aktives» Verhältnis zu den Farben bekommen. Das Augentier Mensch hat mit Farbstoffen natürlicher Herkunft aus dem Bereich der Pflanzen, Tiere und Mineralien nicht nur seine Umgebung, sondern auch sich selbst seiner Umwelt gegenüber verändert. So sei an die im Altertum klassischen Farbstoffe Purpur aus Drüsen der Purpur- und Stachelschnecken, Indigo aus *Indigofera*-Arten, Safran aus den Narbenschenkeln des Crocus oder an den Krapp-Farbstoff der Färberröte erinnert; Pigmente, die ihrer Kostbarkeit wegen teilweise in Gold und Edelsteinen aufgewogen wurden. Daß auch heute biogene Farbstoffe von großer wirtschaftlicher Bedeutung sein können, mögen zwei Beispiele zeigen: das inzwischen zwar ökonomischer synthetisch herstellbare β-Carotin, das aus dem Bereich der Lebensmittelchemie als gelb färbendes Pigment und Provitamin A nicht mehr wegzudenken ist, oder das berühmte Kosmeticum des Morgenlandes, der Henna-Farbstoff aus *Lawsonia*-Arten, ohne den unsere modernen Haarpflegemittel nicht möglich wären.

Ohne Frage stehen für uns die psychischen Momente der Farbe und Farbstoffe im Vordergrund ihrer Wirkung. Nicht nur im täglichen Leben spielt die Psychologie der Farbe eine wichtige Rolle. In der Malerei, aber auch in der Poesie hat sie ihre höchste Ausdruckskraft gefunden. Die Empfindung der Farbe gegenüber ist eine Empfindung sui generis, die darüber hinaus mit Erlebnissen der uns immer als Farbstoffträger gegenübertretenden Natur eng verknüpft ist. Diese psychischen Effekte der Farbe wurden bereits von Goethe in seiner Farbenlehre untersucht. Allerdings bewertete man in der Folgezeit häufig das Farberlebnis als bloße Lust- bzw. Unlustempfindung. Erst Stefanescu-Goanga (1912) rückte von dieser Subjektivierung des Farbempfindens ab, «um auf dem Boden der Tatsachen des Einzelbewußtseins Analysen der Farben als reine Empfindungsqualitäten» zu gewinnen. Diese Empfindungsqualität der Farbe gegenüber, die zur Zeit ein aktuelles Forschungsgebiet der Physiologie ist und bis zu Farbtests zur Persönlichkeitsbestimmung führt, ist eng verbunden mit der Erscheinungsweise, in der die Farbe uns in unserer Umwelt, z.B. in gefärbten Organismen gegenüber tritt. So differenzieren wir Farben, die an Pigmente, Farbstoffe gebunden sind, von solchen, die von bestimmten Strukturen abhängen. Strukturfarben, durch Interferenz des Lichtes an dünnen Schichten entstanden, geben uns die wechselnden Farbeindrücke von tierischen Schuppen und Feldern wieder.

Gibt es eine Begründung dafür, pflanzliche Farbstoffe oder Pigmente in einer Zusammenfassung herauszugeben? Gibt es über das den Menschen psychisch beeinflussende Moment noch weitere Gemeinsamkeiten der in ihrer chemischen und biochemischen Struktur recht unterschiedlichen Farbstoffe biogener und hier insbesondere pflanzlicher Herkunft? Diese Frage ist sicherlich auf Grund unseres heutigen Wissens nur mangelhaft zu beantworten. Im Vordergrund der Gemeinsamkeiten der meisten biogenen Farbstoffe steht ohne Frage ihr Effekt, der mit ihren optischen Eigenschaften zusammenhängt. Sind es bei den Tieren Funktionen des Abschreckens, des Imponierens und des Anlockens, ist ihre Aufgabe im Bereich der Blüten und Früchte offensichtlich eng verknüpft mit der Verbreitung der Art. Aber schon hier erschöpft sich das Gemeinsame aller Farbstoffe. Während einzelne Gruppen für die sie produzierenden Organismen zweifelsohne wichtige, ja lebenswichtige physiologische Funktionen haben (z.B. die Chlorophylle), steht für andere Pigmente, u.a. auch für die als Lichtschutzsubstanzen oder Photoakzeptoren diskutierten Carotinoide der endgültige Nachweis einer metabolischen Rolle noch aus. Offen bleibt, inwieweit es sich häufig bei diesen Farbstoffen mit Sekundärstoffcharakter um biochemische Fossilien handelt. Andererseits hat gerade das möglicherweise für Pflanzen «nutz»lose β-Carotin als Provitamin A wichtige physiologische Funktionen als Effektor im Metabolismus vieler Tiere und des Menschen.

Trotz weniger Gemeinsamkeiten und vieler Unterschiede dieser Stoffgruppen in

Biochemie, Biogenese und in möglichen Funktionen regen sie immer wieder zu Versuchen an, das Problem biogener Farbstoffe von der biologisch-biochemischen Seite abzutasten. Sehr wohl bewußt, daß nicht nur manche Aspekte, auch manche Farbstoffgruppen zu wenig oder gar nicht behandelt wurden, lege ich diese Zusammenfassung mit dem «Mut zu Lücke» vor, den heutzutage jeder Herausgeber bei der Publikation einer Übersicht haben muß. Wenn diese Referate neben ihrer Darstellung des Standes der Farbstoff-Forschung vor allem die noch vorhandenen weißen Flecken unseres Wissens um die Biologie, vor allem aber um die mögliche Funktion und Bedeutung dieser Stoffklassen für die Organismen selbst aufzeigen, ist der Sinn dieses Bandes erfüllt. Daß darüber hinaus die Beschäftigung mit pflanzlichen Farbstoffen von ihrer Biochemie bis zur Physiologie und Psychologie dem Ästheten unter ihren Erforschern – und welcher Wissenschaftler ist das eigentlich nicht? – Befriedigung gibt, mag ein persönlicher Wunsch des Herausgebers sein.

Light and Pigments

H. Mohr

Biologisches Institut II, Universität Freiburg i. Br.

1. Introduction

The photoautotrophic higher plant is optimised with respect to the factor light. The process of optimisation – an integral aspect of development – is naturally controlled by endogenous factors (gene-controlled developmental homeostasis). However, the environmental factor light also plays a decisive role in regulating the optimization of a developing plant with respect to the light factor. Tht term *photomorphogenesis* has been used to designate the fact that light controls development (that is, growth, differentiation and morphogenesis) of a plant independently of photosynthesis. Naturally, light does not carry any *specific* information with regard to development. It is the specific reception and the correct *interpretation* of photosignals from the environment which permit the regular development of plants. Phenomenologically, light is an «elective factor» which deeply influences the manner in which the genes present in a particular plant organism are used during development (Mohr, 1972). Beyond its significance for plant biology, photomorphogenesis can be considered as a model process in our efforts to understand the control of gene expression by environmental factors during development of eucaryotic organisms, including man (Mohr and Sitte, 1971). Fig. 1 shows seedlings of the mustard (*Sinapis alba*) plant illustrating the basic phenomena of photomorphogenesis. Without the light factor, normal development cannot occur; rather, the plant will etiolate until it dies. Thus, when we study the mechanism of photomorphogenesis, we study in fact the mechanism of *normal development* in higher plants.

The *formation of pigments* is a conspicuous feature accompanying photomorphogenesis. Control by light of pigment formation has in fact been widely used as a model system to study the mechanism of photomorphogenesis. The principal insights obtained during the analysis of the biochemical model systems constitute a solid basis in our present efforts to understand (in molecular terms) the appearance of structural and functional traits during photomorphogenesis.

2. Mass and sensor pigments

Those pigments which colour the higher plant have the following obvious functions: 1. absorption, transfer, and conversion of light energy to drive the endergonic process of photosynthesis; 2. establishment of the communication between plants and animals (e.g., pigments in flowers and in fruits); 3. absorption of undesired light (light filter function). As an example, the nearly ubiquitous appearance of anthocyanin in seedlings is explained by their function of protecting the flavins and cytochromes of the seedling against the photooxidative action of high light intensities. 4. quenching in the case of carotenoids. It is established knowledge that in photosynthetic membranes carotenoids serve as antenna pigments, absorbing light and transforming the energy with considerable efficiency to chlorophyll (see 1.). Even more importantly, carotenoids serve a protective function by efficiently quenching potentially dangerous triplet chlorophyll and singlet oxygen (Burnett, 1976). Self-photooxidation of chlorophyll and the photosensitizing action of chlorophyll in the absence of carotenoids are well documented in the literature (see Frosch et al., 1979).

To satisfy these functions a relatively high concentration of the pigments is required: *mass pigments*. As representatives we chose the chlorophylls and the anthocyanins of the juvenile plant.

Besides mass pigments the higher plant contains *sensor pigments*. Usually their concentration is so low that this class of pigments does not contribute to the colour of the plant. For this reason the sensor pigments of the higher plant have been overlooked until recently by the plant biochemists concerned with natural products. The functions of plant sensor pigments can be characterised as follows: 1. Control of plant development within the limits given by developmental homeostasis (photomorphogenesis, photoperiodism; see Mohr, 1972). 2. Photomodulation of plant behaviour (photonastic movements, phototropic movements). 3. Control of intracellular movements of organelles (e.g., chloroplasts). The dominating sensor pigment in the life cycle of the higher plant is the chromoprotein *phytochrome*. Besides phytochrome, *flavoproteins* («cryptochrome») play a part, in particular in the phototropic responses.

A characteristic feature of the function of sensor pigments is that the free energy required for the observable changes in metabolism, form, and structure of a plant is not obtained from the light signal itself. Rather, the operation of the sensor pigment is part of a switching mechanism; the free energy used up in the signal-response-chain is derived from dissimilatory processes, as a rule from cell respiration.

In this chapter, we deal with the question of the extent to which synthesis and contents of pigments in the higher plant are regulated by the environmental factor light. We will consider *chlorophyll*, *anthocyanin* (in the juvenile plant), and *phytochrome*. As an experimental system – representative by and large for the dicotyledonous angiosperms – we chose the mustard seedling (*Sinapis alba*) during the developmental period between sowing of the seed and 84 h after sowing (25 °C; see Mohr, 1966).

We anticipate an important experimental result: whereas the formation of the sensor pigment phytochrome is *not* influenced by light (Schäfer et al., 1972; Drumm et al., 1975); the formation of the mass pigments, chlorophyll a, b and anthocyanin, depends on light. Without light these mass pigments cannot be synthesised by the mustard seedling. A priori there are two alternatives to explain the effect of light on the formation of mass pigments: 1. The mass pigments are synthesised in the dark in minute concentrations. Light absorbed by this initial amount activates the further formation of the mass pigments (positive feedback or *autocatalysis*). 2. Formation of the mass pigments is regulated by a sensor pigment such as phytochrome («heterogenous *photocatalysis*»).

Even Georg Klebs (1917), who developed the concept of heterogenous photocatalysis, interpreted light-dependent chlorophyll formation as an instance of autocatalysis. Today we consider chlorophyll as well as anthocyanin syntheses as paradigms of heterogenous photocatalysis (Steinitz et al., 1976; Mohr and Kasemir, 1977).

3. The experimental system

Over the years we have investigated the development of mustard seedling cotyledons under strictly standardized growing conditions at 25 °C (Mohr 1966, 1972; Mohr and Schopfer, 1977). These epigeal cotyledons are peculiar organs (Fig. 1). As long as the seedling develops exclusively in the dark, the cotyledons function as compact storage organs. Filled with fat and storage protein they serve the requirements of the rapidly growing axis system. The cotyledons themselves do not grow or develop significantly as long as the seedling is kept in complete darkness. However, when the seedling is illuminated with white light of sufficiently high irradiance we observe a strong expansion of the cotyledons and a rapid transition into photosynthetic organs which is very similar in internal structure and in function to a normal photosynthetically active leaf (Lovell and Moore, 1970). We may assume that the light-mediated development of an epigeal cotyledon with a high expansion factor represents the corresponding events in young, true leaves in almost every respect, in particular in the development of the photosynthetic apparatus. However, since the cotyledons contain a nonlimiting supply of storage molecules, the molecular events leading to a photosynthetically active leaf can be studied much more conveniently in these organs than, e.g., in primary leaves which have been used preferentially in the past (e.g. Bradbeer and Montes, 1976). The biological unit «mustard cotyledon» offers further advantages: while the fresh weight of the organ increases under the influence of light, cell number and DNA content remain virtually constant and are not significantly affected by light. Thus, phototransformation of the mustard cotyledon from a storage organ to a photosynthetically active leaf with a particularly high assimilatory capacity involves a cell population (approximately $5.6 \cdot 10^5$ cells per cotyledon) which already exists in the seed (Lovell and Moore, 1970; Mösinger, 1976).

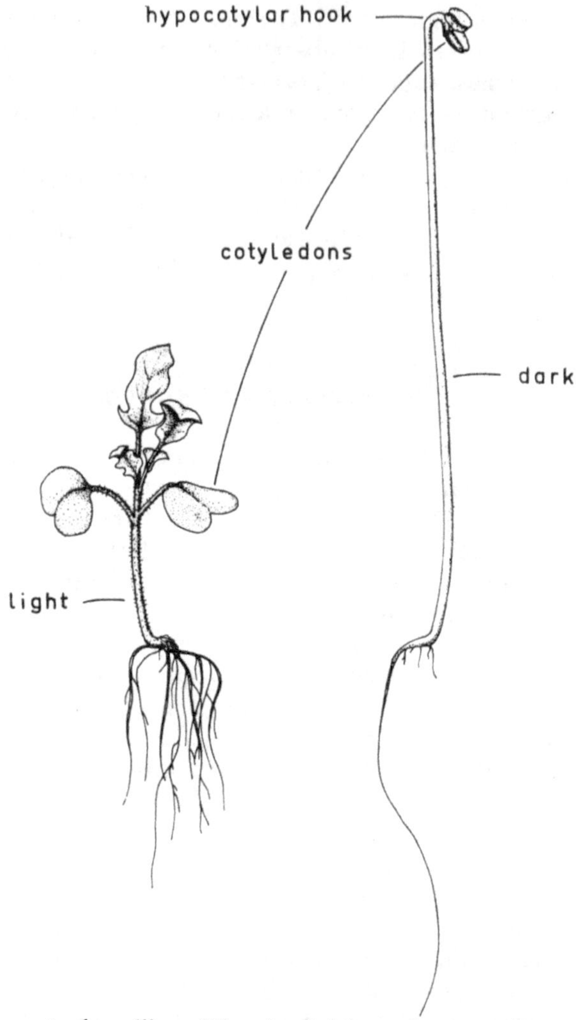

Fig. 1: These two mustard seedlings (*Sinapis alba*) have the same chronological age and are virtually identical genetically. The differences in morphogenesis are due to light. The drawings emphasize the point that light causes the cotyledons to develop from storage organs (right) to photosynthetically active leaves (left). Unlike, for instance, pea cotyledons, the mustard cotyledon is not devoid of function after depletion of lipid and protein reserves, since in the light it expands, becomes green and persists as a photosynthetic organ. (Based on Mohr, 1972)

4. Phytochrome

In higher plants phytochrome is the most important sensor pigment for the detection of photosignals from the environment and for making use of this information to regulate the orderly development of the living system. The photoreceptor involved in the phototransformation of the mustard cotyledons is phytochrome. This bluish chromoprotein is characterized by photochromic properties, i.e., it changes its absorption properties after exposure to light (see Mitrakos and Shropshire, 1972).

In this review, we use the phytochrome model as shown in Fig. 2. This model

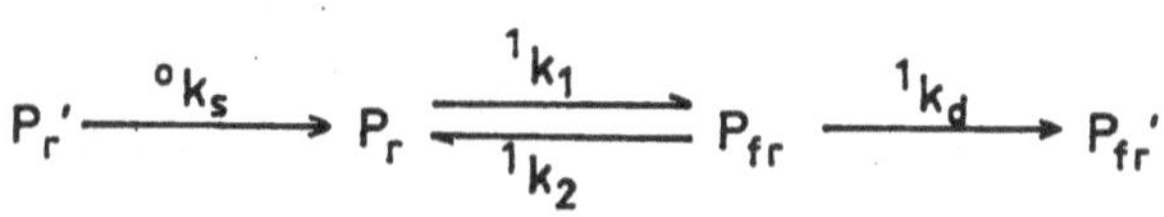

Fig. 2: A model of the phytochrome system as it occurs in the hypocotylar hook (and very probably also in the cotyledons) of the mustard seedling. The symbols ^0k_s, $^1k_1$, $^1k_2$, ^1k_d represent the rate constants of de novo synthesis of P_r, back and forth phototransformations and P_{fr} destruction, respectively. Only $^1k_1$ and $^1k_2$ are light-dependent. (Based on Oelze-Karow et al., 1976). We have ignored the so-called dark reversion of P_{fr} ($P_{fr} \xrightarrow{\text{dark}} P_r$) in Fig. 2, since its significance and physiological relevance are not established (see Oelze-Karow et al., 1976)

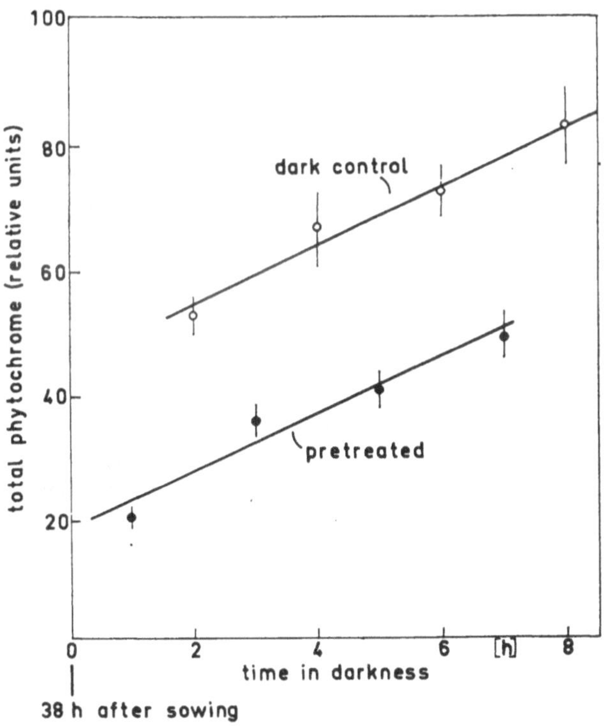

Fig. 3: Increase of total phytochrome in the mustard seedling cotyledons in the dark. The increase is due to de novo synthesis of P_r (Quail et al., 1973). ○, control (no light treatment); ●, treatment as follows: 36 h dark – 1 h red – 5 min far-red. Obviously, the rate of synthesis of P_r (^0k_s) does not depend on the amount of P_{tot}. (Based on Drumm et al., 1975)

quantitatively describes the properties of the phytochrome system as it occurs in the cotyledons and in the hypocotylar hook of the mustard seedling during the period of our experimentation. All observations have led to the concept that the P_{fr} form of the chromoprotein is the physiologically active form, the effector molecule of the phytochrome system; while the P_r form has no physiological effect. The sum of the amounts of P_r and P_{fr} is called «total phytochrome» or P_{tot}. In a dark-grown seedling, only P_r is present. Synthesis de novo of P_r is a zero order process (during the period of our experimentation at least) which neither depends on light nor on the amount of phytochrome already present (Fig. 3). The physiologically active P_{fr} originates from P_r through a first-order phototransformation which is photoreversible. The effector molecule P_{fr} is not stable. It disappears through a first-order destruction process which is independent of light. The half-life of P_{fr} in the mustard cotyledons at 36 h after sowing is 45 min at 25 °C. *Biochemically*, phytochrome is a chromoprotein consisting of a protein moiety (possibly with a small amount of carbohydrate) and a chromophoric group. The phytochrome chromophore is a phycobilin (an open-chained tetrapyrrole, similar to the algal chromophore phycocyanobilin) (see Rüdiger's chapter in the present volume). *Biophysically* speaking, an important feature of the phytochrome system is that the absorption spectra of P_r (peak in the red spectral range at 665 nm) and P_{fr} (peak in the far-red spectral range at 725 nm) overlap throughout the visible spectrum. This overlap is the reason why photostationary states are a characteristic of the phytochrome system, under conditions of saturating irradiations (Hanke et al., 1969). The photostationary state of the phytochrome system as defined by $\varphi_\lambda = [P_{fr}]_\lambda/[P_{tot}]$ is rapidly established in the red and far-red range of the spectrum even at a moderate photon fluence rate; that is, only «light pulses» are required to virtually establish photostationary states in the phytochrome system.

On the basis of this information, the operational criteria for the involvement of phytochrome in a particular photoresponse can be defined as follows: a photoresponse can be induced by a standard red light pulse (φ red = 0.8)*; the induction by red light can be fully reversed by immediately following with a far-red light pulse (φ standard far-red = 0.023*; φ 756 nm < 0.01*). In the case of phytochrome involvement, the extent of the response following the irradiation sequence of red plus far-red will be identical to the extent of the response following a brief treatment with far-red light alone.

Another important property of the phytochrome system is that it develops a

---

* The φ values given are those for the hypocotylar hook (Hanke et al., 1969; Oelze-Karow and Mohr, 1973; Schäfer et al., 1975). For technical reasons, the precision of the corresponding data obtained from the cotyledons is less satisfactory, but in principle the properties of the phytochrome system are undoubtedly the same in both organs (see Schäfer et al., 1973). Since, however, the absolute values of the rate constants are different in hook and cotyledons (Schäfer et al., 1973), the φ_λ-values may not be identical. Moreover, the φ_λ-values are subject to changes during development (Schäfer et al., 1972).

steady state under continuous light. Based on the model in Fig. 2, any change of total phytochrome can be described by the equation

$$\frac{d[P_{tot}]}{dt} = {}^0k_s - {}^1k_d \cdot [P_{fr}].$$

Under steady state conditions, i.e., «no change of P_{tot}», the rate of P_r synthesis is equal to the rate of P_{fr} destruction, ${}^0k_s = {}^1k_d \cdot [P_{fr}]$. Therefore,

$$[P_{fr}] \text{ steady state} = {}^0k_s/{}^1k_d.$$

In words: the steady state concentration of the effector molecule P_{fr} is only a function of the rate constants for P_r synthesis and P_{fr} destruction, which are both light-independent. This means that the steady state concentration of the effector molecule P_{fr} does not depend on the wavelength of the light incident on the system. The only prerequisite is that the incident light is absorbed by both phytochrome forms to an extent sufficient to establish a steady state. On the other hand, the concentration of P_{tot} is *dependent* on the wavelength of light under steady state conditions as shown by the following equation

$$\frac{[P_{fr}]}{[P_{tot}]_\lambda} = \varphi_\lambda \quad \text{or} \quad [P_{tot}]_\lambda = \text{constant} \cdot \frac{1}{\varphi_\lambda}.$$

In words: the ratio φ between the concentrations of P_{fr} and P_{tot} under steady state conditions is a function of wavelength since ${}^1k_1$ and ${}^1k_2$ are wavelength dependent. This implies that $[P_{tot}]$ is a function of φ_λ.

The existence of photosteady states of the phytochrome system at least up to about 730 nm has been used in the experimental approach to photomorphogenesis; namely, this property of the phytochrome system offers the opportunity to run the steady state of the phytochrome system at a wavelength which does not cause significant chlorophyll synthesis or photosynthesis. For long term irradiation we have been using a high irradiance (fluence rate) (3.5 Wm^{-2}) standard far-red light source which is equivalent, as far as the phytochrome system is concerned, to the 718 nm wavelength. Since the formation of chlorophyll is negligible under this kind of light we can study phytochrome-mediated photomorphogenesis without interference of photosynthesis.

The mathematical treatment of the phytochrome model (Fig. 2) can, of course, be performed at a more sophisticated level. I would like to consider only one further aspect, however. Under steady state conditions («no change of P_{tot}») the systems analytical treatment of the phytochrome model leads to a linear relationship between the term

$$\frac{P_{tot}^{N_\lambda} - P_{tot}^{\infty}}{P_{tot}^{\infty}}$$

and the reciprocal of the applied irradiance (fluence rate), $1/N_\lambda$ (Fig. 4). Since P_{tot} can be measured *in situ* with considerable accuracy, the theoretical prediction can

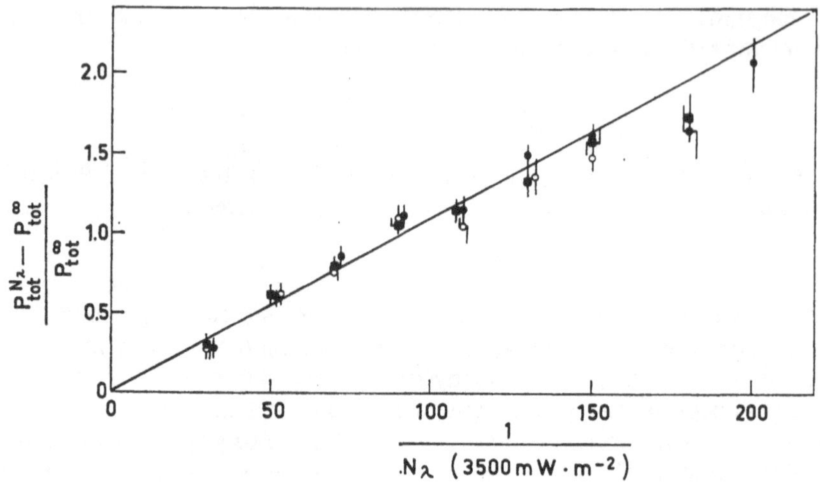

Fig. 4: Irradiance (fluence rate) dependency of the amount of total phytochrome P_{tot} present under steady state conditions in continuous standard far-red in cotyledons of the mustard seedling. Term $P_{tot}^{N_\lambda} - P_{tot}^{\infty}/P_{tot}^{\infty}$ is plotted against the reciprocal of the irradiance (fluence rate) N_λ. *Straight line:* theoretical (systems analytical) result; *points:* empirical results obtained with 3 independent experimental approaches. $P_{tot}^{N_\lambda}$, total phytochrome ($[P_r]$ plus $[P_{fr}]$) under steady state conditions at irradiance N_λ; P_{tot}^{∞}, total phytochrome under steady state conditions at infinitely high irradiance. (Based on Schäfer and Mohr, 1974)

be tested. Fig. 4 shows that the photometric measurement of the P_{tot} term in the cotyledons of the mustard seedling agrees with the prediction and thus confirms the *quantitative* validity of the phytochrome model (Fig. 2). The experimental data, obtained with three different approaches fit well into the theoretical straight line.

The model (Fig. 2) can be extended. The so-called *High Irradiance Responses* (HIR) of photomorphogenesis (see Hartmann, 1966; Mohr, 1972) can only be explained quantitatively by the model if it is extended by some elements (Schäfer, 1975). In particular, the refined model considers the interaction between phytochrome and its intracellular receptor(s) which must be postulated to account for the physiological evidence (see Mohr, 1972). Since the direct *measurements* of phytochrome-receptor binding in situ and in vitro are still a matter of controversy (see Marmé, 1977; Pratt, 1978) this aspect will not be followed any further. In any case, the simple phytochrome model as illustrated in Fig. 2 is adequate for an explanation of photomorphogenesis as long as the law of reciprocity holds («induction conditions», «light pulse experiments»), and for the justification of the use of continuous far-red light to operate the phytochrome system under steady state conditions (see Mohr, 1972; Schäfer et al., 1973).

5. Phytochrome control of Chl-a * formation

Chl-a is a characteristic marker molecule of the plastid. Its synthesis takes place within the plastid compartment (Rebeiz and Castelfranco, 1973). In continuous white light, plastogenesis in the mustard cotyledons (Fig. 5) follows in principle the

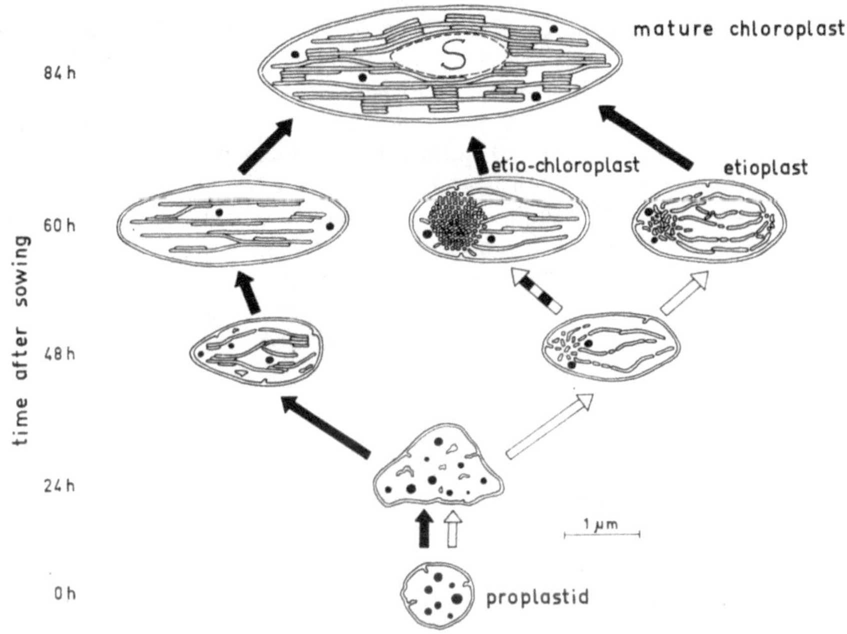

Fig. 5: A scheme showing the time course of plastid development in mustard cotyledons. White arrows: development in complete darkness; dark arrows: development in continuous white light (7,000 lx); white/dark arrow: treatment of the seedling with 4 red light pulses given 36, 40, 44 and 48 h after sowing. S, starch. (Based on Girnth et al., 1978.)

same temporal pattern of ultrastructural changes as described for other species, e.g., bean (*Phaseolus vulgaris*) or corn (*Zea mays*) (Bradbeer and Montes, 1976; Whatley, 1977). In mustard cotyledons, the stage of a fully mature chloroplast is reached approximately 84 h after sowing (25 °C). In complete darkness, the plastids of the angiosperms develop only to the stage of the etioplasts which differ conspicu-

---

* Abbreviations: Chl-a, chlorophyll a; Chl-b, chlorophyll b; Chlide-a, chlorophyllide a; PChl, protochlorophyll(ide). PChl will be used as a collective term including protochlorophyllide ester and protochlorophyllide. However, there is experimental evidence that protochlorophyllide ester is not photoconverted at all or at a much slower rate than protochlorophyllide (Wolff and Price, 1957; Virgin, 1960). This implies that the primary product of the photoconversion is chlorophyllide a rather than chlorophyll a.

ously from chloroplasts (Fig. 5). A characteristic of the etioplast is the prolamellar body, a network of branched tubuli which are sometimes arranged in the form of a highly ordered, «paracristalline» lattice. The etioplasts do not contain chlorophyll, but rather, only a very small amount of protochlorophyll which is localized within the prolamellar body (see Bogorad, 1966). In the angiosperms, the step from protochlorophyll to chlorophyll is light dependent: PChl → Chlide-a* photoconversion (Smith, 1960; for details see Schneider's article in the present volume). The effective light absorption is performed by the PChl-holochrome (a PChl-protein complex). Thus, Chl-a accumulation only occurs in the light, provided that the light can be absorbed by protochlorophyll attached to the protein ($\lambda < 700$ nm).

We should first consider the «capacity» of the mustard cotyledons to synthesise Chl-a. By «capacity» we mean the maximum flux in the biosynthetic channel leading to Chl-a. This capacity can only be measured if we saturate the PChl → Chlide-a photoconversion. We do this with standardised fluorescent white light. The rate of Chl-a accumulation at saturated PChl photoconversion serves as a gauge for the «capacity» of the biosynthetic channel leading to Chl-a. Fig. 6 shows that the capacity at 60 h after sowing strongly depends on the light pretreatment which was received by the seedling before the onset of white light. The light pretreatment acts exclusively

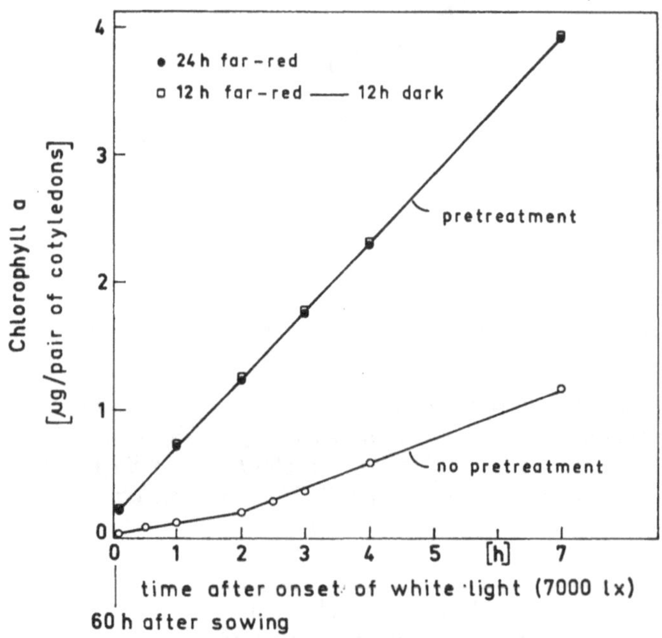

Fig. 6: Accumulation of Chl-a in the mustard seedling cotyledons under continuous standard white light (7,000 lx). The seedlings were either pretreated with standard far-red light between 36 and 48 h (or 60 h) after sowing (upper curve) or kept in complete darkness until the beginning of the white light (lower curve) (Based on Kasemir et al., 1973.)

through phytochrome (Kasemir et al., 1973). It is obvious that the light pretreatment has two different effects: 1. The so-called lag phase – 2 h in the present case – is eliminated. 2. The capacity is strongly increased. The capacity at a given time in etiolated seedlings and in seedlings pretreated with phytochrome (operationally, with light pulses, or with long-term far-red light) was measured by determining the slope of the Chl-a accumulation curve during the linear phase of Chl-a accumulation under saturating white light after the initial lag phase (2 h) was overcome (see Fig. 6). To determine the capacity in continuous white light, the first derivative of the Chl-a accumulation curve (Fig. 7) was determined graphically.

It is obvious from Fig. 8 that the capacity depends on the physiological age of the seedling and that it is strongly increased by phytochrome, operationally under continuous far-red light. The time course of the capacity under continuous *white* light

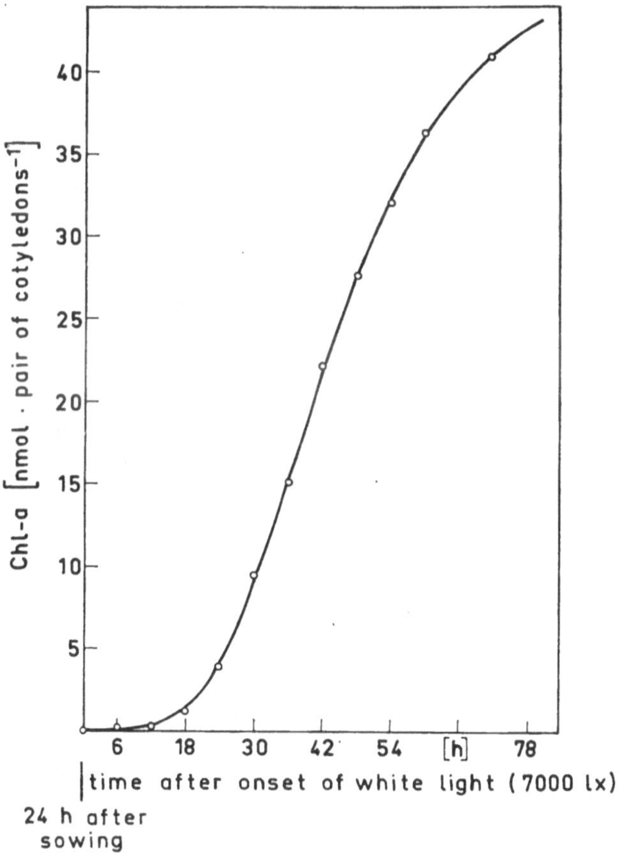

Fig. 7: Time course of Chl-a accumulation in continuous standard white light (7,000 lx) which is saturating for the PChl → Chlide-a photoconversion in the cotyledons of the mustard seedling. Onset of light: 24 h after sowing. (Based on Gehring et al., 1977.)

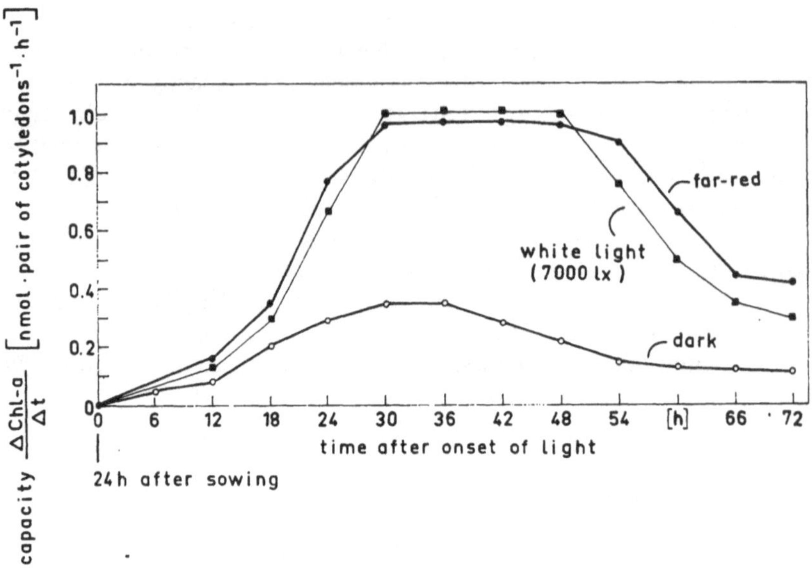

Fig. 8: Time course of the capacity of the Chl-a synthesizing metabolic pathway in the mustard seedling cotyledons in darkness and under continuous far-red or white light. In the latter case, capacity is defined as the first derivative of the Chl-a accumulation curve in saturating white light (see Fig. 7). In darkness and under far-red light (under which conditions no significant Chl-a accumulation takes place), the capacity is operationally defined as the increase of Chl-a between 2 and 4 h after the onset of standard white light (see Fig. 6). (Based on Gehring et al., 1977.)

(which permits Chl synthesis and photosynthesis) deviates only slightly from the time course of the «capacity» under continuous standard far-red light (which does not permit significant Chl accumulation). It is concluded that the capacity for Chl-a synthesis is determined by phytochrome irrespective of whether or not Chl-a is actually being synthesised. A comparison of the time courses of the capacity in light and dark suggests that the *temporal pattern* is the same. While the actual capacity is indeed determined by phytochrome, the developmental pattern – the temporal pattern of the capacity – is not shifted significantly by light. Moreover, it is clear from Fig. 8 that photosynthesis, which takes place in white light at a high rate, exerts no positive influence on the capacity. Instead, ongoing photosynthesis leads to a somewhat earlier decrease of the capacity.

In continuous darkness or in continuous light we do not see any manifestation of an endogenous circadian rhythm. However, when we turn off the light after 12 h the physiological clock (Gehring et al., 1977) comes into play (Fig. 9). It manifests itself in a strong overshoot and in a conspicuous circadian oscillation of the capacity. The induction (or release) of the physiological clock is caused by a light → dark transition. The photoreceptor involved is phytochrome.

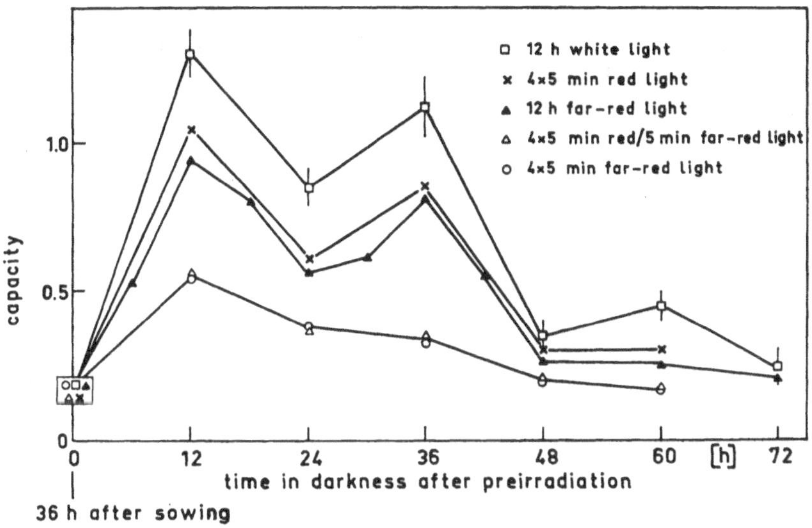

Fig. 9: Time course of the capacity of the Chl-a synthesizing pathway in the mustard seedling cotyledons in darkness following a 12 h light pretreatment (between 24 and 36 h after sowing). The light pulses (5 min each) were given at 24, 28, 32 and 36 h after sowing. Operational definition of capacity, see Fig. 8. (Based on Gehring et al., 1977.)

Experiments designed to localise the site of regulation of the capacity have led to the result that both phytochrome and the physiological clock operate at the level of 5-aminolevulinate (ALA) synthesis (for biochemical details of chlorophyll synthesis see Schneider's chapter in the present volume). It is obvious from Fig. 10 that the capacity of the ALA-forming system and the capacity of the Chl-a-synthesising system almost coincide under continuous far-red light, as well as under conditions in which the physiological clock plays a major role. It is concluded that phytochrome and the physiological clock determine the capacity of Chl-a biogenesis through the capacity of ALA synthesis.

The capacity of the ALA-forming system was determined analogously to the Chl-a capacity from the slope of the linear ALA accumulation curve under conditions in which ALA-dehydratase (EC 4.2.1.24) in the mustard cotyledons was inhibited by the application of levulinate ($8.5 \cdot 10^{-2}$ mol $\cdot$ l^{-1}) (for details see Masoner and Kasemir, 1975 and Gehring et al., 1977).

6. Phytochrome control of Chl-b* formation

It is agreed that Chl-b is formed from Chlide-a in some way. However, debate continues with regard to the *immediate* precursor of Chl-b. Shlyk (1971) concluded that «in contrast to Chl-a, Chl-b is not formed directly from Chlide-a but needs an intermediate». Virgin (1977) suggests «that Chl-b is only synthesised in connection with a phototransformation of PChl to Chlide-a». He thinks that Chl-b formation takes place from freshly formed Chl-a molecules and «that «old» molecules are not reactive in this respect».

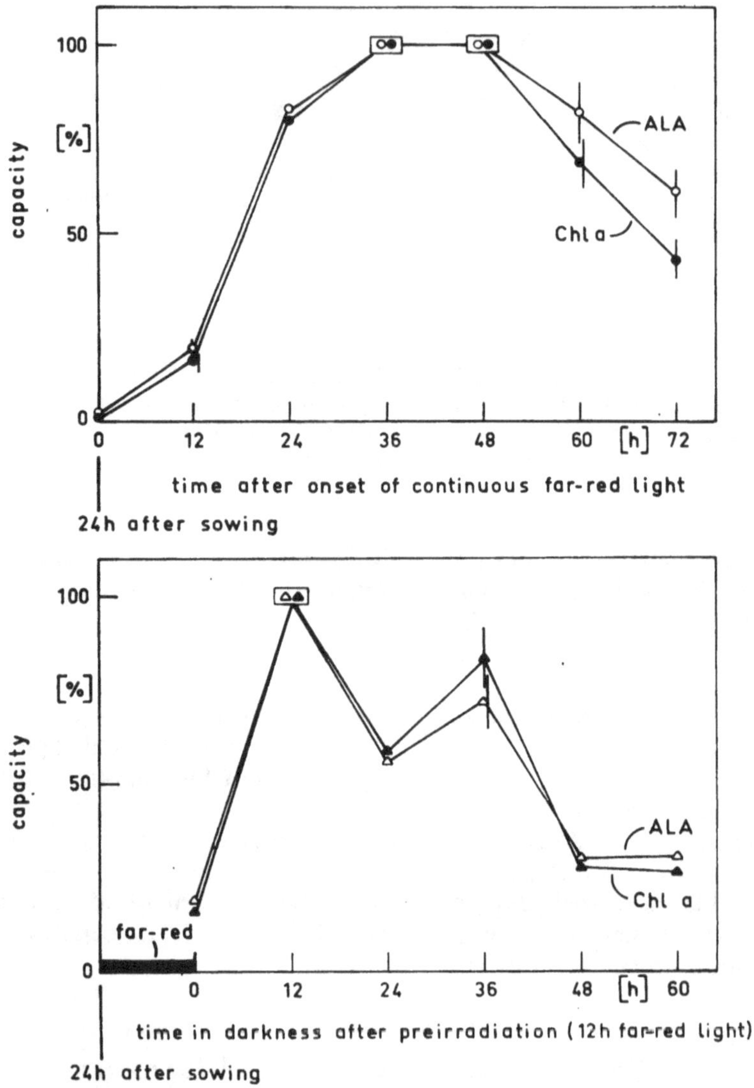

Fig. 10: Time course of the capacities for Chl-a synthesis and 5-aminolevulinate (ALA) synthesis in the mustard seedling cotyledons in continuous far-red light and in darkness following a 12 h pretreatment with far-red light (between 24 and 36 h) after sowing. (Based on Gehring et al., 1977.)

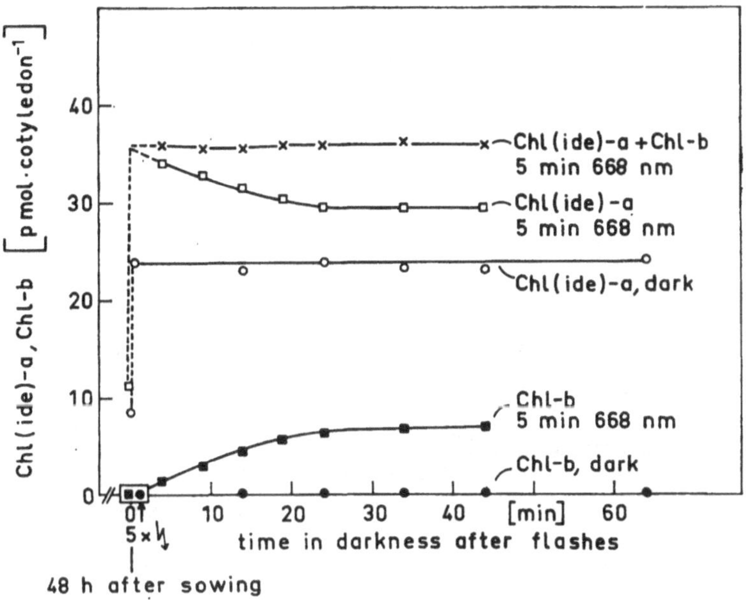

Fig. 11: Time courses of decrease of Chl(ide)-a (□, ○) and formation of Chl-b (■, ●) in mustard seedling cotyledons in darkness after saturating photoconversion of PChl → Chl(ide)-a by white light flashes at 48 h after sowing. Symbols indicating light pretreatment at 36 h after sowing: (□, ■) 5 min 668 nm light; (×) sum of Chl(ide)-a plus Chl-b, 5 min 668 nm light. (○, ●) no light pretreatment. (Based on Oelze-Karow and Mohr, 1978.)

Figs. 11 and 12 show that without a red light pretreatment, which acts through phytochrome, Chl-b will not be formed at all in the mustard cotyledons during the first 60 min after the onset of white light, even though a relatively large amount of Chl-a is in fact formed from Chlide-a during this time period (Kasemir and Prehm, 1976). If a red light pulse pretreatment is given, formation of Chl-b takes place stoichiometrically at the cost of Chl(ide)-a (Fig. 11). Chl-b is formed only as long as the esterification of Chlide-a → Chl-a is not yet completed. Our present data (see Oelze-Karow and Mohr, 1978) are only compatible with the concept that Chlide-a is the common precursor for Chl-a and Chl-b.

A striking difference between the effect of phytochrome on Chl-a and Chl-b accumulation is that in the case of Chl-a, the action of phytochrome can be considered as a quantitative modulation of a capacity which already exists in the dark-grown seedling (see Fig. 8). On the other hand, formation of Chl-b (at least during the first 60 min after the onset of white light) depends on phytochrome pretreatment in an obligatory, qualitative manner (no phytochrome action, no Chl-b).

A *different* dependency of Chl-a and Chl-b accumulation on phytochrome has also been observed during the steady state period of Chl accumulation (between 90 and 180 min after the onset of white light in our experiments, Fig. 12). The

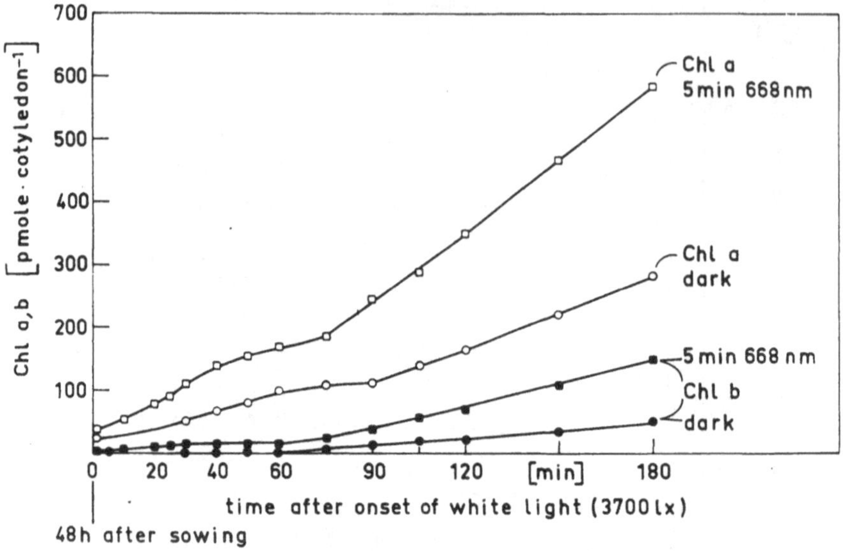

Fig. 12: Time courses of the formation of Chl(ide)-a (□, ○) and Chl-b (■, ●) in mustard seedling cotyledons in continuous white light (3,700 lx) after a saturating photoconversion of PChl → Chl(ide)-a at 48 h after sowing. Symbols with regard to pretreatment: (□, ■) 5 min 668 nm light given at 36 h after sowing; (○, ●) no light pretreatment. The pulse pretreatments act exclusively through phytochrome (P_{fr}). Continuous white light is saturating for PChl → Chlide-a photoconversion. (Based on Oelze-Karow and Mohr, 1978.)

relative effect of the red light pulse pretreatment (which operates exclusively through phytochrome) on Chl-b accumulation is considerably stronger than on Chl-a accumulation.

7. Phytochrome control of anthocyanin formation

The mustard seedling produces 5 anthocyanins all of which contain cyanidin as the aglycone (Havelange and Schumacker, 1966). Anthocyanin synthesis in the mustard cotyledons takes place only in the *epidermal* cells (Wagner and Mohr, 1966). There is no anthocyanin synthesis without P_{fr} (Bühler, 1977). However, even if P_{fr} is present in the cotyledons from the time of sowing onwards, anthocyanin synthesis only starts at 27 h after sowing (at 25 °C) (Steinitz et al., 1976). This «starting point» (i.e., the point in time at which the epidermal cells become capable of forming anthocyanin as a response to P_{fr} formation) cannot be shifted by any light treatment or by the application of nutrients. At a given temperature parameter, it is a *system's constant*.

Steinitz et al. (1976) have shown that the epidermal cells of the mustard cotyledons become *competent* for P_{fr} with respect to anthocyanin synthesis only at 27 h after sowing. Before this time the phytochrome system (which can be demonstrated immediately after full imbibition of the seed) is not yet coupled or connected to those cell functions which lead to anthocyanin synthesis. In other words: with regard to anthocyanin synthesis, the red light signal originally received by P_r and passed to P_{fr} has not yet been irreversibly passed to some further constituent in the signal-response-chain connecting the phytochrome system with anthocyanin synthesis. Even a 27 h treatment with red light (starting with sowing of the seeds) is totally ineffective with regard to anthocyanin synthesis, provided that P_{fr} is removed from the cotyledons after termination of the red light by a saturating pulse with 756 nm light which lowers the level of P_{fr} to probably less than 1/10 of a percent ($\varphi_{756\text{ nm}} < 0.001$, see Schäfer et al., 1975) (Fig. 13).

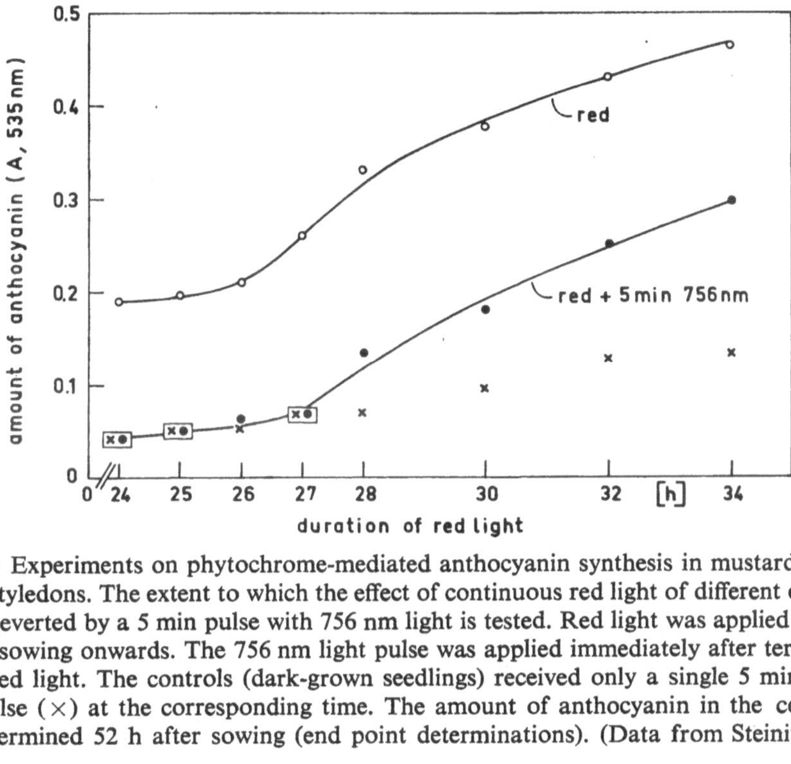

Fig. 13: Experiments on phytochrome-mediated anthocyanin synthesis in mustard (*Sinapis alba*) cotyledons. The extent to which the effect of continuous red light of different durations can be reverted by a 5 min pulse with 756 nm light is tested. Red light was applied from the time of sowing onwards. The 756 nm light pulse was applied immediately after termination of the red light. The controls (dark-grown seedlings) received only a single 5 min 756 nm light pulse (×) at the corresponding time. The amount of anthocyanin in the cotyledons was determined 52 h after sowing (end point determinations). (Data from Steinitz, 1977.)

At 36 h after sowing all epidermal cells potentially capable of performing anthocyanin synthesis have become competent for P_{fr} with regard to anthocyanin synthesis. At this developmental stage, the dose-response-function (amount of anthocyanin as a function of the amount of the effector molecule P_{fr}) is linear for low amounts of P_{fr} and extrapolates through point zero (Fig. 14). Thus, the simple equation

$$[\text{amount of anthocyanin}] = c \cdot [P_{fr}]$$

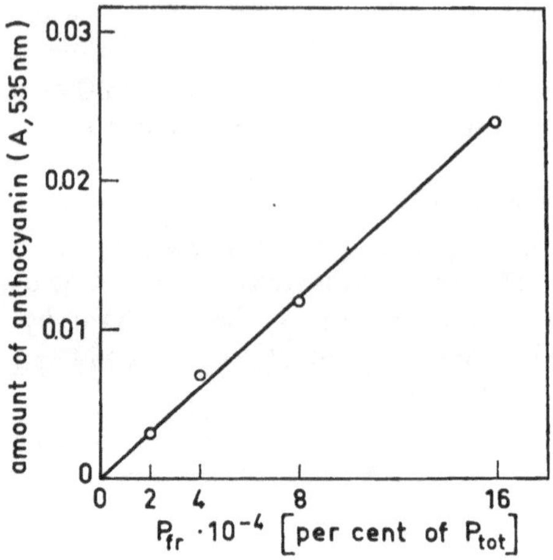

Fig. 14: Quantitative relationship between amount of effector molecule P_{fr} (made by a brief light pulse given 36 h after sowing) and amount of anthocyanin produced within 24 h by a given amount of P_{fr}. System: mustard seedling (*Sinapis alba*). The dose-response-curve is linear in the neighbourhood of zero and extrapolates through zero. (Based on Drumm and Mohr, 1974.)

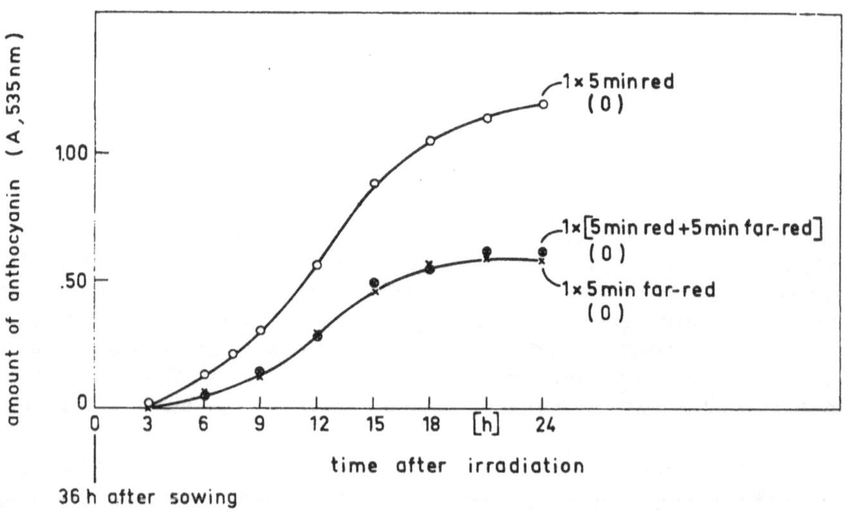

Fig. 15: Time-courses of anthocyanin accumulation in the mustard seedling after pulse treatment with red and/or far-red light (5 min each). Pulses were given at point zero, i.e. 36 h after sowing. Full reversibility of the effect of a red light pulse by following immediately with a far-red light pulse documents the involvement of phytochrome in the light-mediated response. (Based on Lange et al., 1971.)

Table 1: Anthocyanin and phenylalanine ammonia-lyase (PAL) from mustard seedling cotyledons are compared with regard to starting points and points of loss of full reversibility. For comparison, a marker enzyme of the plastid compartment (ribulosebisphosphate carboxylase, abbreviated Carboxylase) is also included.

	starting point [h]	point of loss of full reversibility [h]
Anthocyanin	27	27
PAL	27	27
Carboxylase	48	18

Starting point: the point on the time scale at which a phytochrome-mediated increase of the enzyme or end product level is detectable. The *point of loss of full reversibility* is the point on the time scale up to which the inductive effect of continuous red light (given from the time of sowing) remains fully reversible by a saturating 756 nm light pulse. The time scale starts with the sowing of the seeds (time after sowing). (Data from Steinitz et al., 1976; Frosch et al., 1977.)

describes the relationship between P_{fr} and anthocyanin whereby c is a constant, specific for the system. The problem has been to study the «mechanism» behind this equation, i.e., the sequence of individual molecular or biophysical steps between the formation of P_{fr} and the appearance of anthocyanin. Knowledge about the time course (kinetics) of the response following inductive light pulses is an indispensable prerequisite for any discussion about «mechanisms» (Fig. 15). When these data are subject to probit transformation, straight lines are found. This means that the time course of anthocyanin accumulation shows the characteristics of a Gaussian curve, irrespective of the amount of P_{fr} established by the inductive light pulse. The details of the time courses of anthocyanin synthesis (following a light pulse as well as under continuous light) are difficult to reconcile with idea that the rate of anthocyanin synthesis is *controlled* by enzymes whose appearance is mediated by P_{fr} (see Mohr, 1972). On the other hand, the notion that the induction by P_{fr} of anthocyanin synthesis is *related* to the induction of enzymes in the phenylpropanoid pathway has been supported by experimental evidence in the case of phenylalanine ammonia-lyase (PAL) more than 10 years ago (Durst and Mohr, 1966). We know today that phytochrome-mediated photomorphogenesis (i.e., normal development, see Fig. 1) is generally related to differential gene expression (see Mohr and Schopfer, 1977) and that several enzymes of the phenylpropanoid pathway are under light control (Hahlbrock et al., 1976). In the case of PAL it was shown (e.g., by Frosch et al., 1977) that the light induction of this enzyme and the light induction of anthocyanin in the mustard seedling cotyledons are in fact closely linked. As an example, PAL becomes detectable in the mustard cotyledons at 27 h after sowing; this «starting point» cannot be shifted by light; the inductive effect of continuous red light (given from the time of sowing) remains fully reversible by 756 nm-light up to the starting point (27 h after sowing); PAL induction is a graded response over a wide range of P_{fr} doses and depends strongly on the photon fluence rate (High Irradiance Response) as does anthocyanin induction, etc. In brief: there can hardly be any doubt that in the mustard cotyledons the induction of PAL by phytochrome (which involves a corresponding increase of the rate of de novo synthesis of the enzyme protein, see Schopfer, 1977) is closely connected to phytochrome-mediated anthocyanin synthesis (Table 1). However, in spite of all coincidences it is not proved that PAL is in fact the site at which the rate of anthocyanin synthesis is *controlled.*

8. Light Control of anthocyanin formation in the mesocotyl of *Sorghum vulgare*

Experience with the mustard seedling (*Sinapis alba*) suggests that anthocyanin synthesis in this dicotyledonous seedling must be attributed exclusively to the action of phytochrome (see Lange et al., 1971; Drumm and Mohr, 1974; Drumm et al., 1975). In particular, we could not find any indication for the involvement of a blue light photoreceptor other than phytochrome (Steinitz et al., 1976). On the other hand, there are angiosperms such as buckwheat (Mohr and van Nes, 1963; Scherf and Zenk, 1967), corn (Duke and Naylor, 1976), and milo (Downs and Siegelman, 1963) in which a predominant action of blue light – in addition to phytochrome – has been shown with regard to the formation of seedling anthocyanin.

We have chosen the milo (*Sorghum vulgare*) seedling to investigate the kind of interaction – if any – between phytochrome and the blue light photoreceptor. Accumulation of anthocyanin, predominantly a red anthocyanin previously identified as an acylated cyanidin-3-glucoside (see Stafford, 1965, 1966), was measured in the first internode (mesocotyl). Without light, no formation of the red anthocyanin was observed in this organ. The major results are summarised in Table 2: Phytochrome (P_{fr}) can only act once a blue (or UV) light effect has occurred; on the other hand, the expression of the blue (or UV) light effect is controlled by P_{fr}. It is concluded that there is an obligatory dependency (or *sequential* interaction) between the blue (or UV) light effect and the light effect occurring through phytochrome. In dichromatic experiments, i.e., simultaneous, high fluence rate irradiation with two kinds of light (see Hartmann, 1966; Drumm and Mohr, 1978) it was shown that the blue-light-mediated photoreaction *per se* is not affected by the presence or near absence of P_{fr}. The data obtained substantiate the conclusion that the initial action of the blue (or UV) light is totally independent of phytochrome, even though the blue light effect can only manifest itself in the presence of P_{fr}.

Milo and mustard differ with regard to anthocyanin synthesis insofar as mustard can perform those reactions which are blue-(or UV-)light-dependent in milo, even in complete darkness. Buckwheat (Mohr and van Nes, 1963; Scherf and Zenk, 1967) can be considered as an example of those plants which respond slightly to phytochrome without pretreatment. However, a high fluence rate, blue light pretreatment strongly increases the responsiveness towards P_{fr}.

Table 2: Induction (or noninduction) of anthocyanin synthesis in the mesocotyl (1st internode) of *Sorghum vulgare* (milo) seedlings by light of different qualities (white light: xenon light, similar to sunlight, 250 W · m^{-2}; blue light: interference filter, λ_{max} 450 nm, 9 W · m^{-2}; UV: λ_{max} 350 nm, 7.5 W · m^{-2}). After the 3 h light treatment the seedlings were kept in the dark for 24 h before extraction of anthocyanin. (Based on Drumm and Mohr, 1978.)

light treatment (onset, 60 h after sowing)	amount of anthocyanin (measurement, 87 h after sowing) [relative units]
27 h dark	0
27 h white light	115
27 h red light	0
27 h far-red light	0
3 h white light	19
3 h blue light	7
3 h UV	19
3 h white light + 5 min red light*	19
3 h white light + 5 min 756 nm-light*	6
3 h white light + 5 min 756 nm-light + 5 min red light	20
3 h UV + 5 min red light	19
3 h UV + 5 min 756 nm-light	5
3 h UV + 5 min 756 nm-light + 5 min red light	19

* Photoequilibria of the phytochrome system (see Schäfer et al., 1975): $\varphi_{red} \approx 0.8$; $\varphi_{756} < 0.01$. A 5 min light pulse suffices to virtually establish the photoequilibrium.

With regard to the identity of the blue light photoreceptor we probably deal with a very common flavin-type photoreceptor pigment (more recently called «cryptochrome») which controls a large number of physiological responses in a wide diversity of photosensitive organisms (for reviews see Haupt, 1977; Mohr, 1972, chapter 21).

There are other reports on the obligatory dependency of phytochrome action on a pretreatment with short wavelength light. Wellmann (1971, 1976) has described the phenomenon that the expression of an inductive effect of UV light ($\lambda_{max} < 300$ nm) on flavonoid and concomitant enzyme biosynthesis in cell suspension cultures of parsley (*Petroselinum hortense*) as well as intact parsley seedlings is controlled by phytochrome. Phytochrome (P_{fr}) without a short wavelength light pretreatment does not mediate any of the responses. In contrast to the work with milo, visible light is totally ineffective in the light pretreatment with parsley. The peak of the action spectrum is below 300 nm. Thus, the photoreceptor pigment in Wellmann's system differs from the one responsible for the blue light effect in the milo mesocotyl. The question of whether or not the blue light and the UV light effects in milo are due to the same photoreceptor pigment cannot be answered conclusively at present.

Recently, control by light of *chlorophyll* synthesis in the shoot of milo seedlings was investigated (Sawhney et al., 1980). Surprisingly, it was found that at least in short term experiments (30 min) no specific blue light effect exists in light control of Chl-a and Chl-b syntheses. Rather, evidence suggests that chlorophyll synthesis in *Sorghum vulgare* is exclusively under the control of phytochrome. Thus, light control of anthocyanin synthesis (Drumm and Mohr, 1978) and light control of chlorophyll synthesis (Sawhney et al., 1980) *in the same system* are totally different.

9. General significance

Theoretical aspect. Control of mass pigment synthesis by the sensor pigment phytochrome is an integral part of photomorphogenesis. These responses may be considered as paradigms of differentiation processes in which the effector (P_{fr}) and in part the «mechanism» are known. It is realized, of course, that biochemical model systems of differentiation which include only simple, water-soluble end products (such as anthocyanins) will probably not lead to a satisfactory solution of the basic problem of developmental biology, i.e., the relationship between enzymes and form («Gestalt») or, in other words, the relationship between enzyme specificity and structural pattern. However, we believe that studies of simple biochemical model systems of differentiation will increase our understanding of the physiological «laws» and molecular foundations of differentiation and thus contribute to the efforts of elucidating the connection in multicellular, higher organisms between modulation of enzyme synthesis and development of structural and functional specificity in space and time.

Practical aspect. In modern plant industry, commercial growers either grow or supplement the growth of high value crops in controlled greenhouses using excessive amounts of artificial light for the production of their plants. The phytochrome-mediated photomorphogenic responses, including chlorophyll synthesis, are essential for an optimum regulation of the efficiency of photosynthesis. If the precise light requirements of a developing plant were known, the grower would probably need *far less* electrical energy per unit of fresh mass of produced crop. It is a practical aim of phytochrome research to predict the best combination of photomorphogenic lighting for producing a desired crop in a given time at minimum cost.

Acknowledgement. Support by the Deutsche Forschungsgemeinschaft (SFB 46) is gratefully acknowledged. I thank my research associates (see references) and my secretary, Barbara Hoffmann, for fruitful collaboration.

References

Bradbeer, J. W., and G. Montes, 1976: In: Light and Plant Development (H. Smith, ed.). Butterworths, London.

Bogorad, L., 1966: In: The Chlorophylls (L. P. Vernon and G. R. Seely, eds.). Academic Press, New York.

BÜHLER, B., 1977: Doctoral Thesis, University of Freiburg.
BURNETT, J. H., 1976: In: Chemistry and Biochemistry of Plant Pigments, Vol. 1 (T. W. Goodwin, ed.). Academic Press, London.
DOWNS, R. J., and H. W. SIEGELMAN, 1963: Plant Physiol. **38**, 25–30.
DRUMM, H., and H. MOHR, 1974: Photochem. Photobiol. **20**, 151–157.
DRUMM, H., and H. MOHR, 1978: Photochem. Photobiol. **27**, 241–248.
DRUMM, H., A. WILDERMANN, and H. MOHR, 1975: Photochem. Photobiol. **21**, 269–273.
DUKE, S. O., and W. W. NAYLOR, 1976: Physiol. Plant. **37**, 62–68.
DURST, F., and H. MOHR, 1966: Naturwiss. **53**, 531–532.
FROSCH, S., H. DRUMM, and H. MOHR, 1977: Planta **136**, 181–186.
FROSCH, S., M. JABBEN, R. BERGFELD, H. KLEINIG, and H. MOHR, 1979: Planta **145**, 497–505.
GEHRING, H., H. KASEMIR, and H. MOHR, 1977: Planta **133**, 295–302.
GIRNTH, C., R. BERGFELD, and H. KASEMIR, 1978: Planta **141**, 191–198.
HAHLBROCK, K., K. H. KNOBLOCH, F. KREUZALER, J. R. M. POTTS, and E. WELLMANN, 1976: Eur. J. Biochem. **61**, 199–207.
HANKE, J., K. M. HARTMANN, and H. MOHR, 1969: Planta **86**, 235–249.
HARTMANN, K. M., 1966: Photochem. Photobiol. **5**, 349–366.
HAUPT, W., 1977: Bewegungsphysiologie der Pflanzen. Thieme, Stuttgart.
HAVELANGE, A., and R. SCHUMACKER, 1966: Bull. Soc. Roy. Sci. Liège **35**, 125–141.
KASEMIR, H., U. OBERDORFER, and H. MOHR, 1973: Photochem. Photobiol. **18**, 481–486.
KASEMIR, H., and G. PREHM, 1976: Planta **132**, 291–295.
KLEBS, G., 1917: Sitzungsberichte Heidelberger Akad. Wiss., Math.-naturwiss., Kl., Abt. B, 3. Abh.
LANGE, H., W. SHROPSHIRE, and H. MOHR, 1971: Plant Physiol. **47**, 649–655.
LOVELL, P. H., and K. G. MOORE, 1970: J. Exptl. Bot. **21**, 1,017–1,030.
MARMÉ, D., 1977: Ann. Rev. Plant Physiol. **28**, 173–198.
MASONER, M., and H. KASEMIR, 1975: Planta **126**, 111–117.
MÖSINGER, E., 1976: Diploma Thesis, University of Freiburg.
MOHR, H., 1966: Z. Pflanzenphysiol. **54**, 63–83.
MOHR, H., 1972: Lectures on Photomorphogenesis. Springer, Berlin-Heidelberg-New York.
MOHR, H., and H. KASEMIR, 1977: In: Research in Photobiology (A. Castellani, ed.). Plenum, New York.
MOHR, H., and P. SCHOPFER, 1977: In: Nucleic Acids and Protein Synthesis in Plants (L. Bogorad and J. H. Weil, eds.). Plenum, New York.
MITRAKOS, K., and W. SHROPSHIRE (eds.), 1972: Photochrome. Academic Press, London and New York.
MOHR, H. and P. SITTE, 1971: Molekulare Grundlagen der Entwicklung. BLV, München.
MOHR, H., and E. VAN NES, 1963: Zeitschrift für Botanik **51**, 1–16.
OELZE-KAROW, H., and H. MOHR, 1973: Photochem. Photobiol. **18**, 319–330.
OELZE-KAROW, H., and H. MOHR, 1976: Photochem. Photobiol. **23**, 61–67.
OELZE-KAROW, H., and H. MOHR, 1978: Photochem. Photobiol. **27**, 189–193.
OELZE-KAROW, H., E. SCHÄFER, and H. MOHR, 1976: Photochem. Photobiol. **23**, 55–59.
PRATT, L. H., 1978: Photochem. Photobiol. **27**, 81–105.
QUAIL, P. H., E. SCHÄFER, and D. MARMÉ, 1973: Plant Physiol. **52**, 124–127.
REBEIZ, C. A., and P. A. CASTELFRANCO, 1973: Ann. Rev. Plant Physiol. **24**, 129–172.
SAWHNEY, S., H. OELZE-KAROW, and H. MOHR, 1980: In: Proceedings in Photomorphogenesis Research (J. de Greef, ed.). University of Antwerp Press, Antwerp.
SCHÄFER, E., 1975: J. Math. Biol. **2**, 41–56.

Schäfer, E., T.-U. Lassing, and P. Schopfer, 1975: Photochem. Photobiol. **22**, 193–202.
Schäfer, E., B. Marchal, and D. Marmé, 1972: Photochem. Photobiol. **15**, 457–464.
Schäfer, E., and H. Mohr, 1974: J. Math. Biol. **1**, 9–15.
Schäfer, E., W. Schmidt, and H. Mohr, 1973: Photochem. Photobiol. **18**, 331–334.
Scherf, H., and M. H. Zenk, 1967: Z. Pflanzenphysiol. **57**, 401–418.
Schopfer, P., 1977: Ann. Rev. Plant Physiol. **28**, 223–252.
Shlyk, A. A., 1971: Ann. Rev. Plant Physiol. **22**, 169–184.
Smith, J. H. C., 1960: In: Comparative Biochemistry of photoreactive Systems (M. B. Allen, ed.). Academic Press, New York.
Stafford, H. A., 1965: Plant Physiol. **40**, 130–138.
Stafford, H. A., 1966: Plant Physiol. **41**, 953–961.
Steinitz, B., 1977: Doctoral Thesis, University of Freiburg.
Steinitz, B., H. Drumm, and H. Mohr, 1976: Planta **130**, 23–31.
Virgin, H. I., 1960: Physiol. Plant. **13**, 155–164.
Virgin, H., 1977: Physiol. Plant. **40**, 45–49.
Wagner, E., and H. Mohr, 1966: Planta **71**, 204–221.
Wellmann, E., 1971: Planta **101**, 283–286.
Wellmann, E., 1976: Photochem. Photobiol. **24**, 659–660.
Whatley, J. M., 1977: New Phytol. **78**, 407–420.
Wolff, J. B., and L. Price, 1957: Arch. Biochem. Biophys. **72**, 293–301.

Carotenoid Biosynthesis

Brian H. Davies

Department of Biochemistry and Agricultural Biochemistry,
University College of Wales, Aberystwyth

1. Introduction

It is some 150 years since the first descriptions of what we now know as the carotenoid pigments appeared in the literature. During this time, the carotenoids have been studied in all branches of natural science. The unique combination of their variety of structure and their characteristically intense yellow, orange, red and purple colours ideally suited them for their part in the development of such analytical procedures as adsorption chromatography and absorption spectrophotometry, techniques which, in their turn, had facilitated the recognition of over 90 carotenoids by 1952. Of these (Goodwin, 1952), two-thirds had been isolated from higher plants. This situation resulted less from an early intrinsic interest in higher plants than from the fact that only higher plant tissues were available on a sufficient scale to provide the large amounts of pure carotenoid then necessary for structure determination by traditional chemical methods. Since the advent of smaller-scale preparative (e.g. thin-layer chromatography) and physico-organic methods (e.g. mass spectrometry), and with the ready availability, through the world's culture collections, of a wealth of pure strains and mutants of microorganisms, the interests of carotenoid chemists have broadened to include fungi, bacteria and algae. It is from smaller-scale studies of the pigments of such organisms that many unusual carotenoid structures have emerged in recent years (Straub, 1976) and an almost exponential rate of growth in the number of carotenoids of known structure has taken place. Of nearly 500 carotenoids now known, only about a quarter have been isolated from higher plants (Davies, 1976).

2. Structure and nomenclature

The structures of carotenoids reflect their mode of biosynthesis since most contain 8 isoprene units joined in such a way that the linking of the units is reversed at the centre of the skeletally-symmetrical C_{40} molecule (fig. 1). Certain non-photosynthetic bacteria, however, form carotenoids that are biogenetically C_{30} (Davies, 1977a). The basic feature of carotenoids, on which nearly all their properties and functions depend, is the polyene chain, the sequence of conjugated olefinic bonds running through the backbone of the molecule. This chromophore usually has 9, 10 or 11 conjugated double bonds, but differing degrees of unsaturation may result in as few as 3 or as many as 15 such bonds. Cyclization of the carbon skeleton is limited to the formation in many carotenoids of a 6- (or sometimes 5-) membered ring at one or both ends. Xanthophylls, which are formed (formally and biosynthetically) from the hydrocarbon carotenes, are characterized by the presence of oxygen functions (—OH, —OR, —OCOR, —C=O, —CHO, —COOH, —COOR, epoxides, etc.) and may undergo skeletal modifications involving chain elongation (to C_{45} and C_{50} carotenoids in some non-photosynthetic bacteria) or degradation. Acetylenic and allenic bonds may also occur.

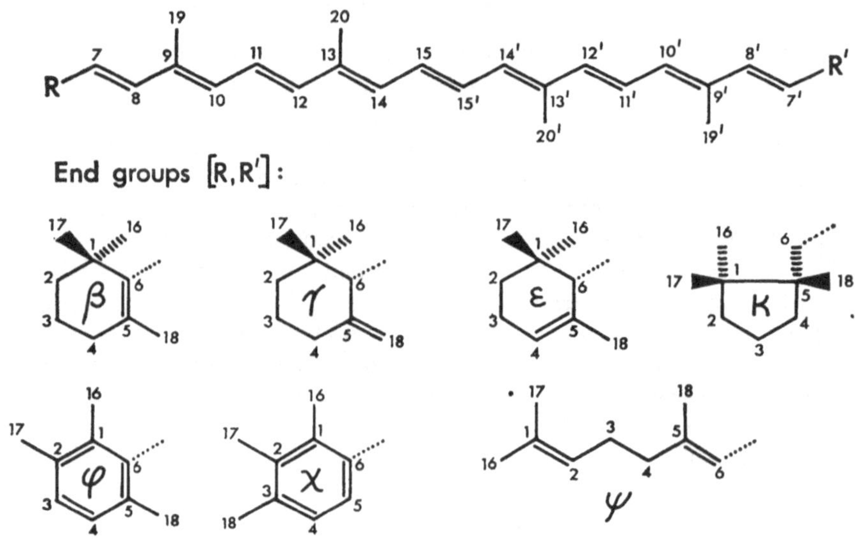

Fig. 1: The carotene end groups and the numbering of the carbon atoms of a carotenoid molecule. The end group first in the Greek alphabet is shown on the left and its carbons take the unprimed numerals

Rules for the semi-systematic nomenclature of carotenoids (Commission on Biochemical Nomenclature, 1971, 1975) describe all such structures by using only seven different end groups (*β*, *γ*, *ε*, *κ*, *φ*, *χ*, *ψ*; see fig. 1) and, of these, only four (*β*, *ε*, *κ*, *ψ*) occur in higher plant carotenoids. The end groups at either extremity of the (fully-unsaturated all-*trans*) polyene chain are used to form the C_{40} carotene stem names (e.g. *β*, *β*-carotene, *β*, *ε*-carotene, *ψ*, *ψ*-carotene, etc.). These names, modified by the appropriate prefixes and suffixes (with locant designation, see fig. 1) to denote changes in the carbon skeleton (nor-, seco-, apo-, homo-), hydrogenation level (hydro-, dehydro-, *retro*-), oxygenation (methoxy-, epoxy-, -ol, -one, -al, -oic acid, etc.), or to denote configuration (*cis*-, *R*, *S*), are used to describe the structures of all known carotenoids. These conventions are used in the Appendix (p. 48) to define the structures of all the carotenoids mentioned by their commoner trivial names in this chapter.

3. Biosynthesis

The carotenoids, with their bewildering variety of structures, clearly present the biochemist with intriguing questions concerning not only the biosynthesis of the basic C_{40} skeleton but also the origin of such diverse features as allenic linkages, acetylenic bonds and the alicyclic and aromatic rings, and the mechanism of such processes as oxygen insertion, ring formation, opening and contraction, chain elongation and degradation.

Two approaches have been used in studying carotenoid biosynthesis, the indirect kinetic approach and the more direct use of isotopes (Davies, 1977b, c). The former relies on quantitative measurements of the carotenoids of a tissue or organism (a) at different stages of development, (b) in different mutant strains, or (c) after growth in the absence and presence, and sometimes after the removal, of an inhibitor. Whereas method (a) can be used with all carotenogenic organisms, the use of the second technique (b) is largely confined to microorganisms (see Davies, 1973) since microbial mutants are so readily available. The use of the inhibitor method (c) is also more restricted in higher plants for not all the known inhibitors of carotene desaturation and cyclization are active in green tissues. Often, the results of studies of these three types are open to ambiguous interpretation and, at best, only allow biosynthetic pathways to be postulated rather than proved.

The second approach uses substrates labelled with ^{3}H, ^{14}C or ^{18}O (and latterly, with ^{2}H and ^{13}C) to prove that individual reactions (or reaction sequences) actually occur. In carrying out such studies on carotenoids, however, the biochemist is faced with serious practical difficulties. Not least among them is the instability of these natural products in solution and on chromatography; they are especially sensitive to heat, light, oxygen, acids and, in some cases, to alkali. Any failure strictly to observe a number of precautions during manipulation (see Davies, 1976) results in low overall recoveries and in the appearance of artifacts. Another major obstacle is the hydrophobic nature of the pigments; this means that they cannot readily be used as substrates for aqueous enzyme preparations and many experiments have had to rely on the use of radioactive substrates which are precursors not only of carotenoids, but also of sterols and other terpenoids. Incorporations into individual carotenoids are low and there may be contamination of samples with other radioactive terpenoids which, although they are not carotenoids, may have similar chromatographic behaviour. In spite of the risks involved in extensive manipulative procedures, particularly rigorous purification of each carotenoid is necessary for the results of such experiments to be conclusive.

Many experimental organisms and tissues show permeability barriers to common carotenoid precursors and these have to be circumvented by the use of cell-free preparations. Once these and other problems are overcome, the use of labelled substrates, especially if labelled stereospecifically (see Britton, 1976a, b), can yield considerable information about reaction sequences and the mechanism of individual reactions. Because of the many techniques available for breaking microbial cells, most carotenogenic microorganisms yield suitable enzyme systems; the most successful of these have been from *Phycomyces blakesleeanus* (Davies, 1973; Gregonis and Rilling, 1974; Bramley and Davies, 1975), a *Mycobacterium* sp. (Gregonis and Rilling, 1974) and from *Flavobacterium* spp. (McDermott et al., 1973). Only two preparations from higher plants have consistently proved successful: a soluble extract of an acetone powder from tomato plastids (the «tomato system»; Jungalwala and Porter, 1969) and the «non-aqueous chloroplast system» (Charlton et al., 1967; Britton and Goodwin, 1971).

Sufficient experiments have now been carried out with such approaches and systems for us to conclude that the general features of carotenoid biosynthesis are essentially the same in higher plants, algae, fungi and bacteria. Some of the important differences are viewed against the background of this common pathway in the following account of carotenoid formation. Particular attention is paid, however, to biosynthesis in higher plants.

3.1. Early steps, up to C_{20} (see figure 2)

In carotenoid biosynthesis, as in the formation of other terpenoids, mevalonic acid (MVA; C_6) is converted via its 5-phosphate and 5-pyrophosphate, and with the loss of C-l as CO_2, into isopentenyl pyrophosphate (IPP; C_5), often referred to as the basic building unit of all isoprenoid compounds. In most organisms, MVA is formed via 3-hydroxy-3-methylglutaryl-CoA (HMG-CoA) from three molecules of acetyl-CoA (reviewed by Beytía and Porter, 1976).

Carotenoid formation in the chloroplast, however, poses peculiar problems. MVA is certainly converted into phytoene (II; the first C_{40} hydrocarbon precursor of carotenoids) in chloroplasts (Charlton et al., 1967; Buggy et al., 1969) but the problem is the obscure origin of endogenous MVA (see Goodwin, 1977 for a full discussion). Greening tissues rapidly incorporate $^{14}CO_2$ but not exogenous [^{14}C]MVA into carotenoids, although MVA is incorporated extrachloroplastically into other terpenoids. Chloroplasts are not known to carry out glycolysis or β-oxidation of fatty acids, the two processes which, in other tissues, normally provide acetyl-CoA for synthesis. Although a stromal-phase acetyl-CoA synthase could activate free acetate, it is more likely that a pyruvate dehydrogenase bound to the lamellar membrane (Stumpf, 1975) forms acetyl-CoA from the pyruvate thought to arise from CO_2 via glycollate, glyoxylate, glycine and serine (Gore et al., 1974). Alternatively, it may be the case that HMG-CoA is formed from leucine and CO_2, for the chloroplast is permeable to leucine which is incorporated into carotenoids in green tissues (Ehrenberg and Faludi-Dániel, 1962). However it is formed in the chloroplast, HMG-CoA can be converted into MVA for the essential enzyme, HMG-CoA reductase, is present within the chloroplast (Brooker and Russell, 1975).

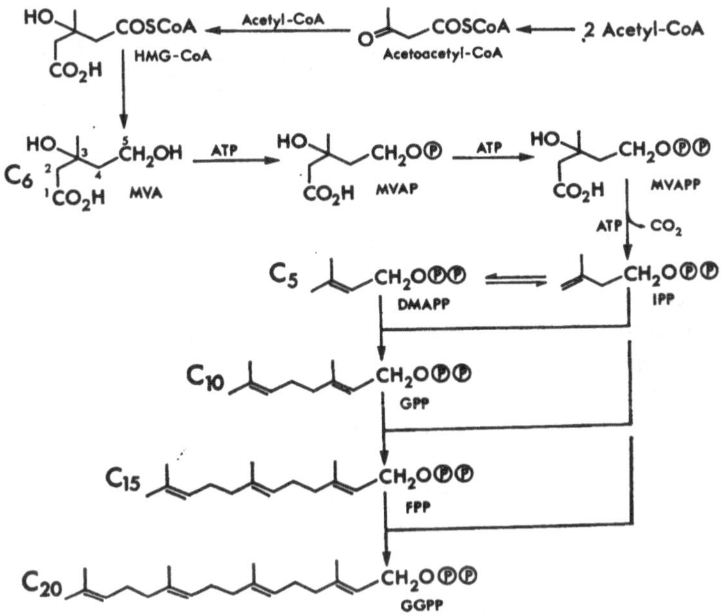

Fig. 2: A general pathway for the biosynthesis of geranylgeranyl pyrophosphate (GGPP) from acetyl-CoA.

Later steps are common to chloroplasts and other tissues. Isomerization of IPP to dimethylallyl pyrophosphate (DMAPP), the condensation of the two C_5 molecules (IPP and DMAPP) to yield geranyl pyrophosphate (GPP; C_{10}) and subsequent additions of C_5 units (IPP) result in the formation of farnesyl pyrophosphate (FPP; C_{15}; dimerizes to C_{30} squalene via presqualene pyrophosphate (PSPP); Epstein and Rilling, 1970) and geranylgeranyl pyrophosphate (GGPP; C_{20}). The carotenes (C_{40}) are products of the dimerization of GGPP; again, a cyclopropylcarbinyl pyrophosphate, prephytoene pyrophosphate (PPPP), is an intermediate (Altman et al., 1972).

3.2. Formation of the first C_{40} hydrocarbon, phytoene

In squalene formation, the cyclopropylcarbinyl cation formed by rearrangement following the loss of pyrophosphate from PSPP is stabilized by the stereospecific transfer of a hydride ion from NADPH (Epstein and Rilling, 1970). An analogous reaction for carotene biosynthesis would involve the formation from PPPP of lycopersene (I), the C_{40} counterpart of squalene. This would then have to lose two hydrogens to yield phytoene (II), a known precursor of the coloured carotenes.

The formation of PPPP from GGPP has been demonstrated (a) in a *Mycobacterium* sp. (Johnson et al., 1974), (b) by using purified yeast squalene synthase and (c) in the «tomato system»; it is converted by the latter systems (b and c) into lycopersene only as long as NADPH is present (fig. 3a; Qureshi et al., 1973; Barnes et al., 1973). There is, however, no such requirement for NADPH in carotene formation by chloroplasts (Charlton et al., 1967) or fungi (Davies, 1973); the addition of NADPH merely stimulates squalene formation (Davies, 1973).

From this evidence, and the fact that the conjugated triene phytoene (II), rather than lycopersene (I), accumulates in microorganisms when dehydrogenation is blocked by mutation (Davies, 1973) or by inhibition with diphenylamine (DPA; Davies, 1973) or Sandoz 6706 (Kleinig, 1974; Kümmel and Grimme, 1975), in an albino maize mutant (Anderson and Robertson, 1960; Treharne et al., 1960) and in maize seedlings treated with Sandoz 6706 (Bartels and McCullough, 1972; Ben-Aziz and Koren, 1974), it may be concluded that lycopersene is not an obligatory intermediate in carotene formation. Its formation from GGPP (in a lower yield than that of squalene from FPP; Beytia et al., 1973) by the yeast and tomato enzymes (Qureshi et al., 1973; Barnes et al., 1973) and by an enzyme system from *Neurospora crassa* (Grob et al., 1961), and even its presence in some carrot strains (Nusbaum-Cassuto and Villoutreix, 1965), may result from a lack of absolute specificity for FPP as the substrate for squalene synthase. The apparent conversion of lycopersene into phytoene by the «tomato system» (Barnes et al., 1973) could again be rationalized in terms of a non-absolute substrate specificity.

The «lycopersene problem» is now being resolved (after nearly 20 years) by the purification of phytoene synthetase from tomato plastids (Maudinas et al., 1977). The enzyme complex*, which has a molecular weight of about 200000, converts either IPP or GGPP into phytoene. There is an absolute requirement for Mn^{2+} but not for $NADP^+$; ATP is stimulatory (Maudinas et al., 1975) while NADPH inhibits phytoene formation and does not stimulate (in the purified enzyme) lycopersene production. The enzyme complex yields a sub-unit, PPPP synthetase (m.wt. 40000) on chromatography with low ionic strength buffers on DEAE cellulose (Islam et al., 1977).

The proposal (Altman et al., 1971) that phytoene could be formed directly from a cyclopropylcarbinyl cation (formed from PPPP on pyrophosphate loss) has now been substantiated (fig. 3). Samples of phytoene from higher plants and fungi have a predominantly 15-*cis* configuration (Jungalwala and Porter, 1965; Aung Than et al., 1972) while those from certain non-photosynthetic bacteria are all-*trans* (II; Kushwaha et al. 1972; Gregonis and Rilling, 1973, 1974; Khan et al., 1975). Studies with tomato slices (Williams et al., 1967) and bean leaf chloroplasts (Buggy et al., 1969) using [2-^{14}C, (5*R*)-5-$^{3}H_1$]MVA and [2-^{14}C, 5-$^{3}H_2$]MVA show that the 1-*pro-S*-hydrogen is lost from each GGPP molecule (5-*pro*-S of MVA) in 15-*cis* phytoene formation. This has been confirmed in microbial systems where no ^{3}H is incorporated from [4-^{14}C, (1-*S*)-1-$^{3}H_1$]GGPP into 15-*cis* phytoene (*Phycomyces*) while half the ^{3}H is retained in all-*trans* phytoene (*Mycobacterium*; Gregonis and Rilling, 1974).

One of the current views of the situation (Davies and Taylor, 1976) is illustrated in fig. 3 (for other possible mechanisms see Nes and McKean, 1977). The cyclo-

* For details, see Porter and Spurgeon (1979)

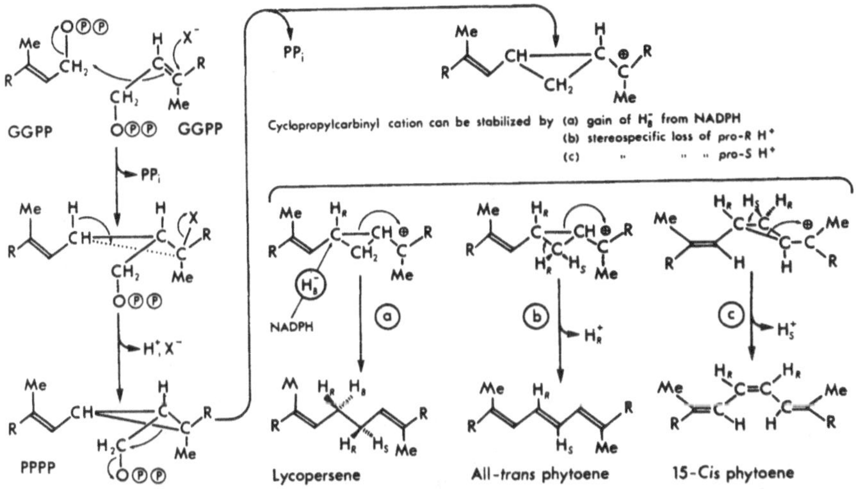

Fig. 3: Formation of prephytoene pyrophosphate (PPPP) from GGPP and its conversion into (a) lycopersene, (b) all-*trans* phytoene and (c) 15-*cis* phytoene. See the text for details

propylcarbinyl cation formed from PPPP can be stabilized in any of three ways, (a) by gaining a hydride ion from NADPH to yield lycopersene, as in a non-specific squalene synthase, (b) by the stereospecific loss of a *pro-R* proton to yield all-*trans* phytoene (as in some bacteria) or (c) by the stereospecific loss of the adjacent *pro-S* proton to form 15-*cis* phytoene, the first C_{40} hydrocarbon precursor of higher plant carotenoids.

3.3. Desaturation of phytoene

As a result of detailed carotenoid analyses on the fruits of mutant tomatoes, Porter and Lincoln (1950) proposed that phytoene is sequentially desaturated to lycopene (VII). Determinations of the structures of the intermediates (Davis et al., 1966) showed that didehydrogenations occur alternately to the left and right of the central chromophore of phytoene (II) to form phytofluene (III), ζ-carotene (IV), neurosporene (VI) and lycopene (VII). The introduction of each new double bond in the dehydrogenation sequence takes place by *trans* elimination of the 2-*pro-S* and 5-*pro-R* hydrogens originating from MVA (Williams et al., 1967; McDermott et al., 1973; Britton, 1976a). Presumably, further didehydrogenations lead to the formation of 3,4-didehydrolycopene (VIII) in some fungi and the fully desaturated 3,4,3′,4′-tetradehydrolycopene in Valencia oranges (Winterstein et al., 1960).

Since higher plant phytoene has the 15-*cis* configuration and ζ-carotene, neurosporene and lycopene generally have all-*trans* geometry, a *cis-trans* isomerization must occur at an early stage in the desaturation sequence. Tomato phytofluene (Jungalwala and Porter, 1965) differs from microbial phytofluene (see Davies and Taylor, 1976) in that it, too, is predominantly 15-*cis*. It is therefore reasonable to

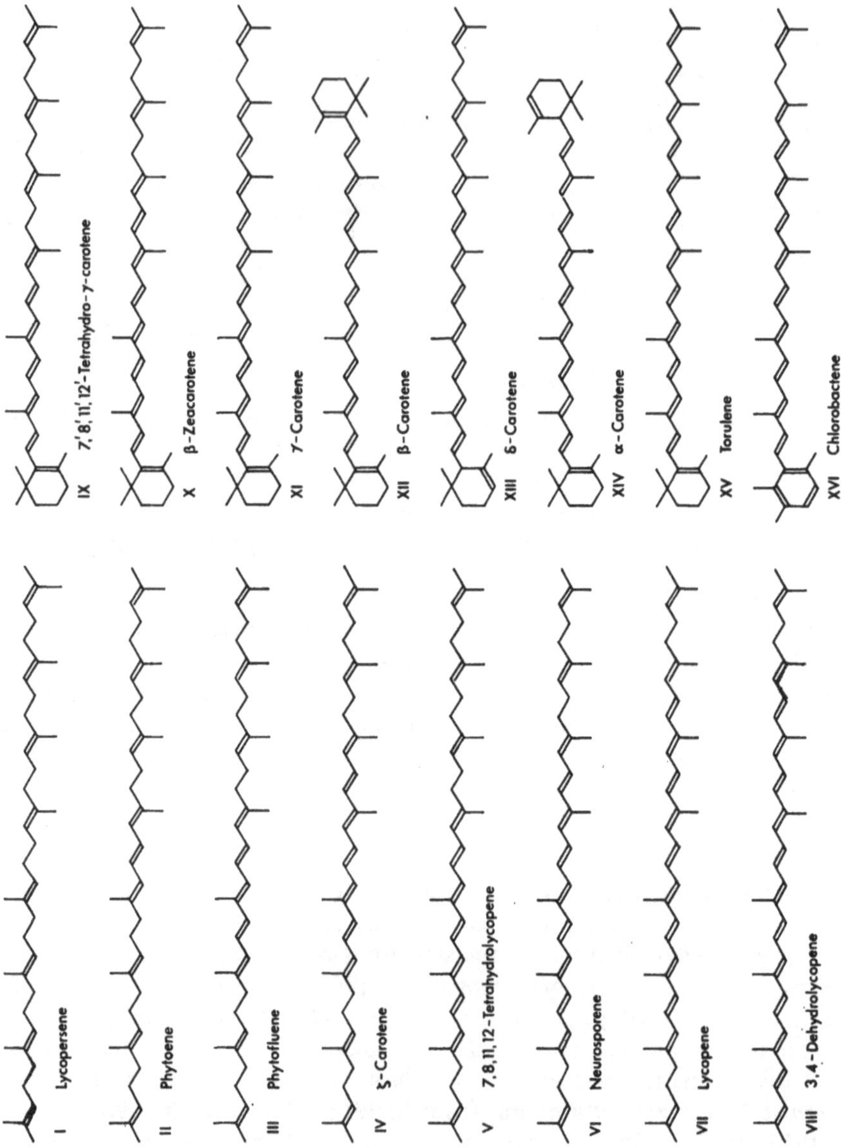

postulate that, in higher plants, 15-*cis* phytoene is dehydrogenated to 15-*cis* phytofluene which then isomerizes to all-*trans* prior to further dehydrogenation.

The conversion of detergent-solubilized 15-*cis* [^{14}C]phytoene into unsaturated acyclic (including 15-*cis* phytofluene) and cyclized carotenes has been demonstrated in «tomato systems» from the fruit of different genetic selections (Kushwaha et al., 1970) and in *Phycomyces blakesleeanus* mutant enzyme systems (see Davies, 1973).

An extract from red tomato fruit converts *cis* phytofluene into *trans* phytofluene, *trans* phytofluene to *trans* ζ-carotene, and *trans* ζ-carotene into neurosporene and lycopene as well as into cyclized carotenes (Qureshi et al., 1974a). A similar system from a tangerine mutant is capable both of converting *trans* ζ-carotene into more unsaturated all-*trans* carotenes and their poly-*cis* isomers* (proneurosporene and prolycopene; Qureshi et al., 1974b) and also of converting *cis* ζ-carotene (Raymundo and Simpson, 1972) into poly-*cis* and all-*trans* carotenes (Qureshi et al., 1974c). ζ-Carotene appears to be the branch point for poly-*cis* carotene formation in the

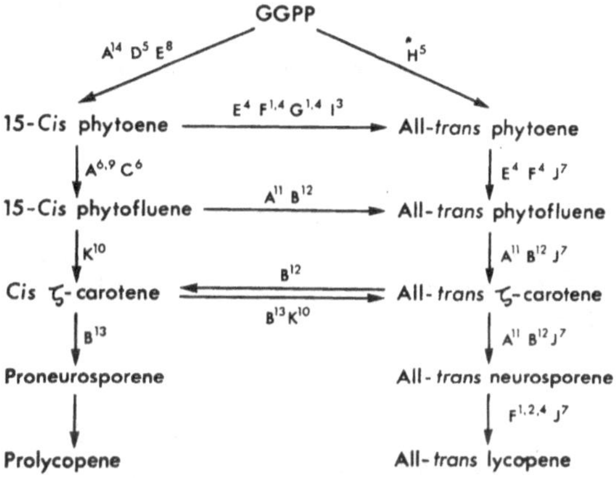

Fig. 4: Biogenetic transformations of acyclic carotenes (after Davies and Taylor 1976). The letters (A–K) indicate those organisms in which *individual reactions* have either been demonstrated *directly* (solubilized labelled substrate) or *specifically* inferred

A. Red tomato fruit (soluble enzyme from plastids)
B. Tangerine tomato fruit (same system)
C. «Hi-β», «Hi-ε» and Golden Jubilee tomato fruit (same system)
D. *Phycomyces blakesleeanus*, C5 mutant (cell-free system)
E. *P. blakesleeanus*, C9 mutant (cell-free system)
F. *P. blakesleeanus*, C115 mutant (cell-free system)
G. *Rhodospirillum rubrum*
H. *Mycobacterium* sp. (soluble enzyme system)
I. *Flavobacterium* R1560 (cell-free preparation)
J. *Halobacterium cutirubrum* (cell-free preparation)
K. *Scenedesmus obliquus*, PGl mutant

The superscript numbers refer to literature sources, as follows:
1. Bramley, 1973 – 2. Bramley et al., 1977 – 3. Brown et al., 1975 – 4. Davies, 1973 – 5. Gregonis and Rilling, 1974 – 6. Kushwaha et al., 1970 – 7. Kushwaha et al., 1976 – 8. Lee and Chichester, 1969 – 9. Papastephanou et al., 1973 – 10. Powls and Birtton, 1977 – 11. Qureshi et al., 1974a – 12. Qureshi et al., 1974b – 13. Qureshi et al., 1974c – 14. Shah et al., 1968.

* Prolycopene is now known to be 7,9,7′,9′-tetra-*cis* lycopene (Englert et al., 1979; Clough and Pattenden, 1979). Its precursors in the tangerine tomato are 15-*cis*-phytoene, 15,9′-di-*cis* phytofluene, 9,9′-di-*cis* ζ-carotene and 9,7′,9′-tri-*cis* neurosporene (Clough and Pattenden, 1979).

tangerine tomato (fig. 4). The final step of the general desaturation sequence, the conversion of neurosporene into lycopene, has been demonstrated *in vitro* in microbial (Kushwaha et al., 1976; Bramley et al., 1977) and higher plant (B. H. Davies and M. A. Flores-Ortega, unpublished work) systems.

Carotene desaturation in other microbial systems is more complex. The first difference is that some photosynthetic bacteria (e.g. *Rhodospirillum rubrum*; Davies, 1970a) use an unsymmetrical isomer of ζ-carotene (7,8,11,12-tetrahydrolycopene, V) exclusively as the intermediate between phytofluene and neurosporene, while some fungi (e.g. *Neurospora crassa*; Davies et al., 1974) have alternative dehydrogenation sequences in which both isomeric conjugated heptaenes are intermediates. Other differences from higher plants are in the configuration of phytoene and phytofluene. In those fungi and photosynthetic bacteria which have been examined, and in the non-photosynthetic bacterium *Flavobacterium* R 1560, a cell-free preparation of which converts labelled 15-*cis* into all-*trans* phytoene (Brown et al., 1975), 15-*cis* phytoene is formed first, isomerized to all-*trans* phytoene (a step which is at least in part photochemical in *Mucor hiemalis*; Herber et al., 1972) and then dehydrogenated to all-*trans* phytofluene. A different situation exists in a mutant of the alga *Scenedesmus obliquus* in which the *cis* to *trans* isomerization is delayed until the ζ-carotene level (Powls and Britton, 1977). In those bacteria which form all-*trans* phytoene directly from PPPP (e.g. *Halobacterium cutirubrum*, Kushwaha et al., 1972, 1975), the all-*trans* configuration is present throughout the dehydrogenation sequence. A detailed comparison of these configurational relationships has been made elsewhere (Davies and Taylor, 1976), but the salient points are summarized in fig. 4.

3.4. Triterpenoid (C_{30}) carotenoids

On investigating the acyclic carotenes of the non-photosynthetic Gram-positive bacterium *Streptococcus faecium* UNH 564 P, Taylor and Davies (1973) made the surprising observation that, although these compounds correspond in their absorption spectra to those of the carotenes (II–VI) described above, they are all triterpenoid (C_{30}) rather than C_{40} molecules. Gas-liquid chromatography of the perhydrocarotenes (Taylor and Davies, 1975) was particularly useful in showing that they had the same carbon skeleton as squalene. The current nomenclature rules permit their designation as 4,4′-diapophytoene, 4,4′-diapophytofluene, 4.4′-diapo-ζ-carotene, 4,4′-diapo-7,8,11,12-tetrahydrolycopene and 4,4′-diaponeurosporene (Taylor and Davies, 1974a). Xanthophylls are also present in the organism; these include 4-hydroxy-4,4′-diaponeurosporene and its D-glucoside (Taylor and Davies, 1974b) and 4,4′-diaponeurosporen-4-al (Taylor and Davies, 1976).

Studies on the biosynthesis of the triterpenoid carotenoids have concentrated on the formation of diapophytoene by a cell-free preparation from *S. faecium*. Labelled MVA, IPP and FPP, the latter prepared using a barley embryo enzyme system (Davies et al., 1975), are all incorporated. That the molecule arises from the dimerization of FPP, rather than by the degradative loss of the terminal C_5 units from C_{40} phytoene, has been demonstrated by following the simultaneous incorporations of [^{14}C]IPP and [^{3}H]FPP (Davies and Taylor, 1976).

3.5. Formation of alicyclic carotenes

It is clear that, in some organisms, lycopene (VII) is cyclized first at one end to form γ-carotene (XI) and then at the other to give β-carotene (XII). It is equally clear that, in other organisms, particularly fungi (see fig. 5), the primary cyclization may occur at one end of the molecule of an acyclic carotene as soon as the 7,8-bond is desaturated and the secondary cyclization at the other end may await corresponding desaturation (Goodwin, 1971; Davies, 1975). Thus in fungi, the monocyclic isomers of 7,8,11,12-tetrahydrolycopene (V), neurosporene (VI) and lycopene (VII), i.e. 7′,8′,11′,12′-tetrahydro-γ-carotene (IX), β-zeacarotene (X) and γ-carotene (XI) respectively, may all be on the route to β-carotene (XII). The torulene (XV) of some fungi may result either from the further dehydrogenation of γ-carotene (XI) or from the cyclization of 3,4-didehydrolycopene (VIII; see fig. 5).

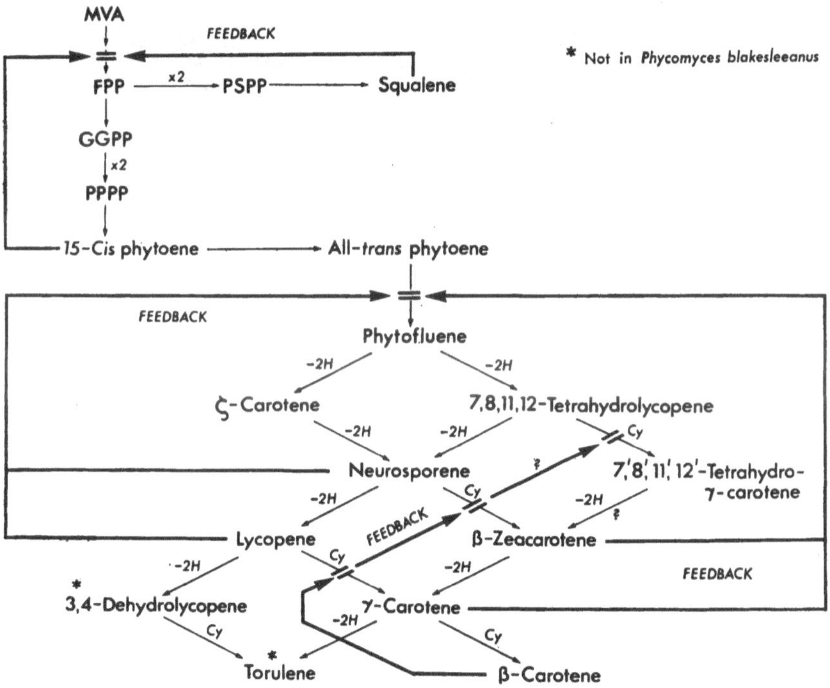

Fig. 5: Carotene biosynthesis and its negative feedback control in fungi

Similar routes, involving δ-carotene (XIII), are thought to lead to α-carotene (XIV) formation in higher plants (Davies 1976). Only in fungi, however, have alternative routes to bicyclic carotenes actually been shown to occur (Bramley et al., 1977).

Unambiguous definitions of the actual *in vivo* substrates for the carotene cyclase(s) must await the purification of the enzymes concerned. What has become clear recently is the actual mechanism of the reaction forming β-rings in carotenoids. Studies with stereospecifically-labelled MVA showed, some years ago (Goodwin and Williams, 1965a, b; Williams et al., 1967), that β- and ε-rings are formed independently but possibly from a common enzyme-bound intermediate (often visualized as a carbonium ion; Britton, 1976b). A cyclized carbonium ion formed (fig. 6a) as a result of proton attack at C-2 could be stabilized by proton loss (fig. 6b) either from C-6 (to yield a β-ring) or from C-4 (to yield an ε-ring). The proton attacking C-2 would be retained in the cyclized product.

The PGl strain of *Scenedesmus obliquus* accumulates ζ-carotene when it is cultured heterotrophically in the dark (Powls and Britton, 1977). Cells were harvested, resuspended in D_2O and their starch reserves depleted, and then the suspension was illuminated in an air-CO_2 atmosphere. Bicyclic carotenoids were formed and α-

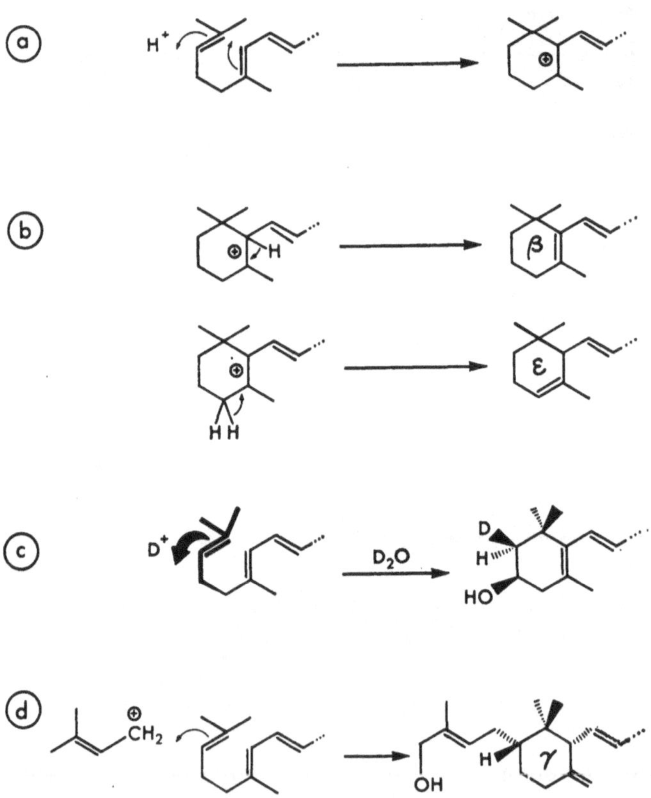

Fig. 6: Mechanisms of carotenoid cyclization reactions visualized as involving carbonium ions. Attack at C-2 by a proton gives (a) a carbonium ion which is (b) the common precursor of β- and ε-rings in C_{40} carotenes. The direction of this attack in zeaxanthin biosynthesis (c) differs from that of an electrophilic C_5 unit in sarcinaxanthin biosynthesis (d). See the text for details

carotene, β-carotene, lutein and zeaxanthin were purified and submitted to mass spectrometric analysis. Each bicyclic molecule formed from the unlabelled acyclic precursor contained 2 atoms of deuterium (Britton et al., 1977a).

The same approach was used with a *Flavobacterium* sp., R 1519, in which nicotine blocks cyclization and causes lycopene to accumulate; removal of the inhibitor and aerobic reincubation results in the formation of zeaxanthin (McDermott et al., 1974). After removal of the nicotine, cells were reincubated aerobically in a medium prepared using D_2O. The resulting dideuteriozeaxanthin was examined by ^{1}H n.m.r. spectroscopy which revealed the location of the deuteriums at C-2 and C-2′ and showed that the sample had the (2*S*, 2′*S*) configuration (fig. 6c). It was concluded that the initial proton attack in the cyclization of lycopene is on the *re,re* face of the C-1,2 double bond (Britton et al., 1977b).

3.6. C_{45} and C_{50} homocarotenoid formation

The C_{45} and C_{50} homocarotenoids are characteristic of certain types of non-photosynthetic bacteria. Structurally, they have C_{40} skeletons formally extended at C-2 (and/or C-2′) by additional isoprene units. The commonest types have, in the central C_{40} core of the molecule, acyclic ψ-end groups (e.g. bacterioruberin; *Halobacterium* spp.) or cyclic γ- (e.g. sarcinaxanthin; *Sarcina lutea* and *S. flava*) or ε-end groups (e.g. decaprenoxanthin; *Flavobacterium dehydrogenans*, *Corynebacterium* spp.).

The C_5 addition reactions leading to bacterioruberin (H. *cutirubrum*; Kushwaha and Kates, 1976) and sarcinaxanthin (*S. flava*; Al-Windawi, 1977; Davies, 1979), like proton-initiated cyclization and 1,2-hydration reactions of C_{40} carotenes, can be inhibited by nicotine. Although the two reaction types, initiated respectively by attack at C-2 of an acyclic carotene by either a proton or an electrophilic C_5 unit, are similar in showing sensitivity to nicotine, they are stereochemically different (fig. 6c, d). Since the configuration of sarcinaxanthin is (2*R*, 2′*R*) (Hertzberg and Liaaen-Jensen, 1977), it must be formed by the C_5 units attacking the *si, si* faces of the 1,2-double bonds of an acyclic carotene, whereas proton attacks are on the *re, re* faces (Britton et al., 1977b; fig. 6).

3.7. Aromatic carotenes

The aromatic carotenes, which have 1,2,5-trimethylphenyl (φ-) or 1,2,3-trimethylphenyl (χ-) end groups, were first encountered in the Japanese sea sponge, *Reniera japonica* (Yamaguchi, 1957, 1958, 1960), but the end groups have since been found in the pigments of both photosynthetic (*Chloropseudomonas, Chlorobium, Chromatium* and *Thiocystis*) and non-photosynthetic (*Mycobacterium* and *Streptomyces*) bacteria. Studies on their biosynthesis have been few, but Moshier and Chapman (1973) demonstrated the incorporation of [2-^{14}C]MVA into chlorobactene (XVI) by cultures of *Chloropseudomonas ethylica*. By oxidatively degrading the chlorobactene to labelled benzenecarboxylic acids, these workers showed that the aromatic ring arises from MVA and decarboxylation of the acids revealed that stereospecific migration of the methyl group derived from C-3′ of MVA must have taken place in the formation of the 1,2,5-trimethylphenyl (φ-) end group. The effects of different concentrations of nicotine on *Chlorobium limicola* var. *sulphatophilum* suggest that lycopene (VII) and γ-carotene (XI) are precursors of chlorobactene (Leutwiler and Chapman, 1978).

3.8. Aspects of the control of carotene formation in *Phycomyces*

While evidence from *Fusarium aquaeductuum* (Lang and Rau, 1972) tends to militate against the operation of a multienzyme carotenogenic complex, the situation in *Phycomyces* is quite different (Davies and Taylor, 1976). Quantitative genetic complementation studies on the C 2 (carotenoidless) and C 9 (lycopene-accumulating) mutants of *P. blakesleeanus* suggest that the conversion of phytoene into β-carotene (fig. 5) is catalyzed by a multienzyme aggregate containing four dehydrogenases and two cyclases (de la Guardia et al., 1971). The operation of such an aggregate in *Phycomyces* (which may be the case only during the exponential phase of growth; Hsu et al., 1974) is consistent with the low conversions (compared with those of MVA) into β-carotene by a cell-free preparation even using efficiently solubilized high specific activity carotene substrates (Davies, 1973, 1975). Further genetic studies on *P. blakesleeanus* mutants (Ootaki et al., 1973) showed that only three genes are involved in normal carotenogenesis, *carA*, *carB* and *carR*. While *carA* and *carB* mutants are all white, only the *carB* mutants accumulate phytoene, the *carA* mutants (e.g. C 2) are carotenoidless and the red *carR* mutants (e.g. C 9) accumulate lycopene. The derivation from a red lycopene-producing strain of a leaky *carB* mutant (S 86) which forms less phytoene than other *carB* mutants but also forms (in decreasing amounts in order of desaturation) phytofluene, ζ-carotene, neurosporene and lycopene showed that the dehydrogenases of the aggregate are four copies of the same enzyme coded by the *carB* gene (Eslava and Cerdá-Olmedo, 1974). Although it is also suggested that the cyclases are two copies of the product of the *carR* gene (Eslava and Cerdá-Olmedo, 1974), they show different sensitivities to nicotine (Davies and Rees, 1973).

The production of carotenes by *P. blakesleeanus* is also influenced by light and, possibly, by sexual activity. While the stimulatory effects of β-ionone on β-carotene formation have been recognized for some time (Goodwin, 1971), it was only recently shown that retinol (vitamin A) and β-ionone, presumably acting as trisporic acid mimics, stimulate the formation of phytoene in *carB*, and lycopene in *carR*, mutants (Eslava et al., 1974). The *carB* mutants contain more phytoene, and the *carR* mutants more lycopene, than the wild-type *Phycomyces* contains β-carotene. This suggests that β-carotene exerts end-product regulation so that a lack of β-carotene induces a greater carotenogenic activity with an increased accumulation of intermediates (Eslava and Cerdá-Olmedo, 1974). Indeed, studies using cycloheximide and 2-(4-chlorophenylthio)triethylamine hydrochloride (CPTA) indicate that β-carotene regulates its own production through regulating enzyme synthesis (Murillo and Cerdá-Olmedo, 1975), End-product regulation of enzyme synthesis and photoinduction may exert their effects through a common mechanism (operating before phytoene, for *carB* and *carR* mutants do not significantly increase carotenogenesis as a result of illumination; Eslava and Cerdá-Olmedo, 1974).

Evidence has appeared from our own biosynthetic studies which indicates an element of negative feedback control in *P. blakesleeanus*. Individual carotenes were added to an enzyme system actively forming β-carotene from [2-^{14}C]MVA (Bramley and Davies, 1976). While the addition of β-carotene itself specifically inhibited cyclase activity, lycopene, γ-carotene, neurosporene or β-zeacarotene, but not β-carotene, each inhibited phytoene dehydrogenation. The addition of phytoene or squalene reduced terpenoid formation in general. The ability of lycopene, and the failure of β-carotene, to inhibit phytoene desaturation in *Phycomyces* is confirmed by the growth characteristics of two mutant strains (Bramley and Davies, 1975). In a high β-carotene strain (C 115), no phytoene but only β-carotene accumulates in the late stages of growth, in contrast to the situation in the C 9 lycopene-accumulating strain where phytoene accumulates and radioactivity from [2-^{14}C]MVA appears in phytoene rather than in lycopene as soon as high lycopene levels are reached. The feedback mechanisms thought to operate in *P. blakesleeanus* are illustrated in fig. 5.

3.9. C_{40} Xanthophyll formation

Much of the evidence that xanthophylls are formed from carotenes in the presence of oxygen has come from studies on algae such as *Chlorella pyrenoidosa* (Claes, 1958, 1959) and *Scenedesmus obliquus* (Britton et al., 1977c; Powls and Britton, 1977); this evidence, however, is indirect. Direct proof of this long-held assumption is as follows. *Flavobacterium* R 1519, which normally forms zeaxanthin, accumulates labelled lycopene when grown in the presence of [^{14}C]MVA and nicotine. When inhibitor and substrate are both removed, anaerobic resuspension results in the lycopene being converted into labelled β-carotene. When air is admitted, the β-carotene is converted into zeaxanthin which carries the radioactive label (McDermott et al., 1974). The significance of this microbial experiment to the higher plant situation is that the chloroplast xanthophylls are, like zeaxanthin, 3-hydroxy-derivatives of α- or β-carotene. The 3-hydroxyl groups originate as molecular oxygen (Yamamoto et al., 1962) and are probably inserted by a mixed function oxygenase, for the hydroxylation is stereospecific (3*R*) and no ketonic intermediate is involved (see Britton, 1976a, b).

A second important feature of chloroplast xanthophylls is that some (antheraxanthin, violaxanthin, neoxanthin) are 5,6-epoxides. These oxygen functions also come from molecular oxygen (Yamamoto and Chichester, 1965) and are inserted stereospecifically (5*R*, 6*S*) and, therefore, enzymically (see Britton, 1976a).

The 3-hydroxy-5,6-epoxy-β-end group may have an important role in the formation of other structural features of higher plant and algal xanthophylls. Studies on the levels of fruit carotenoids of colour varieties of *Capsicum annuum* (Davies et al., 1970a) showed that, in red fruits, capsanthin and capsorubin were formed from antheraxanthin and violaxanthin, presumably by pinacollic rearrangements leading to ϰ-end groups (fig. 7a). In the mature yellow fruit of another variety which lacks the enzymic capability for cyclopentanol ring formation, neoxanthin was formed, again apparently at the expense of violaxanthin. Thus an alternative transformation of the 3-hydroxy-5,6-epoxy-β-end group may form the allenically-linked 3,5-dihydroxy-β-end group of neoxanthin (fig. 7b), although another view is that an intermediate peroxide may be involved (Weedon, 1970). Both allenic and acetylenic bonds are found in xanthophylls and are not known in carotenes, so their formation must be an exclusive feature of the further metabolism of xanthophylls. It is conceivable that the algal acetylenic carotenoids such as diadinoxanthin may be formed from epoxides via allenes. This seemed to be the case in *Euglena gracilis* where diadinoxanthin appeared to be formed at the expense of neoxanthin (fig. 7c; B. H. Davies and A. J. Cutler, unpublished work). The *retro* carotenoid eschscholtzxanthin is formed from antheraxanthin by a mechanism thought to involve the 3-hydroxy-5,6-epoxy-β-end group (fig. 7d; Williams et al., 1966).

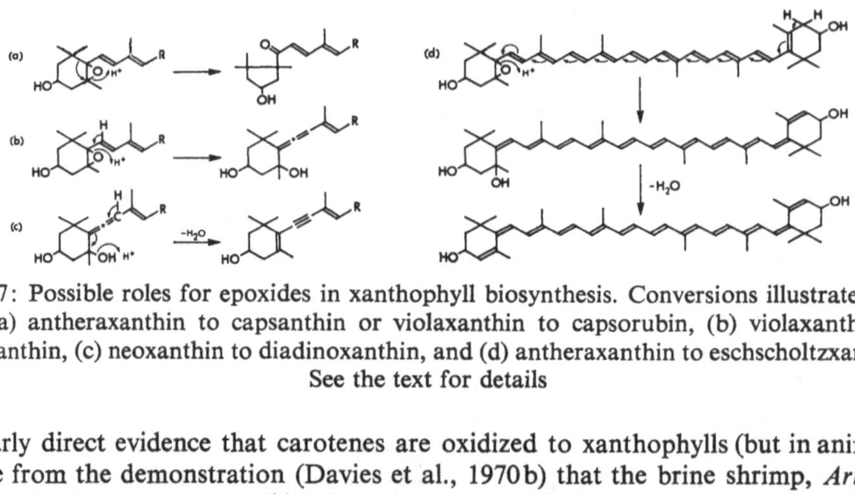

Fig. 7: Possible roles for epoxides in xanthophyll biosynthesis. Conversions illustrated are of: (a) antheraxanthin to capsanthin or violaxanthin to capsorubin, (b) violaxanthin to neoxanthin, (c) neoxanthin to diadinoxanthin, and (d) antheraxanthin to eschscholtzxanthin. See the text for details

Early direct evidence that carotenes are oxidized to xanthophylls (but in animals) came from the demonstration (Davies et al., 1970b) that the brine shrimp, *Artemia salina*, is able to convert [^{14}C]β-carotene, via echinenone (4-keto-β-carotene) into canthaxanthin (4,4'-diketo-β-carotene). Such 4-β-ketocarotenoids are characteristic of certain animals (crustacea, fish, birds; Fox, 1976), are the secondary carotenoids produced by many green algae under nutritional stress (see Czygan, 1968a, b; Donkin, 1976) and may also occur in flowers of *Adonis* species (Egger, 1965; Egger and Kleinig, 1967; Czygan, 1969) and in microorganisms. The evidence from the alga *Dictyococcus cinnibarinus* (Gribanovski-Sassu, 1972) and from a *Brevibacterium* sp. (Hsieh et al., 1974) is that 4-hydroxylation of the β-ring precedes oxidation of C-4 to a ketone; no evidence could be obtained for the participation of the expected 4-hydroxy intermediates in *Artemia*, possibly because they exist only in an enzyme-bound form (Davies et al., 1970b), but such intermediates seem to be involved in some other crustacea (Gilchrist, 1968).

Acyclic ketocarotenoids are formed by some photosynthetic bacteria. The oxygen originates from molecular oxygen in those organisms which form 2-ketocarotenoids aerobically (e.g. *Rhodopseudomonas sphaeroides*; Shneour, 1962) but those organisms which oxidize C-4 (of a ψ-end group) to a ketone under strict anaerobiosis (e.g. *Chromatium okenii*) are more likely to hydrate a 3,4-double bond and then dehydrogenate the intermediate alcohol (Goodwin, 1971).

Of the carotenoids of the four families of photosynthetic bacteria (Rhodospirillaceae, Chromatiaceae, Chlorobiaceae and the Cyanobacteria), the biosynthesis of those of the Rhodospirillaceae (e.g. *Rhodospirillum rubrum* and *Rhodopseudomonas sphaeroides*) is the best understood. These pigments are typically mono- (e.g. spheroidene) or dimethoxy-carotenoids (e.g. spirilloxanthin); the insertion of each methoxyl group is accompanied by a 3,4-didehydrogenation at that end of the molecule. The end group is formed from the ψ-end group (of e.g. lycopene in spirilloxanthin formation or neurosporene in spheroidene biosynthesis) by a sequence of three reactions, namely 1,2-hydration, 3,4-didehydrogenation and O-methylation using S-adenosylmethionine (Singh et al., 1973). While 1,2-hydration is always the first step (Davies and Aung Than, 1974), the other two reactions may occur in either order (Davies, 1970b). This, and the fact that methoxyl group formation can proceed before carotene dehydrogenation (to lycopene) is complete, means that diphenylamine-inhibited cultures of *R. rubrum* may have over 40 different carotenoids as a result of the apparently low substrate specificity of what is assumed to be but a small number of enzymes (Davies, 1975).

3.10. Partially degraded carotenoids

The characteristic carotenoids of some tissues, such as the seco- and apo-carotenoids of citrus fruits, are probably the products of the first stages of carotenoid degradation, formed by ring opening and β-oxidation. Yokoyama and White (1966) suggested that β-apo-8'-carotenal and β-apo-10'-carotenal are natural degradation products of β-carotene while β-citraurin arises from zeaxanthin (or β-cryptoxanthin). A ring-opening mechanism involving an intermediate peroxide (Weedon, 1970) could lead to seco-carotenoid formation. Thus β-carotene would yield semi-β-carotenone and then β-carotenone, α-carotene would yield semi-α-carotenone and triphasiaxanthin would be formed from β-cryptoxanthin. Such reactions would be consistent with the changes in the carotenoid levels of ripening fruit of *Triphasia trifolia* (Yokoyama and White, 1970). Azafrin (C_{27}), bixin (C_{25}) and crocetin (C_{20}) are all thought to be formed from C_{40} carotenoids by oxidative degradation (Weedon, 1971).

Our studies on carotenoid biosynthesis receive financial support from the Science Research Council.

Appendix

Trivial and semi-systematic names and structures of the carotenoids mentioned above.*

Antheraxanthin	5,6-Epoxy-5,6-dihydro-β,β-carotene-3,3′-diol
β-Apo-8′-carotenal	8′-Apo-β-caroten-8′-al
β-Apo-10′-carotenal	10′-Apo-β-caroten-10′-al
Azafrin	5,6-Dihydroxy-5,6-dihydro-10′-apo-β-caroten-10′-oic acid
Bacterioruberin	2,2′-Bis(3-hydroxy-3-methylbutyl)-3,4,3′,4′-tetradehydro-1,2,1′,2′-tetrahydro-ψ,ψ-carotene
Bixin	Methyl hydrogen 9′-*cis*-6,6′-diapocarotene-6,6′-dioate
Canthaxanthin	β,β-Carotene-4,4′-dione
Capsanthin	3,3′-Dihydroxy-β,κ-caroten-6′-one
Capsorubin	3,3′-Dihydroxy-κ,κ-carotene-6,6′-dione
α-Carotene	β,ε-Carotene
β-Carotene	β,β-Carotene
γ-Carotene	β,ψ-Carotene
δ-Carotene	ε,ψ-Carotene
ζ-Carotene	7,8,7′,8′-Tetrahydro-ψ,ψ-carotene
β-Carotenone	5,6,5′,6′-Diseco-β,β-carotene-5,6,5′,6′-tetrone
Chlorobactene	φ,ψ-Carotene
Crocetin	8,8′-Diapocarotene-8,8′-dioic acid
β-Cryptoxanthin	β,β-Caroten-3-ol
Decaprenoxanthin	2,2′-Bis(4-hydroxy-3-methyl-2-butenyl)-ε,ε-carotene
Diadinoxanthin	5,6-Epoxy-7′,8′-didehydro-5,6-dihydro-β,β-carotene-3,3′-diol
4,4′-Diapo-ζ-carotene	4,4′-Diapo-7,8,7′,8′-tetrahydro-ψ,ψ-carotene
4,4′-Diaponeurosporen-4-al	4,4′-Diapo-7′,8′-dihydro-ψ,ψ-caroten-4-al
4,4′-Diaponeurosporene	4,4′-Diapo-7,8-dihydro-ψ,ψ-carotene
4,4′-Diapophytoene	4,4′-Diapo-7,8,11,12,7′,8′,11′,12′-octahydro-ψ,ψ-carotene
4,4′-Diapophytofluene	4,4′-Diapo-7,8,11,12,7′,8′-hexahydro-ψ,ψ-carotene
4,4′-Diapo-7,8,11,12-tetrahydrolycopene	4,4′-Diapo-7,8,11,12-tetrahydro-ψ,ψ-carotene
3,4-Didehydrolycopene	3,4-Didehydro-ψ,ψ-carotene
Echinenone	β,β-Caroten-4-one
Eschscholtzxanthin	4′,5′-Didehydro-4,5′-*retro*-β,β-carotene-3,3′-diol
4-Hydroxy-4,4′-diaponeurosporene	4,4′-Diapo-7′,8′-dihydro-ψ,ψ-caroten-4-ol

* Figure 1 shows end groups and numbering of carbons throughout the carotenoid molecule. A full list of carotenoid names and structures has been prepared by Straub (1976).

Lutein	β,ε-Carotene-3,3'-diol
Lycopene	ψ,ψ-Carotene
Lycopersene	7,8,11,12,15,7',8',11',12',15'-Decahydro-ψ,ψ-carotene
Neoxanthin	5',6'-Epoxy-6,7-didehydro-5,6,5',6'-tetrahydro-β,β-carotene-3,5,3'-triol
Neurosporene	7,8-Dihydro-ψ,ψ-carotene
Phytoene	7,8,11,12,7',8',11',12'-Octahydro-ψ,ψ-carotene
Phytofluene	7,8,11,12,7',8'-Hexahydro-ψ,ψ-carotene
Sarcinaxanthin	2,2'-Bis(4-hydroxy-3-methyl-2-butenyl)-γ,γ-carotene
Semi-α-carotenone	5,6-Seco-β,ε-carotene-5,6-dione
Semi-β-carotenone	5,6-Seco-β,β-carotene-5,6-dione
Spheroidene	1-Methoxy-3,4-didehydro-1,2,7',8'-tetrahydro-ψ,ψ-carotene
Spirilloxanthin	1,1'-Dimethoxy-3,4,3',4'-tetradehydro-1,2,1',2'-tetrahydro-ψ,ψ-carotene
3,4,3',4'-Tetradehydrolycopene	3,4,3',4'-Tetradehydro-ψ,ψ-carotene
7',8',11',12'-Tetrahydro-γ-carotene	7',8',11',12'-Tetrahydro-β,ψ-carotene
7,8,11,12-Tetrahydrolycopene	7,8,11,12-Tetrahydro-ψ,ψ-carotene
Torulene	3',4'-Didehydro-β,ψ-carotene
Triphasiaxanthin	3'-Hydroxy-5,6-seco-β,β-carotene-5,6-dione
Violaxanthin	5,6,5',6'-Diepoxy-5,6,5',6'-tetrahydro-β,β-carotene-3,3'-diol
β-Zeacarotene	7',8'-Dihydro-β,ψ-carotene
Zeaxanthin	β,β-Carotene-3,3'-diol

References

ALTMAN, L. J., R. C. KOWERSKI, and H. C. RILLING, 1971: Synthesis and conversion of presqualene alcohol to squalene. J. Amer. Chem. Soc. **93**, 1782–1783.

ALTMAN, L. J., L. ASH, R. C. KOWERSKI, W. W. EPSTEIN, B. R. LARSEN, H. C. RILLING, F. MUSCIO, and D. E. GREGONIS, 1972: Prephytoene pyrophosphate. A new intermediate in the biosynthesis of carotenoids. J. Amer. Chem. Soc. **94**, 3257–3259.

AL-WINDAWI, H. K., 1977: Studies on bacterial carotenoids. Ph. D. Thesis, University of Wales.

ANDERSON, I. C., and D. S. ROBERTSON, 1960: Role of carotenoids in protecting chlorophyll from photodestruction. Plant Physiol. **35**, 531–534.

AUNG THAN, P. M. BRAMLEY, B. H. DAVIES, and A. F. REES, 1972: The stereochemistry of phytoene. Phytochemistry **11**, 3187–3192.

BARNES, F. J., A. A. QURESHI, E. J. SEMMLER, and J. W. PORTER, 1973: Prelycopersene pyrophosphate and lycopersene, intermediates in carotene biosynthesis. J. Biol. Chem. **248**, 2768–2773.

BARTELS, P. G., and C. MCCULLOUGH, 1972: A new inhibitor of carotenoid synthesis in higher plants: 4-chloro-5-(dimethylamino)-2-α,α,α-(trifluoro-*m*-tolyl)-3(2H)-pyridazinone (Sandoz 6706). Biochem. Biophys. Res. Commun. **48**, 16–22.

BEN-AZIZ, A., and E. KOREN, 1974: Interference in carotenogenesis as a mechanism of action of the pyridazinone herbicide Sandoz 6706. Accumulation of C_{40} carotenoid precursors, inhibition of β-carotene synthesis and enhancement of phytoene epoxidation. Plant Physiol. **54**, 916–920.

BEYTÍA, E. D., and J. W. PORTER, 1976: Biochemistry of polyisoprenoid biosynthesis. Ann. Rev. Biochem. **45**, 113–142.

BEYTÍA, E., A. A. QURESHI, and J. W. PORTER, 1973: Squalene synthetase. III. Mechanism of the reaction. J. Biol. Chem. **248**, 1856–1867.

BRAMLEY, P. M., 1973: Carotenoid biosynthesis in microorganisms. Ph. D. Thesis, University of Wales.

BRAMLEY, P. M., and B. H. DAVIES, 1975: Carotene biosynthesis by cell extracts of mutants of *Phycomyces blakesleeanus*. Phytochemistry **14**, 463–469.

BRAMLEY, P. M., and B. H. DAVIES, 1976: β-Carotene biosynthesis by extracts of the C115 mutant of *Phycomyces blakesleeanus*. Phytochemistry **15**, 1913–1916.

BRAMLEY, P. M., B. H. DAVIES, and AUNG THAN, 1977: Alternative pathways of carotene cyclization in *Phycomyces blakesleeanus*. Phytochemistry **16**, 235–238.

BRITTON, G., 1976a: Biosynthesis of carotenoids. In T. W. Goodwin (ed.), Chemistry and biochemistry of plant pigments, 2nd ed., Vol. 1, 262–327. Academic Press, London, New York and San Francisco.

BRITTON, G., 1976b: Later reactions of carotenoid biosynthesis. Pure Appl. Chem. **47**, 223–236.

BRITTON, G., and T. W. GOODWIN, 1971: Biosynthesis of carotenoids. Methods Enzymol. **18(C)**, 654–701.

BRITTON, G., W. J. S. LOCKLEY, R. POWLS, T. W. GOODWIN, and L. M. HEYES, 1977a: Carotenoid transformations during chloroplast development in *Scenedesmus obliquus* PGl demonstrated by deuterium labelling. Nature **268**, 81–82.

BRITTON, G., W. J. S. LOCKLEY, N. J. PATEL, T. W. GOODWIN, and G. ENGLERT, 1977b: The use of deuterium labelling to elucidate the stereochemistry of the initial step of the cyclization reaction in zeaxanthin biosynthesis in a *Flavobacterium*. J. Chem. Soc., Chem. Commun. 655–656.

BRITTON, G., R. POWLS, and R. M. SCHULZE, 1977c: The effect of illumination on the pigment composition of the ζ-carotenic mutant, PGl, of *Scenedesmus obliquus*. Arch. Microbiol. **113**, 281–284.

BROOKER, J. D., and D. W. RUSSELL, 1975: Subcellular localization of 3-hydroxy-3-methylglutaryl coenzyme A reductase in *Pisum sativum* seedlings. Arch. Biochem. Biophys. **167**, 730–737.

BROWN, D. J., G. BRITTON, and T. W. GOODWIN, 1975: Carotenoid biosynthesis by a cell-free preparation from a *Flavobacterium* species. Biochem. Soc. Trans. **3**, 741–742.

BUGGY, M. J., G. BRITTON, and T. W. GOODWIN, 1969: Stereochemistry of phytoene biosynthesis by isolated chloroplasts. Biochem. J. **114**, 641–643.

CHARLTON, J. M., K. J. TREHARNE, and T. W. GOODWIN, 1967: Incorporation of [2-^{14}C]-mevalonic acid into phytoene by isolated chloroplasts. Biochem. J. **105**, 205–212.

CLAES, H., 1958: Biosynthese von Carotinoiden bei *Chlorella*. 4. Die Carotinsynthese einer *Chlorella*-Mutante bei anaerober Belichtung. Z. Naturforsch. **13B**, 222–224.

CLAES, H., 1959: Biosynthese von Carotinoiden bei *Chlorella*. 5. Die Trennung von Licht-

und Dunkelreactionen bei der lichtabhängigen Xanthophyllsynthese von *Chlorella.* Z. Naturforsch. **14B**, 4–7.

Clough, J. M., and G. Pattenden, 1979: Naturally-occurring poly-*cis* carotenoids. Stereochemistry of poly-*cis* lycopene and its congeners in «tangerine» tomato fruit. J. Chem. Soc., Chem. Commun. 616–619.

Commission on Biochemical Nomenclature, 1971: IUPAC Commission on the Nomenclature of Organic Chemistry and IUPAC-IUB Commission on Biochemical Nomenclature. Tentative rules for the nomenclature of carotenoids. Biochemistry (U.S.A.) **10**, 4827–4837.

Commission on Biochemical Nomenclature, 1975: IUPAC Commission on the Nomenclature of Organic Chemistry and IUPAC-IUB Commission on Biochemical Nomenclature. Nomenclature of carotenoids (Recommendations, 1974). Biochemistry (U.S.A.).

Czygan, F.-C., 1968a: Sekundär-Carotinoide in Grünalgen. I. Chemie, Vorkommen und Faktoren, welche die Bildung dieser Polyene beeinflussen. Arch. Mikrobiol. **61**, 81–102.

Czygan, F.-C., 1968b: Sekundär-Carotinoide in Grünalgen. II. Untersuchungen zur Biogenese. Arch., Mikrobiol. **62**, 209–236.

Czygan, F.-C., 1969: Untersuchungen zum Stoffwechsel der Ketocarotinoide in *Adonis*-Arten. I. Pigment-Zusammensetzung nicht normal gefärbter Blütenblätter. Planta (Berl.) **85**, 35–41.

Davies, B. H., 1970a: A novel sequence for phytoene dehydrogenation in *Rhodospirillum rubrum.* Biochem. J. **116**, 93–99.

Davies, B. H., 1970b: Alternative pathways of spirilloxanthin biosynthesis in *Rhodospirillum rubrum.* Biochem. J. **116**, 101–110.

Davies, B. H., 1973: Carotene biosynthesis in fungi. Pure Appl. Chem. **35**, 1–28.

Davies, B. H., 1975: Carotenoids – aspects of biosynthesis and enzymology. Ber. Deutsch. Bot. Ges. **88**, 7–25.

Davies, B. H., 1976: Carotenoids. In: T. W. Goodwin (ed.), Chemistry and biochemistry of plant pigments, 2nd ed., Vol. 2, 38–165. Academic Press, London, New York and San Francisco.

Davies, B. H., 1977a: C_{30} Carotenoids. In: T. W. Goodwin (ed.), International Review of Biochemistry, Vol. 14, Biochemistry of lipids II, 51–100. University Park Press, Baltimore, London and Tokyo.

Davies, B. H., 1977b: Carotenoids in higher plants. In: M. Tevini and H. K. Lichtenthaler (eds.), Lipids and lipid polymers in higher plants, 199–217. Springer-Verlag, Berlin, Heidelberg and New York.

Davies, B. H., 1977c: Lipid-soluble leaf pigments. Biochem. Soc. Trans. **5**, 1256–1259.

Davies, B. H., 1979: Solved and unsolved problems of carotenoid formation. Pure Appl. Chem. **51**, 623–630.

Davies, B. H., and Aung Than, 1974: Monohydroxycarotenoids from diphenylamine-inhibited cultures of *Rhodospirillum rubrum.* Phytochemistry **13**, 209–219.

Davies, B. H., and A. F. Rees, 1973: 7′, 8′, 11′, 12′-Tetrahydro-γ-carotene: a novel carotene from *Phycomyces blakesleeanus.* Phytochemistry **12**, 2745–2750.

Davies, B. H., and R. F. Taylor, 1976: Carotenoid biosynthesis – the early steps. Pure Appl. Chem. **47**, 211–221.

Davies, B. H., S. Matthews, and J. T. O. Kirk, 1970a: The nature and biosynthesis of carotenoids of different colour varieties of *Capsicum annuum.* Phytochemistry **9**, 797–805.

Davies, B. H., W.-J. Hsu, and C. O. Chichester, 1970b: The mechanism of the conversion of β-carotene into canthaxanthin by the brine shrimp, *Artemia salina* L. (Crustacea: Branchiopoda). Comp. Biochem. Physiol. **33**, 601–615.

DAVIES, B. H., C. J. HALLETT, R. A. LONDON, and A. F. REES, 1974: The nature of «Θ-carotene». Phytochemistry **13**, 1209–1217.

DAVIES, B. H., A. F. REES, and R. F. TAYLOR, 1975: Preparation of labelled terpenyl pyrophosphates using extracts of barley seed embryos (*Hordeum vulgare* Zephyr). Phytochemistry **14**, 717–722.

DAVIS, J. B., L. M. JACKMAN, P. T. SIDDONS, and B. C. L. WEEDON, 1966: Carotenoids and related compounds. Part XIV. The structure and synthesis of phytoene, phytofluene, ζ-carotene and neurosporene. J. Chem. Soc. (C) 2154–2165.

DE LA GUARDIA, M. D., C. M. G. ARAGÓN, F. J. MURILLO, and E. CERDÁ-OLMEDO, 1971: A carotenogenic enzyme aggregate in *Phycomyces:* evidence from quantitative complementation. Proc. Nat. Acad. Sci., U.S.A. **68**, 2012–2015.

DONKIN, P., 1976: Ketocarotenoid biosynthesis by *Haematocuccus lacustris*. Phytochemistry **15**, 711–715.

EGGER, K., 1965: Die Ketocarotinoide in *Adonis annua* L. Phytochem. 4, 609–618.

EGGER, K., and H. KLEINIG, 1967: Die Ketocarotinoide in *Adonis annua* L. III. Vergleich mit synthetischen Substanzen. Phytochem. 6, 903–905.

EHRENBERG, L., and A. FALUDI-DÁNIEL, 1962: On the role of leucine in the carotenoid synthesis of the chloroplasts of higher plants. Acta Chem. Scand. **16**, 1523–1526.

ENGLERT, G., B. O. BROWN, G. P. MOSS, B. C. L. WEEDON, G. BRITTON, T. W. GOODWIN, K. L. SIMPSON, and R. J. H. WILLIAMS, 1979: Prolycopene, a tetra-*cis* carotene with two hindered *cis* double bonds. J. Chem. Soc., Chem. Commun. 545–547.

ESLAVA, A. P., and E. CERDÁ-OLMEDO, 1974: Genetic control of phytoene dehydrogenation in *Phycomyces*. Plant Sci. Lett. **2**, 9–14.

ESLAVA, A. P., M. I. ALVAREZ, and E. CERDÁ-OLMEDO, 1974: Regulation of carotene biosynthesis in *Phycomyces* by vitamin A and β-ionone. Eur. J. Biochem. **48**, 617–623.

EPSTEIN, W. W., and H. C. RILLING, 1970: Studies on the mechanism of squalene biosynthesis. The structure of presqualene pyrophosphate. J. Biol. Chem. **245**, 4597–4605.

FOX, D. L., 1976: Animal biochromes and structural colours, 2nd. ed. University of California Press, Berkeley, Los Angeles and London.

GILCHRIST, B. M., 1968: Distribution and relative abundance of carotenoid pigments in *Anostraca* (Crustacea: Branchiopoda). Comp. Biochem. Physiol. **24**, 123–147.

GOODWIN, T. W., 1952: The comparative biochemistry of the carotenoids. Chapman and Hall, London.

GOODWIN, T. W., 1971: Biosynthesis. In: O. ISLER (ed.), Carotenoids, 577–636. Birkhäuser, Basel.

GOODWIN, T. W., 1977: The prenyllipids of the membranes of higher plants. In: M. TEVINI and H. K. LICHTENTHALER (eds.), Lipids and lipid polymers in higher plants, 29–47. Springer-Verlag, Berlin, Heidelberg and New York.

GOODWIN, T. W., and R. J. H. WILLIAMS, 1965a: A mechanism for the cyclization of an acyclic precursor to form β-carotene. Biochem. J. **94**, 5C–7C.

GOODWIN, T. W., and R. J. H. WILLIAMS, 1965b: A mechanism for the biosynthesis of α-carotene. Biochem. J. **97**, 28C–32C.

GORE, M. G., H. M. HILL, R. B. EVANS, and L. J. ROGERS, 1974: Inhibition by isonicotinyl hydrazide of pigment formation in higher plants. Phytochemistry **13**, 1657–1665.

GREGONIS, D. E., and H. C. RILLING, 1973: Photoinduced prephytoene pyrophosphate synthesis in a *Mycobacterium* sp. Biochem. Biophys. Res. Commun. **54**, 449–454.

GREGONIS, D. E., and H. C. RILLING, 1974: The stereochemistry of *trans*-phytoene synthesis. Some observations on lycopersene as a carotene precursor and a mechanism for the synthesis of *cis-* and *trans* phytoene. Biochemistry (U.S.A.) **13**, 1538–1542.

GRIBANOVSKI-SASSU, O., 1972: Effect of diphenylamine on carotenoid synthesis in *Dictyococcus cinnibarinus*. Phytochemistry **11**, 3195–3198.

GROB, E. C., K. KIRSCHNER, and F. LYNEN, 1961: Neues über die Biosynthese den Carotinoide. Chimia (Switz.) **15**, 308–310.

HERBER, R., B. MAUDINAS, J. VILLOUTREIX, and P. GRANGER, 1972: Formation de phytoene all-*trans* chez *Mucor hiemalis*. Biochim. Biophys. Acta **280**, 194–202.

HERTZBERG, S., and S. LIAAEN-JENSEN, 1977: Bacterial carotenoids. LIII. C_{50}-Carotenoids. 19. Absolute configuration of sarcinaxanthin and sarcinaxanthin mono-β-D-glucoside. Isolation of sarcinaxanthin diglycoside. Acta Chem. Scand. **B 31**, 215–218.

HSIEH, L. K., T.-C. LEE, C. O. CHICHESTER, and K. L. SIMPSON, 1974: Biosynthesis of carotenoids in *Brevibacterium* sp. KY-4313. J. Bact. **118**, 385–393.

HSU, W.-J., D. C. AILION, and M. DELBRÜCK, 1974: Carotenogenesis in *Phycomyces*. Phytochemistry **13**, 1463–1468.

ISLAM, M. A., S. A. LYRENE, E. M. MILLER Jr., and J. W. PORTER, 1977: Dissociation of prelycopersene pyrophosphate synthetase from phytoene synthetase complex of tomato fruit plastids. J. Biol. Chem. **252**, 1523–1525.

JOHNSON, J. H., B. C. REED, and H. C. RILLING, 1974: Early photoinduced enzymes of photoinduced carotenogenesis in a *Mycobacterium* species. J. Biol. Chem. **249**, 402–406.

JUNGALWALA, F. B., and J. W. PORTER, 1965: The configuration of phytoene. Arch. Biochem. Biophys. **110**, 291–299.

JUNGALWALA, F. B., and J. W. PORTER, 1969: Enzymatic synthesis of phytoene in tomato. Methods Enzymol. **15**, 454–460.

KHAN, N., D. E. LOEBER, T. P. TOUBE, and B. C. L. WEEDON, 1975: Carotenoids and related compounds. Part XXX. Stereochemistry and synthesis of phytoene. J. Chem. Soc. Perkin I, 1457–1464.

KLEINIG, H., 1974: Inhibition of carotenoid synthesis in *Myxococcus fulvus* (*Myxobacteriales*). Arch. Microbiol. **97**, 217–226.

KÜMMEL, H. W., and L. H. GRIMME, 1975: The inhibition of carotenoid biosynthesis in green algae by Sandoz H 6706: accumulation of phytoene and phytofluene in *Chlorella fusca*. Z. Naturforsch. **30 C**, 333–336.

KUSHWAHA, S. C., and M. KATES, 1976: Effect of nicotine on biosynthesis of C_{50} carotenoids in *Halobacterium cutirubrum*. Canad. J. Biochem. **54**, 824–829.

KUSHWAHA, S. C., G. SUZUE, C. SUBBARAYAN, and J. W. PORTER, 1970: The conversion of phytoene-^{14}C into acyclic, monocyclic and dicyclic carotenes and the conversion of lycopene-15,15′-^{3}H to mono- and dicyclic carotenes by soluble enzyme systems obtained from the plastids of tomato fruit. J. Biol. Chem. **245**, 4708–4717.

KUSHWAHA, S. C., E. L. PUGH, J. K. G. CRAMER, and M. KATES, 1972: Isolation and identification of dehydrosqualene and C_{40}-carotenoid pigments in *Halobacterium cutirubrum*. Biochim. Biophys. Acta **260**, 492–506.

KUSHWAHA, S. C., M. KATES, and J. W. PORTER, 1976: Enzymatic synthesis of C_{40} carotenes by cell-free preparation from *Halobacterium cutirubrum*. Canad. J. Biochem. **54**, 816–823.

LANG, W., and W. RAU, 1972: Untersuchungen über die lichtabhängige Carotinoidsynthese. IX. Zum Induktionsmechanismus der carotinoid-bildenden Enzyme bei *Fusarium aquaeductuum*. Planta **106**, 345–354.

LEE, T.-C., and C. O. CHICHESTER, 1969: Geranylgeranyl pyrophosphate as the condensing unit for enzymatic synthesis of carotenes. Phytochemistry **8**, 603–609.

LEUTWILER, L. S., and D. J. CHAPMAN, 1978: Biosynthesis of cyclic carotenoids in photosynthetic bacteria. Abstr. 5th Int. Symp. Carotenoids, Madison, p. 34.

MCDERMOTT, J. C. B., G. BRITTON, and T. W. GOODWIN, 1973: Carotenoid biosynthesis in a *Flavobacterium* sp.: stereochemistry of hydrogen elimination in the desaturation of phytoene to lycopene, rubixanthin and zeaxanthin. Biochem. J. **134**, 1115–1117.

MCDERMOTT, J. C. B., D. J. BROWN, G. BRITTON, and T. W. GOODWIN, 1974: Alternative pathways of zeaxanthin biosynthesis in a *Flavobacterium* species. Experiments with nicotine as inhibitor. Biochem. J. **144**, 231–243.

MAUDINAS, B., M. L. BUCHOLTZ, C. PAPASTEPHANOU, S. S. KATIGAR, A. V. BRIEDIS, and J. W. PORTER, 1975: Adenosine-5′-triphosphate stimulation of the activity of a partially-purified phytoene synthetase complex. Biochem. Biophys. Res. Commun. **66**, 430–436.

MAUDINAS, B., M. L. BUCHOLTZ, C. PAPASTEPHANOU, S. S. KATIGAR, A. V. BRIEDIS, and J. W. PORTER, 1977: The partial purification and properties of a phytoene synthesizing enzyme system. Arch. Biochem. Biophys. **180**, 354–362.

MOSHIER, S. C., and D. J. CHAPMAN, 1973: Biosynthetic studies on aromatic carotenoids: biosynthesis of chlorobactene. Biochem. J. **136**, 395–404.

MURILLO, F. J., and E. CERDÁ-OLMEDO, 1975: Regulation of carotenogenesis in *Phycomyces.* 10th FEBS Meeting, Paris, Abstr. 1431.

NES, W. R., and M. L. MCKEAN, 1977: Biochemistry of steroids and other isopentenoids. University Park Press, Baltimore, London and Tokyo.

NUSBAUM-CASSUTO, E., and J. VILLOUTREIX, 1965: Mise en évidence du lycopersène chez les végétaux superieurs. C. R. Acad. Sci. **260**, 1013–1015.

OOTAKI, T., A. C. LIGHTY, M. DELBRÜCK, and W.-J. HSU, 1973: Complementation between mutants of *Phycomyces* deficient with respect to carotenogenesis. Molec. Gen. Genet. **121**, 57–70.

PAPASTEPHANOU, C., F. J. BARNES, A. V. BRIEDIS, and J. W. PORTER, 1973: Enzymatic synthesis of carotenes by cell-free preparations of fruit of several genetic selections of tomatoes. Arch. Biochem. Biophys. **157**, 415–425.

PORTER, J. W., and R. E. LINCOLN, 1950: I. *Lycopersicon* selection containing a high content of carotenes and colourless polyenes; II. The mechanism of carotene biosynthesis. Arch. Biochem. **27**, 390–403.

PORTER, J. W., and S. L. SPURGEON, 1979: Enzymatic synthesis of carotenes. Pure Appl. Chem. **51**, 609–622.

POWLS, R., and G. BRITTON, 1977: The role of isomers of phytoene, phytofluene and ζ-carotene in carotenoid biosynthesis by a mutant strain of *Scenedesmus obliquus*. Arch. Microbiol. **115**, 175–179.

QURESHI, A. A., F. J. BARNES, E. J. SEMMLER, and J. W. PORTER, 1973: Biosynthesis of prelycopersene pyrophosphate and lycopersene by squalene synthetase. J. Biol. Chem. **248**, 2755–2767.

QURESHI, A. A., A. G. ANDREWES, N. QURESHI, and J. W. PORTER, 1974a: The enzymatic conversion of *cis*-[^{14}C]phytofluene, *trans*-[^{14}C]phytofluene and *trans*-[^{14}C]ζ-carotene to more unsaturated acyclic, monocyclic and dicyclic carotenes by a cell-free preparation of red tomato fruits. Arch. Biochem. Biophys. **162**, 93–107.

QURESHI, A. A., M. KIM, N. QURESHI, and J. W. PORTER, 1974b: The enzymatic conversion of *cis*-[^{14}C]phytofluene, *trans*-[^{14}C]-phytofluene and *trans*-[^{14}C]ζ-carotene to poly-*cis* acyclic carotenes by a cell-free preparation of tangerine tomato fruit plastids. Arch. Biochem. Biophys. **162**, 108–116.

QURESHI, A. A., N. QURESHI, M. KIM, and J. W. PORTER, 1974c: The isolation, purification and characterization of *cis*-ζ-carotene and the demonstration of its conversion to acyclic, monocyclic and dicyclic carotenes by a soluble enzyme system obtained from the plastids of tangerine tomato fruit. Arch. Biochem. Biophys. **162**, 117–125.

RAYMUNDO, L. C., and K. L. SIMPSON, 1972: The isolation of a poly-*cis*-ζ-carotene from the tangerine tomato. Phytochemistry **11**, 397–400.

SHAH, D. V., D. H. FELDBRUEGGE, A. R. HOUSER, and J. W. PORTER, 1968: Conversion of ^{14}C-labelled geranylgeranyl pyrophosphate to phytoene by a soluble tomato plastid enzyme system. Arch. Biochem. Biophys. **127**, 124–131.

SHNEOUR, E. A., 1962: The source of oxygen in *Rhodopseudomonas spheroides* carotenoid pigment conversion. Biochim. Biophys. Acta **65**, 510–511.

SINGH, R. K., G. BRITTON, and T. W. GOODWIN, 1973: Carotenoid biosynthesis in *Rhodopseudomonas sphaeroides*. s-Adenosylmethionine as the methylating agent in the biosynthesis of spheroidene and spheroidenone. Biochem. J. **136**, 413–419.

STRAUB, O., 1976: Key to carotenoids; lists of natural carotenoids. Birkhäuser Verlag, Basel and Stuttgart.

STUMPF, P. K., 1975: Biosynthesis of fatty acids in spinach chloroplasts. In: T. GALLIARD and E. I. MERCER (eds.), Recent advances in the chemistry and biochemistry of plant lipids, 95–113. Academic Press, London, New York and San Francisco.

TAYLOR, R. F., and B. H. DAVIES, 1973: Triterpenoid carotenoids from a pigmented *Streptococcus* sp. Biochem. Soc. Trans. **1**, 1091–1092.

TAYLOR, R. F., and B. H. DAVIES, 1974a: Triterpenoid carotenoids and related lipids. The triterpenoid carotenes of *Streptococcus faecium* UNH 564P. Biochem. J. **139**, 751–760.

TAYLOR, R. F., and B. H. DAVIES, 1974b: Triterpenoid carotenoids and related lipids. Triterpenoid monohydroxy- and monoglucosyloxy-carotenoids from *Streptococcus faecium* UNH 564P. Biochem. J. **139**, 761–769.

TAYLOR, R. F., and B. H. DAVIES, 1975: Gas-liquid chromatography of carotenoids and other terpenoids. J. Chrom. **103**, 327–340.

TAYLOR, R. F., and B. H. DAVIES, 1976: Triterpenoid carotenoids and related lipids. Triterpenoid carotenoid aldehydes from *Streptococcus faecium* UNH 564P. Biochem. J. **153**, 233–239.

TREHARNE, K. J., E. I. MERCER and T. W. GOODWIN, 1966: Carotenoid biosynthesis in some maize mutants. Phytochemistry **5**, 581–587.

WEEDON, B. C. L., 1970: Allenic and acetylenic carotenoids. Rev. Pure Appl. Chem. (Australia) **20**, 51–66.

WEEDON, B. C. L., 1971: Occurrence. In: O. ISLER (ed.), Carotenoids, 25–59. Birkhäuser, Basel.

WILLIAMS, R. J. H., G. BRITTON, and T. W. GOODWIN, 1966: A possible mechanism for the biosynthesis of eschscholtzxanthin. Biochim. Biophys. Acta **124**, 200–203.

WILLIAMS, R. J. H., G. BRITTON, and T. W. GOODWIN, 1967: The biosynthesis of cyclic carotenes. Biochem. J. **105**, 99–105.

WINTERSTEIN, A., A. STUDER, and R. RÜEGG, 1960: Neuere Ergebnisse der Carotinoidforschung. Ber. Deut. Chem. Ges. **93**, 2951–2965.

YAMAGUCHI, M., 1957: Chemical constitution of renieratene. Bull. Chem. Soc. Japan **30**, 979–983.

YAMAGUCHI, M., 1958: Renieratene, a new carotenoid containing benzene rings isolated from a sea sponge. Bull. Chem. Soc. Japan **31**, 739–742.
YAMAGUCHI, M., 1960: Total synthesis of renieratene and renierapurpurin. Bull. Chem. Soc. Japan **33**, 1560–1562.
YAMAMOTO, H. Y., and C. O. CHICHESTER, 1965: Dark incorporation of $^{18}O_2$ into antheraxanthin by bean leaf. Biochim. Biophys. Acta **109**, 303–305.
YAMAMOTO, H. Y., C. O. CHICHESTER, and T. O. M. NAKAYAMA, 1962: Biosynthetic origin of oxygen in the leaf xanthophylls. Arch. Biochem. Biophys. **96**, 645–649.
YOKOYAMA, H., and M. J. WHITE, 1966: Citrus carotenoids. VI. Carotenoid pigments in the flavedo of Sinton citrangequat. Phytochemistry **5**, 1159–1173.
YOKOYAMA, H., and M. J. WHITE, 1970: Carotenoid formation in *Triphasia trifolia*. Phytochemistry **9**, 1795–1797.

The reversible, light-induced conversions of xanthophylls in the chloroplast

A. Hager

Institut Biologie I der Universität Tübingen

1. Introduction

Besides chlorophylls, all photosynthetic organisms also contain carotenoids in their thylakoid membranes. These yellow lipophilic pigments have a two-fold function. One type acts as accessory, light-harvesting pigment which can efficiently transfer at least part of the absorbed light energy to the chlorophyll, and the other type functions as a protecting agent which can prevent photooxidative destruction during strong illumination (review: Isler, 1971).

However, a few xanthophylls which exhibit very fast light-induced concentration changes occuring in a few minutes, are an exception. The following review will discuss the causes, molecular mechanisms, and possible functions of these reversible light-dependent xanthophyll conversions in the metabolism of the chloroplasts.

2. General characteristics of the xanthophyll conversions

Since investigations of Sapozhnikov et al., (1957) and Blass et al., (1959), it has been known that an epoxide xanthophyll named Viol * decreases in higher plants and in *Chlorella* upon illumination of the organisms. Yamamoto et al. (1962a and b) have shown that Viol is converted to Zea during this process.

Exact quantitative measurements have clearly demonstrated (Fig. 1) that the content of the di-epoxide compound Viol (Fig. 2) decreases very rapidly at the beginning of strong illumination and that subsequently the concentration of a one-epoxide compound, Anth increases initially and is finally followed by accumulation of the fully epoxide-free compound, Zea.

---

* Abbreviations: Viol = violaxanthin = (3 S, 5 R, 6 S, 3′ S, 5′ R, 6′ S)-5,6,5′,6′-Diepoxy-5,6,5′,6′-tetrahydro-β,β-carotene-3,3′-diol; Anth = antheraxanthin = 5,6-Epoxy-5,6-dihydro-β,β-carotene-3,3′-diol; Zea = zeaxanthin = (3 R, 3′ R)-β,β-Carotene-3,3′-diol; Asc = ascorbic acid; DHAsc = dehydro-ascorbic acid; GSH = glutathione; GSSG = oxidized GSH; DCMU = 3(3′4′-dichlorophenyl)-1,1-dimethylurea; CCCP = m-chlorocarbonylcyanidephenylhydrazon; DNP = 2,4-dinitrophenol; PMS = phenazinemethosulfate

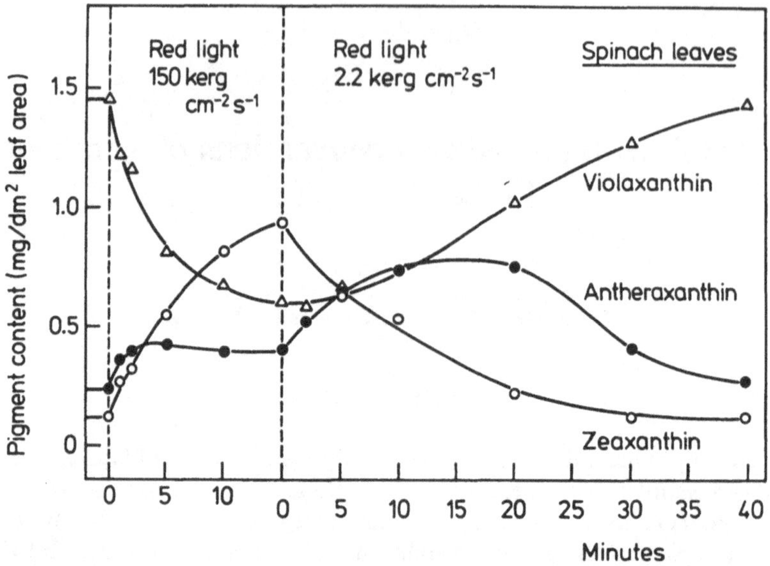

Fig. 1: Kinetics of de-epoxidation (under strong light) and epoxidation (under weak light) in spinach leaves. From Figs. 1–3 of Hager (1967b). Methods of pigment determination: Hager and Meyer-Bertenrath (1966, 1967), Hager and Stransky (1970a)

The de-epoxidation could also be initiated by red light which is not absorbed by the xanthophylls themselves (Fig. 1). This fact shows that the chlorophyll functions as a light acceptor (Hager, 1966; Sapozhnikov et al., 1966; Lee and Yamamoto, 1968).

At lower light intensities (Fig. 1), under special conditions and also in the dark, the epoxide compound is reformed (Hager, 1967b). Therefore, depending on the light intensity, the equilibrium of the reaction system changes. This fact has been documented on a variety of green organisms (see Chap. 3). Fig. 2 shows this dependency in *Lemna gibba*: Zea, which is immediately formed upon illumination, is preserved at high light intensities, but decreases again to zero at low intensities due to the relatively fast back reaction (epoxidation). We must therefore assume that the Zea formed during average light intensities, immediately changes back to Viol. The back reaction compensates the Zea formation. Only at very high light intensities is the rate of the epoxidation insufficient which leads to an accumulation of Zea. As we know today, different enzyme systems are responsible for the de-epoxidation (forward reaction) and for the epoxidation (back reaction) (Hager, 1969, Hager and Perz, 1970, Westerhoff, 1974). Therefore different pathways can be followed, thus forming a cyclic reaction sequence called the xanthophyll cycle (Hager, 1967a) or violaxanthin cycle (Lee and Yamamoto, 1968).

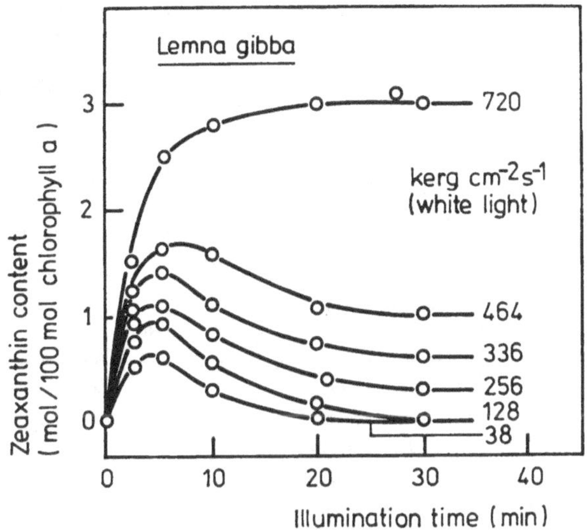

Fig. 2: Kinetics of Zea formation from Viol in *Lemna gibba* under various light intensities. From Fig. 2 of Siefermann (1972)

3. Distribution of the xanthophyll cycle in the plant kingdom

The question arises, as to how widely the cycle is spread in the plant kingdom. An answer to this question could give an indication of the cycle's importance in plant metabolism. We have found (Hager and Stransky, 1970a, b, c; Stransky and Hager, 1970a, b, c) that only photobacteria and three groups of algae do not possess the cycle. These algae contain phycobilines and comprise the Cryptophyceae which are very simply constructed flagellates and the blue and red algae (Figs. 3 and 4). These organisms are supposed to be phylogenetically related and are quite primitive; they have at most double thylakoid membranes and no stacking (Lee, 1972; Toyama, 1974). The green and the brown algae, in addition to the mosses, ferns, and seed plants, contain the previously discussed epoxide cycle with Viol, Anth and Zea. The other algae show a variant of the cycle: A mono-epoxide compound with an alkine bond, which is called diadinoxanthin and is de-epoxidized to diatoxanthin in the presence of light (Figs. 3 and 4). One can see that the xanthophyll cycle has been never forgotten during the phylogenetic development of the plants and this indicates that it probably has an important function.

Repeated attempts at evaluation of this function have been made. Two hypothesis have been primarily considered.

4. First attempts at an interpretation of the xanthophyll conversions

On the bases of tests with $H_2{}^{18}O$, Sapozhnikov et al. (1961, 1965, 1972) and Saakov (1965a, b) assumed that the change of the oxygen content of these xanthophylls is connected to the O_2 liberation during photosynthesis. They found an

No cyclic carotenoid changes	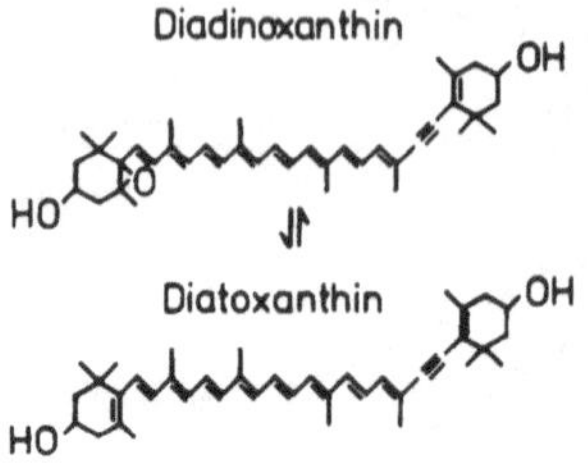	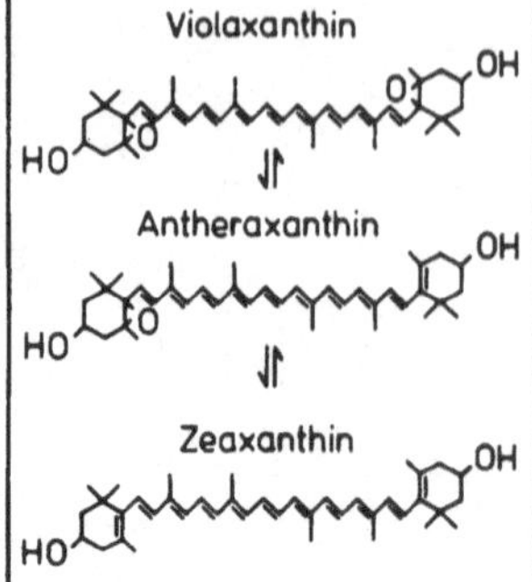
Photobacteria Cyanophyceae Cryptophyceae Rhodophyceae	Euglenophyceae Dinophyceae Chrysophyceae Bacillariophyceae Xanthophyceae Chloromonadophyceae	Chlorophyceae Phaeophyceae Higher plants

Fig. 3: Distribution of the two types of light-dependent xanthophyll conversions in the plant kingdom (Hager, 1975)

introduction into Viol of ^{18}O from $H_2{}^{18}O$ and concluded from this that the photosynthetically liberated oxygen runs at least partly over this epoxide cycle, namely, in the following way:

$$\text{Zeaxanthin} + 2H_2{}^{18}O \xrightarrow{\text{dark}} {}^{18}O\text{-Violaxanthin} + 4H$$

$$^{18}O\text{-Violaxanthin} \xrightarrow{\text{light}} \text{Zeaxanthin} + {}^{18}O_2$$

During this reaction the light energy directly absorbed from Viol should lead to the photooxidation of the epoxide oxygen, e.g., to the liberation of O_2 from this xanthophyll. Red light absorbed by the chlorophyll should also be effective if it is transferred to the Viol via the triplet state of the chlorophyll (Sapozhnikov, 1969, 1972).

These results could be differently interpreted, however, if one assumes that gaseous $^{18}O_2$ is formed during the photosynthetic splitting of $H_2{}^{18}O$ and is then introduced into Zea.

The investigations of Yamamoto and Chichester (1965) and Takeguchi and Yamamoto (1968) show in fact that the epoxy group is only formed from atmospheric O_2. Furthermore, there are still a few other arguments against the assumption presented by Sapozhnikov et al. which are discussed in Chap. 5.

Another interpretation of the xanthophyll cycle was provided by Krinsky (1966). He pointed out that the addition of malate or other reductants to lyophylized cells of *Euglena* in the dark increases the percentage of the epoxide-free compound diato-

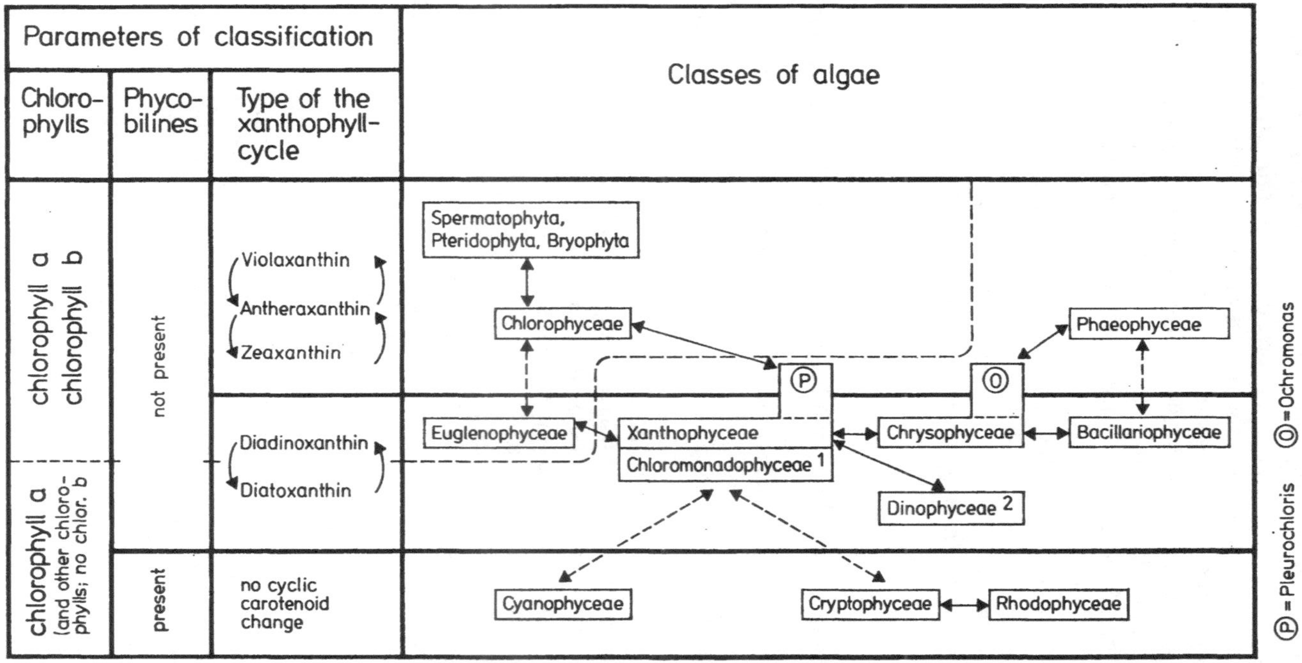

Fig. 4: Relationship between individual classes of algae deduced from the pattern of the pigments and from the type and distribution of light-induced xanthophyll transformations (Stransky and Hager 1970). 1) Chapman and Haxo (1966) 2) Loeblich III (1966)

xanthin (then wrongly called Zea), and that, on the other hand, in the presence of light the percentage of the epoxide-containing diadinoxanthin (then antheraxanthin) is increased. He then concluded that the formation of the epoxide compound protects the cell from photooxidative destruction. In the triplet state light-activated chlorophyll, usually deactivated through the processes of photosynthesis, should also react with molecular oxygen, thereby forming a singlet-activated oxygen (Krinsky, 1971). This $^1O_2$ could produce a lethal oxidation in the cell. Through the introduction of this $^1O_2$ as epoxide oxygen into the xanthophyll damage to the cell should be prevented (Krinsky, 1971). An enzymatic de-epoxidation would reestablish the protective oxygen-free compound diatoxanthin.

This hypothesis has also not been supported by new results which have been found by studying the xanthophyll cycle in isolated chloroplasts.

5. Induction of de-epoxidation in isolated chloroplasts

Chloroplasts isolated by usual methods show no de-epoxidation upon illumination. This reaction takes place, however, if a factor from the supernatant is again added to the thylakoids after centrifugation of the chloroplast suspension (Hager, 1966; 1967a). This factor is ascorbic acid which, according to Franke and Heber (1964) and Gerhardt (1964), is found to be present in relatively high amounts in chloroplasts. Although the volume of the chloroplasts is only 10% of that of the cell, they contain 40–50% of the total ascorbic acid content of the cell.

The dependence of de-epoxidation on photosynthetic electron transport can be studied in the presence of this essential cofactor. It was surprising to find that Zea is only formed if electrons flow through those parts of the electron transport chain which are involved in phosphorylation (Fig. 5) (Hager, 1966; 1969; Westerhoff, 1975). Moreover, the addition of uncouplers fully inhibits the formation of Zea.

These results led to the suggestion that ATP or an energy-rich state other than ascorbic acid is required for the de-epoxidation (Hager, 1966).

On the other hand, the formation of Zea was not necessary for ATP formation or the O_2 evolution. For instance, the *Hill* reaction with ferricyanide as electron acceptor takes place with formation of ATP but without formation of Zea (Hager, 1966, 1975). This result contradicts the assumption made by the Russian school, i.e., that the xanthophyll cycle is necessary for photosynthetic oxygen evolution.

However, there was another observation which made it again doubtful that Zea formation depends on photophosphorylation. In isolated chloroplasts we found that a strong formation of Zea can also occur merely by decreasing the pH of the incubation medium from 7 to 5 without illumination (Fig. 5; Hager, 1969).

Neither light, nor ATP, nor an energy-rich state was necessary for the formation of Zea during this reaction in the chloroplasts. This type of reaction which depends on a relatively low pH of 5, was also found to occur with the enzyme de-epoxidase (isolated from the chloroplasts) after addition of the substrates Viol and Asc.

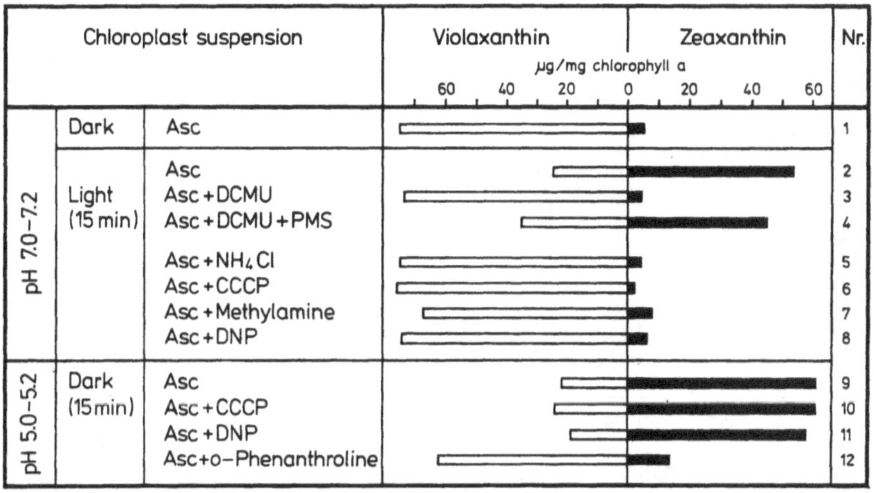

Fig. 5: The change of the Viol and Zea concentration in isolated spinach chloroplasts under different conditions. Explanation given in the text. Phosphate buffer (67 mM); Asc (10 mM), DCMU (10 µM), PMS (50 µM), FMN (1 mM); NH_4Cl (10 mM); CCCP (No 6:5 µM; No 10: 100 µM); Methylamine (5 mM); DNP (3 mM); *o*-phenanthroline (10 mM); red light (150 kerg/cm²/sec). (Hager 1969)

6. Properties of isolated de-epoxidase; mechanism of Zea formation

The enzyme of the de-epoxidation can be isolated from spinach chloroplasts (Hager, 1966; Hager and Perz, 1970; Perz, 1970; Hager, 1975). The water-soluble enzyme has a molecular weight of 54,000. Viol ($K_m = 1.06 \times 10^{-5}$ mol/l), Anth ($K_m = 1.09 \times 10^{-5}$ mol/l) and diadinoxanthin ($K_m = 1.6 \times 10^{-4}$ mol/l) were used as substrates. However, no reaction occurs with *cis*-stereoisomeres of these xanthophylls and with epoxides of carotenes (Perz,1970). Asc is required as a co-substrate. To remain active, the de-epoxidase isolated from chloroplasts of *Lactuca sativa* needs chloroplast lipids (monogalactosyldiglyceride) in addition (Yamamoto et al., 1974). A new and very exact investigation of lipid requirement and substrate specificity of violaxanthin de-epoxidase has been published by Yamamoto and Higashi (1978).

The sharp pH optimum of the enzyme lies at pH 5.2 and is therefore nearly identical with that of 4.8 measured in isolated chloroplasts (Hager, 1969; Perz, 1970) (Fig. 6). No reaction occurs in the neutral pH region. This means that the isolated enzyme transforms the added substrate Viol to Zea under the same conditions as the enzyme in situ. It also means that we have a pH indicator in, or at, the thylakoid membrane, which indicates a pH of 5 by the formation of Zea (Fig. 7).

If Zea is formed upon illumination in chloroplasts buffered at a pH of 7 or >7 (Fig. 8), one can deduce that the pH decreases to 5 in the thylakoid compartment

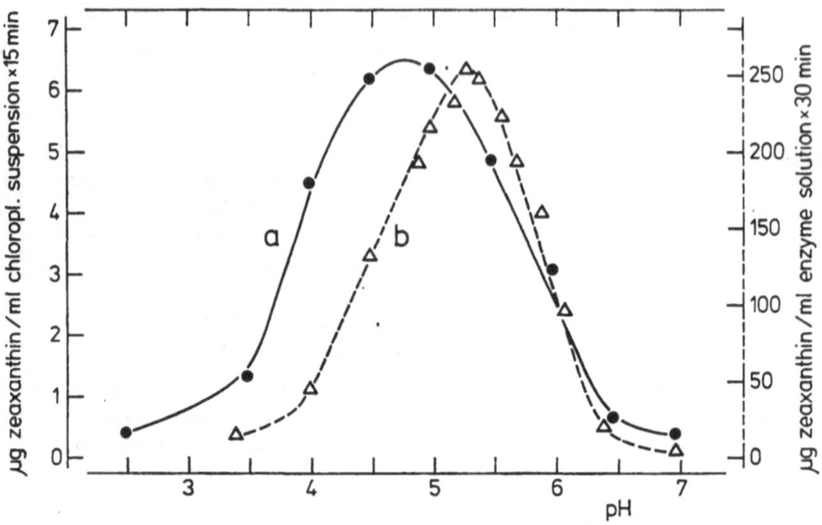

Figs. 6a and b: The pH dependence of the formation of Zea, a) in isolated spinach chloroplasts (formed from the endogenous Viol), b) with the aid of isolated de-epoxidase and added Viol (Hager, 1966, 1969). Citrate-phosphate buffer (0.1 M) and Asc 10 mM; 90 μg chlorophyll a/ml chloroplast-suspension.

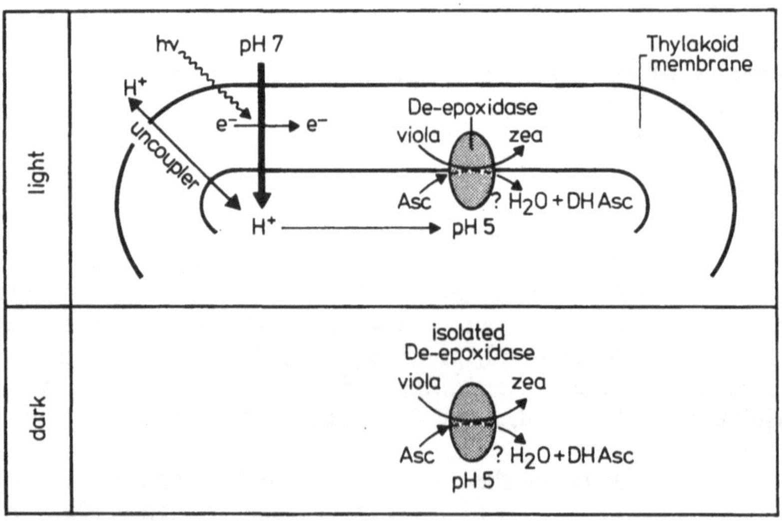

Fig. 7: A model of the Asc-dependent formation of Zea from Viol in the presence of de-epoxidase. The enzyme is active at pH 5 both in the isolated state and in situ in the membrane. This low pH is produced in the thylakoid loculus normally through the light-dependent proton shift. Since the formation of a proton gradient is correlated with the chemiosmotic formation of ATP, it appears that Zea formation depends in many cases on photophosphorylation. As a result, Zea formation is inhibited by uncouplers which prevent an accumulation of H^+ in the loculus (Hager, 1969)

in which the de-epoxidase is located. Furthermore, we can infere that a proton gradient which has a magnitude of 2 to 3 pH units is formed across the thylakoid membrane. With the help of this endogenous pH-dependent enzyme it was possible to prove the existence of the proton gradient across the thylakoid membrane (Hager, 1969) postulated by Mitchell (1966, 1967).

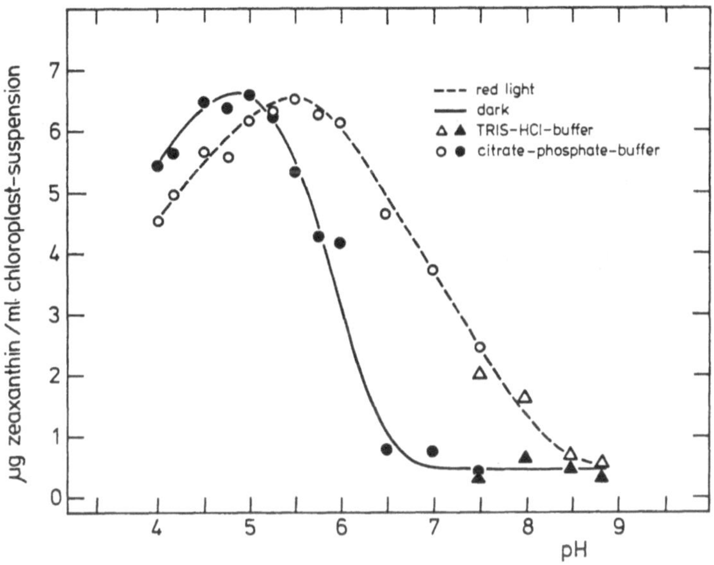

Fig. 8: Effect of pH of the suspension medium on the formation of Zea from Viol (see Fig. 7) in isolated spinach chloroplasts upon illumination (red light 150 kerg/cm²/sec; 15 min) and in the dark (15 min). 90 µg chlorophyll a/ml chloroplast suspension (Hager, 1969)

On the basis of these results, it is now understandable (see Fig. 7) why Zea formation in isolated chloroplasts can occur only under photophosphorylating conditions. Electron transport processes which occur under these conditions lead to a proton gradient across the membrane and at last to an acidification of the thylakoid loculus (review: Trebst and Avron, 1978). As a result of this physiological pH decrease, the constitutive enzyme de-epoxidase is activated and forms Zea. If uncouplers are added under such conditions (Hager, 1969) or the membrane is made permeable to H^+ in another way, an accumulation of protons in the thylakoid loculus upon illumination cannot occur and therefore neither can formation of Zea (Figs. 5 and 7).

To summarize: the condition for an enzymatic de-epoxidation is a pH decrease in the thylakoid loculus. Since this normally leads to (chemiosmotic) ATP formation, Zea formation is often a sensitive indicator of phosphorylation (Westerhoff, 1974); an energy-rich state or ATP is only an apparent requirement for Zea formation.

7. Activation of de-epoxidase in intact cells through a nonphysiological decrease in pH (CO_2 aeration)

The following test should be mentioned as another proof of the rather unusual pH-dependent activation of de-epoxidase. When one exposes spinach leaves to a CO_2 atmosphere in the dark, Zea is formed in the chloroplasts (Hager, 1975). The reason is the formation of carbonic acid which leads to a pH decrease in the thylakoids and thereby to an activation of the enzyme. In addition, this experiment shows that spinach leaves, which have been left several hours in the dark form Zea after exposure to CO_2. An appreciable decrease in Zea formation occurs only after 24 hours. This indicates that leaves held in the dark for a longer period still contain a sufficient amount of the reductant, which we assume to be ascorbic acid. Even so, the question arises as to how it is possible that the oxidised Asc can be reduced again upon illumination, despite the high speed of the xanthophyll cycle reaction.

8. Regeneration of the de-epoxidation reductant

The DHAsc formed during the Viol → Zea conversion can be reduced enzymatically to Asc in chloroplasts (Table) with the aid of two enzymes, one the NADPH-dependent glutathion-reductase which was characterized by Santarius and Stocking (1969) and Jablonski and Anderson (1978), and the other a GSH-dependent dehydro-ascorbic acid reductase (Foyer and Halliwell (1976, 1977), Halliwell and Foyer (1978)). The enzymatic reduction of DHAsc has an optimum value of pH 6.6 (Fig. 9). If one assumes that a stoichiometric amount of Asc is produced from GSH and is completely used up in the de-epoxidation, then 2 Mol NADPH are necessary to convert 1 Mol Viol to Zea (Fig. 12).

The kinetic measurements are consistent with this proposed reaction mechmanis (Fig. 10); DHAsc can function as a cosubstrate for the de-epoxidase if it is reduced

Table: Formation of L-Asc from L-DHAsc in chloroplasts at a pH of 6.5 in different reduction systems. Determination of the Asc produced, by the Asc-dependent Zea-formation at a pH of 5.2 (Perz, 1970; and Hager, 1975)

	Chloroplast suspension: experimental conditions	Added redox-compounds	Zeaxanthin (nMol / mg Chl.)
Dark	pH 5,2 for 20 min	DH Asc + GSH	5,7
Light	pH 7,0 for 20 min	DH Asc + GSH	4,1
Dark	pH 6,5 for 30 min, then pH 5,2 for 20 min	DH Asc + GSH DH Asc + NADPH DH Asc + GSSG + NADPH DH Asc + GSSG + NADH DH Asc + FADH DH Asc + GSSG	9,4 0,5 6,8 3,5 0 0

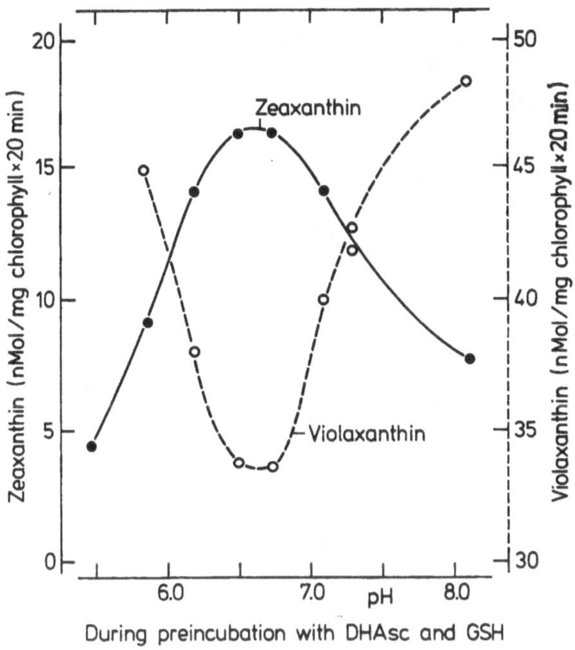

Fig. 9: The pH dependence of the enzymic reduction (after 20 min) of L-DHAsc to L-Asc in isolated chloroplasts. Determination of the L-Asc produced, by the thereby triggered Viol → Zea transformation at pH 5.2.

Samples (2 ml): Chloroplasts with 1.86 mg chlorophyll; 20 μmol GSH; 20 μmol DHAsc; 80 mol potassium phosphate buffer at different pH (from 5.5 to 8.5). Incubation: 20 min at 22 °C; each sample was then adjusted to a pH of 5.2 with a saturated KH_2PO_4 solution and was further incubated for 20 min in the dark (Perz 1970; Hager, 1975)

to Asc with the aid of photosystem I (washed whole chloroplasts) (Siefermann and Yamamoto, 1974a).

In summary: the forward reaction, i.e. the de-epoxidation, depends on a light-induced shift of pH and on the Asc concentration. Another possibility to regulate these xanthophyll conversions has been proposed by Siefermann and Yamamoto (1974 and 1975a). They believe that the availability of Viol for the de-epoxidase depends on the redox reactions between photosystem I and II.

However, to be able to explain the function of the complete xanthophyll cycle it would be necessary to know the details of the back reaction, that is, the introduction of oxygen into Zea to form an epoxide.

9. Epoxidation (back reaction) in whole leaves

It has already been established that it is necessary to introduce atmospheric oxygen for the formation of the epoxide group in algae and whole plants (Yamamoto and Chichester, 1965; Hager, 1967a, b; Yamamoto and Takeguchi, 1972). This intro-

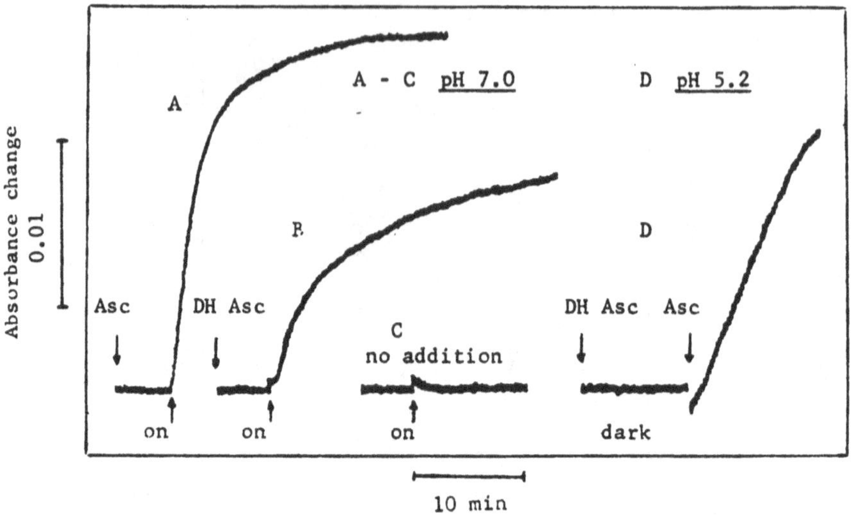

Fig. 10: Effect of Asc and DHAsc on Viol de-epoxidation in light at pH 7.0, and dark at pH 5.2. De-epoxidation is shown as an absorbance increase at 505 nm with 540 nm as reference. The reaction mixtures contain chloroplasts equivalent to 11.1 μg chlorophyll ml^{-1}, 400 mM sorbitol, 10 mM NaCl, and 50 mM Hepes-NaOH buffer, pH 7.0 (curves A—C) and 100 mM sodium citrate buffer, pH 5.2 (curve D). Sodium Asc and DHAsc concentrations were 16 mM and 10 mM respectively. Actinic light at 670 nm 36 kerg $cm^{-2}s^{-1}$ was used for curves A—C. (Siefermann and Yamamoto, 1974b)

duction of oxygen can take place in the dark and is accelerated by increasing the oxygen concentration. In addition epoxidation (i.e., Viol formation) is increased when leaves are exposed to low light intensities (Fig. 1) (Hager, 1967b; Yamamoto et al., 1967). This fact can be explained by the light-induced opening of the stomata and the thereby enhanced availability of atmospheric O_2, and furthermore by photosynthetic oxygen evolution in the proximity of the reactant Zea (Hager, 1967b). Another light-dependent component of epoxidation has been proved in investigations employing isolated chloroplasts and is explained in the following section.

10. Epoxidation (Zea → Viol transformation) in isolated chloroplasts

As early as 1966 it was assumed that the back reaction is catalyzed by a mixed-function oxygenase (Hager, 1966). Later this assumption was confirmed in our laboratory using isolated spinach chloroplasts: Besides oxygen, a reductant, namely, NADPH is necessary for the epoxidation (Perz, 1970; Westerhoff, 1974; Hager, 1975). Fig. 11 illustrates the formation of Viol from Zea as influenced by the NADPH concentration, and Fig. 12 the kinetics of the reaction (Westerhoff, 1974). The reaction can also take place in the presence of NADH, however, it occurs at lower rates. The optimum pH of the reaction is 7.6 (Fig. 13). These results, supported by those

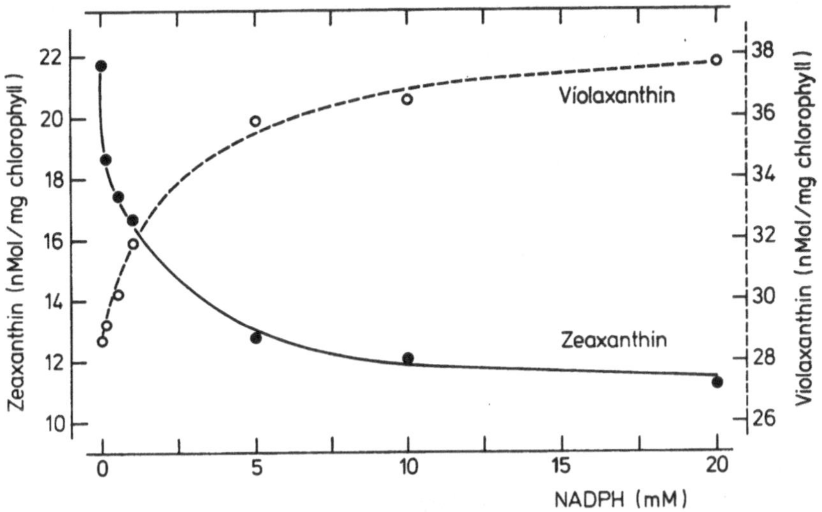

Fig. 11: Dependence of Viol formation from Zea (epoxidation) in isolated chloroplasts on the NADPH concentration. Samples (1 ml): 16 μmol phosphate buffer, pH 7.6; chloroplast suspension containing 1.02 mg chlorophyll; increasing amounts of NADPH (from 0.1 to 20 μmol); continuous oxygen aeration for 45 min at 23 °C (Perz, 1970; Hager, 1975)

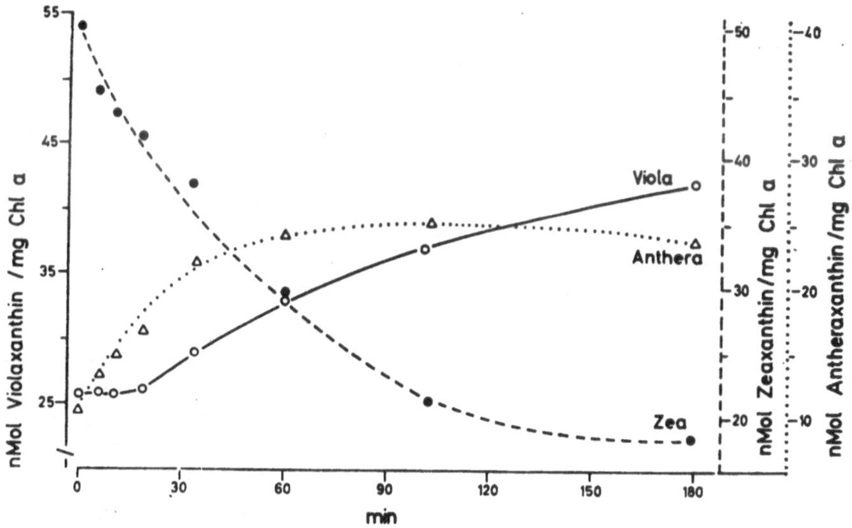

Fig. 12: Kinetics of Zea epoxidation to Anth and Viol in isolated spinach chloroplasts. 800 erg/cm²/sec, 18 °C, FAD 2.5 μM, pH 7.5 (Westerhoff, 1974)

on lettuce chloroplasts (Siefermann and Yamamoto, 1975b, c), show that the epoxidation runs optimally at a pH which is higher than the optimum in de-epoxidation. This surprising phenomenon can only be explained if one assumes that the two

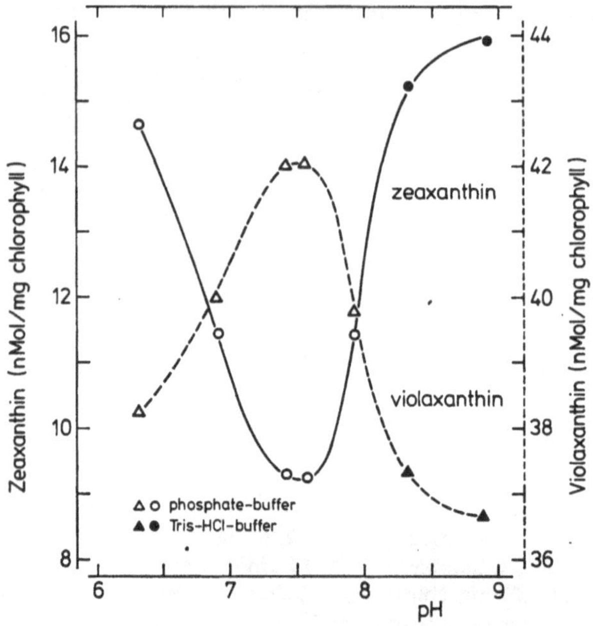

Fig. 13: The pH dependence of the back reaction (Viol formation from Zea) in isolated chloroplasts. Samples (1 ml): 50 μmol potassium-phosphate buffer or 100 μmol Tris/HCl buffer at different pH (pH 6.3 to 8.9); 0.91 mg chlorophyll; O_2 aeration; reaction time 45 min; temperature 24 °C (Perz, 1970; Hager, 1975)

enzymes of the xanthophyll cycle are localized at different sides of the thylakoid membrane (see Chap. 12).

Epoxidation is strongly inhibited by the copper chelate salicylaldoxim (5 mM) (Hager, 1966; Perz, 1970). This observation and the necessity of NADPH and O_2 as co-substrates indicates that the epoxidase is a metal-containing mixed-function oxygenase. The participation of NADPH in the epoxidation reaction already shows that light may stimulate this reaction, not only through the evolution of oxygen, but also through the formation of NADPH in photosystem I. In fact, it has been established that a light-dependent transformation of Zea to Viol occurs in the presence of ferredoxin and NADP in isolated chloroplasts (Westerhoff, 1974). This reaction is already saturated at very low light intensities at which the light-formed cosubstrate NADPH cannot yet be analytically detected (Westerhoff, 1974). This indicates a close proximity and a functional coupling of the NADP-reducing enzyme system with that of the epoxidase.

It is now possible to initiate the forward and back reaction simultaneously in isolated chloroplasts in the presence of NADP, ferredoxin, and ascorbic acid. At low light intensities the NADP-dependent epoxidation predominates and at higher light intensities the ascorbic acid-dependent de-epoxidation takes over (Westerhoff, 1975).

However, epoxidation also takes place in the dark; this is not a contradiction because certain amounts of reduced NADP are always present in the chloroplasts in the intact cell. Heber and Krause (1972) have determined that 30% of NADP is present in the chloroplast in the reduced state in the dark. Thus NADPH could be continuously regenerated when the PGA/HAP shuttle runs in reverse (Heber, 1974).

11. Interdependence of the reaction partners of the xanthophyll cycle

Considering the whole cycle (Fig. 14) one can see that one reductant is necessary for the light-dependent forward reaction, and one for the light-stimulated back reaction. It has been pointed out that the NADPH is oxidized directly during epoxidation of Zea, and indirectly during de-epoxidation of Viol through the sequence NADPH → GSSG → DHAsc.

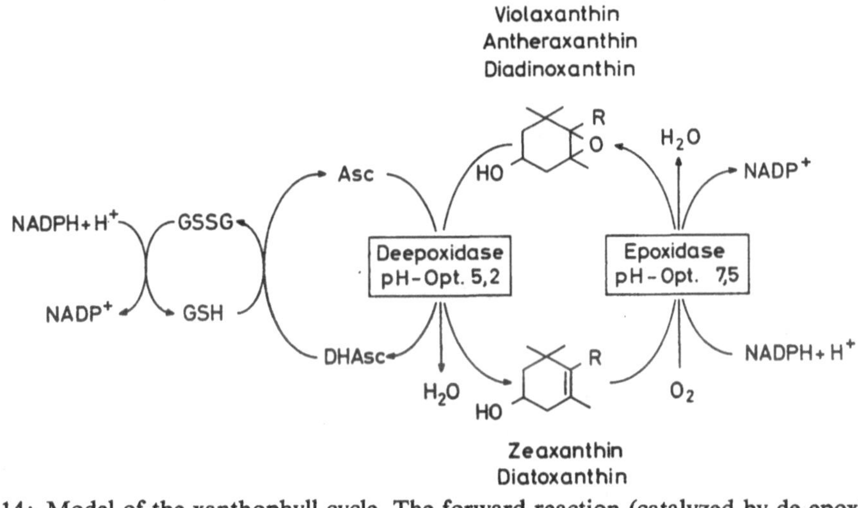

Fig. 14: Model of the xanthophyll cycle. The forward reaction (catalyzed by de-epoxidase) and the back reaction (epoxidase) have very different pH optima and require different reductants (Hager, 1975)

It is therefore likely that we have a system in which reduction energy is used up continuously through the formation of H_2O. From this fact one can make a number of assumptions concerning the functions of the cycle; these assumptions will be discussed in more detail in Chap. 13.

12. Localization of the enzymes of the xanthophyll cycle in chloroplasts

Two different enzyme systems can be postulated based on the very different pH optimum and the entirely different cofactor requirement of the forward and back

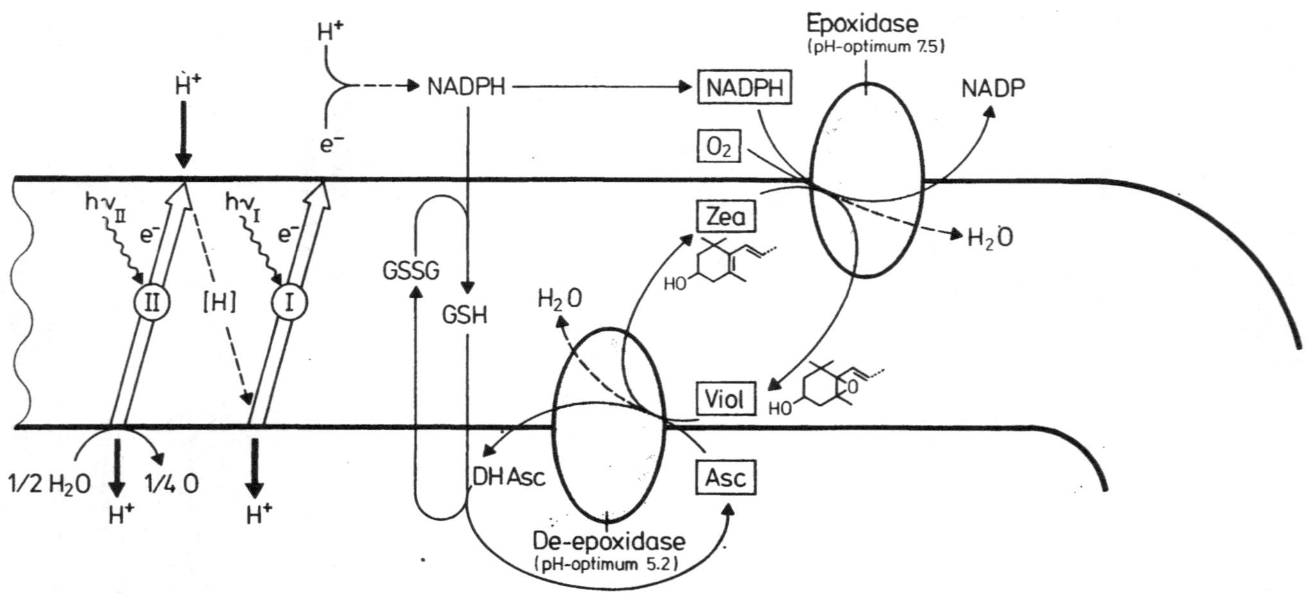

Fig. 15: Localization and mechanisms of the xanthophyll cycle in the thylakoid membrane. The cyclic reaction can occur only because of the different pH optima of the de-epoxidase and epoxidase, when a light-dependent proton gradient is formed across the thylakoid membrane (further explanation in the text) (Hager, 1975)

reaction. Furthermore, the isolated de-epoxidase (Hager and Perz, 1970) cannot catalyze the back reaction (epoxidation) under any circumstances.

Both enzymes should be located in the vicinity of their substrates in the thylakoid-membrane (Fig. 15). Since the activity of the de-epoxidase (pH optimum 5.2) is very sensitive to the light-induced pH decrease in the thylakoid loculus, the enzyme has to be directly connected to this compartment. On the other hand, there is hardly any inhibition of the epoxidase, which has a sharp pH optimum of 7.5, if the pH is about 5 during illumination in the thylakoid loculus. Therefore, the epoxidase can be only localized at the outside of the membrane. This situation is very likely in view of the previous explanation that the NADPH-induced epoxidation in isolated chloroplasts (in the presence of ferredoxin and NADP) is already saturated at low intensities at which no or very low quantities of reduced NADP are observed (Westerhoff, 1974, 1975). The NADPH formed must therefore be used up immediately for the epoxidation at the site of its formation. Since it has been suggested that the ferredoxin-NADP-reductase is located at the outside of the thylakoid-membrane (Berzborn, 1968, 1969; Böhme, 1978), it is logical to assume that the epoxidase is localized in its immediate vicinity.

Due to the presence of the de-epoxidase and epoxidase on different sides of the membrane the substrates Viol and Zea have to migrate from one side to the other in a type of shuttle. Since the de-epoxidase, and the epoxidase have very different pH optima, the xanthophyll cycle can only occur if a light -dependent proton gradient is established across the thylakoid membrane, i.e., under phosphorylating conditions. From these and the above mentioned findings one can make a number of assumptions concerning the function of the cycle.

13. Considerations concerning the function of the light-dependent xanthophyll conversions

a) It is possible that the ATP/NADPH ratio can be improved through a continuous decrease of the reduction energy in chloroplasts with the help of the xanthophyll cycle. This formation of extra ATP can be useful for a sufficient energy supply for CO_2 fixation, for transport mechanisms in chloroplasts, and for the synthesis of biopolymeres and proteins, etc. The following can be said about the magnitude of the improvement. If one considers that the quantity of Zea formed during the first 15–30 sec after illumination is the measure of the rate of the cycle, then one would expect a turnover of 2 to 3 μmol Viol/mg chlorophyll/h to result. Since 2 molecules NADPH per molecule Zea are used up during the back reaction and since it is probable that the same number is used up in the forward reaction (through the reduction of DHAsc), a total of 8–12 μmol NADPH/mg chlorophyll/h are oxidized. However, there is not yet any evidence that the cycle occurs at these measureably high starting velocities during illumination.

Furthermore, it is possible that the xanthophyll cycle is only important at given times, for example, during the change from dark to light. From the investigation of Heber and Krause (1972) it is known that a third of the total NADP amount in intact chloroplasts in the dark is still present in the reduced form. In addition, it must be assumed that a considerable amount of the other electron carriers are also present in the reduced state. This follows from the above-mentioned results showing that the de-epoxidation, which requires a reductant of $E'_0 = -0{,}058$ V, can also be started after long lasting darkness (Hager, 1975). A fast oxidation of the electron acceptors would provide good starting conditions for triggering photosynthetic electron transport upon illumination and forming ATP. This can occur by the participation of the fast starting xanthophyll cycle.

In this case, the relative increase in the ATP level gives the cycle an effect similar to that of cyclic photophosphorylation. There is one difference, however: during ATP formation from cyclic photophosphorylation NADPH remains accumulated; thus the noncyclic electron transport cannot be initiated. In the case of xanthophyll transformation NADPH is oxidized, and thus normal noncyclic electron transport and oxygen evolution can start.

b) The modulation of the activity of enzymes of the reductive pentosephosphate cycle, both in the photosynthetic procaryonts and in higher plants, appears to occur through light via the SH/SS relation (Duggan and Anderson, 1975; Anderson, 1974a). Foyer and Halliwell (1976) assume that glutathione stabilizes the enzymes of the Calvin cycle. Since coupling of the Asc/DHAsc and GSH/GSSG system occurs during de-epoxidation, it is possible that the xanthophyll cycle regulates photosynthetic carbon metabolism by means of these systems.

c) Yamamoto and Takeguchi (1972) suggest that the xanthophyll cycle could be part of a regulatory system for photosynthesis which functions by influencing membrane properties. However, to date there are no experimental indications which would support this suggestion.

This work has been supported by a grant from the Deutsche Forschungsgemeinschaft.

Summary

The known mechanisms inducing the light-dependent conversion of xanthophylls are reviewed and the importance of the xanthophyll cycle for the metabolism of chloroplasts is discussed.

Two types of the xanthophyll have been demonstrated in photosynthetic organisms. The reversible violaxanthin $\rightleftharpoons$ antheraxanthin $\rightleftharpoons$ zeaxanthin reaction occurs in seed plants, ferns, bryophytes, Chlorophyceae, and Phaeophyceae (Figs. 3 and 4). In most of the other algae the reversible diadinoxanthin $\rightleftharpoons$ diatoxanthin reaction was proven. Phycobiline-containing algae such as Cyanophyceae, Cryptophyceae, Rhodophyceae, as well as photosynthetic bacteria, do not have xanthophyll cycle.

Both of the reactions violaxanthin $\rightarrow$ zeaxanthin and diadinoxanthin $\rightarrow$ diatoxanthin represent a light-induced de-epoxidation (Figs. 1 and 3); and both of the reactions zeaxanthin → violaxanthin and diatoxanthin → diadinoxanthin represents an epoxidation by molecular oxygen.

Recently, cofactors of the xanthophyll cycle have been discovered, and some of the concerned enzymes have been characterized. These results have provided new insights into the mechanism of the xanthophyll cycle.

The enzyme catalyzing the *de-epoxidation* has a molecular weight of 54.000 and a pH optimum of 5.2 (corresponding to the pH optimum in vivo) (Figs. 5 and 6). The enzyme is inactive at pH values ≥ 7. The de-epoxidase is located on the inner surface of the thylakoid. During irradiation of chloroplasts the enzyme is activated by the drop in pH within the thylakoid due to photosynthetic electron transport (Figs. 7 and 8). Therefore, under certain conditions the formation of zeaxanthin (diatoxanthin) indicates a drop in pH within the thylakoid and the formation of a proton gradient across the thylakoid membrane. Since such a proton gradient can produce ATP (according to the chemiosmotic theory), zeaxanthin formation can also be used as an indicator for photophosphorylation. Uncouplers such as CCCP inhibit zeaxanthin formation (and ATP formation) by destroying the pH gradient (Figs. 5 and 7). Besides the light-induced drop in pH within the thylakoid, the de-epoxidation reaction needs ascorbate as reductant. The dehydroascorbic acid formed during this reactions is re-reduced to ascorbic acid by the coupled reaction of a NADPH-dependent glutathione reductase and a glutathione-dependent de-hydro-ascorbic acid reductase (Table, Figs. 9 and 14).

In contrast to the de-epoxidation, the *epoxidation* also occurs at pH values $\geq$pH 7 (pH optimum at pH 7.5; Fig. 13). The epoxidation is catalysed by a mixed-function oxygenase located on the outer surface of the thylakoid (Fig. 15). The reaction depends on oxygen and NADPH (Figs. 11, 14). It is independent of light, since chloroplasts also contain some reduced NADP in the dark. But the epoxidation can be stimulated by light due to the photosynthetic formation of NADPH and O_2.

Since de-epoxidase and epoxidase are located on opposite sides of the thylakoid membrane, the simultaneous occurence of de-epoxidation and epoxidation requires transport of the violaxanthin and zeaxanthin substrates, which are located within the membrane, from the inner side to the outer side of the membrane (Fig. 15). Due to the different pH optima of the two enzymes, maximum activity of the xanthophyll cycle depends on a light-induced proton gradient across the thylakoid membrane. This corresponds to phosphorylating conditions.

It is suggested that one *function of the xanthophyll cycle* is the destruction of reducing equivalents at the site of their formation within the chloroplast. This improves the final ATP/NADPH ratio. In addition, a (temporary) overreduction around the thylakoid can be prevented, which should inhibit noncyclic electron transport and O_2 evolution. Such a local accumulation of NADPH can be considered mainly in highly organized chloroplast systems which need ATP for active transport, metabolic and biosynthetic processes, and CO_2 fixation. Other possible functions of the xanthophyll cycle are discussed.

References

ANDERSON, L. E., T.-C. L. Ng, K.-E. Y. PARK, 1974: Inactivation of pea leaf chloroplastic and cytoplasmic glucose-6-phosphate dehydrogenases by light and dithiothreitol. Plant Physiol. **53**, 835–839.

BERZBORN, R., 1968: Nachweis der Ferredoxin-NADP-Reduktase in der Oberfläche des Chloroplasten-Lamellarsystems mit Hilfe spezifischer Antikörper. Z. Naturforschg. **23 b**, 1096–1104.

–, 1969: Untersuchungen über die Oberflächenstruktur des Thylakoidsystems der Chloroplasten mit Hilfe von Antikörpern gegen die Ferredoxin-NADP-Reduktase. Z. Naturforschg. **24 b**, 436–446.

BLASS, U., J. M. ANDERSON, and M. CALVIN 1959: Biosynthesis and possible relations among the carotenoids and between chlorophyll a and b. Plant Physiol. **34**, 329–333.

BÖHME, H., 1978: Reactions of antibodies against ferredoxin, ferredoxin-NADP$^+$ reductase and plastocyanin with spinach chloroplasts. Eur. J. Biochem. **84**, 87–93.

CHAPMAN, D. J., and F. T. HAXO, 1966: Chloroplast pigments of Chloromonadophyceae. Phycol. **2**, 89–91.

DUGGAN, J. X., L. E. ANDERSON, 1975: Light-regulation of enzyme activity. Planta **122**, 293–297

FOYER, C. H., and B. HALLIWELL, 1976: The presence of glutathione and glutathione reductase in chloroplasts: a proposed role in ascorbic acid metabolism. Planta **133**, 21–25.

–, –, 1977: Purification and properties of dehydroascorbate reductase from spinach leaves. Phytochemistry **16**, 1347–1350.

FRANKE, W., und U. HEBER, 1964: Über die quantitative Verteilung der Ascorbinsäure innerhalb der Pflanzenzelle. Z. Naturforschg. **19 b**, 1146–1149.

GERHARDT, B., 1964: Untersuchungen über Beziehungen zwischen Ascorbinsäure und Photosynthese. Planta **61**, 101–129.

GRUMBACH, K. H., H. K. LICHTENTHALER, and K. H. ERISMANN, 1978: Incorporation of $^{14}CO_2$ in photosynthetic pigments of *Chlorella pyrenoidosa*. Planta **140**, 37–43.

HAGER, A., 1966: Die Zusammenhänge zwischen lichtinduzierten Xanthophyll-Umwandlungen und Hill-Reaktion. Ber. Deutsch. Bot. Ges. **79**, 94–107.

–, 1967a: Untersuchungen über die lichtinduzierten, reversiblen Xanthophyllumwandlungen an *Chlorella* und *Spinacia oleracea*. Planta **74**, 148–172.

–, 1967b: Untersuchungen über die Rückreaktionen im Xanthophyll-Cyclus bei *Chlorella*, *Spinacia* und *Taxus*. Planta **76**, 138–148.

–, 1969: Lichtbedingte pH-Erniedrigung in einem Chloroplasten-Kompartiment als Ursache der enzymatischen Violaxanthin → Zeaxanthin-Umwandlung; Beziehungen zur Photophosphorylierung. Planta **89**, 224–243.

–, 1975: Die reversiblen, lichtabhängigen Xanthophyllumwandlungen im Chloroplasten. Ber. Deutsch. Bot. Ges. **88**, 27–44.

–, und T. MEYER-BERTENRATH, 1966: Die Isolierung und quantitative Bestimmung der Carotinoide und Chlorophylle von Blättern, Algen und isolierten Chloroplasten mit Hilfe dünnschichtchromatographischer Methoden. Planta **69**, 198–217.

–, und –, 1967: Die Identifizierung der an Dünnschichten getrennten Carotinoide grüner Blätter und Algen. Planta **76**, 149–168.

–, und H. PERZ, 1970: Veränderung der Lichtabsorption eines Carotinoids im Enzym (De-Epoxidase)-Substrat (Violaxanthin)-Komplex. Planta **93**, 314–322.

–, und H. STRANSKY, 1970a: Das Carotinoidmuster und die Verbreitung des lichtinduzierten

Xanthophyll-Cyclus in verschiedenen Algenklassen. I. Methoden zur Identifizierung der Pigmente. Arch. Mikrobiol. **71**, 132–163.
–, und –, 1970b: Das Carotinoidmuster und die Verbreitung des lichtinduzierten Xanthophyll-Cyclus in verschiedenen Algenklassen. III. Grünalgen. Arch. Mikrobiol. **72**, 68–83.
–, und –, 1970c: Das Carotinoidmuster und die Verbreitung des lichtinduzierten Xanthophyll-Cyclus in verschiedenen Algenklassen. V. Einzelne Vertreter der Cryptophyceae, Euglenophyceae, Bacillariophyceae, Chrysophyceae and Phacophyceae. Arch. Microbiol. **73**, 77–89.
HALLIWELL, B., and C. H. FOYER, 1978: Properties and physiological function of a glutathione reductase purified from spinach leaves by affinity chromatography. Planta **139**, 9–17.
HEBER, U., 1974: Metabolic exchange between chloroplasts and cytoplasm. Ann. Rev. Plant Physiol. **25**, 393–421.
–, and G. H. KRAUSE, 1972: Hydrogen and Proton transfer across the chloroplast envelope, 1023–1033. Proc. IInd Intern. Congr. Photosynth. Res. Stresa 1971, Dr. W. Junk N. V. Publ., The Hague.
ISLER, O., 1971: Carotenoids. Basel, Stuttgart: Birkhäuser.
JABLONSKI, P. P., and J. W. ANDERSON, 1978: Light-dependent reduction of oxidized glutathione by ruptured chloroplasts. Plant Physiol. **61**, 221–225.
KRINSKY, N. J., 1966: The role of carotenoid pigments as protective agents against photosensitized oxidations in chloroplasts. In: T. W. GOODWIN (ed.), Biochemistry of Chloroplasts I, 423–430. Academic Press, New York.
–, 1971: Function. In: O. Isler (ed.), Carotenoids, 669–716. Verlag Birkhäuser, Basel und Stuttgart.
LEE, R. E., 1972: Origin of plastids and the phylogeny of algae. Nature **237**, 44–46.
LEE, K. H., and H. Y. YAMAMOTO, 1968: Action spectra for light-induced de-epoxidation and epoxidation of xanthophylls in spinach leaf. Photochem. Photobiol. **7**, 101–107.
LOEBLICH, A. R. III, 1966: Aspects of the physiology and biochemistry of the Pyrrhophyta. Phykos, **5**, 216–255.
MITCHELL, P., 1966: Chemiosmotic coupling in oxidative and photosynthetic phosphorylation. Biol. Rev. **41**, 445–502.
–, 1967: Proton-translocation phosphorylation in mitochondria, chloroplasts and bacteria – natural fuel cells and solar cells. Fed. Proc. **26**, 1370–1379.
PERZ, H. F., 1970: Die enzymatische Umwandlung der Carotinoide des Xanthophyllzyklus im zellfreien System. Diss. Univ. München.
SAAKOV, V. S., 1965a: Metabolism of violaxanthin ^{14}C in the leaf and its role in photosynthesis reaction. Dokl. Akad. Nauk SSSR **165**, 230–233.
–, 1965b: On the possible role of xanthophylls in oxygen transfer during photosynthesis. Fiziol. Rastenij **12**, 377–385.
SANTARIUS, K. A., and C. R. STOCKING, 1969: Intracellulare localization of enzymes in leaves and chloroplast membrane permeability to compounds involved in amino acid synthesis. Z. Naturforschg. **24b**, 1170–1179.
SAPOZHNIKOV, D. J., 1969: Transformations of xanthophylls in the chloroplasts. In: H. Metzner (Ed.), Progress in Photosynthesis Research, Vol. II, 694–700.
–, 1972: Investigation of the violaxanthin cycle. In: 3. Intern. Symp. Carotenoids, Cluij, Romania, 1972. Butterworths, London.
–, D. G. ALKHAZOV, Z. M. EIDELMAN, N. V. BAZHANOVA, J. K. LEMBERG, T. G. MASLOVA, A. B. GIRSHIN, I. A. POPOVA, V. S. SAAKOV, O. F. POPOVA, and G. A. SHIRYAEVA, 1961:

The assimilation of ^{18}O from heavy-oxygen water into violaxanthin when the plant is exposed to the action of light. Bot. Zhur. **46**, 673–676.

–, Z. M. EIDELMAN, N. V. BAZHANOVA, and O. F. POPOVA, 1959a: The inhibitory effect of hydroxylamin on the photoreactions in the course of xanthophyll transformation. Dokl. Akad. Nauk SSSR **127**, 1128–1131.

–, L. R. KOLOTOVA, and Y. E. GILLER, 1966: Action spectrum of de-epoxidation of violaxanthin. Dokl. Akad. Nauk SSSR **171**, 740–741.

–, T. A. KRASOVSKAYA, and A. N. MAYEVSKAYA, 1957: Changes observed in the relation between the main carotenoids in the plastids of green leaves exposed to light. Dokl. Akad. Nauk SSSR **113**, 465–467.

–, T G. MASLOVA, N. V. BAZHANOVA, and O. F. POPOVA, 1965: Kinetics of inclusion of ^{18}O from $H_2{}^{18}O$ into Violaxanthin. Biofizika **10**, 349–351.

SIEFERMANN, D., 1972: Kinetik studies on the xanthophyll cycle of *Lemna gibba* L. – Influence of photosynthetic oxygen and supplied reductor. In: FORTI et al. (Eds.), IInd Intern. Congr. on Photosynthesis, Stresa 1971, 629–635. Dr. Junk N. V., The Hague.

–, and H. YAMAMOTO, 1974a: Light-induced de-epoxidation of violaxanthin in lettuce chloroplasts. III. Reaction kinetics and effect of light intensity on de-epoxidase activity and substrate availability. Biochim. Biophys. Acta **357**, 144–150.

–, –, 1974b: Light-induced de-epoxidation of violaxanthin in lettuce chloroplasts. V. Dehydroascorbate, a link between photosynthetic electron transport and de-epoxidation. In: Proc. 3rd Int. Congr. Photosynth. Res. Avron, M. (ed.), Amsterdam: Elsevier, 1975, Vol. 3, pp. 1991–1998.

–, –, 1975a: Light-induced de-epoxidation of violaxanthin in lettuce chloroplasts. IV. The effect of electron transport conditions on violaxanthin availability. Biochim. Biophys. Acta **387**, 149–158.

–, –, 1975b: NADPH and oxygen-dependent epoxidation of zeaxanthin in isolated chloroplasts. Biochem. Biophys. Res. Com. **62**, 456–461.

–, –, 1975c: Properties of NADPH and oxygen-dependent zeaxanthin epoxidation in isolated chloroplasts. A transmembrane model for the violaxanthin cycle. Arch. Biochem. Biophys. **171**, 70–77.

STRANSKY, H., 1976: Studies on carotenoid metabolism in higher plants. In: Abstracts of the Symposium on lipids and lipid polymers in higher plants, Botanical Institut of Karlsruhe, W-Germany, p. 89.

–, und A. HAGER, 1970a: Das Carotinoidmuster und die Verbreitung des lichtinduzierten Xanthophyll-Cyclus in verschiedenen Algenklassen. II. Xanthophyceae. Arch. Mikrobiol. **71**, 164–190.

–, und –, 1970b: Das Carotinoidmuster und die Verbreitung des lichtinduzierten Xanthophyll-Cyclus in verschiedenen Algenklassen. IV. Cyanophyceae und Rhodophyceae. Arch. Mikrobiol. **72**, 84–96.

–, und –, 1970c: Das Carotinoidmuster und die Verbreitung des lichtinduzierten Xanthophyll-Cyclus in verschiedenen Algenklassen. VI. Chemosystematische Betrachtung. Arch. Mikrobiol. **73**, 315–323.

TAKEGUCHI, D. A., and H. Y. YAMAMOTO, 1968: Light-induced $^{18}O_2$ uptake by epoxy xanthophylls in New Zealand spinach leaves (*Tetragonia expansa*). Biochim. Biophys. Acta **153**, 459–465.

TÔYAMA, S., 1974: Electron microscope studies on the morphogenesis of plastids. IX. The phylogeny of photosynthetic apparatus with special reference to lamellar structure. Jap. J. Bot. 20, 413–438.

TREBST, A., and M. AVRON, 1977: Photosynthesis I, Encyclopedia of Plant Physiol. Vol. 5. Berlin: Springer.

WESTERHOFF, N., 1974: Beziehungen zwischen den lichtabhängigen Xanthophyll-Umwandlungen und dem photosynthetischen Elektronentransport bzw. der Photophosphorylierung Ber. Deutsch. Bot. Ges. 87, 545–551.

–, 1975: Beziehungen zwischen den lichtabhängigen, reversiblen Pigmentumwandlungen des Xanthophyllcyclus und dem photosynthetischen Elektronentransport sowie der Photophosphorylierung. Dissertation Univ. Münster.

YAMAMOTO, H. Y., J. L. CHANG, and M. S. AIHARA, 1967: Light-induced interconversion of violaxanthin and zeaxanthin in New Zealand spinach-leaf segments. Biochim. Biophys. Acta **141**, 342–347.

–, –, CHENCHIN, E. E., YAMADA, D. K., 1974: Effect of chloroplast lipids on violaxanthin de-epoxidase activity. Proceeding Third Intern. Congr. Photosynthesis, 1974, Rehovot, Israel, p. 1999–2006, Elsevier Scientific Publ. Comp., Amsterdam.

–, H. Y., and C. O. CHICHESTER, 1965: Dark incorporation of oxygen into antheraxanthin by bean leaf. Biochim. biophys. Acta (Amsterdam) **109**, 303–305.

–, –, and T. O. M. NAKAYAMA, 1962a: Xanthophylls and the Hill reaction. Photochem. Photobiol. **1**, 53–57.

–, H. Y., HIGASHI, R. M., 1978: Violaxanthin de-epoxidase; lipid composition and substrate specificity. Arch. Biochem. Biophys. **190**, 514–522.

–, T. O. M. NAKAYAMA, and C. O. CHICHESTER, 1962b: Studies on the light and dark interconversions of leaf xanthophylls. Arch. Biochem. **97**, 168–173.

–, and C. A. TAKEGUCHI, 1972: Concepts on the role of epoxy carotenoids in plants. In: Forti et al. (Eds.), Proc. IInd Intern. Congr. Photosynthesis, Stresa 1971, 621–627. Dr. Junk N. V., The Hague.

Photoregulation of carotenoid biosynthesis: an example of photomorphogenesis.

Werner Rau

Botanisches Institut der Universität München

1. Introduction

Morphosis is defined as the specific morphological properties of a plant that have developed and differentiated under the influence of a particular exogenous (environmental) factor. Thus the term «morphosis» characterizes the final result of the development while the sequence of developmental steps leading to this result is designated as «morphogenesis». If light is the exogenous factor, we are dealing with «photomorphogenesis».

The major problem in photomorphogenesis is the question of how light regulates and controls the expression of competence provided by the genetic information of a particular organism, or, in other words, the question of the mechanism of photoregulation. Hence, the aim of research in this field is to elucidate the sequence and biochemical nature of the processes in a given organism that intervene between the absorption of irradiation and the responses thereto, i.e., the ultimate morphogenic phenomena. In a classical sense the term morphogenetic response would be defined as an alteration of the plant's form («formative Wirkung»). However, since the basis of such morphoses are specific alterations of metabolism, such as the production of cellular compounds and of the enzymes catalyzing such syntheses, we presently extend the term photomorphogenesis to all processes occuring during a photoregulated development of a plant. This definition of photomorphogenesis will be used in the following chapters.

Consequently, in this article our present knowledge of the mechanism of photomorphogenesis in general will be briefly summarized; more attention will be given to a particular response, i.e., the «blue-light»-induced biosynthesis of carotenoids in fungi which can serve as an example of photomorphogenesis in a broader sense.

2. On the mechanism of photoregulation of morphoses

The sequence of reactions of the regulatory chain may be divided essentially into two parts:

1. Absorption of light by a photoreceptor for the particular responses and the mechanism of the photoreceptor's transformation by irradiation.
2. Subsequent processes evoked by the transformations of the photoreceptor.

2.1. Photoreceptors and primary reactions

The usual technique employed to identify a photoreceptor which is responsible for a certain response includes comparison of the action spectrum of a response to the absorption spectra of putative pigments. For plant responses to illumination three important types of action spectra are known, except for several special cases. An example of each group is shown in Fig. 1.

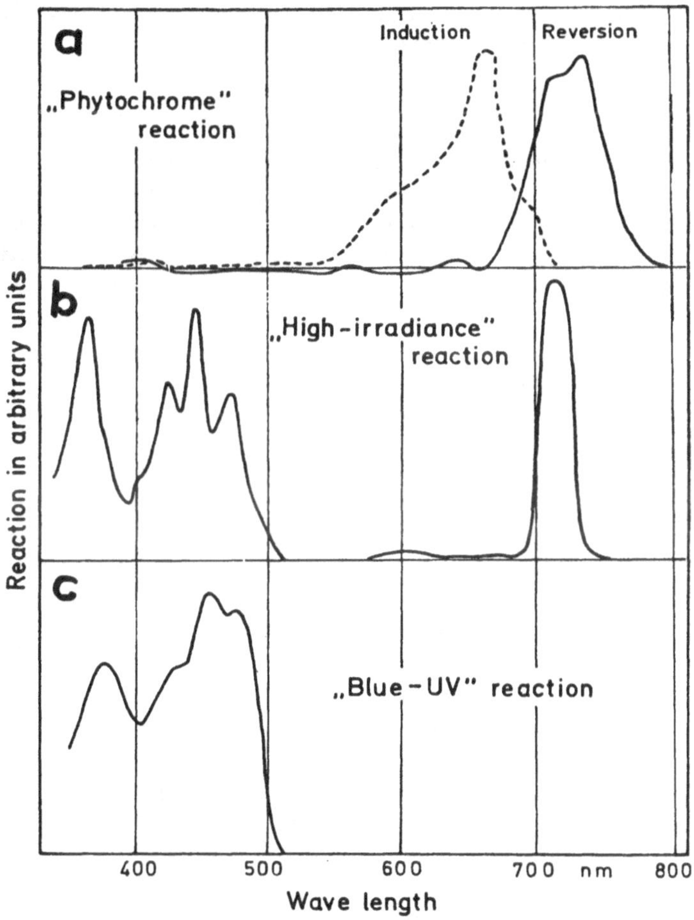

Fig. 1: Comparison of three types of action spectra of photomorphoses. a: «Phytochrome» action spectrum (from Withrow et al., 1957). b: Action spectrum of «high irradiance response» of growth inhibition in hypocotyles of *Lactuca sativa* (from Hartmann, 1967). c: Action spectrum of photoinduction of carotenoid biosynthesis in *Fusarium aquaeductuum* (from Rau, 1967a)

2.1.1. Action spectrum of phytochrome

Many responses are induced by a brief illumination with red light showing the well-known action spectrum (Fig. 1a) with maximum effect at 660 nm (R). Further characteristics are saturation of the response by low irradiation energies and reversion of the effect by a subsequent short exposure to far-red light (Fig. 1a). In these cases the acting photoreceptor is undoubtedly phytochrome. The primary reaction is the transformation of P_{660} to the P_{730} form. P_{730} is the morphogenetically «active» form of phytochrome triggering all subsequent processes. Typical examples of this mode of action, which are referred to as the «low energy phytochrome system», are photoregulation of seed germination and photoperiodically controlled induction of flowering, which exclusively depends on periodical light breaks.

2.1.2. Action spectrum of «high irradiance reaction» (HIR)

Numerous photomorphoses are controlled by a prolonged irradiation. The corresponding action spectrum reveals two major spectral regions with high quantum efficiency, namely, the far-red light (λ = 710–720 nm) and the blue-to-near-UV light as shown in detail in Fig. 1b; but not all action spectra of presumed HIR show such a sharp peak in the far-red light (e.g., Wagner and Mohr, 1966), some even have a maximum in the red part of the spectrum (for ref. see Jose and Vince-Prue, 1978). Since these action spectra are obtained only with prolonged or continuous illumination and with relatively high irradiation energy, the corresponding responses were designated as «high energy reactions» or «high irradiance reactions (HIR)». In contrast to the «classical» phytochrome reaction, the HIR is irreversible and is irradiance-dependent. The identification of the photoreceptor responsible for the HIR is complicated by the fact that – besides the interference with the low-energy reaction – to date no naturally occuring plant pigment has been identified that has an absorption which is identical to the HIR action spectrum. Although phytochrome was demonstrated to absorb light also of the UV-blue region (Siegelman and Butler, 1965), it does not show a corresponding complex pattern of absorption. Results obtained by dual-wavelength irradiation (Hartmann, 1967) provide convincing evidence that phytochrome is the photoreceptor for the far-red region of the spectrum; for the UV-blue region, however, phytochrome is – despite a proposal by Hartmann and Menzel (1969) and the calculations of Schäfer (1975) – a most unlikely candidate. An alternative suggestion will be discussed in the following chapter.

The mode of «primary action» in the HIR is still unknown, but several models and hypotheses to explain it have been proposed. With the two exceptions mentioned above, these explanations only refer to phytochrome and neglect the «blue light photoreceptor». The assumptions are more or less based on, or partially take into account, the following observations:

- Continuous irradiation with far-red (or even with red light) maintains a low but constant P_{730} level over a long period of time («photostationary state»).
- P_{730} may be subject to destruction (or «decay») indicated by the loss of photoreversible spectral changes.

- The physiological responses usually depend, over a wide range, on the irradiance.
- Some HIR responses show a «threshold» reaction in an all-or-none manner.
- Phytochrome undergoes permanent «cycling», i.e., an interconversion between P_{660} and P_{730} during continuous irradiation with light of suitable wavelengths.
- A significant proportion of the phytochrome present in vivo exists in intermediate forms under high-intensity light; these intermediates (or an «activated» P_{730}) are thought to play a predominant role in photoinduction.
- Active phytochrome molecules are associated with membranes or are photoreversibly membrane-bound and may exert a positive reaction only in such a form.

Since the problems have been discussed thoroughly in many valuable reviews which emphasize the different points of view (Hartmann, 1966; Mohr, 1972; Shropshire, 1972; Mitrakos and Shropshire, 1972; Deutch and Deutch, 1974; Schäfer, 1975; Smith, 1975; Vince-Prue, 1975; Schopfer, 1977; Jose and Vince-Prue, 1978) primary action should not be stressed here in more detail.

2.1.3. Action spectrum of blue light

Several photoregulated phenomena are triggered exclusively by near UV and blue light;.in the following chapters they will be covered by the term «blue light responses». For such responses only light with wavelengths of 350–500 nm is effective; this spectral region was first examined in detail for photoinduced movements, namely phototropism and chloroplast movement. Action spectra of some of these processes are compiled in Fig. 2. They show major peaks of effectiveness at 370–380 nm and three peaks, or at least shoulders, between 400 and 500 nm; light of wavelengths longer than 520 nm is entirely ineffective. Further examples of such reactions induced exclusively by blue light are the induction of two-dimensional growth of fern gametophytes (Mohr, 1956), development of chloroplasts (e.g., Berger and Bergmann, 1967), maintenance of photosynthetic activity in chloroplasts (Clauss, 1970; Schmid and Clauss, 1977), the effect of light on algal respiration (Kowallik, 1967), and in a broader sense the control of photosynthetic products (for ref. see Voskresenskaya, 1972; Kandeler, 1972). The nature of the photoreceptor as well as possible primary reactions will be discussed along with photoinduced carotenoid synthesis in the second part of this article.

Comparison of the action spectra of blue light responses with those of HIR (Fig. 1) reveals the striking similarity in the spectral region of 350–500 nm which has already been pointed out by others, for example, by Shropshire (1972). As mentioned in the preceding chapter, there have been previous attempts to explain blue light responses in terms of phytochrome (Hendricks and Borthwick, 1959; Hartmann, 1966; Hartmann and Menzel, 1969). For the same purpose Schäfer (1975) recently calculated an action spectrum for continuous irradiation responses based on his model of phytochrome action; the shape of this spectrum is in good agreement with the action spectra of HIR responses in the far-red part but is obviously different in the UV-blue region. We therefore assumed (Rau, 1975), based on suggestions of other authors, (cf. Downs and Siegelman, 1963; Grill and Vince, 1965;

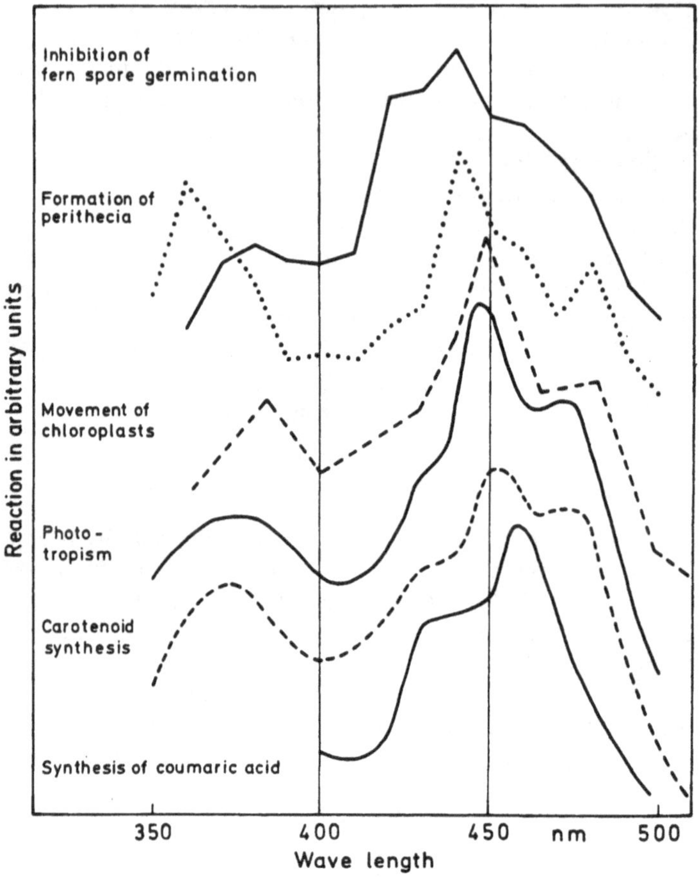

Fig. 2: Comparison of the action spectrum of carotenoid biosynthesis in *Fusarium aquaeductuum* with action spectra of different developmental and movement responses in various plants induced by «blue light». From top to bottom: Inhibition of phytochrome-mediated germination of spores of *Pteris vittata* (after Sugai, 1971); induction of perithezia in *Nectria haematococcus* (after Curtis, 1972); positive phototaxis of chloroplasts in *Lemna trisulca* (after Zurzycki, 1972); phototropism in *Phycomyces* (after Curry and Gruen, 1959); carotenoid synthesis (after Rau, 1967a); biosynthesis of coumaric acid (after Engelsma, 1967)

Engelsma, 1967) that the blue and the far-red activities in plant photomorphogenesis are mediated by distinct photochemical systems; the proposal of two photoreceptors is widely accepted (for ref. see Smith, 1975; Jose and Vince-Prue, 1978). In this connection certain photoresponses have to be referred to for which the involvement of the two photoreceptors has been unequivocally demonstrated, namely, the photoregulation of anthocyanin accumulation in *Sorghum* (Downs and Siegelman, 1963), *Fagopyrum* (Mohr and van Nes, 1963; Scherf and Zenk, 1967), gherkin (Engelsma and Meijer, 1965) and corn (Duke and Naylor, 1976), and development of plastids in roots (Björn, 1967). These data had been neglected for a long time in previous discussions; only recently have they been revived and served to prompt further investigations (Drumm and Mohr, 1978).

2.2. Processes following photoinduction

Many photoregulated responses, especially photoinduced activity changes of various enzymes, are curtailed or at least reduced by inhibitors of protein synthesis. From these results it has been concluded that de novo synthesis of enzymes may be a consequence of irradiation; this assumption was confirmed for two enzymes by density-labelling experiments (for ref. and discussion see Schopfer, 1977). Mainly based on data indicating inhibition of phytochrome-mediated responses by substances known to inhibit nucleic acid and protein synthesis, Mohr (1966) postulated the hypothesis of «differential gene expression», i.e., photoregulation of transcription was proposed, which, as a consequence, implies the involvement of de novo synthesis of specific mRNAs. Several arguments against this hypothesis as a general mechanism

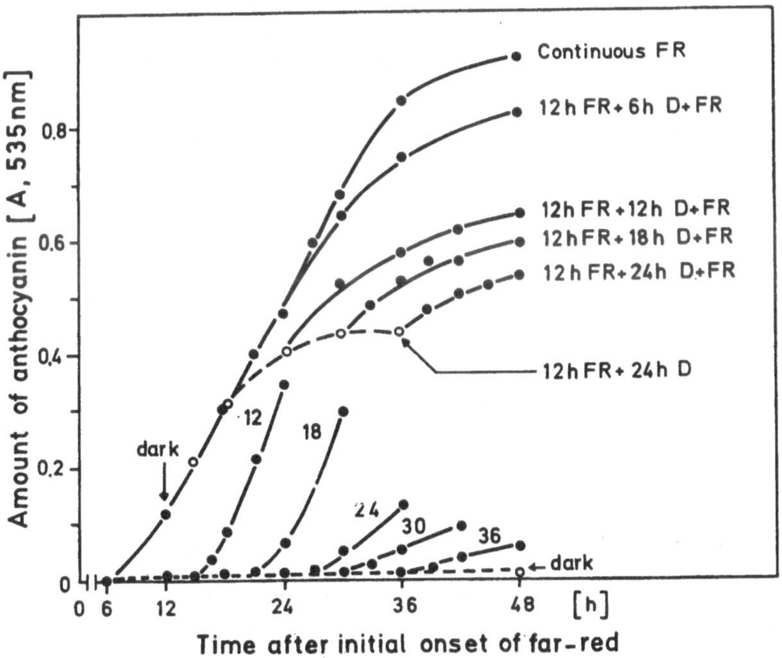

Fig. 3: Initial lag phase (lower part) and secondary lag phase (upper part) of far-red mediated anthocyanin synthesis in *Sinapis alba*. To determine the secondary lag phase the seedlings were irradiated for 12 hours with far-red, placed in darkness and re-irradiated with far-red after dark intervals of different duration as indicated in the figure. The numbers (lower part of the figure) indicate the time of onset of far-red light (after Lange et al., 1967)

of photoregulation have been raised. Detailed and basic discussions of the problem have recently been published elsewhere (Smith, 1975; Schopfer, 1977), but major points of criticism should briefly be mentioned here. Inhibitors have been questioned as valuable tools because of their possible unspecificity. Also, several phytochrome-mediated photoresponses are short-term activation or inactivation processes, i.e., rapid, light-dependent changes without a lag phase, an example of which is shown in Fig. 3. For this mode of action, Schopfer (1973) proposed the model of «photomodulation», rather than an induction mechanism. Other responses happen so rapidly that it is inconceivable by reasons of timing that the chain of events connecting them to the initial stimulus should include the synthesis of RNA and protein molecules. Such results were interpreted to indicate that light mediated changes in membrane properties trigger the subsequent reactions (for ref. see Briggs and Rice, 1972; Smith, 1975).

Hence, in contrast to the gene expression hypothesis, it has been argued that light might regulate mRNA translation during protein synthesis, rather than regulating mRNA synthesis through transcription, or even might control the activation of pre-existing inactive enzyme molecules rather than be involved in enzyme synthesis. Finally, it should be emphasized that – as many authors have pointed out – more than one «primary reaction» of the photoreceptors as well as multiple levels of regulation of the subsequent processes could be involved in photomorphogenesis.

3. Photoregulation of carotenoid biosynthesis in fungi

In several species of bacteria and fungi, biosynthesis of carotenoids is strictly photoregulated. The investigations in our laboratory were carried out mainly with the imperfect fungus *Fusarium aquaeductuum*, partly also with *Neurospora crassa*. The following chapter will essentially deal with the results obtained from these two

Table 1. Species of fungi in which carotenogenesis is strictly photoregulated

Organism	Authors
Aspergillus giganteus mut. alba	Zurzycka (1963)
Cephalosporium diospyri	Codner and Platt (1959)
Dacryopinax spathularia	Goldstrohm and Lilly (1965)
Fusarium aquaeductuum	Rau and co-workers (1959–1978)
Fusarium coeruleum	Rau et al. (unpubl. data)
Fusarium oxysporum	Carlile (1956); Rau et al (unpubl. data)
Neurospora crassa	Went (1901); Haxo (1949); Zalokar (1954); Harding and Mitchell (1968); Rau et al. (1968)
Neurospora sitophila	Ishii and Akagi (1948)
Pyronema confluens	Carlile and Friend (1956)
Sphaerobolus stellatus	Friederichsen and Engel (1957)
Syzygites megalocarpus	Wenger and Lilly (1966)
Verticillium agaricinum	Valadon and Mummery (1971)

species of fungi. But the phenomenon of photoinduced carotenogenesis is known to occur in numerous other fungal species as listed in Table 1. The presentation of our progress in elucidating the regulatory mechanism is arranged according to the sequence of events that presumably follow illumination of the organisms.

3.1. Photoreceptor

The action spectrum of the induction of carotenoid synthesis is shown in Fig. 4 in comparison with absorption spectra of a flavoprotein and of a carotenoid dissolved in a mixture of ethanol/H_2O; on account of the similarities of the spectra both

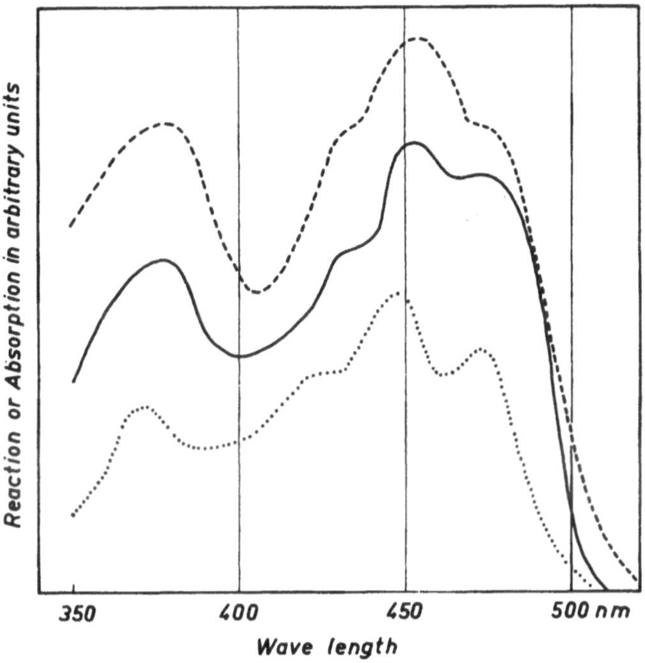

Fig. 4: Comparison of the action spectrum for photoinduced carotenoid synthesis: —— (Rau, 1967a) with absorption spectra of a flavoprotein: ---- (from Robinson et al., 1962) and of lutein dissolved in a mixture of ethanol/H_2O: · · · · · (from Hager, 1970)

pigments might be considered to act as a photoreceptor. Especially regarding phototropism, many arguments for or against either of the two candidates have been discussed for a long time (for ref. see Song et al., 1972), and this debate need not be repeated here in detail. Song and Moore (1974) presented convicing evidence ruling out carotenoids as photoreceptors. Therefore, in accordance with previous suggestions by many other authors, they conclude that the photoreceptor for phototropism is a flavin or flavoprotein; we favour this interpretation and would like to extend it to all other «blue-light» responses.

3.2. «Primary reaction»

Only a brief exposure to light is sufficient to induce carotenoid synthesis in fungi, and also in some bacteria, tested for photoregulation to date (for ref. see Batra, 1971). The amount of pigment produced is proportional to the logarithm of the incident light quantity (Rau, 1967a). Saturation of photoinduction was found only with *Neurospora crassa* (Zalokar, 1955; Rau et al., 1968), while in all examined *Fusarium* species saturation of the photoreaction could not be reached either under continuous light (Rau, 1967a) or with high illumination (Theimer 1968). Following a brief irradiation, carotenoid synthesis starts only after a «lag phase» and continues for a period of 8 to 36 h in the dark (Fig. 9). The length of both the lag phase and the period of pigment accumulation is distinctively different for the various organisms.

The finding that the photoreaction is independent of the temperature (Zalokar, 1956; Rau, 1962; Harding, 1974) indicates that the «primary reaction» is a photochemical process. What kind of reaction can we expect? The following results are taken for evidence that a photooxidative step is involved in the «primary reaction».

a) Oxygen is essential for optimum photoinduction, but to some extent induction also takes place without O_2. The data compiled in Fig. 5 show that under anaerobic conditions saturation of the photoreaction is reached at a relatively low light dosage and proved to be independent of both light intensity and time of illumination. Furthermore, mycelia which are subsequently aerated with O_2 are susceptible to an additional photoinduction (Rau, 1969). Similar results demonstrating the induc-

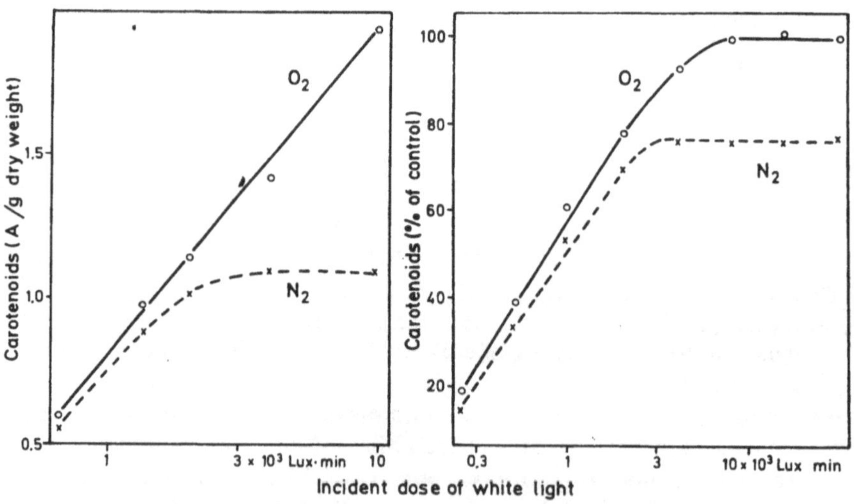

Fig. 5: Dose-response relationships of photoinduced carotenogenesis for irradiation in O_2 or N_2 atmosphere. The amount of carotenoids was determined after completion of pigment accumulation (under continuous aeration). Left: *Fusarium aquaeductuum*. Right: *Neurospora crassa*; amount of carotenoids calculated in % of pigment content of a sample irradiated with the highest dosage under O_2 (from Rau, 1969)

tion of pigment synthesis under anaerobic conditions were reported by Zalokar (1954) and Trinci and Banbury (1969). In contrast, Howes et al. (1969) found no photo-induction of carotenoid synthesis in a *Mycobacterium* sp. under a N_2 atmosphere. But one may conclude at least for the fungal system that oxygen is not involved directly in the photochemical reaction of the photoreceptor, but rather functions as an electron acceptor keeping the photoreceptor in a proper state of oxidation.

b) Dithionite, a strong reducing substance, specifically prevents photoinduction when applied to the mycelia during or up to a few minutes after illumination, but it is completely ineffective when added approximately 20 min after the onset of irradiation (Fig. 6). On the other hand, treatment of the mycelia with hydrogen peroxide in the dark may, at least to a certain degree, substitute for light in the induction of carotenogenesis (Theimer and Rau, 1970).

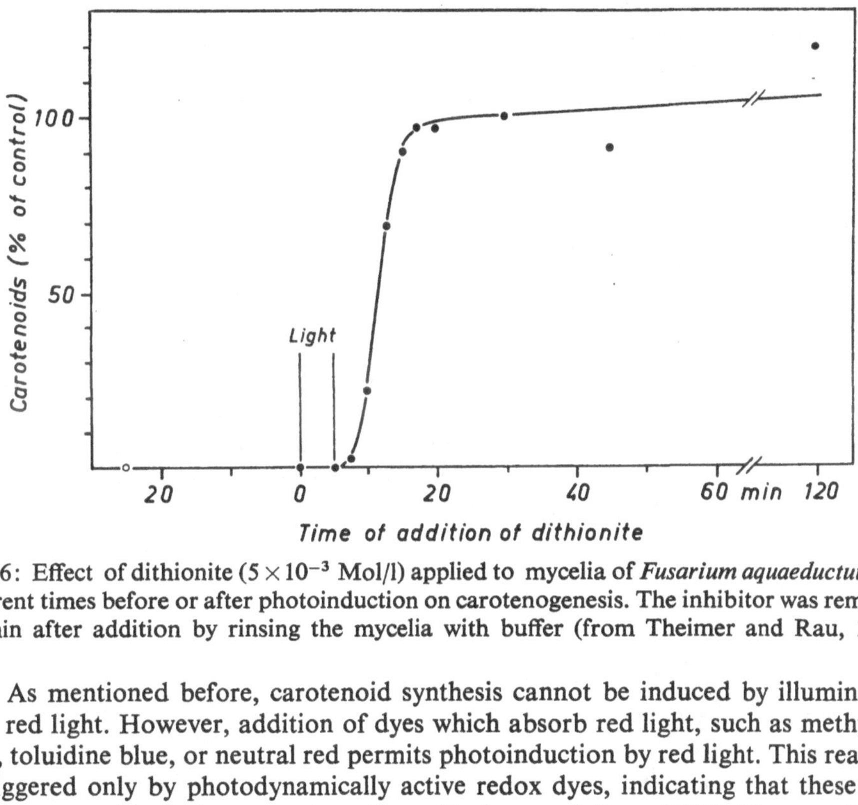

Fig. 6: Effect of dithionite (5×10^{-3} Mol/l) applied to mycelia of *Fusarium aquaeductuum* at different times before or after photoinduction on carotenogenesis. The inhibitor was removed 30 min after addition by rinsing the mycelia with buffer (from Theimer and Rau, 1970)

c) As mentioned before, carotenoid synthesis cannot be induced by illumination with red light. However, addition of dyes which absorb red light, such as methylene blue, toluidine blue, or neutral red permits photoinduction by red light. This reaction is triggered only by photodynamically active redox dyes, indicating that these dyes may act as artifical photoreceptors (Lang-Feulner and Rau, 1975). The question is still open as to whether the redox dyes are able to replace the natural photoreceptor, or whether the dyes act as an additional receptor by transferring the energy to the natural photoreceptor; however, recent results (Britz et al., 1979) tend to support the former assumption.

In the past few years direct evidence for redox reactions as a consequence of irradiation has been accumulated for some «blue-light»-sensitive systems. Blue light irradiation of a partially anaerobic mycelium of Neurospora crassa yielded reduction of a b-type cytochrome; the initial reaction was shown to be the photoreduction of a flavin moiety (Muñoz et al., 1974; Muñoz and Butler, 1975). A similar electron transfer was also reported for different *in vitro* systems (for ref. see Ninnemann et al., 1977). These light-inducible reductions were shown to be localized in particulate fractions obtained from *Neurospora* mycelium and corn coleoptiles (Brain et al., 1977). Most recently, Briggs and co-workers found that illumination with red light also effected the reduction of a b-type cytochrome in particulate fractions from corn coleoptiles in the presence of methylene blue. They discussed their results and the current literature in some detail (Britz et al., 1979). Summarizing the data from our laboratory, and considering the results from other groups, we may draw a hypothetical scheme illustrating the events during the «primary reaction» of the «blue-light» – photoreceptor (Fig. 7). For this scheme we assume that irradiation of the photoreceptor causes oxidation of a yet unknown »substance», X_{red}. The resulting »photooxidation product» is rapidly stabilized by subsequent reactions, presumably within 20 min in Fusarium (Fig. 6). The reduced photoreceptor is then re-oxidized by transferring electrons to cytochrome and/or finally to O_2. Two «key-substances» in the reaction chain may be considered as «triggers» for further induction steps: the reduced component (cytochrome ?) or the oxidized component (X_{ox}). Our data on the effect of dithionite and H_2O_2 support the assumption that X_{ox} is the more likely candidate.

A striking parallelism of these events to redox reactions involved in phytochrome transformation should be mentioned: Mumford and Jenner (1971) found that the dark reversion $P_{730} \rightarrow P_{660}$ of isolated phytochrome is markedly enhanced by adding reducing agents, both dithionite or reduced ferredoxin being especially effective. But at the moment, an assumption of a similarity between, or even an identity, of primary reactions during phytochrome-mediated photoinductions to that of the blue light receptor is mere speculation.

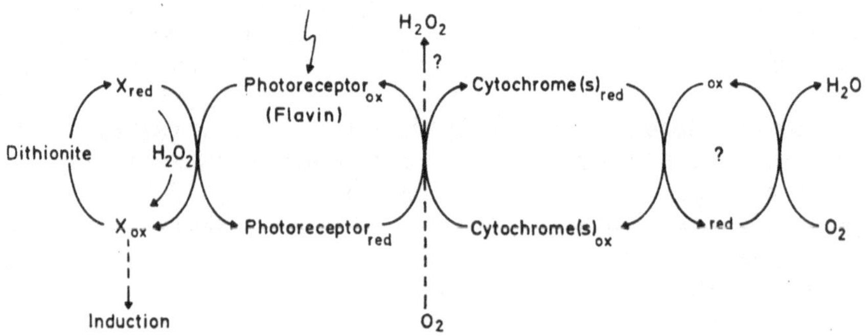

Fig. 7: Scheme illustrating the proposed events during the «primary reaction» of the «blue light» photoreceptor

3.3. Acetylcholine

Since the detection of rapid phytochrome-mediated responses (i.e., leaf movements, Fondeville et al., 1966) a change of the permeability of membranes – as already mentioned in chapter 2. 2. – has been considered one of the primary steps. After Jaffe (1970) had discovered that the mammalian neurohumour acetylcholine is also present in plant tissues, and that its level is regulated by phytochrome, this substance was subsequently shown to regulate a number of photomorphoses (for ref. see Briggs and Rice, 1972; Kandeler, 1972). For example, Gressel et al. (1971) investigated the blue-light-dependent sporulation of the fungus *Trichoderma* and reported on the substitution of acetylcholine for light; but simultaneous inhibition of acetylcholine-esterase with eserin proved to be essential. We also investigated the effect of acetylcholine on the induction of carotenoid synthesis in *Fusarium aquaeductuum*. The mycelia were kept in the dark and acetylcholine was added up to a concentration of 10^{-4} Mol/l both without eserin or together with eserin at various concentrations. Although the experiments were repeated many times, we were neither able to induce pigment synthesis by acetylcholine under aerobic, nor under anaerobic conditions (Jochum-Felbermeir, 1977). Similarly, Kasemir and Mohr (1972) could not find any effect of acetylcholine on phytochrome-controlled processes in *Sinapis*.

3.4. Involvement of nucleic acids

Is there any evidence for the involvement of gene de-repression and synthesis of specific mRNAs in photoinduced carotenoid synthesis? Rather indirect evidence comes from the following results. Light dependent pigment synthesis is completely blocked by cycloheximide, a potent inhibitor of cytoplasmic protein synthesis when applied prior to illumination. However, this inhibition is reversible: when cycloheximide was washed out of the mycelium at various times in subsequent darkness the data compiled in Fig. 8 were obtained after a sufficient period of time for enzyme and pigment synthesis. We drew two conclusions from these results: 1. Photoinduction takes place even under conditions of blocked protein synthesis; this is demonstrated by carotenoid production in the dark simply after removing the protein synthesis block. Therefore, irradiation must induce the formation of an «induction product» which ranges prior to translation and enzyme synthesis in the sequence of photoinduced reactions; 2. The «photoinduction product» appears to be rather stable, which is indicated by the finding that the same amount of carotenoids was synthesized after withdrawal of cycloheximide as would have been produced after a prolonged treatment with the drug (compare both curves in Fig. 8). Therefore, the induction product cannot be identical with the «photooxidation product» (see 3.2) which is completely stabilized 20 min after irradiation (see Fig. 6) (Rau, 1971; see also Harding and Mitchell, 1968).

In contrast to anthocyanin synthesis (Fig. 3) and to phytochrome-mediated carotenoid synthesis in *Sinapis* (Schnarrenberger and Mohr, 1970), the blue-light-dependent carotenoid synthesis in fungi shows a secondary lag phase after a second

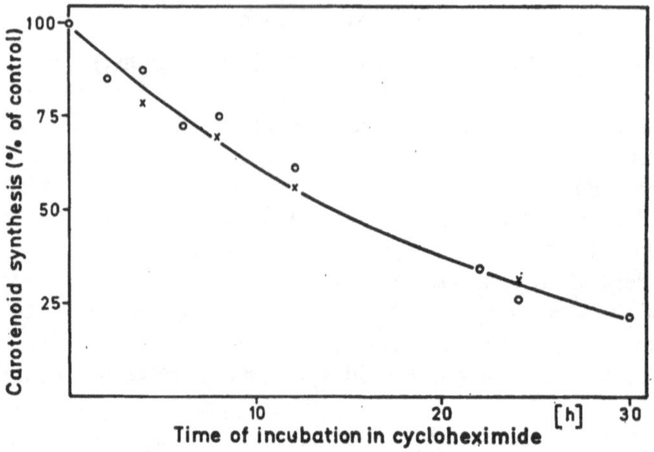

Fig. 8: Stability of photoinduction during blocked protein synthesis (○—○): cycloheximide applied prior to irradiation, washed out at different times prior to (control sample) or after irradiation, and the amount of carotenoids determined after completion of pigment accumulation. «Ability for synthesis» of the mycelium (×—×): Irradiation only after removing cycloheximide at different times (from Rau, 1971)

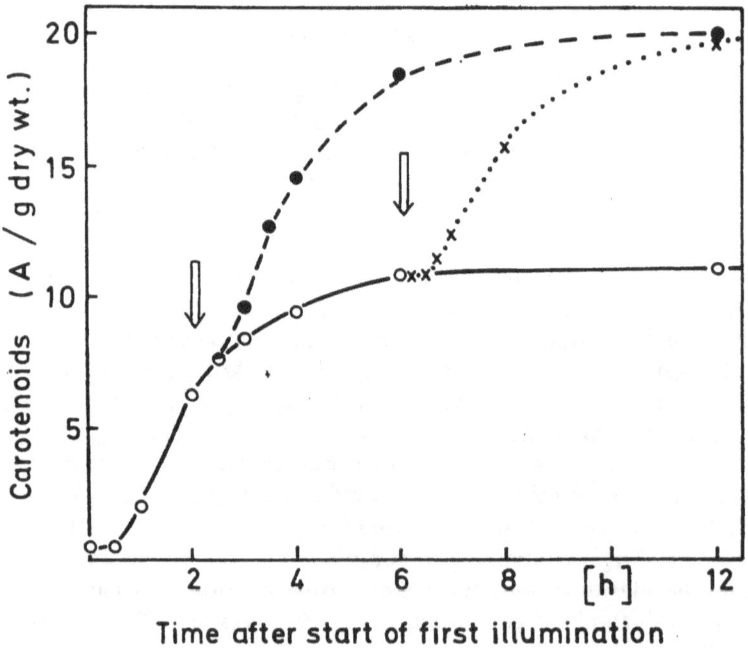

Fig. 9: Time course of carotenoid accumulation in *Neurospora crassa* after first and second illumination (10 min, white light). ○—○: first illumination only; ●--●: second illumination after 2 h; ×...×: second illumination after 6 h. Arrows indicate time of second illuminations (from Rau and Rau-Hund, 1977)

light treatment (Fig. 9) (Rau and Rau-Hund, 1977). Consequently, the mode of regulation in the latter case seems to be different from the «modulation» proposed for the other systems.

Until recently an inhibitor of transcription which completely and specifically blocks photoinduced carotenogenesis in fungi had not been found; transcription is only partially inhibited by actinomycin D in *Verticillium agaricinum* (Mummery and Valadon, 1973) and *Neurospora crassa* (Subden and Bobowski, 1973). We finally succeeded in finding a potent inhibitor. Incubation of the mycelia with distamycin A added prior to illumination in a concentration of 3×10^{-4} Mol/l fully blocked photoinduced pigment production without severe reduction of oxygen uptake by the mycelia. The specificity of distamycin A was tested by an examination of the incorporation of radioactively labelled uridine and leucine. The data indicate a differential inhibition of the synthesis of both nucleic acids and proteins (Felbermeir and Rau, 1974). Several attempts to obtain direct evidence for an involvement of de novo synthesis of mRNA in photoregulatory processes have so far been unsucessful (Dittes and Mohr, 1970; Harel and Bogorad, 1973). Despite some advantages of our system («step down» conditions to reduce other syntheses; short periods of irradiation and labelling with specific precursors) we also failed, although a presumable enrichment of a nucleic acid fraction with mRNA was attempted by isolating polyribosomes (Schrott and Rau, 1975). However, using affinity chromatography of double labelled polysomal RNA on poly(U)-sepharose, we were recently able to demonstrate different $^3H/^{14}C$ ratios for the bound poly (A) containing fraction (mRNA) when compared to the unbound fraction: when [^3H] uridine was present in the light-induced sample and [^{14}C] uridine in the dark control, the $^3H/^{14}C$ ratio in the bound material was found to be higher than that of the unbound fraction, and vice versa. No such shift was observed when both sample and control were kept in the dark. Thus our data may be interpreted to provide conclusive evidence for a photoinduced de novo synthesis of mRNA (Schrott and Rau, 1977). Similar evidence for light-dependent formation of mRNA by some nonfungal plant systems has recently accumulated (i.e., Verdier, 1975; Rosner et al., 1975; Gressel, 1978).

3.5. De novo synthesis of enzymes

As mentioned before (see 3.4.) photoinduced carotenoid accumulation in fungi is blocked completely by cycloheximide, an inhibitor of cytoplasmic protein synthesis; corresponding results have been obtained with bacteria using chloramphenicol (for ref. see Batra, 1971). From these results it was concluded that irradiation induces the de novo production of carotenogenic enzymes. This kind of conclusion has been critisized, however, (cf., Schopfer, 1977). Prerequisites for the accurate estimation of enzyme activities are the isolation of crude or purified enzyme solutions and a functional in vitro assay system. In the case of carotenogenesis, the in vitro determination of enzyme activities is hampered by the fact that only few cell-free systems from different organisms have been established to date (Yokoyama et al., 1962; Porter, 1969; Bramley and Davies, 1975; Kushwaha et al., 1976).

Furthermore, in the organism which served as source for the cell-free systems, carotenoid synthesis is not strictly photoregulated; the only exception is the *Mycobacterium* sp. in which Johnson et al. (1974) were able to demonstrate that the activity of prephytoen synthetase is absent in homogenates of cells grown in the dark and the enzyme may be synthesized de novo. Therefore, the proof of photoinduced de novo synthesis of carotenogenic enzymes must await further progress in this field of research.

Essentially two alternative proposals may be discussed for the mechanism involved in photoinduction of carotenogenesis: Either light triggers a concurrent synthesis (or activation?) of each of the different enzymes of the pathway, or only the synthesis of the first enzyme – the «key-enzyme» – is photoregulated, and the enzymes catalyzing subsequent steps are induced successively by their substrates.

When synthesis (or activation?) of the enzymes is inhibited by addition of cycloheximide at different times after irradiation and the mycelium is subsequently given sufficient time to complete carotenoid synthesis in the dark, the amount of each of the different carotenoids produced may be taken as a relative indicator of the enzyme activity already present at the time of addition of the inhibitor. The results of such experiments (Fig. 10) show that the different enzyme activities of the biosynthetic pathway are formed sequentially. Furthermore, under strict anaerobic conditions carotenoid synthesis is blocked reversibly, while enzymes must have been produced to a certain degree (Rau, 1971). This finding led to an experiment designed to provide data decisive for one of the above mentioned alternatives. It is based upon the following consideration: In the case of «substrate induction», no carotenogenic enzymes should be produced under anaerobic conditions, because no substrates had been synthesized; after subsequently restoring aerobic conditions pigments should accumulate again in sequence, and each with a lag phase. Since this is not the case (Fig. 11), «substrate induction» seems not to be involved. We therefore conclude that light induces carotenogenic enzymes as a «coupled group». A coordinated photoinduction of enzyme activity has also been found in the flavonoid glycoside pathway (for ref. see Hahlbrock et al., 1976).

Due to the lack of appropriate evidence it is highly speculative to suggest the existence of a «carotenoid operon» involved in this regulatory mechanism similar to the «PAL-operon» introduced by Zucker (1972) for the photoregulation of phenolic compounds. In my opinion, the occurrence of a multi-enzyme aggregate (La Guardia et al., 1971) is equally hypothetical on the basis of the presently available data.

3.6. Carotenoid synthesis

Similar to the photoinduced synthesis of carotenogenic enzymes (Fig. 10), the individual carotenoids are also synthesized in a sequence (Fig. 12) closely following the proposed biosynthetic pathway. In *Fusarium aquaeductuum* lycopene is evidently not converted further by a cyclase but rather seems to be an end product (Bindl et al., 1970).

Results and problems of the biosynthesis of carotenoids have been reviewed in detail in the last few years (Goodwin, 1971; see also the presentation by Davies in this volume).

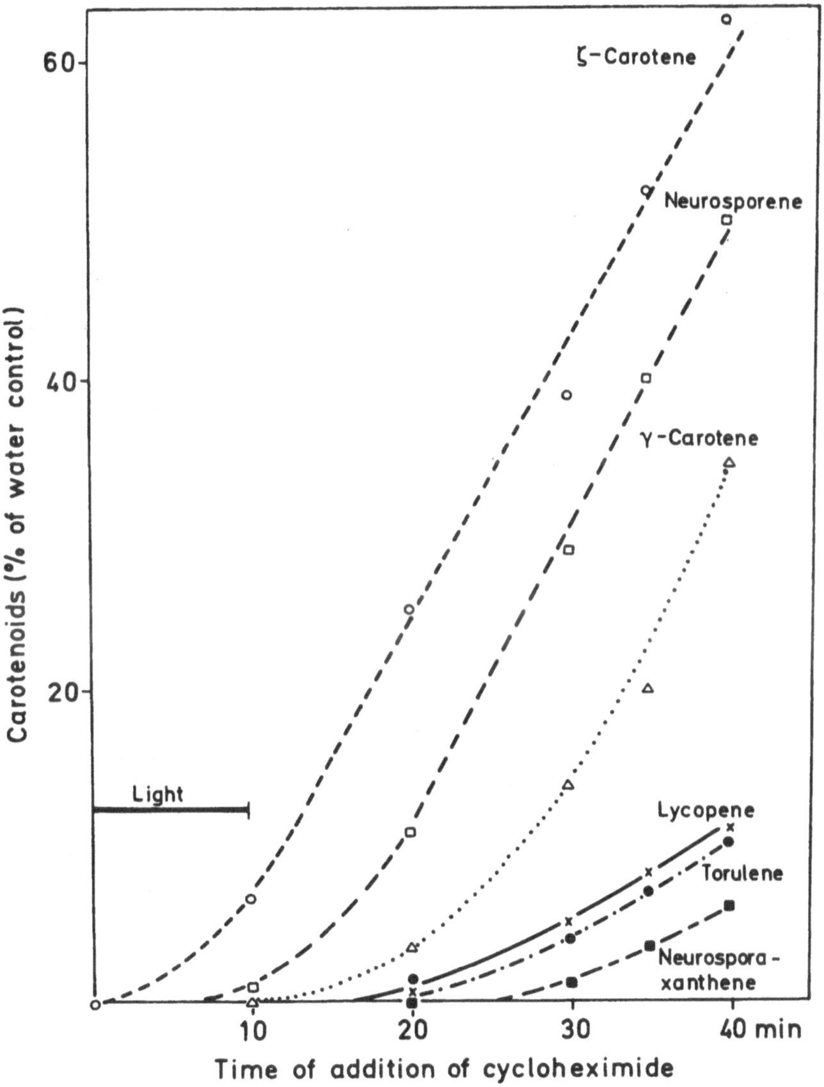

Fig. 10: The amount of individual carotenoids synthesized in *Fusarium aquaeductuum* during a 36 hr incubation as a function of time of addition of cycloheximide after illumination (from Bindl, Lang and Rau, 1970)

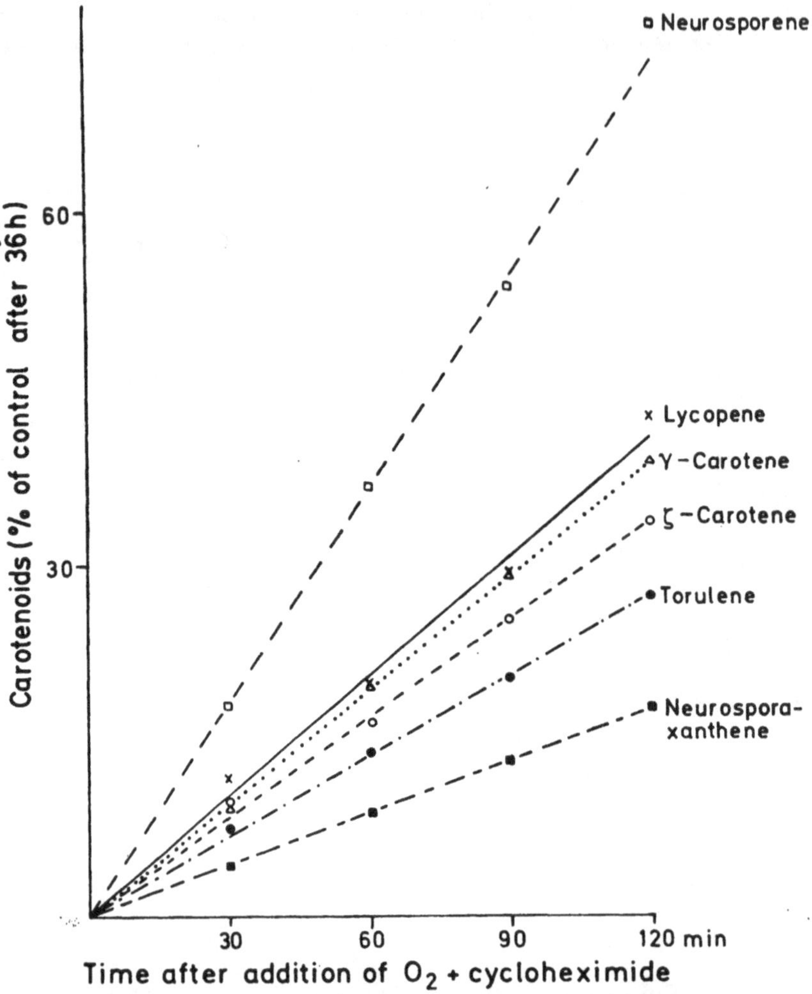

Fig. 11: Kinetics of synthesis of the individual carotenoids in ***Fusarium aquaeductuum*** after the following pretreatment: 10 min illumination under oxygen, 3 hr incubation under anaerobic conditions, subsequent simultaneous addition of oxygen and cycloheximide (from Lang and Rau, 1972)

Summary

The mechanism of photoregulation of morphoses involves a sequence of events which is triggered by illumination and finally leads to a visible response. The sequence of events may be divided into 2 main sections: 1. photoreactions induced by absorption of radiation by the acting photoreceptor and 2. dark reactions following photo

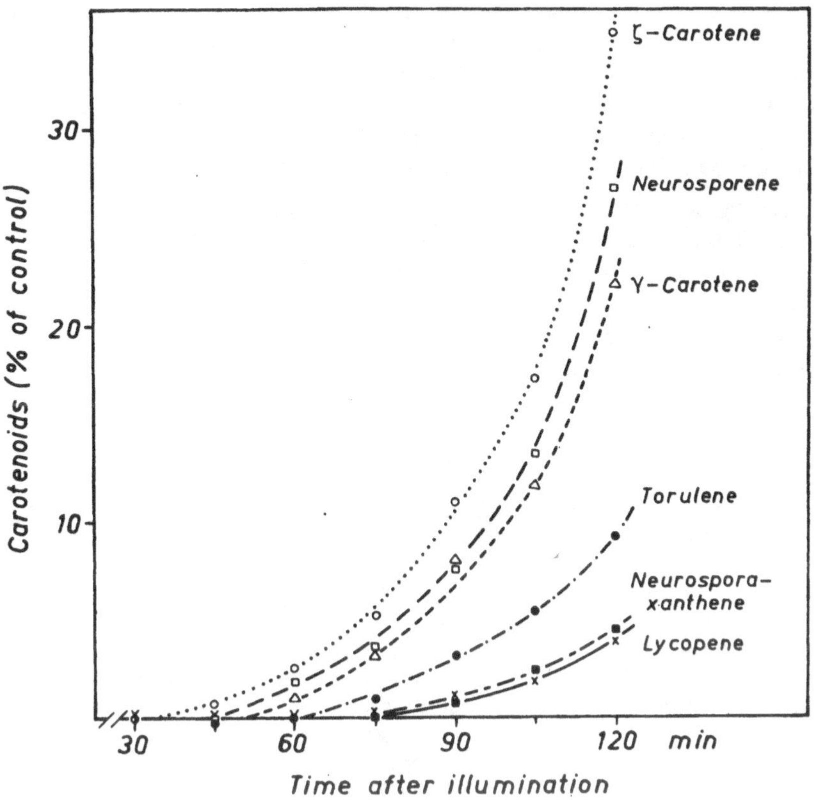

Fig. 12: Time course of synthesis of individual carotenoids in *Fusarium aquaeductuum* after a brief (10 min) exposure to light. Carotenoid content of control samples was determined when carotenogenesis has finished (36 hr) (from Bindl, Lang and Rau, 1970)

induction. Three types of action spectra are characteristic for plant systems: The «classical» phytochrome action spectrum, the «high irradiance» action spectrum, and the «blue light» action spectrum. Properties of both the photoreactions and the corresponding dark reactions are briefly discussed.

In several species of bacteria and fungi carotenoid biosynthesis is strictly photoregulated. According to our investigations on the fungi *Fusarium aquaeductuum* and *Neurospora crassa* the mechanism of photoregulation involves the following sequence of reactions. A flavin or flavoprotein acts as photoreceptor. The data presented are taken as evidence that the primary event is a redox reaction leading to the oxidation of a yet unknown component which might induce the subsequent steps. Acetylcholine was found to be unable to mimic photoinduction. Experiments designed to study the effect of an inhibitor of transcription (distamycin A) and of the reversible inhibition of translation (cycloheximide) yielded results which led to the conclusion that synthesis of specific RNA is involved in photoregulation; data are presented to support this assumption. From further results it is concluded that carotenogenic enzymes are synthesized de novo and are induced as a coupled group rather than individually by their substrates. The results available to date consistently support the hypothesis suggesting that the mechanism of photoinduction of carotenoid biosynthesis in fungi involves photoregulated differential gene activations.

Acknowledgements. I thank most warmly all co-workers who have contributed ideas and results to the work of our laboratory, and also all colleagues, for their helpful and stimulating discussions. In particular I should like to mention Dr. Theimer, Dr. Schrott and my wife, Dr. Rau-Hund, and thank them also for their help in preparing this manuscript. My thanks are also due to the Deutsche Forschungsgemeinschaft for generous financial support of our work.

References

BATRA, P. P., 1971: Mechanism of light-induced carotenoid synthesis in nonphotosynthetic plants. In: A. C. GIESE (ed.), Photophysiology, Bd. VI, 47–76. Acad. Press, New York and London.

BERGER, CH., und L. BERGMANN, 1967: Farblicht und Plastidendifferenzierung im Speichergewebe von *Solanum tuberosum* L. Z. Pflanzenphysiol. **56**, 439–445.

BINDL, E., W. LANG und W. RAU, 1970: Untersuchungen über die lichtabhängige Carotinoidsynthese: VI. Zeitlicher Verlauf der Synthese der einzelnen Carotinoide bei *Fusarium aquaeductuum* unter verschiedenen Induktionsbedingungen. Planta (Berlin) **94**, 156–174.

BJÖRN, L., O. 1967: The light requirement for different steps in the development of chloroplasts in excised wheat roots. Physiol. Plantarum **20**, 483–499.

BRAIN, R. D., J. A. FREEBERG, C. V. WEISS and W. R. BRIGGS, 1977: Blue light-induced absorbance changes in membrane fractions from corn and *Neurospora*. Plant Physiol. **59**, 948–952.

BRAMLEY, P. M., and B. H. DAVIES, 1975: Carotene biosynthesis by cell extracts of mutants of *Phycomyces blakesleeanus*. Phytochem. **14**, 463–465.

BRIGGS, W. R., and H. V. RICE, 1972: Phytochrome: Chemical and physical properties and mechanism of action. Ann. Rev. Plant Physiol. **23**, 293–334.

BRITZ, ST. J., E. SCHROTT, S. WIDELL and W. R. BRIGGS, 1979: Red light-induced reduction of a particle-associated b-type cytochrome from corn in the presence of methylene blue. Photochem. Photobiol. **29**, 359–362.

CARILILE, M. J., 1956: A study of the factors influencing non-genetic variation in a strain of *Fusarium oxysporum*. J. gen. Microbiol. **14**, 643–654.

CARILILE, M. J., and J. FRIEND, 1956: Carotenoids and reproduction in *Pyronema confluens*. Nature (London) **178**, 369.

CLAUSS, H., 1970: Der Einfluß von Rot- und Blaulicht auf die Hillaktivität von *Acetabularia*-Chloroplasten. Planta (Berlin) **91**, 32–37.

CODNER, R. C., and B. C. PLATT, 1959: Light-induced production of carotenoid pigments by *Cephalosporia*. Nature (London) **184**, 741.

CURRY, G. M., and H. E. GRUEN, 1959: Action spectra for the positive and negative phototropism of *Phycomyces* sporangiophores. Proc. Nat. Acad. Sci. (Washington) **45**, 797–804.

CURTIS, C. R., 1972: Action spectrum of the photoinduced sexual stage in the fungus *Nec-*

taria haematococca Berk. and Br. var. *cucurbitae* (Snyder and Hansen) Dingley. Plant Physiol. **49**, 235–239.

DEUTCH, B. and B. I. DEUTCH, 1974: A kinetic theory of first order cyclical processes – phytochrome controlled red light induced cereal leaf unfolding compared with theory. Physiol. Plant, **32**, 273–281.

DITTES, H., and H. MOHR, 1970: MAK-Chromatographie der RNS im Zusammenhang mit der phytochromgesteuerten Morphogenese. Z. Naturforschg. **25b**, 708–710.

DOWNS, R. J., and H. W. SIEGELMAN, 1963: Photocontrol of anthocyanin synthesis in milo seedlings. Plant Physiol. **38**, 25–30.

DRUMM, H. and H. MOHR, 1978: The mode of interaction between blue (UV) light photoreceptor and phytochrome in anthocyanin formation of the *Sorghum* seedling. Photochem. Photobiol., **27**, 241–248.

DUKE, S. O. and W. NAYLOR, 1976: Light control of anthocyanin biosynthesis in *Zea* seedlings. Physiol. Plant. **37**, 62–68.

ENGELSMA, G., 1967: Phenol synthesis and photomorphogenesis. Philips Techn. Rev. **28**, 101–110.

ENGELSMA, G., and G. MEIJER, 1965: The influence of light of different spectral regions on the synthesis of phenolic compounds in gherkin seedlings in relation to photomorphogenesis. Acta Bot. Neerl. **14**, 54–72.

FELBERMEIR, M., und W. RAU, 1974: Hemmung der lichtabhängigen Carotinoidsynthese durch Distamycin. Zusammenfassung der Vorträge. Deutsch. Bot. Ges. und Verein. Angew. Bot., Tagung Würzburg 1974, S. 147.

FONDEVILLE, J. C., H. A. BORTHWICK, and S. B. HENDRICKS, 1966: Leaflet movement of *Mimosa pudica* L. indicative of phytochrome action. Planta (Berlin) **69**, 357–364.

FRIEDERICHSEN, I., und G. ENGEL, 1957: Beiträge zur Kenntnis des Abschußrythmus und des Farbstoffs von *Sphaerobolus stellatus* (Tode) pers. Planta (Berlin) **49**, 578–587.

GOLDSTROHM, D. D., and V. G. LILLY, 1965: The effect of light and the survival of pigmented and nonpigmented cells of *Dacryopinax spathularia*. Mycologia **57**, 612–623.

GOODWIN, T. W., 1971: Biosynthesis. In: O. ISLER (ed.) Carotenoids, 577–636.

GRESSEL, J., 1978: Light requirements for the enhanced synthesis of a plastid mRNA during *Spirodela* greening. Photochem. Photobiol. **27**, 167–169.

GRESSEL, J., L. STRAUSBAUCH, and E. GALUN, 1971: Photomimetic effect of acetylcholine on morphogenesis in *Trichoderma*. Nature (London) **232**, 648.

GRILL, R., and D. VINCE, 1965: Photocontrol of antocyanin formation in turnip seedlings. II. The possible role of phytochrome in the response to prolonged irradiation with far-red or blue light. Planta (Berlin) **67**, 122–135.

HAGER, A., 1970: Ausbildung von Maxima im Absorptionsspektrum von Carotinoiden im Bereich um 370 nm; Folgen für die Interpretation bestimmter Wirkungsspektren. Planta (Berlin) **91**, 38–53.

HAHLBROCK, K., K.-H. KNOBLOCH, F. KREUZALER, J. R. M. POTTS and E. WELLMANN, 1976: Coordinated induction and subsequent activity changes of two groups of metabolically interrelated enzymes (Light-induced synthesis of flavonoid glycosides in cell suspension cultures of *Petroselinum hortense*). Eur. J. Biochem. **61**, 199–206.

HARDING, R. W., 1974: The effect of temperature on photo-induced carotenoid biosynthesis in *Neurospora crassa*. Plant Physiol. **54**, 142–147.

HARDING, R. W., and H. K. MITCHELL, 1968: The effect of cycloheximide on carotenoid biosynthesis in *Neurospora crassa*. Arch. Biochem. **128**, 814–818.

HAREL, E., and L. BOGORAD, 1973: Effect of light on ribonucleic acid metabolism in greening maize leaves. Plant Physiol. **51**, 10–16.

HARTMANN, K. M., 1966: A general hypothesis to interpret «High energy phenomena» of photomorphogenesis on the basis of phytochrome. Photochem Photobiol. **5**, 349–366.

HARTMANN, K. M., 1967: Ein Wirkungsspektrum der Photomorphogenese unter Hochenergiebedingungen und seine Interpretation auf der Basis des Phytochroms (Hypokotylwachstumshemmung bei *Lactuca sativa* L.). Z. Naturforschg. **22b**, 1172–1175.

HARTMANN, K. M., and H. MENZEL, 1960: Proc. Intern. Bot. Congr., 11th, Abstr. p. 85.

HAXO, F., 1949: Studies on the carotenoid pigments of *Neurospora*. I. Composition of the pigment. Arch. Biochem. **20**, 400.

HENDRICKS, S. B., and H. A. BORTHWICK, 1959: Photocontrol of plant development by the simultaneous excitation of two interconvertible pigments. II. Theory and control of anthocyanin synthesis. Bot. Gaz. **120**, 187–193.

HOWES, C. D., P. P. BATRA, and C. F. BLAKELEY, 1969: Absolute requirement for oxygen during illumination for photoinduced carotenoid synthesis. Biochim. Biophys. Acta **189**, 298.

ISHII, R., and M. AKAGI, 1948: *Monilia* (*Neurospora*) *sitophila*. Propagation and pigment production. J. Ferment. Techn. (Japan) **26**, 198–203.

JAFFE, M. J., 1970: Evidence for the regulation of phytochrome-mediated processes in bean roots by the neurohumor Acetylcholine. Plant Physiol. **46**, 768–777.

JOCHUM-FELBERMEIR, M., 1977: Untersuchungen zum Induktionsmechanismus der lichtregulierten Carotinoidsynthese bei *Fusarium aquaeductuum*. Ph. D. Thesis, Universität München.

JOHNSON, J. H., B. C. REED and H. C. RILLING, 1974: Early photoinduced enzymes of photoinduced carotenogenesis in a *Mycobacterium* species. J. biol. Chem. **249**, 402–406.

JOSE, A. M. and D. VINCE-PRUE 1978: Phytochrome action: A reappraisal. Photochem. Photobiol. **27**, 209–216.

KANDELER, R., 1972: Photoregulation. Fortschr. Bot. **34**, 211–226.

KASEMIR, H., and H. MOHR, 1972: Involvement of acetylcholine in phytochrome-mediated processes. Plant Physiol. **49**, 453–454.

KOWALLIK, W., 1967: Action spectrum for an enhancement of endogenous respiration by light in *Chlorella*. Plant Physiol. **42**, 672–676.

KUSHWAHA, S. C., M. KATES and J. W. PORTER, 1976: Enzymatic synthesis of C_{40} carotenes by cell-free preparation from *Halobacterium cutirubrum*. Can. J. Biochem. **54**, 816–823.

LA GUARDIA, M. D. DE, C. M. G. ARAGON, F. J. MURILLO, and E. CERDA-OLMEDO, 1971: A carotenogenic enzyme aggregate in *Phycomyces*: Evidence from quantitative complementation. Proc. nat. Acad. Sci. (Washington) **68**, 2012–2015.

LANG, W., und W. RAU, 1972: Untersuchungen über die lichtabhängige Carotinoidsynthese. IX. Zum Induktionsmechanismus der carotinoidbildenden Enzyme bei *Fusarium aquaeductuum*. Planta (Berlin) **106**, 345–354.

LANG-FEULNER, J., and W. RAU, 1975: Redox dyes as artificial photoreceptors in light-dependent carotenoid synthesis. Photochem. Photobiol. **21**, 179–183.

LANGE, H., I. BIENGER, und H. MOHR, 1967: Eine neue Beweisführung für die Hypothese einer differentiellen Genaktivierung durch Phytochrom 730. Planta (Berlin) **76**, 359–366.

MITRAKOS, K. und W. SHROPSHIRE (ed.), 1972: Phytochrome. Academic Press, London.

MOHR, H., 1956: Die Beeinflussung der Keimung von Farnsporen durch Licht und andere Faktoren. Planta (Berlin) **46**, 534–551.

MOHR, H., 1966: Differential gene activation as a mode of action of phytochrome 730. Photochem. Photobiol. **5**, 469–483.

MOHR, H., 1972: Lectures on Photomorphogenesis. Springer, Berlin, Heidelberg, New York.

MOHR, H., and E. VAN NES, 1963: Der Einfluß sichtbarer Strahlung auf die Flavonoidsynthese und Morphogenese der Buchweizenkeimlinge (*Fagopyrum esculentum* Moench). Z. Bot. **51**, 1–16.

MUMFORD, F. E., and E. L. JENNER, 1971: Catalysis of the phytochrome dark reaction by reducing agents. Biochemistry **10**, 98–101.

MUMMERY, R. S. and L. R. G. VALADON, 1973: Effect of certain nucleic acid and protein inhibitors on carotenogenesis in *Verticillium agaricinum*. Physiol. Plant. **28**, 254–258.

MUÑOZ, V. and W. L. BUTLER, 1975: Photoreceptor pigment for blue light in *Neurosporaa crassa*. Plant Physiol. **55**, 421–426.

MUÑOZ, V., S. BRODY and W. L. BUTLER, 1974: Photoreceptor pigment for blue light responses in *Neurospora crassa*. Biochem. Biophys. Res. Commun. **58**, 322–327.

NINNEMANN, H., R. J. STRASSER and W. L. BUTLER, 1977: The superoxide anion as electron donor to the mitochondrial electron transport chain. Photochem. Photobiol. **26**, 41–47.

PORTER, J. W., 1969: Enzymatic synthesis of Carotene and related compounds. Pure Appl. Chem. **20**, 449.

RAU, W., 1962: Über den Einfluß der Temperatur auf die lichtabhängige Carotnioidbildung von *Fusarium aquaeductuum*. Planta (Berlin) **59**, 123–137.

RAU, W., 1967a: Untersuchungen über die lichtabhängige Carotinoidsynthese. I. Das Wirkungsspektrum von *Fusarium aquaeductuum*. Planta (Berlin) **72**, 14–28.

RAU, W., 1967b: Untersuchungen über die lichtabhängige Carotinoidsynthese II. Ersatz der Lichtinduktion durch Mercuribenzoat. Planta (Berlin) **74**, 263–277.

RAU, W., 1969: Untersuchungen über die lichtabhängige Carotinoidsynthese. IV. Die Rolle des Sauerstoffs bei der Lichtinduktion. Planta (Berlin) **84**, 30–42.

RAU, W., 1971: Untersuchungen über die lichtabhängige Carotinoidsynthese. VII. Reversible Unterbrechung der Reaktionskette durch Cycloheximid und anaerobe Bedingungen. Planta (Berlin) **101**, 251–264.

RAU, W., 1975: Zum Mechanismus der Photoregulation von Morphosen am Beispiel der Carotinoidsynthese. Ber. Deutsch. Bot. Ges., **88**, 45–60.

RAU, W., I. LINDEMANN und A. RAU-HUND, 1968: Untersuchungen über die lichtabhängige Carotinoidsynthese III. Die Farbstoffbildung von *Neurospora crassa* in Submerskultur. Planta (Berlin) **80**, 309–316.

RAU, W., and A. RAU-HUND, 1977: Light-dependent carotenoid synthesis. X. Lag-phase after a second illumination period in *Fusarium aquaeductuum* and *Neurospora crassa*. Planta (Berlin) **136**, 49–52.

ROBINSON, J. C., L. KEAY, R. MOLINARI, and I. W. SIZER, 1962: L-α-Hydroxy acid oxidases of hog renal cortex. J. biol. Chem. **237**, 2001–2010.

ROSNER, A., K. M. JACOB, J. GRESSEL and D. SAGHER, 1975: The early synthesis and possible function of $0{,}5 \times 10^6$ M rRNA after transfer of darkgrown *Spirodela* plants to light. Biochem. Biophys. Res. Commun. **67**, 383–391.

SCHÄFER, E., 1975: A new approach to explain the «high irradiance responses» of photomorphogenesis on the basis of phytochrome. J. Math. Biol. **2**, 41–56.

SCHERF, H., and M. H. ZENK, 1967: Der Einfluß des Lichtes auf die Flavonoidsynthese und die Enzyminduktion bei *Fagopyrum esculentum* Moench. Z. Pflanzenohysiol. **57**, 401–418.

SCHMID, R., and H. CLAUSS, 1977: Is the nucleus involved in the blue light mediated photomorphoses in *Acetabularia mediterranea*? In: C. L. F. Woodcock (ed.), Progress in Acetabularia Research. 255–269. Acad. Press, New York and London.

SCHNARRENBERGER, C., and H. MOHR, 1970: Carotenoid synthesis in mustard seedlings as controlled by phytochrome and inhibitors. Planta (Berlin) **94**, 296–307.

SCHOPFER, P., 1973: Modulation und Determination der Enzymbildung als Regulationsprinzipien bei der phytochromabhängigen Photomorphogenese. Ber. Deutsch. Bot. Ges. **86**, 271–286.

SCHOPFER, P., 1977: Phytochrome control of enzymes. Ann. Rev. Plant Physiol. **28**, 223–252.

SCHROTT, E. L. und W. RAU, 1975: Versuche zum direkten Nachweis einer Beteiligung von m-RNS an der Photoinduktion der Carotinoidsynthese. Ber. Deutsch. Bot. Ges. 88, 233–243.

SCHROTT, E. L. and W. RAU, 1977: Evidence for a photoinduced synthesis of poly (A) containing mRNS in *Fusarium aquaeductuum*. Planta (Berlin) **136**, 45–48.

SHROPSHIRE, W. jr., 1972: Phytochrome, a photochromic sensor. In: A. C. GIESE (ed.) Photophysiol. Bd. VII, 33–72. Acad. Press, New York and London.

SIEGELMAN, H. W., and W. L. BUTLER, 1965: Properties of phytochrome. Ann. Rev. Plant Physiol. **16**, 383–393.

SMITH, H. 1975: Phytochrome and Photomorphogenesis. Mc Graw-Hill, London.

SONG, P.-S., and TH. A. MOORE, 1974: On the photoreceptor pigment for phototropism and phototaxis: Is a carotenoid the most likely candidate? Photochem. Photobiol. **19**, 435–441.

SONG, P.-S., TH. A. MOORE, and M. SUN, 1972: Excited states of some plant pigments. In: C. O. CHICHESTER (ed.) The Chemistry of Plant Pigments, 33–74. Acad. Press, New York and London.

SUBDEN, R. E., and G. BOBOWSKI, 1973: Evidence for actinomycin D inhibition of transcription of carotenoid loci in *Neurospora*. Experientia **29**, 965–967.

SUGAI, M., 1971: Photomorphogenesis in *Pteris vittata* IV. Action spectra for inhibition of phytochrome-dependent spore germination. Plant Cell Phys. **12**, 103–109.

THEIMER, R. R., 1968: Untersuchungen zum Wirkungsmechanismus von Licht und Mercuribenzoat bei der Carotinoidsynthese in *Fusarium aquaeductuum*. Ph. D. THESIS, Universität München.

THEIMER, R. R., und W. RAU, 1970: Untersuchungen über die lichtabhängige Carotinoidsynthese. V. Aufhebung der Lichtinduktion durch Reduktionsmittel und Ersatz des Lichts durch Wasserstoffperoxid. Planta (Berlin) **92**, 129–137.

TRINCI, A. P. J., and G. H. BANBURY, 1969: Effect of light on growth and carotenogenesis of the tall conidiophores of *Aspergillus giganteus*. Trans. Brit. mycol. Soc. **52**, 1–14.

VALADON, L. R. G., and R. MUMMERY, 1971: Effect of light on nucleic acids, protein and carotenoids of *Verticillium agaricinum*. Microbios **4**, 227–240.

VERDIER, G., 1975: Synthesis and translation site of light-induced m-RNS in etiolated *Euglena gracilis*. Biochim. Biophys. Acta **407**, 91–98.

VINCE-PRUE, D. 1975: Photoperiodism in plants. Mc. Graw-Hill, London.

VOSKRESENSKAYA, N. P., 1972: Blue light and carbon metabolism. Ann. Rev. Plant Physiol. **23**, 219–234.

WAGNER, E., and H. MOHR, 1966: Kinetic studies to interpret «high-energy phenomena» of photomorphogenesis on the basis of phytochrome. Photochem. Photobiol. **5**, 397–406.

WENGER, C. J., and V. G. LILLY, 1966: The effects of light on carotenogenesis, growth, and sporulation of *Syzygites megalocarpus*. Mycologia **58**, 671–680.

WENT, F. A. F. C., 1901: *Monila sitophilia* (Mont.) Sacc. ein technischer Pilz Javas. Centr. Bakt. Parasitenkde. **7**, 544.

WITHROW, R. B., W. H. KLEIN and V. ELSTAD, 1957: Action spectra of photomorphogenic induction and its photoinactivation. Plant Physiol. **32**, 453–462.

YOKOYAMA, H., T. O. M. NAKAYAMA and C. O. CHICHESTER, 1962: Biosynthesis of β-carotene by cell-free extracts of *Phycomyces blakesleeanus*. J. biol. Chem. **237**, 681–686.

ZALOKAR, M., 1954: Studies on biosynthesis of carotenoids in *Neurospora crassa*. Arch. Biochem. **50**, 71–80.

ZALOKAR, M., 1955: Biosynthesis of carotenoids in *Neurospora*. Action spectrum of photo-activation. Arch. Biochem. **56**, 318–325.

ZUCKER, M., 1972: Light and enzymes. Ann. Rev. Plant Physiol. **23**, 133–156.

ZURZYCKA, A., 1963: Studies on photomorphosis in *Aspergillus giganteus* mut. alba II. Acta. Biol. Cracov. Ser. Bot. VI, 103–113.

ZURZYCKI, J. 1962: The action spectrum for the light dependent movements of chloroplasts in *Lemna trisulca*. Acta. Soc. Bot. Pol. **31**, 489–538.

Some remarks on the usefulness of algal carotenoids as chemotaxonomic markers

Adolf Weber[1] and Michael Wettern[2]

[1] Institut für Allgemeine Botanik und Botanischer Garten der Universität Hamburg
[2] Botanisches Institut der Technischen Universität Braunschweig

It is the merit of Tswett to have first succeeded in separating plastid pigments by column chromatography in 1906. The great number of papers published since then in this field have made it progressively clearer that some of these pigments, the carotenoids, are present if not in all, then at least in most plants or plant parts. To this day several hundreds of terpene derivatives have been isolated (cf., Straub, 1971; Weedon, 1971), many of them from algae. In the future even more may be expected since the analytical methods (thin layer chromatography, UV-IR and -NMR-spectroscopy) have been continuously refined and allow a better identification in this class of substances.

Due to their widespread presence in the plant kingdom, the carotenoids have been used respectively as chemotaxonomic and chemosystematic criteria, especially where distinctive morphological features of plant groups were not always sufficiently available, i.e., mainly in the fungi and the algae.

Concerning the taxonomy of the above mentioned groups, one is easily tempted to assume that pigments are the most important characteristics as one differentiates on this basis between red, green, blue-green, brown algae etc. And yet we know that a «correct» classification of algae is not possible solely on the basis of colour. Instead, up to date plant systems including the algae are based primarily on morphological classification principles. They represent a «systematical hypothesis» (cf., Heywood, 1966; Hegnauer, 1971), the accuracy of which should be tested with the aid of further criteria such as phytochemistry, cytology and ultrastructure, etc. Finally, it might become possible to describe the «natural system» through a comparison of the maximum number of available characteristics.

Some examples in the following pages might elucidate whether the systematist is or could be brought closer to this goal by the analysis of algal carotenoids. In other words, does such an analysis give information on phylogenetic interrelationships, or are the results only of importance to the taxonomist and thus merely suitable for classification purposes (cf., Merxmüller, 1967). It was a prerequisite for this definitely subjective experiment that the authors had to select one of the up-to-date algal systems. Preference has been given to Fott's system (1971). It is certainly not the authors' aim to prove, with the aid of the distribution of algal carotenoids, that the system of Fott is «better» than the systems of Round (1975), Bourrelly (1968–1970), or Christensen (1962), to name only a few examples. Fott's system shall be merely the basis and reference for the authors' observations on whether the algal carotenoids are only of chemotaxonomic, or also of chemosystematic importance.

It is taken for granted that the algal carotenoids can be considered as «characteristics», i.e., fulfill the following conditions:

1) Their presence must be as genetically stable and their development as independent of invironmental conditions as possible (cf., Heywood, 1971).

2) Sufficient data on the specific characteristics should be available; insufficient knowledge leads to the danger of incorrect taxonomic evaluation (Hegnauer, 1971).

For the carotenoids this also means that they must be correctly identified. Otherwise apparent deviations or similarities in algal taxa could be based on false pigment analyses.

3) The distribution of a characteristic must be sufficiently well known. In this regard, insufficient knowledge determines the usefulness of a characteristic for taxonomic purposes.

4) Finally, the more complex the knowledge of a characteristic, the more valuable it is (Heywood, 1971). The metabolic pathway of a carotenoid should therefore be known. For just as morphological characteristics of plants may have originated in very different ways, identical chemical substances may be produced in completely different biochemical pathways (Reznick, 1964). Apart from this, it is well known that some polyenes, – e.g., the secondary carotenoids – occur only under certain ecological conditions, as will be shown later. It is therefore surely acceptable to follow Merxmüller's view that chemical characteristics are important: «... niemals aber a priori bedeutsamer als die übrigen Merkmalskategorien ...» («... but never a priori more significant than the other characteristics ...», Merxmüller, 1967). Finally a particulary important problem for the chemotaxonomy of the algae is the following: assertions can only make sense if not only the chemical substances, but also the examined species, are identified precisely and are available for a cross-check (cf., Merxmüller, 1967) What this implies for «coccal green algae», for example, is familiar to everyone who has ever worked with this plant group. However, even quite recently papers have been published which fail to identify even the species of the alga examined. Frequently scientists are interested primarily in a physiological problem and make only incidental use of the «taxonomic» results.

After these preliminary observations let us attempt to illustrate by means of some examples the importance of algal carotenoids as an attribute used in evaluating algae systems.

Although this characteristic has long since been included by taxonomists, and, in addition, the algal carotenoids have been dealt with in monographs and synoptic articles (Goodwin, 1971, 1974; Nakayama, 1964; Liaaen-Jensen, 1977; Hager and Stransky, 1970a–b; Stransky and Hager, 1970a–c; Wettern and Weber, 1979; see also the legend of Table 1), a summarizing table is indispensable.

Table 1 includes:

1) a rough classification of the carotenoids according to their chemical character.

Table 1: Distribution pattern of carotenoids in algae

		Cyanophyta	Chromophyta					Rhodophyta	Chlorophyta			flagellates of uncertain taxonomic position			
		Cyanophyceae	Chrysophyceae	Xanthophyceae	Bacillariophyceae	Phaeophyceae	Dinophyceae	Rhodophyceae	Chlorophyceae	Conjugatophyceae	Charophyceae	Euglenophyceae	Cryptophyceae	Chloromonadophyceae	colorless flagellates
a	myxoxanthophyll	$+_{3}$													
	4-keto-myxoxanthophyll	$-_{6}$													
	aphanizophyll	$-_{12}$													
	oscillaxanthin	$-_{12}$													
	myxol-2′-O-methyl-methyl-pentoside	$\bigcirc_{6}$													
	4-keto-myxol-2′-O-methyl-methylpentoside	$\bigcirc_{6}$													
	oscillol-2,2′-di(O-methyl)-methylpentoside	$\bigcirc_{6}$													
b	phytoene						$\bigcirc_{6}$								
	phytofluene						$\bigcirc_{6}$								
	lycopene	$-_{6}$							$\bigcirc_{5}$		$-_{5}$				

c	γ-carotene	$\bigcirc_6$	$-_5$				$\bigcirc_{13}$		$\bigcirc_5$		$-_2$				
d	α-carotene		$+_5$				$\bigcirc_6$	$-_3$	$+_5$	$\bigcirc_5$			$+_4$		
	β-carotene	$+_3$	$+_4$	$+_1$	$+_4$	$+_5$	$+_5$	$+_3$	$+_2$ $\times_5$	$+_2$	$+_2$	$+_4$	$-_5$	$+_5$	$-_5$
	ε-carotene				$-_4$	$-_4$							$\bigcirc_5$		
e	crocoxanthin												$+_4$		
	monadoxanthin												$-_5$		
	diatoxanthin		$-_4$	$+_1$	$+_4$	$\bigcirc_5$	$-_5$					$+_4$			
	diadinoxanthin		$-_4$	$+_1$	$+_4$	$\bigcirc_5$	$-_5$					$+_4$		$?_5$	
	heteroxanthin			$+_1$								$\bigcirc_6$		$?_5$	
	alloxanthin												$+_4$		
	manixanthin												$-_6$		
	pyrrhoxanthin						$\bigcirc_6$								
	pyrrhoxanthinol						$\bigcirc_6$								
f	β-carotene-epoxide		$-_6$												
	β-carotene-diepoxide			$-_1$											
	cryptoxanthin-epoxide		$-_4$	$-_1$					$\bigcirc_2$			$-_4$			
	cryptoxanthin-diepoxide			$-_1$											
	antheraxanthin		$-_4$	$\bigcirc_1$		$-_4$		$\bigcirc_5$	$+_2$	$+_2$	$+_2$				
	violaxanthin		$-_4$	$\bigcirc_1$		$+_5$		$\bigcirc_6$	$+_2$	$+_2$	$+_2$				
	lutein-epoxide							$\bigcirc_6$	$\bigcirc_2$					$?_5$	
	mutatochrome	$\bigcirc_6$													
	aurochrome	$\bigcirc_6$													
	auroxanthin	$\bigcirc_6$													
g	4-hydroxy-3′,4′-diketo-β-carotene								$\times_5$						
	3-hydroxy-3′,4,4′-triketo-β-carotene								$\times_5$						
	3-hydroxy-canthaxanthin								$\times_5$						
	4-hydroxy-4′-keto-β-carotene	$\bigcirc_3$													
	echinenone	$+_5$	$-_4$						$-_2$ $\times_5$			$-_4$			$-_5$
	canthaxanthin	$+_3$							$-_2$ $\times_5$			$-_5$			$-_5$

Table 1: (Continuation)

		Cyanophyta	Chromophyta					Rhodophyta	Chlorophyta			flagellates of uncertain taxonomic position			
		Cyanophyceae	Chrysophyceae	Xanthophyceae	Bacillariophyceae	Phaeophyceae	Dinophyceae	Rhodophyceae	Chlorophyceae	Conjugatophyceae	Charophyceae	Euglenophyceae	Cryptophyceae	Chloromonadophyceae	colores flagellates
g	3′-hydroxy-echinenone	$-_{3}$										$-_{5}$			
	4′-hydroxy-echinenone								$\times_{5}$						
	astaxanthin						$\bigcirc_{6}$		$\times_{7}$			$+_{4}$			
	siphonaxanthin								$-_{8}$						
	fritschiellaxanthin								$\times_{14}$						
	micronone								$\bigcirc_{5}$						
	phoenicopterone								$\bigcirc_{11}$						$-_{5}$
h	neoxanthin	$\bigcirc_{6}$	$-_{4}$	$+_{1}$	$-_{4}$	$\bigcirc_{6}$		$-_{5}$	$+_{2}$	$+_{2}$	$+_{2}$	$+_{4}$			
	desepoxy-neoxanthin		$-_{6}$												
	dinoxanthin		$\bigcirc_{12}$				$-_{12}$								
	vaucheriaxanthin			$+_{1}$											
	fucoxanthin		$+_{4}$	$+_{4}$	$+_{4}$	$+_{5}$	$\bigcirc_{5}$	$\bigcirc_{6}$							
	fucoxanthinol		$-_{6}$			$\bigcirc_{6}$									
	19′-hexanoyl-fucoxanthin		$-_{6}$												
	peridinin						$+_{5}$								

h	peridiniol						○6								
	9′-hexanoyloxy-paracentrone-3-acetate		−6												
i	cryptoxanthin	○3	−4					+3	○2			−4			
	zeaxanthin	−3	−4	○1		−4		+3	+2	+2	+2	+12			
	caloxanthin	○3													
	nostoxanthin	○3													
	β-carotene-2-ol								○6						
	β-carotene-2,2′-ol								○6						
	α-carotene-2-ol								○6						
	α-cryptoxanthin							○6	○8						
	lutein							−3	+2	+2	+2				
	loroxanthin								−12						
	crustaxanthin								○11						
	trihydroxy-α-carotene								−9	−10					
	trihydroxy-β-carotene								×5						
	adonixanthin								×7						

pigment occurs: + ≙ frequently — ≙ occasionally ○ ≙ only rarely × ≙ as secondary carotenoid

The carotenoids were roughly grouped according to chemical aspects:

a: glycosidic-, b: acyclic-, c: monocyclic-, d: bicyclic-, e: acetylenic-, f: epoxidic-, g: ketonic-, h: allenic-, i: hydroxy-carotenoids

Distribution pattern of carotenoids in algae according to: 1) Stransky and Hager (1970a); 2) Hager and Stransky (1970a); 3) Stransky and Hager (1970b); 4) Hager and Stransky (1970b); 5) Goodwin (1971); 6) Liaaen-Jensen (1977); 7) Weber (1975); 8) Weber and Czygan (1972); 9) Weber (1973); 10) Weber (1969); 11) Thommen (1971); 12) Goodwin (1974); 13) Jeffrey et al. (1975); 14) Buchecker et al. (1978)

2) a classification of the carotenoids related respectively to the individual algal phyla or classes based on Fott's system (1971), see table 1.

The disadvantage lies in the fact that differences between taxonomic units ranking below the orders are not taken into consideration at all. On the other hand, it can easily be shown that carotenoids, or carotenoid patterns, exist which are characteristic of an algal class or even an algal phylum.

One or more glycosidic carotenoids are present in almost all Cyanophyta (Hertzberg et al., 1971; Stransky and Hager, 1970b). So far, their presence has not been established in any other algal group. These results show, for example, that the presence of myxoxanthophyll in an organism (see table 1) is sufficient basis for classification as a blue-green alga. Conversely, the absence of this glycosidic carotenoid, as is the case in *Anacystis nidulans* (Stransky and Hager, 1970b), does not mean that this alga should not be included in the blue-green algae, since it contains two carotenoids specific for cyanophytes: caloxanthin and nostoxanthin.

On the whole, it seems possible to separate blue-green algae from all other algae solely on the basis of their specific carotenoids. However, this knowledge does not solve the well known problem of whether the endocyanells of, for example, *Cyanophora paradoxa*, *Glaucocystis nostochinearum* and *Gloeochaete wittrockiana* should be regarded as blue-green algae or as red algae. Their carotenoids and carotenoid patterns are of no use as «chemotaxonomic markers». The cyanelles of these three organisms contain neither carotenoid glycosids nor caloxanthin. Only *Glaucocystis* possibly contains traces of nostoxanthin. Another carotenoid typical for blue-green algae (even though it is also present in other algal groups), namely, the ketocarotenoid echinenone is also lacking (Chapman, 1966; Schmidt et al., 1979). Is it then possible that endocyanelles are really primitive red algae, as assumed by some investigators (Schnepf and Brown, 1971; Robinson and Preston, 1971)? In any case the carotenoids would correspond very well to those of other Bangiales (Stransky and Hager, 1970b). But perhaps this similarity is due to a close relationship with the «common ancestors» of red and blue-green algae (Skuja, 1954).

We must admit that these are all hypotheses that cannot even be proved with the aid of the chemical characteristic «phycobilins». Here too, some of these pigments are present in blue-green algae and cyanelles, as well as in simple red algae. However, in some of the cyanelles examined (Chapman, 1966) the phycobilins that are typical for blue-green algae clearly prevail (Chapman et al., 1967; Rüdiger, 1975). It might be easier to solve the taxonomic problem discussed here if the biochemical pathways of the pigments named above were known and therefore comparable.

If we take a closer look at table 1, further polyenes can be found, which, like the glycosidic carotenoids of the blue-green algae, characterize whole algal classes; i.e., their presence cannot be proven in other algae. As an example let us take peridinin of the Dinophyceae. It represents the main xanthophyll of this group of flagellates (Strain et al., 1971; Whittle and Casselton, 1968) and is apparently not present in other algal classes. However, in this plant group, the obvious consistency in morphological characteristics does not repeat itself in the «characteristic» carotenoids. Several species of the genus *Gymnodinium* contain fucoxanthin instead of peridinin

(Riley and Wilson, 1965; Loeblich III, 1976); the same holds true for *Peridinium foliaceum* and *Peridinium balticum* (Withers and Haxo, 1975; Jeffrey et al., 1975; Mandelli, 1968). In *Gymnodinium micrum* neither the existence of fucoxanthin nor peridinin can be proved (Whittle and Casselton, 1968).

For *P. foliaceum* and *P. balticum* the observed variations in the pigment patterns can be explained: both organisms are binucleate and contain an endosymbiont (Tomas and Cox, 1973; Jeffrey and Vesk, 1976). According to some authors (i.e., Loeblich III, 1976), the endosymbionts probably belong to the Chrysophyta, and it is a well known fact that this algal group (along with the brown algae and some Cryptophyceae) is characterized by the presence of fucoxanthin. The existence of fucoxanthin in the uninucleate *Gymnodinium* species remains unclear, although there is reason for the formulation of evolutionary hypotheses and interrelationships.

An additional example will illustrate the fact that the results of pigment analyses support or help interpret morphological findings. Based on electron microscopic studies, Hibberd and Leedale (1970) separated the Eustigmatophyceae from the Xanthophyceae as a new class. In the same year, pigment analysis of Stransky and Hager (1970a) showed that this was justified: *Pleurochloris commutata*, included among the Eustigmatophyceae by Hibberd and Leedale (1970), contains vaucheriaxanthin, a characteristic pigment for many Xanthophyceae (Nitsche and Egger, 1970) but does not contain diatoxanthin, diadinoxanthin and heteroxanthin (along with other carotenoids), which would be expected in this algal group (Strain et al., 1970; Egger et al., 1969; Stransky and Hager, 1970a).

Without the morphological findings of Hibbert and Leedale (1970), the polyene pattern of *Pleurochloris* – which is more typical of green algae – would have led to an exceptional position of this alga within the Xanthophyceae. Thus, inclusion in a «false» algal class on the basis of insufficient (morphological) characterization would have been inevitable. It is not yet clear whether the pigment pattern found in *Pleurochloris* also applies to the other Eustigmatophyceae.

Up to this point we have mainly compared the presence of individual carotenoids in our observations. Taking the green algae as an example, it is interesting to know the conclusions one reaches when using the whole spectrum of pigments (of one algal species) for taxonomic investigation.

Many green algae contain the same plastid pigments as the higher plants (Hager and Stransky, 1970a). Most of their carotenoids, however, are also present in other algae, e.g., in red algae. If one were to distinguish these from the green algae solely on the basis of their pigment patterns, it would succeed only if all pigments, i.e., also the chlorophylls or the phycobilins were taken into account. The carotenoid composition alone would not be sufficient. And yet the green algae also contain yellow pigments whose taxonomic importance is undisputed. One example is siphonaxanthin: Kleinig (1969) and other investigators (see the synopsis of Goodwin, 1974) were able to prove its presence (often together with its ester, siphonein) in many siphoneous algae (Bryopsidales). A few families do not seem to be able to synthesize this pigment: it is lacking in all Dasycladineae. On the other hand, it can be found in some Siphonocladales algae which differ considerably in their organization from the Siphonales (Kleinig, 1969; Goodwin, 1974). Siphonaxanthin was also isolated from *Microthamnion kuetzingianum* (Weber and Czygan, 1972), a filamentous alga of the order Ulotrichales and from several genera of Volvocales (*Asteromonas*, *Heteromastix* and *Pyramimonas*). Siphonaxanthin is therefore also not a «key carotenoid», and for this reason is only of limited taxonomic value. In examining problems of evolution, however, its distribution is interesting and it is not surprising that the zoospores of the chaetophoraceaen alga *Microthamnion kuetzingianum* show considerable similarity to those of some representatives from the order of the Siphonales (Melkonian, unpubl.).

Loroxanthin, a polyene very similar to the ketocarotenoid siphonaxanthin, has also only been found in green algae up to now. It is identical with pyrenoxanthin, trollein and trihydroxy-α-carotene (Nitsche, 1974) Its presence was proved in many coccal green algae (Hager and Stransky, 1970a; Aitzetmüller et al., 1969) as well as in *Cladophora* and *Ulva* species (Aitzetmüller et al., 1969), and it is not uncommon in many representatives of the Ulotrichales (Weber, 1973). It seems to be of taxonomic relevance in Conjugatophyceae as well. Weber (1969) was able to prove the existence of this pigment in more than ten Desmidiales, not however, in the examined Zygnematales. In the Mesotaeniales, only *Roya obtusa* seems to be able to synthesize loroxanthin.

In the morphologically uniform coccalean green algae the capacity to form secondary carotenoids (Czygan, 1968a, b) is often used for characterization of certain groups. These pigments occur mainly in older cultures of some Chlorococcales (Dersch, 1960; Czygan, 1968a; Kessler, 1967; Kessler and Czygan, 1967; McLean, 1968; Kleinig and Czygan, 1969) but also in Volvocales (Czygan, 1970; Sprey, 1970; Aasen et al., 1969), in some Ulotrichales (Czygan and Kalb, 1966; Weber, 1975; Tupa, 1974) and even in *Acetabularia* (Kleinig, 1966). They are stored either inside or outside of the plastid membrane in lipid globules and their synthesis usually parallels a decomposition of chlorophylls. In most cases derivatives of β-carotene, namely, astaxanthin, canthaxanthin, echinenone and sometimes adonixanthin are synthesized as secondary carotenoids. Up to now only two cases are known in which derivatives of the α-carotene have been found: phoenicopterone in *Haematococcus pluvialis* (Czygan, 1970b) and fritschiellaxanthin in *Fritschiella tuberosa* (Weber, 1975; Buchecker et al., 1978). In *Trentepohlia aurea* (Czygan and Kalb, 1966) and *Dunaliella* (Aasen et al., 1969), on the other hand, an increase of the β-carotene content was noticed. The cause of this phenomenon is believed to be a decrease of available nitrogen in the culture medium. How can this phenomenon be used in taxonomy? Since it is limited neither to specific orders nor even to specific families, it can be of taxonomic relevance only in lower taxa, but only if the ability to form secondary carotenoids represents a stable characteristic and is independent of ecological factors. At this point it seems to the authors that slight doubts are justified. The results of Czygan (1970c) are opposed to the opinion of Kessler et al. (1963) and Kessler and Soeder (1962), who use the formation of secondary carotenoids (along with other chemical characteristics) to characterize different species in the genus *Chlorella*. Czygan observed that this characteristic can be instable:

ultraviolet-irradiated cultures of *Selenastrum gracilis* temporarily lost the ability to synthesize secondary carotenoids, whereas *Chlorella fusca* irretrievably lost this faculty under the same conditions. Deason et al. (1977), who have tested 23 species of *Neospongiococcum*, also voiced certain doubts as to the taxonomic relevance of this characteristic.

Scepticism seems justified not only because of a certain instability, but also because the modificational amplitude is considerable and can therefore easily lead to incorrect evaluations. Christlieb (1977) showed that, in spite of a considerable reduction of the nitrate concentration in the nutrient medium of *Fritschiella tuberosa*, the usual intense red colour of the alga failed to appear when the magnesium concentration was increased simultaneously. However, Dersch already showed in 1960 that more than one factor influences the formation of secondary carotenoids.

Let us venture the following observation as a conclusion. There are sufficient examples of the usefulness of individual algal carotenoids as aids to the taxonomist, though these are frequently only relevant in lower taxa. An examination of the complete pigment pattern usually gives better results. Single polyenes are hardly suitable as the only basis for illuminating the degree of relationship, unless their exact biosynthesis is known. Development of a phylogenetic tree of the algae solely on the basis of pigment distribution is pure speculation.

References

AASEN, A. J., K. E. EIMHJELLEN, and S. LIAAEN-JENSEN, 1969: An Extreme Source of β-Carotene. Acta chem. Scand. **23**, 2544–2545.

AITZETMÜLLER, K., H. H. STRAIN, W. A. SVEC, M. GRANDOLFO, and J. J. Katz, 1969: Loroxanthin, a Unique Xanthophyll from *Scenedesmus obliquus* and *Chlorella vulgaris*. Phytochem. **8**, 1761–1770.

BOURELLY, P., 1966–1970: Les Algues d'eau Douce. T. 1–3. Boubée & Cie, Paris.

BUCHECKER, R., C. H. EUGSTER, and A. WEBER, 1978: Absolute Konfiguration von α-Doradexanthin und von Fritschiellaxanthin, einem neuen Carotinoid aus *Fritschiella tuberosa* Iyeng. Helv. Chim. Acta **61**, 1962–1968.

CHAPMAN, D. J., 1966: The Pigments of the Symbiotic Algae (Cyanomes) of *Cyanophora paradoxa* and *Glaucocystis nostochinearum* and two Rhodophyceae, *Porphyridium aerugineum* and *Asterocystis ramosa*. Arch. Mikrobiol. **55**, 17–25.

CHAPMAN, D. J., W. J. COLE, and H. W. SIEGELMAN, 1967: Phylogenetic Implications of Phycocyanobilin and C-Phycocyanin. Amer. J. Bot. **55**, 314–316.

CHRISTENSEN, T., 1962: Systematisk Botanik, Alger. University of Copenhagen, Copenhagen.

CHRISTLIEB, T., 1977: Pigmentbildung und Morphologie der Grünalge *Fritschiella tuberosa* Iyeng. (Chaetophorineae) in Abhängigkeit von äußeren Ernährungsbedingungen sowie unterschiedlichen Lichtverhältnissen. Diplomarbeit, Hamburg.

CZYGAN, F.-C., 1968a: Sekundär-Carotinoide in Grünalgen. I. Chemie, Vorkommen und Faktoren, welche die Bildung dieser Polyene beeinflussen. Arch. Mikrobiol. **61**, 81–102.

CZYGAN, F.-C., 1968b: Sekundär-Carotinoide in Grünalgen. II. Untersuchungen zur Biogenese. Arch. Mikrobiol. **62**, 209–236.

CZYGAN, F.-C., 1970a: Blutregen und Blutschnee: Stickstoffmangel-Zellen von *Haematococcus pluvialis* und *Chlamydomonas nivalis*. Arch. Mikrobiol. **74**, 69–76.

CZYGAN, F.-C., 1970b: On the Occurrence of Crustaxanthin and Phoenicopterone in Aplanospores of *Haematococcus pluvialis* Flotow em. Wille. Flora Abt. A. **159**, 339–345.
CZYGAN, F.-C., 1970c: Untersuchungen über die Bedeutung der Biosynthese von Sekundär-Carotinoiden als Artmerkmal bei Grünalgen. Arch. Mikrobiol. **74**, 77–81.
CZYGAN, F.-C., and K. KALB, 1966: Untersuchungen zur Biogenese der Carotinoide in *Trentepohlea aurea*. Z. Pflanzenphysiol. **55**, 59–64.
DEASON, T. R., F.-C. CZYGAN, and C. J. SOEDER, 1977: Taxonomic Significance of Secondary Carotenoid Formation in *Neospongiococcum* (Chlorococcales, Chlorophyta). J. Phycol. **13**, 176–180.
DERSCH, G., 1960: Mineralsalzmangel und Sekundär-Carotinoide in Grünalgen. Flora **149**, 566–603.
EGGER, K., H. NITSCHE and H. KLEINIG, 1969: Diatoxanthin und Diadinoxanthin – Bestandteile des Xanthophyllgemisches von *Vaucheria* und *Botrydium*. Phytochem. **8**, 1583–1586.
FOTT, B., 1971: Algenkunde. G. Fischer, Stuttgart.
GOODWIN, T. W., 1971: Algal Carotenoids. In: Goodwin, T. W. (ed.) Aspects of Terpenoid Chemistry and Biochemistry. Acad. Press, London + New York, 315–356.
GOODWIN, T. W., 1974: Carotenoids and Biliproteins. In: Stewart, W. D. P. (ed.) Algal Physiology and Biochemistry. Blackwell Sci. Publ., Oxford, London, Edinburgh, Melbourne, 176–205.
HAGER, A., and H. STRANSKY, 1970a: Das Carotinoidmuster und die Verbreitung des lichtinduzierten Xanthophyllcyclus in verschiedenen Algenklassen. III. Grünalgen. Arch. Mikrobiol. **72**, 68–83.
HAGER, A., and H. STRANSKY, 1970b: Das Carotinoidmuster und die Verbreitung des lichtinduzierten Xanthophyllcyclus in verschiedenen Algenklassen. V. Einzelne Vertreter der Cryptophyceae, Euglenophyceae, Bacillariophyceae, Chrysophyceae und Phaeophyceae. Arch. Mikrobiol. **73**, 77–89.
HEGNAUER, R., 1971: Pflanzenstoffe und Pflanzensystematik. Naturwiss. **58**, 585–598.
HERTZBERG, S., S. LIAAEN-JENSEN, and H. W. SIEGELMAN, 1971: The Carotenoids of Blue-Green Algae. Phytochem. **10**, 3121–3127.
HEYWOOD, V. H., 1971: Taxonomie der Pflanzen. G. Fischer, Stuttgart.
HIBBERD, D. J., and G. F. LEEDALE, 1970: Eustigmatophyceae – a New Algal Class with Unique Organization of the Motile Cell. Nature **225**, 758–760.
JEFFREY, S. W., M. SIELICKI and F. T. HAXO, 1975: Chloroplast Pigment Patterns in Dinoflagellates. J. Phycol. **11**, 374–384.
JEFFREY, S. W., and M. VESK, 1976: Further Evidence for a Membrane-Bound Endosymbiont within the Dinoflagellate *Peridinium foliaceum*. J. Phycol. **12**, 450–455.
KESSLER, E., 1967: Physiologische und biochemische Beiträge zur Taxonomie der Gattung *Chlorella*. III. Merkmale von 8 autotrophen Arten. Arch. Mikrobiol. **55**, 346–357.
KESSLER, E., and C. J. SOEDER, 1962: Biochemical Contribution to the taxonomy of the Genus *Chlorella*. Nature **194**, 1096–1097.
KESSLER, E., W. LANGNER, J. LUDEWIG and H. WIECHMANN, 1963: Bildung von Sekundär-Carotinoiden bei Stickstoffmangel und Hydrogenase-Aktivität als taxonomische Merkmale in der Gattung *Chlorella*. In: Studies on Microalgae and Photosynthetic Bacteria. Japanese Soc. Plant Physiologists. Tokyo 7–20.
KESSLER, E., and F.-C. CZYGAN, 1967: Physiologische und biochemische Beiträge zur Taxonomie der Gattungen *Ankistrodesmus* und *Scenedesmus*. I. Hydrogenase, Sekundär-Carotinoide und Gelatineverflüssigung. Arch. Mikrobiol. **55**, 320–326.

KLEINIG, H., 1966: Die Bildung von Sekundär-Carotinoiden in *Acetabularia mediterranea*. Ber. Dtsch. Bot. Ges. **79**, 126–130.

KLEINIG, H., 1969: Carotenoids of Siphonous Green Algae: a Chemotaxonomic Study. J. Phycol. **5**, 281–284.

KLEINIG, H., and F.-C. CZYGAN, 1969: Lipids of *Protosiphon* (Chlorophyta). I. Carotenoids and Carotenoid Esters of Five Strains of *Protosiphon botryoides* (Kütz.). Z. Naturforsch. **24b**, 927–930.

LIAAEN-JENSEN, S., 1977: Algal Carotenoids and Chemosystematics. In: Faulkner, D. J., and Fenical, W. H. (eds.) Marine Natural Products Chemistry. Plenum Press, New York + London, 239–259.

LOEBLICH III, A. R., 1976: Dinoflagellate Evolution, Speculation and Evidence. J. Protozool. **23**, 13–28.

MANDELLI, E. F., 1968: Carotenoid Pigments of the Dinoflagellate *Glenodinium foliaceum* Stein. J. Phycol. **4**, 347–348.

McLEAN, R., 1967: Primary and Secondary Carotenoids of *Spongiochloris typica*. Physiol. Plant. **20**, 41–47.

MERXMÜLLER, H., 1967: Chemotaxonomie? Ber. Dtsch. Bot. Ges. **80**, 608–620.

NAKAYAMA, T. O. M., 1962: Carotenoids. In: Lewin, R. A. (ed.) Physiology and Biochemistry of Algae. Acad. Press, London, New York, 409–420.

NITSCHE, H., 1974: Die Identität von Loroxanthin mit Pyrenoxanthin, Trollein und Trihydroxy-α-Carotin. Arch. Microbiol. **95**, 79–90.

NITSCHE, H., and K. EGGER, 1970: Die Struktur von Vaucheriaxanthin. Tetrahedron Letters 1435–1438.

REZNIK, H., 1964: Chemische Merkmale als taxonomisches Kriterium. Ber. Dtsch. Bot. Ges. **77**, 54–61.

RILEY, J. P., and T. R. S. WILSON, 1965: The Use of Thin-Layer Chromatography for the Separation and Identification of Phytoplankton Pigments. J. mar. biol. Ass. U. K. **45**, 583–591.

ROBINSON, D. G., and R. D. PRESTON, 1971: Studies on the Fine Structure of *Glaucocystis nostochinearum* Itzigs. II. Membrane Morphology and Taxonomy. Br. phycol. J. **6**, 113–128.

ROUND, F. E., 1975: Biologie der Algen. G. Thieme, Stuttgart.

RÜDIGER, W., 1975: Phycobiliproteide. Ber. Dtsch. Bot. Ges. **88**, 125–139.

SCHMIDT, B., L. KIES and A. WEBER, 1979: Die Pigmente von *Cyanophora paradoxa*, *Gloeochaete wittrockiana* und *Glaucocystis nostochinearum*. Arch. Protistenk. **122**, 164–170.

SCHNEPF, E., and R. M. BROWN Jr., 1971: On Relationships Between Endosymbiosis and the Origin of Plastids and Mitochondria. In: Reinert, J., and Ursprung, H. (eds.) Origin and Continuity of Cell Organelles. Springer-Verlag, Berlin, Heidelberg, New York, 299–322.

SKUJA, H., 1954: Glaucophyta. In: Engler, H., Syllabus der Pflanzenfamilien. Bd. 1. Borntraeger, Berlin, 56–57.

SPREY, B., 1970: Die Lokalisierung von Sekundär-Carotinoiden von *Haematococcus pluvialis* Flotow em. Wille. Protoplasma **71**, 235–250.

STRAIN, H. H., F. L. BENTON, M. C. GRANDOLFO, K. AITZETMÜLLER, W. A. SVEC and J. J. KATZ, 1970: Heteroxanthin, Diatoxanthin and Diadinoxanthin from *Tribonema aequale*. Phytochem. **9**, 2561–2565.

STRAIN, H. H., W. A. SVEC, K. AITZETMÜLLER, M. C. GRANDOLFO, J. J. KATZ, H. KJØSEN, S. NORGARD, S. LIAAEN-JENSEN, F. T. HAXO, P. WEGFAHRT and H. RAPOPORT, 1971:

The Structure of Peridinin the Characteristic Dinoflagellate Carotenoid. J. Am. chem. Soc. **93**, 1823–1825.

Stransky, H., and A. Hager, 1970a: Das Carotinoidmuster und die Verbreitung des lichtinduzierten Xanthophyllcyclus in verschiedenen Algenklassen. II. Xanthophyceae. Arch. Mikrobiol. **71**, 164–190.

Stransky, H., and A. Hager, 1970b: Das Carotinoidmuster und die Verbreitung des lichtinduzierten Xanthophyllcyclus in verschiedenen Algenklassen. IV. Cyanophyceae und Rhodophyceae. Arch. Mikrobiol. **72**, 84–96.

Stransky, H., and A. Hager, 1970c: Das Carotinoidmuster und die Verbreitung des lichtinduzierten Xanthophyllcyclus in verschiedenen Algenklassen. VI. Chemosystematische Betrachtung. Arch. Mikrobiol. **73**, 315–323.

Straub, O., 1971: Lists of Natural Carotenoids. In: Isler, O. (ed.) Carotenoids. Birkhäuser Verlag, Basel, Stuttgart, 771–850.

Thommen, H., 1971: Metabolism. In: Isler, O. (ed.) Carotenoids. Birkhäuser Verlag, Basel, Stuttgart, 637–668.

Tomas, R. N., and E. R. Cox, 1973: The Symbiosis of *Peridinium balticum* (Dinophyceae). I. Ultrastructure and Pigment Analysis. J. Phycol. (Suppl.) **9**, 16.

Tupa, D. D., 1974: An Investigation of Certain Chaetophoralean Algae. Beihefte zur Nova Hedwigia **46**, 1–155.

Twsett, M., 1906: Adsorptionsanalyse und chromatographische Methode. Anwendung auf die Chemie des Chlorophylls. Ber. Dtsch. Bot. Ges. **24**, 384–393.

Weber, A., 1969: Über die Chloroplastenfarbstoffe einiger Conjugaten. Flora Abt. A **160**, 457–473.

Weber, A., 1973: Über die Chlorophylle und Carotinoide einiger Ulotrichales. Mitt. Staatsinst. Allg. Bot. Hamburg **14**, 25–29.

Weber, A., 1975: Chlorophylle und Carotinoide der Chaetophorineae (Chlorophyceae, Ulotrichales). II. Der Einfluß unterschiedlicher Stickstoffkonzentrationen auf die Pigmentgarnitur und die Morphogenese der Grünalge *Fritschiella tuberosa* Iveng. Arch. Microbiol. **102**, 45–52.

Weber, A., and F.-C. Czygan, 1972: Chlorophylle und Carotinoide der Chaetophorineae (Chlorophyceae, Ulotrichales). I. Siphonaxanthin in *Microthammnion kuetzingianum* Naegeli. Arch. Mikrobiol. **84**, 243–253.

Weedon, B. C. L., 1971: Occurence. In: Isler, O. (ed.) Carotenoids. Birkhäuser Verlag, Basel, Stuttgart, 29–59.

Wettern, M., and A. Weber, 1979: Some Remarks on Algal Carotenoids and their Interconversion into Animal Carotenoids. In: Hoppe, H. A., et al. (eds.) Marine Algae in Pharmaceutical Science. Walter de Gruyter & Co., Berlin, New York, 551–568.

Whittle, S. J., and P. J. Casselton, 1968: Peridinin as the Major Xanthophyll of the Dinophyceae. Br. Phycol. Bull. **3**, 602–603.

Withers, N., and F. T. Haxo, 1975: Chlorophyll c_1 and c_2 and Extraplastidic Carotenoids in the Dinoflagellate, *Peridinium foliaceum*. Plant Sci. Lett. **5**, 7–15.

Chromoplasts

Peter Sitte, Heinz Falk and Bodo Liedvogel

Institute of Biology II, Cell Biology, University of Freiburg/Br.

1. Introduction

The term «chromoplast» was coined by A. Meyer in 1883 to designate yellow to red pigmented plastids which act as visual attractants («Lockorgane»). A. F. W. Schimper, who had created the general term «plastids» («Plastiden») in the same year, adopted Meyer's term two years later but warned against connecting it with a definite function of the organelles in question. He rightly pointed out that in many instances no attracting function can be assigned to yellowish, orange, or red pigmented plastids. Since then, Schimper's terminology has been in general use in scientific papers and textbooks (see, e.g., Linsbauer, 1917; Küster, 1956) *. In modern terms, as a *chromoplast* is designated *any carotenoid-bearing, photosynthetically inactive plastid, irrespective of whatever its biological function might be* (if there is one at all). Chromoplasts occur primarily in flowers and fruits, but they can also be found in vegetative organs (Rothert, 1912; Savelli, 1937).

Chromoplasts were extensively studied by the famous old light microscopists (cf., Meyer, 1883a, b; Schimper, 1883, 1885; Courchet, 1888; Rothert, 1912). This is due in part to the pronounced pigmentation of these organelles which facilitates their observation in living cells without staining, and also to their remarkable structural diversity. In fact, the variability of chromoplasts in form and size surpasses that of chloroplasts and leucoplasts by far. (An impressive colored print of Schimper's work of 1885, showing many peculiar forms of chromoplasts, has been included in Kirk's and Tilney-Bassett's handbook [1967].) A new wave of interest in chromoplasts has been aroused by the introduction of chromatographic methods allowing for easy separation and identification of carotenoids (see Isler, 1971; Straub, 1976). After the advent of the electron microscope it became apparent that diversity in size

---

Abbreviations: CIMs = chromoplast internal membranes; DGDG = digalactosyl diacylglycerol; EM = electron microscope; GA = glutaraldehyde; MGDG = monogalactosyl diacylglycerol; ptDNA = plastid DNA, plastome; SDS-PAGE = sodium dodecylsulfate polyacrylamide gel electrophoresis; SL = sulfolipid

* The term «chromoplast» has been used by certain American authors in a quite different sense, i.e. to denote *any* pigmented form of a plastid (see, e.g., Wilson, 1896; Sharp, 1926; Granick, 1961). This would correspond to the old term «chromatophore» as coined by Schmitz in 1883 and soon adopted by Strasburger (1884) and others.

and shape is paralleled by pronounced differences in the fine structure of chromoplasts. Numerous studies of their internal structure have since appeared. In some of these relationships have been sought between chromoplast fine structure and genetical data and/or corresponding pigment complements. From these studies, which are to be dealt with here, certain principles of chromoplastology regarding mainly questions of terminology, development, and the molecular build-up of chromoplasts have emerged in recent years (see, e.g., Sitte, 1974; Simpson, 1976; Sitte, 1977a). However, in spite of all these efforts knowledge of chromoplasts is still fragmentary, particularly when compared with the vast amount of information available on chloroplasts. This is exemplified by the fact that not more than 4% of Kirk's and Tilney-Bassett's handbook, *The Plastids*, is devoted to chromoplasts.

This situation is quite unsatisfactory. Chromoplasts are the principal carotenoid-bearing organelles in higher plants. They serve an important ecological function in mediating pollination and fruit dispersal in so many flowering plants (cf., Savidge, 1976), and they probably could provide some clues to discoveries in topics such as natural senescence, subcellular morphogenesis, and evolutionary aspects of the cell. Although it is generally accepted that dramatic changes, both in terms of structure and metabolism, take place within plastids during chromoplast formation, it is not at all understood, e.g., how far lytic enzymes from outside the plastids are involved (see Matile, 1975; Choe and Thimann, 1975), to what extent new synthetic capabilities are engaged, or whether this transformation process is reversible or monotropic. What is the fate of the plastome during chromoplast transformation? Are chromoplasts metabolically inactice «marker organelles» of cell senescence, as has often been stated since the days of Lamarck (1778)?

At present there are, of course, more questions than answers. This makes it perhaps on the more desirable to survey some prominent findings of studies on chromoplasts.

2. Chromoplasts and Gerontoplasts

Schimper's definition of a chromoplast was based on pigmentation as the sole criterion. Neither development, nor metabolic activity, nor function, the study of which appeared much more tiresome, has been considered. Yet any such «simple» definition, however handy it may be, can lead to wrong interpretations, and exactly this has happened in the case of chromoplasts. Certain observations of chloroplast degeneration in senescent leaves have been extrapolated to the development of chromoplasts in flowers and fruits. Justification for doing so was based on the unproven assumption that all carotenoid-bearing plastids, devoid of chlorophylls, belonged to the very same type of organelle. It became a widely accepted bias that chromoplasts in general were indicators of late cell senescence, that they were without active metabolism, and without any ability to divide or to return to «real» life. It has been stated that chromoplast transformation is an irreversible process (Frey-Wyssling et al., 1955; Frey-Wyssling und Kreutzer, 1958a; Frey-Wyssling, 1959; Tôyama and Funazaki, 1971). The «monotropic» transformation itself has been described as a more or less chaotic process, characterized by breakdown of nearly all formerly existing internal structures.

At first sight this concept of chromoplasts as being premortal organelles of senescent cells seems not to be applicable to young flowers and fruits. However, eleven years before the outbreak of the French Revolution, J.-B. Lamarck offered a plausible explanation to this apparent contradiction. He rightly noted that many petals or carpels which turn yellow or red in the course of their development are green in their earliest stages. He then supposed that the autumnal conditions of cell and plastid senescence were simply anticipated in these short-lived organs. When it became known one and a half centuries later that acylated carotenoids (never found in chloroplasts) are consistently present, not only in yellow autumnal leaves, but equally well in the plastids of many flowers and fruits (Seybold, 1954), this was taken as a confirmation of Lamarck's idea.

Recent studies (to be reported below) have shown, however, that there are many differences between the degenerating chloroplasts of senescent mesophyll cells and the plastids of anabolic cells of young flowers or fruits (Table I). These differences are too important to be neglected. We consequently propose to denote senescent chloroplasts not as chromoplasts, but as *gerontoplasts*. In contrast, one should only speak of a (true) *chromoplast* in the case of nonphotosynthetic, carotenoid-bearing plastids which are able to perform anabolic processes and carotenoid synthesis in particular. Schimper's old concept nowadays appears misleading insofar as it comprised a heterogeneous population of organelles. Many, if not all, of the discrepancies in the literature on «chromoplasts» disappear when one accepts the terminology as proposed here.

Table I: Distinction of Chromoplasts proper and Gerontoplasts

	Chromoplasts *s. str.*	Gerontoplasts
occurring in cells of	flowers and fruits	leaves
status of the cells	anabolic	catabolic, senescent
derived from	young chloroplasts, leukoplasts, proplastids	fullgrown chloroplasts
net synthesis of carotenoids	+	—
division	+	—
fine structure type	diverse	exclusively globulous
function	attraction of animals	—

3. Fine Structure

Comparative morphological studies on chromoplasts, starting around the middle of the 19th century (e.g., Trécul, 1858), soon resulted in attempts to classify these organelles mainly according to their different shapes (see Meyer, 1883a, b; Schimper, 1883, 1885; for historical surveys see Schürhoff, 1924; Straus, 1953). Birefringence

Table II. Occurrence of the Various Types of Chromoplasts in Plant Organs as Observed by Electron Microscopy

GLOBULOUS TYPE

Coniferophytina

Taxus baccata L. (aril) CAMEFORT 1964; WUTTKE 1975 (unpubl.)

Magnoliatae (Dicots)

Allamanda neriifolia Hook. (sep.) FALK 1974 (unpubl.)

Alyssum saxatile L. (pet.) FALK 1974 (unpubl.)

Brassica rapa L. (pet.) FALK 1974 (unpubl.)

Berberis vulgaris L. (pet.) SITTE 1974 (unpubl.)

Caltha palustris L. (pet.) SCHMIDT 1973 (unpubl.); WUTTKE 1975 (unpubl.)

Capsicum annuum L. (yellow and orange fr.) KIRK & JUNIPER 1967; SIMPSON et al. 1977b

Cheiranthus cheiri L. (pet.) FALK 1974 (unpubl.)

Chimonanthus praecox (L.) Link (pet.) FALK 1977 (unpubl.)

Chrysanthemum segetum L. (pet.) LANCE-NOUGARÈDE 1960

Chrysosplenium alternifolium L.
" *oppositifolium* L. (sep.) SITTE 1974

Citrus sinensis (L.) Pers. (fr.) OSUMI 1961; SALEMA 1969; THOMSON 1966, 1969; THOMSON et al. 1967

Cucurbita pepo L. (fr.) DEVIDÉ & LJUBEŠIĆ 1972, 1974; GRILLI 1965a,b; LJUBEŠIĆ 1972; MATIENKO 1964

Doronicum orientale Hoffm. (pet.) FALK 1974 (unpubl.)

Eranthis hiemalis (L.) Salisb. (pet.) FALK 1974 (unpubl.)

Euphorbia cyparissias L. (honey-leaves) SITTE 1974 (unpubl.)

Forsythia × intermedia Zab. (pet.) SITTE 1974 (unpubl.)

Forsythia ovata Nakai (pet.) LEDBETTER & PORTER 1970

Gerbera jamesonii H.Bolus ex Hook f. (pet.) GRÖNEGRESS 1974

Hamamelis mollis Oliv. (pet.) FALK 1974 (unpubl.)

Jasminum nudiflorum Lindl. (pet.) FALK 1974 (unpubl.)

Lamium galeobdolon (L.) Nath. (pet.) FALK 1974 (unpubl.)

Lycopersicon esculentum Mill. (pet.) LABORDE & SPURR 1973

Lycopersicon esculentum Mill. cv. cèrise (fr.) LAVAL-MARTIN 1974

Lycopersicon esculentum Mill. cv. Golden Jubilee (fr.) HARRIS 1967; HARRIS & SPURR 1969a

Malus sylvestris Mill. (fr.) CATESSON 1970

Mahonia aquifolium (Pursh) Nutt. (pet.) SITTE 1974 (unpubl.)

Melampyrum nemorosum L. (pet.) SITTE 1974 (unpubl.)

Mimulus luteus L. (pet.) FALK 1977 (unpubl.)

Nuphar lutea (L.) Sm. (sep.) GRÖNEGRESS 1974

Oenothera tetragona Roth. (pet.) FALK 1977 (unpubl.)

Orobanche fuliginosa Reut. ex Jord. (leaves) KOLLMANN et al. 1969

Oxalis succulenta Barnéoud (pet.) SITTE 1974 (unpubl.)

Physalis alkekengi L. (fr.) LICHTENTHALER 1969

Physalis franchetii Mast. (fr.) SALEMA 1969

Primula elatior (L.) Hill (pet.) FALK 1974 (unpubl.)

Prunus domestica L.ssp.*syriaca* (Borkh.) Janchen (fr.) FALK 1974 (unpubl.)

Pyrus communis L. (fr.) BAIN & MERCER 1964

Ranunculus acer L. (pet.) SITTE 1974 (unpubl.)

Ranunculus ficaria L. (pet.) FALK 1974 (unpubl.)

Ranunculus repens L. (pet.) FREY-WYSSLING & KREUTZER 1958a SITTE 1974 (unpubl.)

Rhinanthus alectorolophus (Scop.) Poll. (pet.) SITTE 1974 (unp.)

Sarothamnus scoparius (L.) Wimm. (pet.) LICHTENTHALER 1970a

Solanum luteum Mill. (fr.) SIMPSON et al. 1975a

Spartium junceum L. (pet.) NOUGARÈDE 1964

Tagetes erecta L. (pet.) STEFFEN 1964a

Taraxacum officinale F.Weber ex Wigg. (pet.) FALK 1974 (unpubl.)

Tussilago farfara L. (pet.) FALK 1974 (unpubl.)

Viola tricolor var.*maxima* hort. (pet.) WIEHLER 1972 (unpubl.); WUTTKE 1975 (unpubl.)

GLOBULOUS TYPE (continued)

Liliatae (Monocots)

Aloe plicatilis (L.) Mill. (tep.) STEFFEN & WALTER 1958

Convallaria majalis L. (fr.) STEFFEN 1964a,b

Hemerocallis flava (L.) L. (tep.) FALK 1975 (unpubl.)

Iris pseudacorus L. (tep.) SITTE 1974 (unpubl.)

Kniphofia Moench. spec. (tep.) FALK 1974 (unpubl.)

Lilium martagon L. (tep.) SITTE 1974 (unpubl.)

Musa × paradisiaca L. (fr.) FALK 1974 (unpubl.)

Tulipa gesneriana L. cv. Olympiade (tep.) LICHTENTHALER 1970b

Tulipa sylvestris L. (tep.) WUTTKE 1975 (unpubl.)

Zantedeschia elliottiana (W.Wats.) Engl. (spatha) GRÖNEGRESS 1974

TUBULOUS TYPE

Magnoliatae (Dicots)

Agrimonia eupatoria L. (pet.) SITTE 1974 (unpubl.)

Capsicum annuum L. (fr.) FREY-WYSSLING & KREUTZER 1958b; KIRK & JUNIPER 1967; LABORDE & SPURR 1973; LICHTENTHALER 1969; SALEMA 1969; SCHÄFER 1971; SIMPSON & LEE 1974,1976,1977; SIMPSON et al. 1974, 1977a,b; SPURR & HARRIS 1968; SUZUKI 1974

Celastrus scandens L. (aril) BORNMAN 1968

Chelidonium majus L. (pet.) SITTE 1974 (unpubl.)

Cotoneaster serotinus Hutchins. (fr.) SIMPSON & LEE 1974

Crataegus ellwangeriana Sarg. (fr.) SIMPSON & LEE 1974

Cucumis sativus L. (pet.) SMITH & BUTLER 1971

Cucurbita pepo L. (pet.) GRILLI 1965c

Euonymus europaeus L. (aril) WUTTKE 1977

Impatiens noli-tangere L. (pet.)
" *parviflora* DC. (pet.) SITTE 1974 (unpubl.)

Oxalis europaea Jord. (pet.) SITTE 1974 (unpubl.)

Physalis pubescens L. (fr.) SALEMA 1969

Physalis peruviana L. (fr.) SIMPSON & LEE 1974; SIMPSON et al. 1977c

Rhododendron flavum (Hoffmgg.) G.Don non Pall. (pet.) LICHTENTHALER 1969

Rosa canina L. (fr.) SITTE 1975 (unpubl.); STEFFEN & WALTER 1955

Rosa rugosa Thunb. (fr.) WUTTKE 1976a

Rosa L. spec. (fr.) GRILLI 1965c; SIMPSON & LEE 1974

Solanum capsicastrum L. (sep., fr.) STEFFEN & WALTER 1955,1958

Solanum pseudocapsicum L. (fr.) SALEMA 1969

Sorbus aucuparia L. (fr.) SIMPSON & LEE 1974; WUTTKE 1977

Sorbus intermedia (Ehrh.) Pers. (fr.) WUTTKE 1977

Tropaeolum majus L. (pet.) FALK 1976; WINKENBACH et al. 1976

Liliatae (Monocots

Asparagus officinalis L. (fr.) SIMPSON et al. 1977c

Hemerocallis fulva L. (tep.) FALK & SITTE 1974 (unpubl.)

Sauromatum guttatum (Wall.) Schott (spadix append.) BERGER & SCHNEPF 1970

Strelitzia reginae Ait. (sep.) GRÖNEGRESS 1974; SIMPSON & LEE 1974; SIMPSON et al. 1975b; FALK 1975 (unpubl.)

RETICULO-TUBULOUS TYPE

Magnoliatae (Dicots)

Liriodendron tulipifera L. (pet.) LJUBEŠIĆ 1978

Liliatae (monocots)

Typhonium divaricatum Decne. (spadix append.) SCHNEPF & CZYGAN 1966

MEMBRANEOUS TYPE

Magnoliatae (Dicots)

Capsicum annuum L. (fr., several cultivars) SPURR & HARRIS 1968

Citrus sinensis (L.) Pers. (fr.) SALEMA 1969

Cucurbita pepo L. cv.Small Sugar (fr.) GRILLI 1965a

Lycopersicon esculentum Mill. cv. Golden Jubilee (fr.) HARRIS & SPURR 1969b; ROSSO 1967b

Liliatae (Monocots)

Corbularia bulbocodium (L.) Haw. (corona) FALK 1977 (unpubl.); MESQUITA 1976

Narcissus pseudonarcissus L. (corona) KOWALLIK & HERRMANN 1972; LIEDVOGEL & SITTE 1974; LIEDVOGEL et al. 1976; MOLLENHAUER & KOGUT 1968; NICHOLS et al. 1967

Sternbergia lutea (L.) Ker-Gawl. ex Schult.f. (corona) FALK 1977 (unpubl.)

CRYSTAL-CONTAINING TYPE

Magnoliatae (Dicots)

Citrullus lanatus (Thunb.) Matsum. et Nakai (fr.) FALK 1975 (unpubl.)

Citrus paradisi Macf. (fr.) PURCELL et al. 1963

Cucurbita pepo L. cv. ovifera (fr., deep orange areas) DEVIDÉ 1970

Daucus carota L. (root) BEN-SHAUL & KLEIN 1965; BEN-SHAUL et al. 1968; FREY-WYSSLING & SCHWEGLER 1965; GRÖNEGRESS 1971; TÔYAMA 1967; TREFFRY et al. 1966; WRISCHER 1972

Lycopersicon esculentum Mill. (fr.) BEN-SHAUL & NAFTALI 1969; GRILLI 1965c; HARRIS 1970; HARRIS & SPURR 1969a,b; LAVAL-MARTIN 1974; MOHR & STEIN 1969; ROSSO 1967a,b, 1968

Liliatae (Monocots)

Clivia nobilis Lindl. (tep.) SITTE 1974 (unpubl.)

Hemerocallis fulva L. (tep., late stage) FALK & SITTE 1974 (unpubl.)

Narcissus × incomparabilis Mill. (corona) FALK 1977 (unpubl.)

Narcissus poeticus L. (corona) KUHN 1970; KUHN et al. 1969

Zea mays L. (leaves, 'lycopenic' mutant) MARZIANI & ORSENIGO 1965; WALLES & HUDAK 1975

Abbreviations:

append. - appendage; fr. - fruit; pet. - petal; sep. - sepal; tep. - tepal

and dichroitic behavior, as exhibited by certain chromoplast types, have also been used in classification.

After a pioneering effort to apply the still inadequate EM techniques of the fourties to chromoplasts (Roberts and Southwick, 1948), it eventually became possible to deal properly with chromoplasts despite their reputation as very delicate organelles. Chromoplasts are typical plastids in that they invariantly possess an envelope consisting of two separate membranes, and a stroma matrix containing ribosomes which are smaller than the cytoplasmic ones, as well as less dense regions containing slender, uranophilic filaments consisting presumbly of DNA. There is great diversity, however, concerning the internal fine structure of chromoplasts. Therefore, contemporary classification of chromoplasts rests entirely on characteristics of fine structure (Steffen und Walter, 1955, 1958; Frey-Wyssling und Kreutzer, 1958a; Kirk and Tilney-Bassett, 1967; Sitte, 1974). Most of the chromoplasts investigated so far by EM methods (see Table II) belong to the globulous, tubulous, membraneous, or crysstallous type, respectively, according to the carotenoid-bearing fine structural elements prevailing in them in their final state of development. These four main chromoplast types can be characterized as follows:

1) *Globulous chromoplasts* are lens-shaped or spheroidal, and contain many carotenoid-carrying lipid droplets in their stroma (Fig. 1). These droplets, termed plastoglobules (Lichtenthaler und Sprey, 1966), vary greatly in diameter, from 0.05

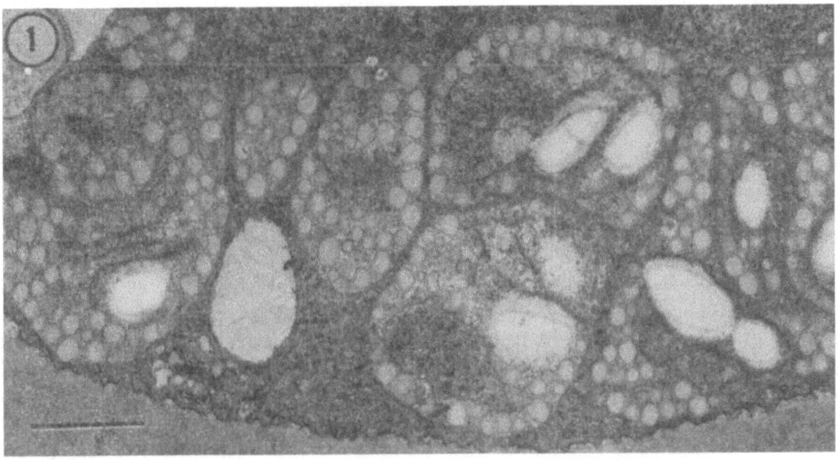

Fig. 1: Globulous chromoplasts in an epidermal cell of an unfolding petal of rape, *Brassica rapa*. In addition to numerous plastoglobules several starch grains can be seen. GA/OsO_4 fixation. The bar in this and the following Figs. corresponds to 1 μm, if not otherwise indicated

up to more than 1.0 μm. Occasionally, remnants of thylakoids are also present which may (cf., Fig. 11), or may not, be associated with the plastoglobules. The globules often are concentrated in the peripheral stroma beneath the envelope. This holds true, for example, for the flower chromoplasts in Asteraceae (Steffen, 1964a), and also for those of Ranunculaceae, in which the center of the chromoplasts is often occupied by other inclusions. The globulous type is by far the commonest among chromoplasts, and it may be regarded as the oldest type in evolution (see Chapter 7).

2) *Tubulous chromoplasts* are characterized by extended carotenoid-carrying internal elements which are more or less circular in cross section (Figs. 2–4). Their diameter is most often about 20 nm, but values of up to 60 nm have been found (e.g., in *Hemerocallis fulva*). These anisometric elements have been termed differently in the literature: «filaments» (Frey-Wyssling and Kreutzer, 1958b; Gunning and Steer, 1975), «fibrils» or «microfibrils» (Steffen und Walter, 1955, 1958; Bornman,

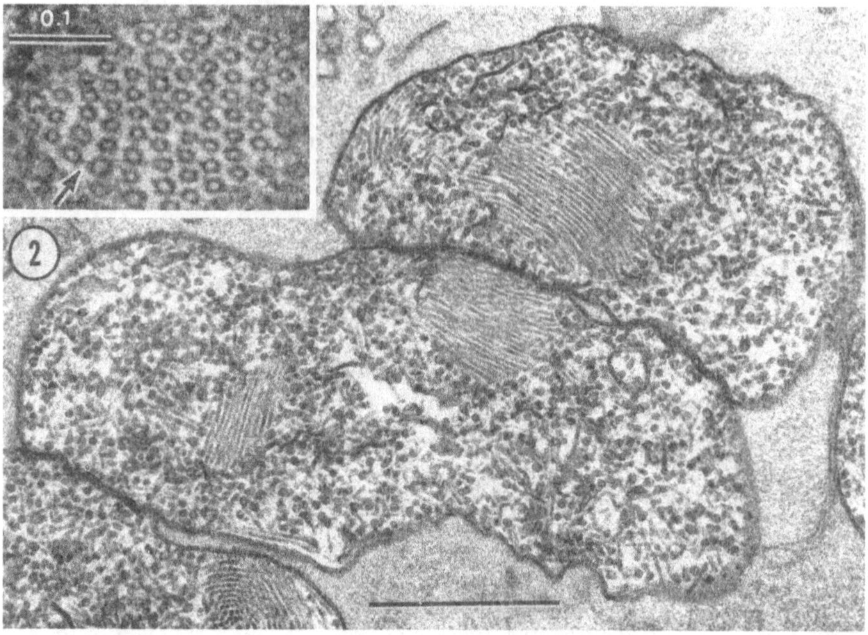

Fig. 2: Chromoplasts of the tubulous type as characteristic for the flowers of nasturtium (*Tropaeolum majus*) show loosely bundled tubules as well as single tubules arranged in various directions. In this $KMnO_4$ fixed preparation the isodiametric lipid bodies are clearly discernible. Inset: Transversely sectioned tubules at higher magnification. Note the cross-bridges between several tubules, preserved by GA/OsO_4 fixation

1968; Simpson and Lee, 1976), »tubuläre Filamente» (tubular filaments) (Lichtenthaler, 1969), or finally «(chromoplast) tubules» (Grilli, 1965c; Salema, 1969; Smith and Butler, 1971; Grönegress, 1974; Sitte, 1974; Falk, 1976; Wuttke, 1977)*. Their denotion as «tubules» corresponds best to their appearance in ultrathin sections. Furthermore, as may be inferred from chemical analyses, chromoplasts seem to be composed of a strongly hydrophobic core and a less hydrophobic superficial layer (see below).

Among these chromoplast tubules, different «subclasses» can be distinguished. In the vast majority of tubulous chromoplasts, the tubules are unbranched. They form conspicuous paralleled aggregates which appear as birefringent and dichroitic areas under the polarizing microscope. Among these tubules two types have been found which differ in a number of characteristic features (see Table III). Whereas many tubulous chromoplasts – preferentially found in flowers – are spheroidal in outline,

* They have also been referred to as «microtubules» (Salema, 1969). This is misleading in more than one respect, however. Chromoplast tubules are lipoprotein structures, are well preserved by fixation with $KMnO_4$, and do not contain tubulin. Consequently, they have nothing to do with cytoplasmic microtubules.

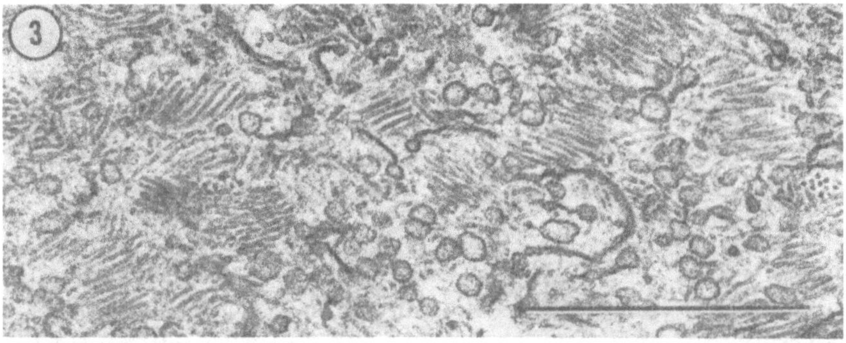

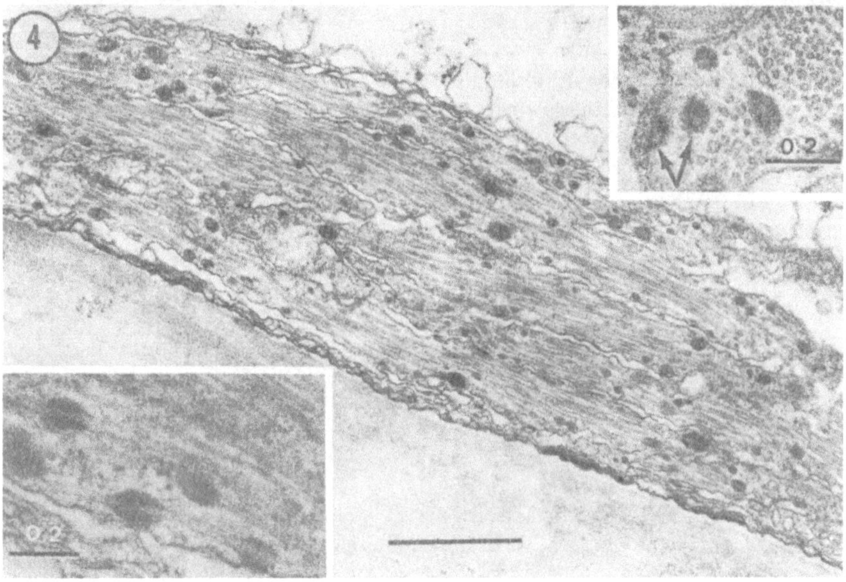

Fig. 3: Chromoplast tubules of *Chelidonium majus* (upper epidermal cell of a petal). Plastoglobules are totally absent. $KMnO_4$ fixation

Fig. 4: Chromoplast of the peripheral mesocarp of a ripe rose hip. Bunched, straight profiles of tubules pierce electron-dense plastoglobules associated with them. Insets: Higher magnification of transverse (upper right) and longitudinal (lower left) sections. OsO_4 fixation. (Kindly provided by Dr. H.-G. Wuttke)

those containing globule-associated tubules are elongated and often spindle-shaped, These latter, which found also in fruits (the hip is a well-known example, Fig. 4). resemble chromoplasts of the crystallous type under the light and polarizing microscope.

Tubule-associated globules are pierced by the tubules (see Fig. 4, insets). According to several authors, these tubules even originate within the globules (Salema, 1969;

Table III: Differences Among Tubules in Chromoplasts (cf., also Simpson and Lee, 1977)

Globule-associated Tubules (cf., Fig. 4)	Tubules Devoid of Globules (cf., Figs. 2, 3)
piercing plastoglobules	not associated with plastoglobules
occurring throughout stroma	confined to distinct areas in the stroma
long, differing in length	shorter, uniform in length
straight	slightly bent
no interconnecting bridges	with interconnecting bridges
extremely stable	often disintegrating in fully developed petals
formation inhibited by CPTA[1]	formation not inhibited by CPTA[1]

[1] CPTA = 2-(4-chlorophenylthio)ethyldiethylammonium chloride. Effect of this inhibitor of carotenoid biosynthesis on chromoplast tubules tested only in mutants of red pepper so far (Simpson and Lee, 1977)

Simpson et al., 1975b). During an early developmental stage the initially spherical and homogeneous globules become lemon-shaped and contain a tubule-like core. Later on, the core elongates and becomes a tubule. Remnants of the globules remain associated with the tubules («nodulated» aspect).

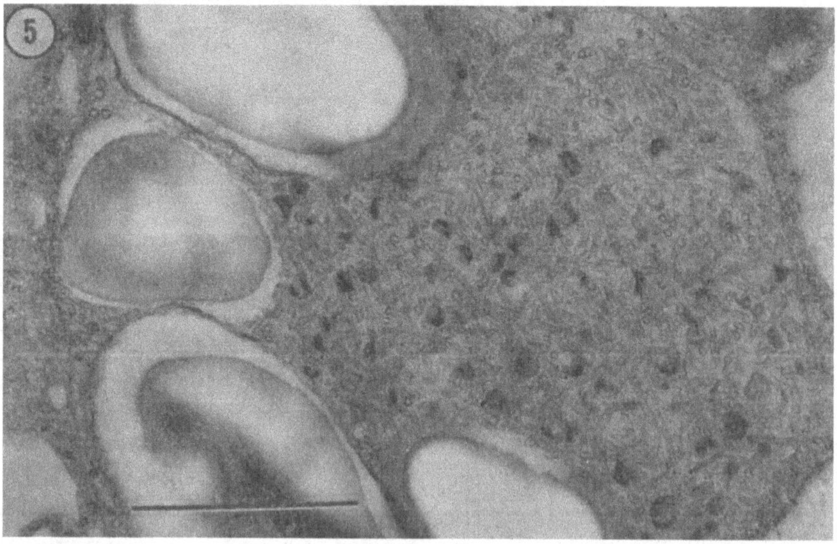

Fig. 5: Chromoplast of the reticulo-tubulous type as contained in cells of the basal, colored area of *Liriodendron* petals. Some prominent starch grains can be seen (left half of the micrograph) in addition to plastoglobules of the same appearance as in Fig. 4. GA/OsO_4 fixation. (Kindly provided by Dr. N. Ljubešić, Zagreb, Yugoslavia)

A third subclass of tubulous chromoplasts with ramified and anastomosing, sinuous tubules has been encountered in rare cases (Fig. 5; Schnepf und Czygan, 1966; Ljubešić, 1978). This type has been termed «reticulo-tubulous» (Sitte, 1974).

3) *Membraneous chromoplasts* contain a set of up to 20 concentric (double) membranes as their most prominent internal structure (Fig. 6). The overall shape of such chromoplasts is spherical, but heavy deformations are not unusual *in situ*. Under the polarizing microscope a faint spheritic birefringence can be detected. It is positive in character and has been attributed to radially oriented lipid molecules (Liedvogel et al., 1976). The occurrence of membraneous chromoplasts in their *typical* form seems to be restricted to flowers – and coronae in particular – of certain Amaryllidaceae (daffodil and some of its close relatives).

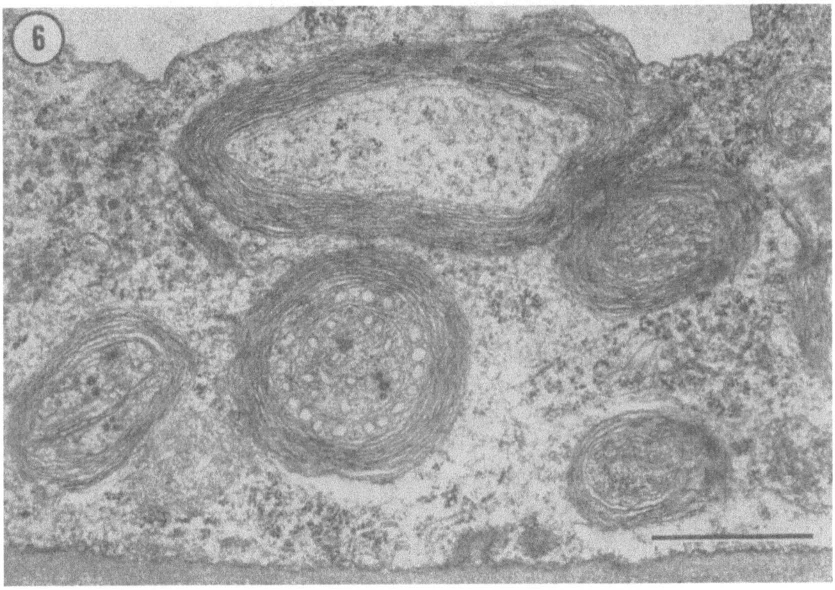

Fig. 6: Chromoplasts in a corona cell of the daffodil with peripherally located, concentric membranes (CIMs). GA/OsO_4 fixation. (From Liedvogel *et al.*, 1976, with permission)

4) *Crystallous chromoplasts* are known, owing to the work of the old light microscopists of more than a century ago. They contain crystalline inclusions of pure carotene (β-carotene, lycopene). Wherever these crystals grow to light-microscopic dimensions, they dictate the shape of the chromoplasts. Well-known examples of this kind are the chromoplasts of carrot roots *, or the fruits of tomatoes and watermelons. Yet carotene crystals can also occur in flower chromoplasts (Fig. 7; Kuhn et al., 1969; Kuhn, 1970) and even, in leaf chloroplasts – in certain lycopene-accumulating mutants of corn and barley – (Marziani and Orsenigo, 1965; Nielsen, 1974;

* Cultivated forms only; the roots of the wild type are unpigmented.

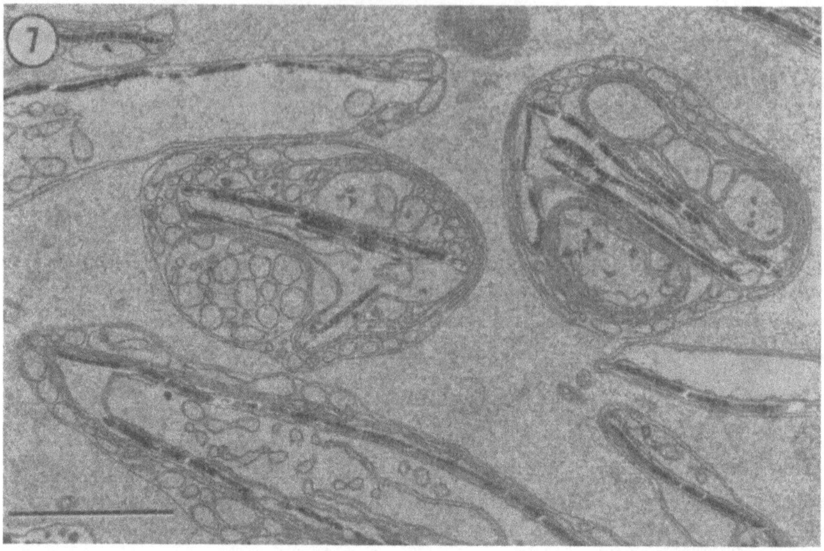

Fig. 7: Crystal-containing chromoplasts of *Narcissus incomparabilis* (deep orange corona) after $KMnO_4$ fixation. Carotene crystals appear as electron-dense, needle-shaped profiles located inside membrane-surrounded compartments (remnants of thylakoids)

Walles and Hudák, 1975). In contradiction to earlier reports (e.g., Ben Shaul and Klein, 1965) it is now well established that the carotene crystals remain surrounded by the chromoplast envelope during all stages of development (Frey-Wyssling and Schwegler, 1965; Rosso, 1968; Grönegress, 1971). Furthermore, it has been shown, that carotene crystals invariably originate, not in the stroma matrix, but within thylakoids, that is, «intralocularly». In ordinary GA-OsO_4 fixation the dense crystals are not sufficiently fixed to withstand extraction by dehydrating and embedding media. In this case, empty spaces are left over within which undulating membranes can be observed (e.g., Harris and Spurr, 1969a). The carotene crystals do not always grow to light-microscopical dimensions. They sometimes appear as submicroscopic crystallites, often in great numbers of up to more than 100 per chromoplast (e.g., flowers of *Clivia*; fruits of *Physalis*).

So far about 100 plant species habe been investigated electronmicroscopically to elucidate the fine structure of their chromoplasts. Some general statements can now be made on the basis of the results obtained (Table II).

a) Many chromoplasts contain additional inclusions apparently devoid of carotenoids. Examples of this kind are starch grains (e.g., in *Brassica* and *Chelidonium*) crystals of protein (*Ranunculus, Musa*) or spherical, strongly osmiophilic bodies of unknown composition (*Mahonia, Caltha*).

b) In many chromoplasts more than just one type of pigment-carrying structures are found. For example, plastoglobules occur not only in globulous but also in membraneous as well as in certain tubulous chromoplasts. Membraneous chromoplasts may contain carotene crystals (*Narcissus incomparabilis*) or tubular elements (e.g., high-β tomato: Harris and Spurr, 1969a). At least in certain cases this makes classification of chromoplasts according to fine structure somewhat arbitrary.

c) It is a well-documented fact that the chromoplasts of different tissues of a given plant do not always belong to the same structural type. In tomatoes, for example, the petal chromoplasts are globulous, the ones in fruits crystallous. In the cucumber and pumpkin, flowers contain tubulous chromoplasts, whereas their fruits bear globulous ones. The situation is reversed in roses with yellow petals, (for ref. see Table 2). Remarkable differences of chromoplast fine structure can be found even within the same organ. Thus, the chromoplasts of the berries of *Asparagus* belong to the globulous type in epidermal cells, and to the tubulous type in the mesocarp (Simpson et al., 1977c; for cherry tomatoes see also: Laval-Martin, 1974). In contrast it should be noted that gerontoplasts are invariantly globulous.

d) A taxonomy based upon chromoplast fine structure does not appear to be promising under these circumstances. This outlook is supported by the fact that in many instances closely related species differ fundamentally with regard to respective chromoplast fine structures in corresponding organs and tissues.

e) A still open question concerns a possible correlation between fine structure and the pigment complement of chromoplasts. As will be seen in section 5.1., the carotenoids do not account for high percentages of the content of plastoglobules, tubules, and membranes. It is not surprising, therefore, that leucoplasts sometimes mimic chromoplast internal fine structure. For example, globulous plastids, however devoid of pigments, have been found in albino fruits of mutants of red pepper (Kirk and Juniper, 1967), of pumpkin (Ljubešić, 1973), in the pith of the barrel cactus (Thomson and Platt, 1973). The leucoplasts of a cultivar of the daffodil with white flowers (cv. «Mount Hood») are indistinguishable in fine structure from wild type chromoplasts, and – with the exception of carotenoids – their internal membranes have been found to contain the same lipids and proteins (Liedvogel and Falk, 1980, Z. Pflanzenphysiol. **98**, 371–375). Tubulous leucoplasts have been described from embryonic suspensor cells in *Phaseolus* (Schepf und Nagl, 1970). In contrast chromoplasts with corresponding internal structure often differ significantly in the main pigments they contain (Simpson et al., 1974; cf., Laborde and Spurr, 1973). For example, the tubulous chromoplasts of hips (*Rosa*), of *Solanum capsicastrum*, of the nasturtium, and of *Strelitzia* contain as their main carotenoids: lycopene, cryptoxanthin, lutein diesters, and β-carotene plus cryptoxanthin, respectively.

In all these cases *fine structure seems not to be correlated with pigmentation.* However, instances are also known in which the existence of such a correlation has been postulated. Examples of this kind have been reported by Harris and Spurr (1969a) from work with carotenoid mutants of tomato.

f) If not only the carotenoids, but also the other constituents of chromoplast substructures (see Section 5.1.) are considered, striking relationships between total chemical composition and fine structure become apparent (Sitte, 1977a, 1978). The percentage of apolar constituents (such as carotenoids and neutral lipids) is highest in plastoglobules (>60% of dry weight), less in tubules (ca. 20%), and least in membranes (below 10%). Conversely, the relative amount of glyco- and phospholipids and of proteins is highest in membranes and lowest in globules. This corresponds to differences in specific surface (that is, surface per volume) of the respective substructures which amounts to less than 0.05 nm^{-1} in the case of plastoglobules, approximately 0.2 nm^{-1} in chromoplast tubules, and well above 0.3 nm^{-1} for membranes. According to well-known laws of physical chemistry it is to be assumed that the apolar constituents in all the different chromoplast substructures are concentrated in the inner part, whereas the surface, being in direct contact with the watery stroma matrix, is covered by moderately hydrophilic compounds. Corresponding models of the molecular build-up of tubules and globules have in fact been proposed (Winkenbach et al., 1976; Wuttke, 1977). On the basis of such considerations a hypothesis has been put forward (Sitte, 1977a, 1978) which postulates that the relative amount of «semipolar» and/or amphiphilic compounds as contained in chromoplast substructures would *determine* their respective geometry. This hypothesis, the validity of which can be tested experimentally in several ways, relies in part on experience with lyotropic mesophases. It is among the first to possibly explain morphogenesis of structures within a living cell by self-assembly.

4. Biochemistry

4.1. Chromoplast Isolation; Characterization of Substructures

Contrary to morphological investigations, reliable biochemical data on chromoplasts and their characteristic substructures have been obtained only recently. Isolation of sufficiently pure chromoplast fractions proved rather difficult. Some early attempts were either totally unsuccessful, or have led to inadequate results (not to be reported here).

Around 1970, isolation techniques were improved to a level high enough to allow certain pioneering analytical investigations. A fraction enriched in crystallous chromoplasts of tomato fruits was used by Jungalwala and Porter (1967) in a study of carotenoid biosynthesis *in vitro*. The pigment crystals of chromoplasts from carrot roots and from coronae of *Narcissus poeticus*, respectively, were isolated by Ben-Shaul et al. (1968) and by Kuhn (1970) and shown not to consist of a chromoprotein as previously postulated, but of pure β-carotene. Plastoglobules isolated from chromoplasts of *Sarothamnus* have been shown to contain lipophilic plastoquinones and carotenoids (Lichtenthaler, 1970a). Herrmann (1972) was the first to isolate membraneous chromoplasts from coronae of the daffodil (*Narcissus pseudonarcissus*) and to show them to contain DNA.

In recent years, the mass isolation of intact chromoplasts in very pure form (as ascertained by EM, see Fig. 8), has been achieved in different instances (membraneous chromoplasts: Falk et al., 1974; Liedvogel et al., 1976; tubulous chromoplasts: Winkenbach et al., 1976; Liedvogel et al., 1978; Thompson, 1978; globulous chromoplasts: Wuttke, 1976b; see also Wuttke, 1977). In spite of certain differences in detail, some general requirements for chromoplast isolation can be summarized as follows. The isolation media used consist of buffered (pH 7.5–8.0), isotonic solutions of sucrose, sorbitol, or mannitol which contain macromolecules of high adsorbing capacity, such as polyvinylpyrrolidone or bovine serum albumin. Tissue homogenization is achieved best by rotating-knife homogenizers. Crude chromoplast fractions are obtained either by flotation (*Tulipa*) or sedimentation (*Tropaeolum*, *Narcissus*). Further purification is achieved by density gradient ultracentrifugation. The different internal substructures of chromoplasts are then set free either by osmotic shock, gentle sonication, or careful use of detergents, and collected by appropriate density gradients (Fig. 8).

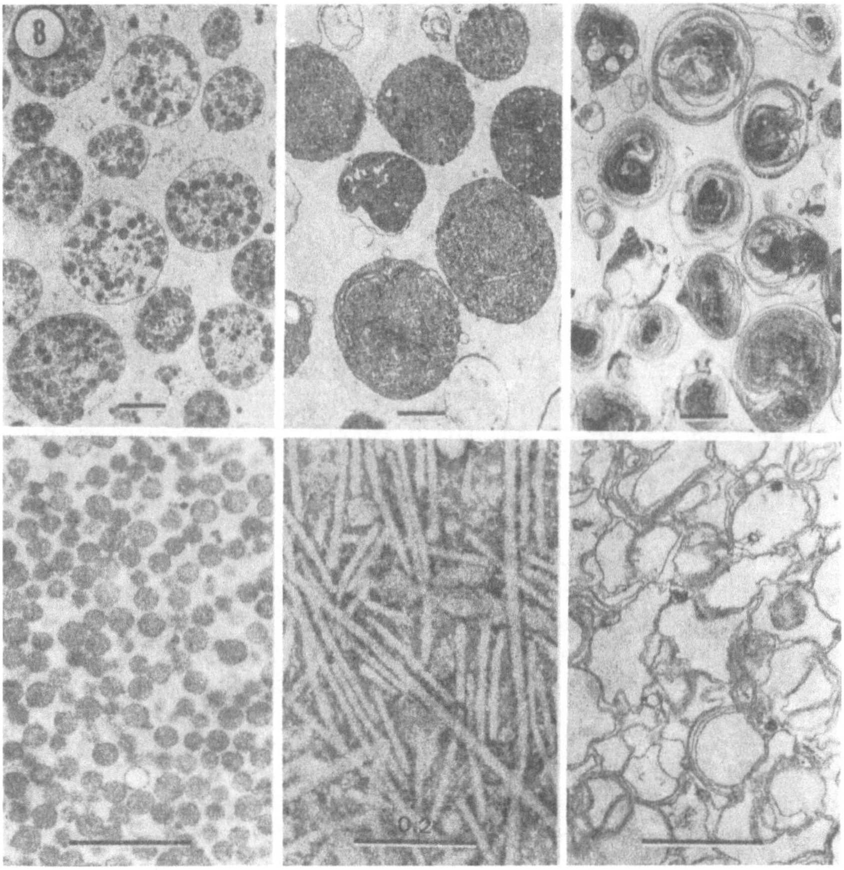

Fig. 8: Upper series: Thin section through fractions of isolated chromoplasts of the globulous (*Tulipa*, left) tubulous (*Tropaeolum*, middle), and membraneous (*Narcissus*, right) type. – Lower series: Corresponding purified substructures in thin sectioned (globules, CIMs) or negatively stained (tubules) preparations. Micrographs of the *Tulipa* material kindly provided by Dr. H.-G. Wuttke

Some pertinent data on the *chemical composition of chromoplast substructures* (not including carotene crystals) are compiled in Table IV.

Chromoplast globules exhibit a buoyant density of less than 1.0 g cm^{-3}, indicating that they consist mainly of lipids (Sitte, 1977b). This expectation is borne out by the analytical data presented in Table IV. For example, about 3/4 (dry weight basis) of the globules from flower chromoplasts of *Caltha palustris* are apolar acylglycerols, about 1/6 are carotenoids, and carotenoid esters. Galactolipids and phospholipids

Table IV: Chemical Composition of Chromoplast Substructures (percent of dry weight)

	Globules[(a)]	Tubules[(b)]	Membranes[(c)]	Thylakoids[(d)]
carotenoids (+ esters)	14	13	4.5	2
chlorophylls	—	—	—	12
galactolipids:				
MGDG	4	28	12	13
DGDG	3	21	40	10
SL	+	+	3	3
phospholipids	+	10	17	7
isoprenoid quinones	+	+	+	++
steryl components	+	+	+	+
diacylglycerols	6	+	+	+
triacylglycerols	68	+	+	+
proteins	++	28	22	52

Chromoplasts from (a) *Caltha palustris*, petals; (b) *Tropaeolum majus*, petals; (c) *Narcissus pseudonarcissus*, corona. (d) For comparison, chloroplasts of *Narcissus pseudonarcissus*, green leaf.
+ means less than 1.5% of dry weight.

comprise not more than 10% of the dry weight. Isoprenoid quinones and sterol lipids are present in small amounts. Chlorophylls are entirely absent.*

Chromoplast globules invariably also contain traces of protein. As shown by SDS-PAGE, different species always contain a main component with a molecular weight of about 30,000 among these globule-associated proteins, which indicates that isolated globules are not merely contaminated by stroma proteins.

Chromoplast tubules have been isolated from flower chromoplasts of *Tropaeolum*

* It seems to be a general rule that plastoglobules never contain chlorophylls. They may, or may not, contain carotenoids. In chromoplasts belonging to the globulous type and in gerontoplasts they are the principal carotenoid-bearing substructures, a fact already well known to the old light microscopists. In chloroplasts, plastoglobules are normally devoid of carotenoids, as can be seen directly in the case of magnoglobules (Falk, 1960) and has been corroborated by analysis of isolated plastoglobules (Greenwood et al., 1963; Lichtenthaler und Sprey, 1966). However, eyespots in some algae chloroplasts are built of plastoglobules which are deeply red from carotenoids.

majus (Winkenbach et al., 1976; Liedvogel et al., 1978) and, in less pure form, from the hips of *Rosa rugosa* (Wuttke, 1976a). In both cases the chemical composition is very similar. In contrast to chromoplast globules, polar lipids (with plastid-specific galactolipids prevailing) predominate in the tubules (60% of dry weight). There are also appreciable amounts of carotenoids, more than 80% of them being lutein diester in the case of the nasturtium. Other apolar lipidic components like acylglycerols, sterols, and plastoquinones amount to only very low percentages. Chlorophyll is entirely absent. A little less than 30% of the tubule dry weight consist of protein. A single polypeptide with a mol. weight of 30,000 constitutes ca. 80% of the tubule protein.

In addition to tubules, flower chromoplasts of the nasturtium also contain nearly isodiametric lipid bodies which prevail in certain developmental stages (Falk, 1976). Their diameter scatters around values of 30–40 nm. They are characterized by an extraordinarily high carotenoid content which is near one-half of the dry weight. The rest is mainly galactolipids (40%) and protein (Winkenbach et al., 1976; Liedvogel et al., 1978). There is evidence indicating that these lipid bodies can be transformed into chromoplast tubules.

Chromoplast internal membranes (CIMs) have been investigated with particular intensity (Liedvogel and Sitte, 1974; Liedvogel et al., 1976; Liedvogel and Kleinig, 1976, 1977) and may be regarded as the best known chromoplast structures at present. They are exceptionally rich in lipid (about 4/5 of dry weight), among which galactolipids (and DGDG in particular) predominate. Phospholipids are well represented, whereas carotenoids with but minimal esterification form only a minor component. As in chromoplast tubules, only a small amount of acylglycerols, sterols, and plastoquinones is to be found in CIMs. Among the proteins, comprising 22% of the dry weight, there are many different species as can be judged from SDS-PAGE analyses. The polypeptide pattern shows no similarity to that of thylakoids (Liedvogel et al., 1976).

4.2. Synthetic Capabilities of Isolated Chromoplasts

As can be inferred from the extensive and dramatic structural changes during chromoplast differentiation, there must be a remarkable metabolic activity in this phase of development. It is characterized not only by the breakdown of structures and compounds accumulated in earlier phases of plastid development, but also by formation of new components. For example, Lichtenthaler (1969, 1970a) has found the accumulation of plastoquinones and secondary carotenoids to strongly correlate with chromoplast development in certain flowers and fruits. Some of the enzymes involved are plastid-specific, as has been shown by studies on fractions enriched in plastids (Kushwaha et al., 1969; Hill et al., 1971; Thomas and Threlfall, 1974).

Recently, work with pure chromoplast fractions has convincingly shown that isolated chromoplasts posses synthetic capabilities which, in certain respects, are even comparable to those of chloroplasts (as reviewed, e.g., by Ellis, 1976; Leech and Murphy, 1976). CIMs have been demonstrated to possess very active galactosyl transferases (Liedvogel and Kleinig, 1976, 1977). Thus, they are able to form galactolipids much like the envelope of chloroplasts (Douce, 1974) to which they apparently are related ontogenetically. As Liedvogel and Kleinig (1977) have shown, CIMs are also endowed with the capabilities of sterol glycosylation, and of acylation of sterol glycosides as well as of MGDG and of phospholipids. In a most recent study, the same authors (Kleinig and Liedvogel, 1978) presented convincing evidence of the existence of a complete machinery for fatty acid synthesis in the chromoplast stroma of the daffodil, paralleling the situation encountered in chloroplasts (for review, see Stumpf, 1976). Furthermore, it was shown that the daffodil flower chromoplasts are not unique in these synthetic activities; since tubulous chromoplasts isolated from petals of *Tropaeolum* likewise synthesize fatty acids from C_2-units in their stroma, whereas the envelope membranes of these organelles appear fitted with acyl transferases as well as glycosyl transferases (Liedvogel et al., 1978).

5. Development

5.1. General Survey

As numerous EM studies have shown, every type of plastid can be transformed into a chromoplast – in perfect agreement with Schimper's postulate. Occasionally tissue-specific differences are observed in the very same organ. In the perianth of *Lilium tigrinum*, for example, the chromoplasts arise from proplastids in the epidermis, and from chloroplasts in the mesophyll (Steffen, 1964a; for an analogous situation, cf., Steffen, 1964b; Grilli, 1965a, b). That even etioplasts can differentiate into chromoplasts has been shown by Walles and Hudák (1975).

Light seems not to play a decisive role in chromoplast development. Tomato fruits turn red in the dark as well (Smith and Smith, 1931; see, however, Khudairi and Arboleda, 1970). Petals of *Forsythia*, when grown in total darkness from their primordia, likewise show significant differences neither in pigmentation nor in chromoplast fine structure when compared with light-grown controls (Sitte, unpubl.). In other instances light seems to influence the amount of carotenoids accumulated, but not the pigment pattern (Simpson and Lee, 1977; Simpson et al., 1977a).

Almost nothing is known about *cellular regulation* and *genetic factors* involved in chromoplast development. Genetics of carotenoid biosynthesis have been investigated primarily in cultivated plants (see ,e.g., Goodwin, 1971, 1973; Britton, 1976). All the genes necessary for carotenoid biosynthesis are located in the nucleus. However, protein synthesis on plastidal ribosomes in *Chlamydomonas* has been found to be necessary for carotenoid accumulation (Sirevåg and Levine, 1973). Mutants of red pepper with white (instead of orange or red) fruits nevertheless possess the usual carotenoid composition in their green leaves (Kirk and Juniper, 1967). Likewise, the different carotenoid composition of color mutants of tomatoes arises only during ripening (Harris and Spurr, 1969).

Not even the *structural changes* accompanying chromoplast formation can be regarded as adequately known. This matter is complicated by an astonishing diversity of the respective processes as observed in different plants. It is difficult to deduce general rules from the data presented in the literature. As many (but by no means all) chromoplasts originate from chloroplasts or juvenile forms thereof, it has been repeatedly proposed that the carotenoid-carrying structures of chromoplasts are derived from decomposing or transforming thylakoids (e. g., Frey-Wyssling and Kreutzer, 1958a; Grönegress, 1974). This hypothesis is, in its simple and general form, no longer tenable. In too many cases chromoplasts are not derived from

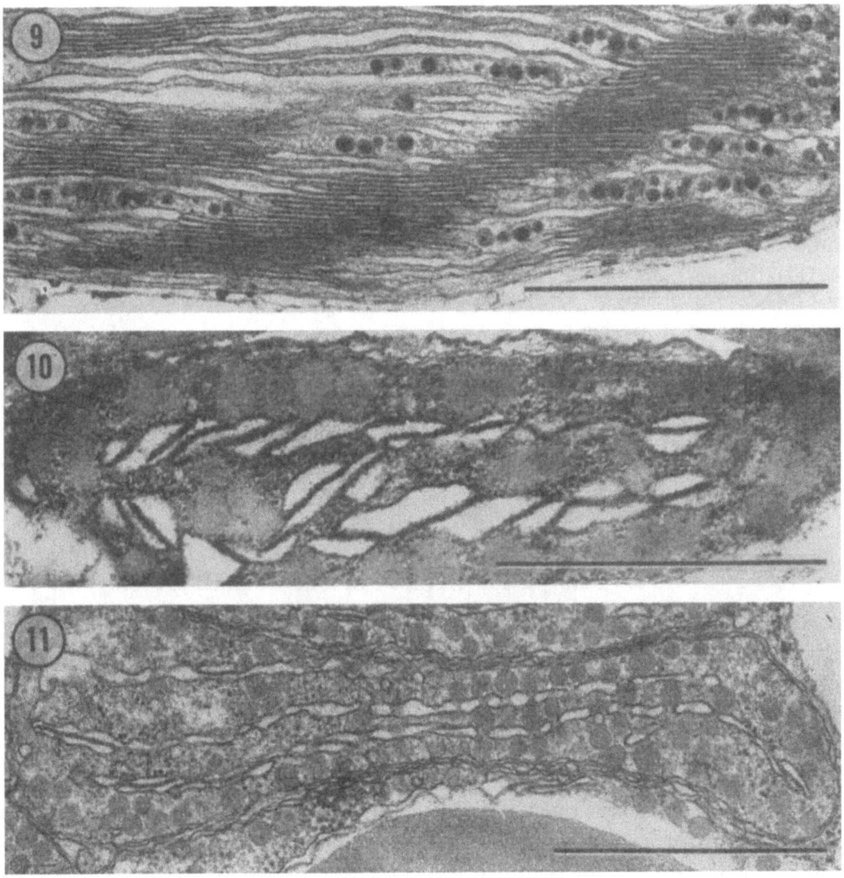

Fig. 9: Early stage of chloroplast-chromoplast transformation in a slightly orange colored rose hip showing a prominent «oblique granum» with staircase-like appearance. Note the small plastoglobules

Fig. 10: More advanced stage of thylakoid displacement in a differentiating chromoplast from a petal cell of *Rhinanthus alectorolophus*

Fig. 11: Young chromoplast in the petal epidermis of *Alyssum saxatile* showing plastoglobules in close contact with single remaining thylakoids. – Fig. 9 to 11: GA/OsO_4 fixation

chloroplasts at all, or the appearance of fine structural elements specific for chromoplasts is not correlated in time with thylakoid breakdown, or finally, these elements originate spatially distant from degrading thylakoids or can be shown to be the result of synthetic activities of the developing chromoplasts. In certain extreme cases the thylakoids of a former chloroplast remain intact even in the fully differentiated chromoplast (Laborde and Spurr, 1973). In the vast majority of examples, in which chromoplast development starts from green plastids, a gradual decomposition of the thylakoid system which is often accompanied by the transitional formation of «oblique grana» actually occurs (see Figs. 9 and 10; cf. Steffen und Walter, 1958; Harris and Spurr, 1969a; Smith and Butler, 1971; Ljubešić, 1972; Falk, 1976). In no case, however, has it been convincingly shown that material freed from thylakoids is directly (that is, without chemical transformation) used again to build chromoplast-specific structures. Membraneous chromoplasts offer an object of special interest in this connection, since a direct transformation of thylakoids into CIMs could be imagined particularly well in this case. However, thylakoids and CIMs differ greatly in their chemical composition, and the enzymatic activities associated with the CIMs resemble those of the plastid envelope, rather than those of the thylakoids (see section 4.2). In accordance with morphological data it therefore appears likely that the CIMs are derived not from thylakoids, but from invaginations of the envelope's inner membrane. This hypothesis has recently been corroborated by EM histochemistry (M. Wrischer, pers. comm.): After incubation of young daffodil flowers with diaminobenzidine (which acts as an electron donor in photosystem I) the reaction product was restricted to the thylakoids, whereas the newly formed, peripherally located CIMs have been found to be entirely free of the reaction product.

5.2. Is There a «Point of No Return» in Chromoplast Development?

According to Schimper, plastids of any form are strictly *sui generis*. Schimper has also postulated that there was no principal restriction as to the particular kind of a plastid from which a certain other could arise. This means that chromoplasts can not only originate from any other form of a plastid, but also that the transformation to a chromoplast is not necessarily irreversible. Schimper himself reported some instances in which transformation of chromoplasts into chloroplasts was observed. Many such cases have been described explicitly in recent years (Thomson et al., 1967; Matienko et al., 1969; Grönegress, 1971, 1974; Devidé and Ljubešić, 1972, 1974; Wrischer, 1972; Sitte, 1974).

Yet Schimper's postulate of an unrestricted interchangeability of the different forms of plastids in a given plant did not remain unchallenged. Particularly Frey-Wyssling's group stressed the fact that a «monotropic» succession of proplastids, leucoplasts, chloroplasts, and finally chromoplasts is to be observed in the vast majority of cases in higher plants (Frey-Wyssling et al., 1955; Frey-Wyssling und Kreutzer, 1958a, Frey-Wyssling, 1959; Tôyama and Funazaki, 1971). In contradiction to Schimper's ideas, these authors postulated that the formation of chromoplasts in particular cannot be reverted, due to an extensive breakdown of internal structures which cannot be restored.

From Section 2 (p. 118) of this review it would seem to be clear that the idea of a « monotropic » transformation should apply primarily to gerontoplasts (see, however, Ljubešić, 1968) but not to chromoplasts *s. str.* Furthermore, the problem of *reversibility* of chromoplast transformation must not be confused with the question of actual *reversion:* In cases where reversion never occurs under « natural » conditions, reversibility still could possibly be demonstrated under certain experimental conditions. And finally, it may be expected that the present question is to be answered differently in different cases, and that no general solution of the problem can be given. That even true chromoplasts can behave as very delicate structures which may already disintegrate in a living cell has been reported in several instances (see, e.g., Steffen, 1964a; Nougarède, 1964; Lichtenthaler, 1969; Falk, 1976). Most impressive, rapid Brownian movement of plastoglobules can be observed in chromoplasts of living cells from petals of *Eranthis* (Falk, unpubl.) and *Eschscholtzia* (Sitte, unpubl.). This indicates a liquefaction of the stroma matrix. Indeed, how any internal structure (except the lipid globules) or the ptDNA could remain intact under such conditions is hardly conceivable. At least in these cases, a reversion to, e.g., chloroplasts seems to be excluded.

5.3. Controlled Multiplication of Chromoplasts

Several forms of plastids, particularly proplastids and chloroplasts, have been shown to multiply by fission (see Küster, 1956; Green, 1964; Stubbe, 1971). Gerontoplasts apparently do not divide, whereas chromoplasts are able to divide*, as has been demonstrated recently in a combined light and EM investigation (Sitte, 1976). It has been found that there is a strict correlation between cell size and the number of chromoplasts per cell in the petal lower epidermis during flower development in *Forsythia.* Although the cells enlarge dramatically, a mean of 5×10^4 chromoplasts per mm^2 of the lower epidermis is maintained throughout all developmental stages. Since chromoplast differentiation is, according to EM evidence, synchronous in these cells, it follows that chromoplast division must occur in them, and that chromoplast multiplication does not come about by chaotic fragmentation. Rather, this process is strictly correlated to cell enlargement, by some unknown regulating mechanism.

5.4. DNA in Chromoplasts

Plastids contain a limited amount of genetic information. This fact, first discovered by geneticists, has recently been corroborated by isolation and EM demonstration of circular DNA from pure chloroplast fractions. (Cell fractionation is an indispensable prerequisite since the buoyant density of ptDNA is only slightly higher than that of nuclear bulk DNA, and only insignificantly lower than that of mitochondrial

* According to Wrischer (1972), chromoplasts in young callus cells of carrot roots divide concomitantly with (and not only after) their transformation into proplastids.

DNA.) In higher plant chloroplasts, the « plastome » has been shown to consist of ring-shaped DNA molecules with a contour length of a little less than 45 μm, which corresponds to a mol. weight of about 92×10^6 or 130 kilobases. Since an average chloroplast contains some 10^{-14} g of DNA, each organelle is likely to contain about 50 copies of the plastome (for review, see Ellis and Hartley, 1974; Tewari et al. 1977; Boschetti, 1978).

Very recently, DNA has also been isolated from chromoplasts and characterized in terms of molecular size, hydrated density, melting temperature, and endonuclease fragmentation. After pioneering work of Kowallik and Herrmann (1972; Herrmann, 1972), further studies in chromoplast DNA of different sources (Falk et al. 1974; Liedvogel, 1976: *Narcissus pseudonarcissus;* Wuttke, 1976b: *Tulipa;* Thompson, 1980: *Tropaeolum*) have revealed the following results:

1) Chromoplast DNA is indistinguishable from chloroplast DNA in that it is circular and has an identical contour length and therefore the same mol. weight. Restriction endonuclease attack (Eco R I) leads to identical fragment patterns in both chloroplast and chromoplast DNA of *Narcissus pseudonarcissus* (Thompson, unpubl.).

2) In *Tropaeolum*, chromoplast as well as chloroplast DNA has a Tm of about 83 °C and a buoyant density of 1.696 g cm^{-3}. A molar G + C content of 35% can be calculated from these data.

Taken together, these findings strongly indicate that chromoplast differentiation is not accompanied by qualitative changes in ptDNA, or by selection of a specific DNA which differs from the bulk of plastome copies present in a chloroplast.

3) The number of plastome copies per average organelle is significantly lower in chromoplasts than in chloroplasts. Since the chromoplasts of the two investigated species are smaller than the corresponding chloroplasts, DNA content and the volume of the organelle possibly correlate with one another (Liedvogel, 1976).

The extent to which chromoplast DNA is transcribed is not yet known. However, the fact that fully developed chromoplasts still possess a number of undegraded copies of the plastome may provide a basis for reversibility of chromoplast transformation.

6. Phylogenetic Considerations

Unlike chromoplasts, *gerontoplasts* by definition are senescent plastids of premortal cells. Both the cells and their plastids are characterized by prevalence of catabolic processes, the physiological significance of which apparently lies in a retraction of nitrogen-containing compounds before leaf fall. In this connection, it may be noted that decomposition of chlorophylls and thylakoids as well as formation of xanthophyll esters and plastoglobules can be induced in green algae by « stress conditions », among them nitrogen deficiency (see, e.g., Czygan, 1968a, b; Kleinig und Wrischer, 1968; Mayer und Czygan, 1969; Sprey, 1970).

If the formation of gerontoplasts in mesophyll cells can in fact be envisaged as an adaptation to periodical leaf-shed, then it becomes understandable that the occurrence of gerontoplasts on a large scale is restricted to the spermatophytes. True *chromoplasts*, on the other hand, appear – as shown in Section 2 – in connection with the evolution of flowers and fruits, and the concomitant evolution of animal pollination (zoophily) and animal-mediated seed dispersal (zoochory). It is tempting to assume that flowering plants, after having «learned» to transform chloroplasts into yellow gerontoplasts, *secondarily* made use of this ability to attract animals by optical signals. According to this line of reasoning, the formation of true chromoplasts not in the vegetative but in the generative domain of the plant became a feature of growing importance in angiosperm phylogeny. In the course of these evolutionary events, the cell biology of this very same organelle was totally reverted: gerontoplasts are functionless remnants of formerly active devices for energy transformation, whereas in chromoplasts there is an amazing formation of new chemical and structural components which serve an important new function in propagation and the survival of the species.

Summary

Chromoplasts are yellow to red plastids, pigmented by – partly secondary – carotenoids. In spite of the obvious ecological, economical, and scientific importance of chloroplasts in higher plants, they still are insufficiently studied. A survey is presented on some recent findings and ideas in chromoplastology.

1) The classic term «chromoplast» which is based on pigmentation, comprises two physiologically and functionally different categories of organelles (Table I). Therefore, it is proposed to distinguish between anabolic, «true» chromoplasts and catabolic gerontoplasts (senescent chloroplasts).

2) Inside chromoplasts, the carotenoids are accumulated within different fine structural elements, namely, lipid globules, tubules or fibrils, membranes, and crystals. Accordingly, several chromoplast types can be defined, the globulous one being the commonest.

3) Since different tissues of the same plant often contain different types of chromoplasts, epigenetic factors seem to predominate in chromoplast differentiation.

4) Chromoplasts and their carotenoid-bearing substructures have been isolated and characterized biochemically (Table IV). With the exception of carotene crystals these substructures are lipoproteinic in nature, carotenoids constituting but a minor part.

5) Chromoplasts possess a remarkable diversity of biosynthetic capabilities, particularly in lipid metabolism.

6) The formation of chromoplasts is a monotropic process in certain cases, but reversible in others. Even full-grown chromoplasts contain undegraded copies of the plastome. They are able to divide.

7) In the evolution of flowering plants it is hypothesized that gerontoplasts appeared as an adaptation to annual leaf shedding, and that the ability to form true chromoplasts was achieved secondarily.

Supported by Sonderforschungsbereich 46.

Note Added in Proof

The manuscript for this article has been completed in the middle of 1978. Since then, several publications on chromoplasts appeared, some of which need brief consideration.

The fine structure of plastids in the leaves of *Euonymus japonica* has been investigated by T. Ikeda (Bot. Mag. Tokyo **92**, 23–30, 1979). The perennial leaves of this plant turn yellow during the winter and show regreening in the spring. Yellowing is accompanied by a complete breakdown of the thylakoid system and the appearance of numerous globules of moderate size (up to 170 per plastid). The appearance of these reversible »gerontoplasts» in resting, but not senescent, leaf cells resembles that of globulous flower chromoplasts.

Fine structure, development, and pigment content of diverse solanacean species has been further studied by the Lee group at Kensington (cf. D. J. Simpson, M. R. Baqar, T. H. Lee: Aust. J. Bot. **26**, 783–792, and 793–806, 1978). In fruit chromoplasts of *Solanum aviculare*, crystalloids of β-carotene grow out from plastoglobules which are formed previously and eventually disappear. A similar situation has been encountered in *S. pseudocapsicum* and in *Physalis alkekengi*, where however not only crystalloids but also «fibrils» (tubules) are observed. Consequently, these authors postulate the existence of two different types of plastoglobules, one type generating crystalloids and the other fibrils. – CPTA (see p. 126), known to inhibit β-carotene accumulation, prevents the formation of both the crystalloids and the fibrils. (For a recent study of CPTA action [and also the influence of elevated temperature] on carotenoid synthesis, compare M. R. Baqar and T. H. Lee: Z. Pflanzenphysiol. **88**, 431–435, 1978.) – The accumulation of certain carotenoids in dependence of distinctive genetic conditions in tomato lines has been investigated by W. P. Mohr (Ann. Bot. **44**, 427–434, 1979). The results indicate that lycopene forms long and narrow crystalloids, whereas β-carotene apparently remains dissolved in plastoglobules.

Sitte (work unpublished so far) succeeded in the experimental formation of nematic liquid crystals of diverse carotenoids. These fibrillar mesophases exhibit the same optical properties as bundles of chromoplast tubules do, among them an extraordinarily high positive anisotropy of up to 0.17. From these model experiments, and in accordance with the hypotheses mentioned on p. 130, it is postulated that chromoplast tubules contain a carotenoid core in the form of a nematic mesophase.

The biosynthetic capabilities of the membraneous chromoplasts of the daffodil have been studied further (B. Liedvogel and H. Kleinig: Planta **144**, 467–471, 1979). Isolated chromoplasts contain all enzymatic activities necessary for the synthesis of diacylgalactosylglycerol and diacylgalabiosylglycerol from acetate, HCO_3^-, *sn*-glycerol 3-phosphate, and UDP-D-galactose. However, the galactolipids mentioned are synthesized from these precursors to only a very low extent in an *in vitro* system. This is attributed to a specificity of diacylglycerol galactosyltransferase for highly unsatured diacylglycerols. Desaturation of saturated (or monounsaturated) fatty acids cannot be brought about by isolated chromoplasts. Intact cells of *Narcissus* coronae are however able to synthesize polyunsaturated fatty acids (H. Kleinig and B. Liedvogel: Planta **144**, 473–477, 1979).

References

BAIN, J. M., and F. V. MERCER, 1964: Organization resistance and the respiration climateric. Aust. J. Biol. Sci. **17**, 78–85.

BEN-SHAUL, Y., and S. KLEIN, 1965: Development and structure of carotene bodies in carrot roots. Bot. Gaz. **126**, 79–85.

—, and Y. NAFTALI, 1969: The development and ultrastructure of lycopene bodies in chromoplasts of *Lycopersicum esculentum*. Protoplasma **67**, 333–344.

—, T. TREFFRY, and S. KLEIN, 1968: Fine structure studies of carotene bodies development. J. Microscopie **7**, 265–274.

BERGER, C., und E. SCHNEPF, 1970: Entwicklung und Altern der Spadix-Appendices von *Sauromatum guttatum* Schott und *Arum maculatum* L. I. Veränderungen der Feinstruktur. Protoplasma **69**, 237–251.

BORNMAN, C. R., 1968: Observations of the ultrastructure of chromoplasts of *Celastrus scandens*. J. S. Afr. Bot. **34**, 295–301.

BOSCHETTI, A., 1978: Biogenese der Chloroplasten und Mitochondrien. G. Fischer, Stuttgart, New York.

BRITTON, G., 1976: Biosynthesis of carotenoids. In: Chemistry and Biochemistry of Plant Pigments (T. W. Goodwin, Ed.). 2nd Ed., Vol. **1**. pp. 262–327. Academic Press, New York, San Franscisco.

CAMEFORT, H., 1964: Evolution de la structure des plastes pendant la maturation de l'arille de l'If (*Taxus baccata* L.). C.R. Acad. Sci. (Paris) **258**, 1017–1020.

CATESSON, A.-M., 1970: Evolution des plastes du péricarp de pomme au cours de la maturation du fruit. Modifications ultrastructurale et accumulation de ferritine. J. Microscopie **9**, 949–974.

CHOE, H. T., and K. V. THIMANN, 1975: The metabolism of oat leaves during senescence. III. The senescence of isolated chloroplasts. Plant Physiol. **55**, 828–834.

COURCHET, M., 1888: Recherches sur les chromoleucites. Ann. Sci. Nat. Bot. Sér. VII, **7**, 263–374.

CZYGAN, F.-C., 1968a: Sekundär-Carotinoide in Grünalgen. I. Chemie, Vorkommen und Faktoren, welche die Bildung dieser Polyene beeinflussen. Arch. Mikrobiol. **61**, 81–102.

CZYGAN, F.-C., 1968b: Sekundär-Carotinoide in Grünalgen. II. Untersuchungen zur Biogenese. Arch. Mikrobiol. **62**, 209–236.

DEVIDÉ, Z., 1970: Ultrastructural changes of plastids in ripe fruit of *Cucurbita pepo* cv. ovifera. Acta Bot. Croat. **29**, 57–62.

—, and N. LJUBEŠIĆ, 1972: Plastid transformations in pumpkin fruits. Naturwissenschaften **59**, 39–40.

—, —, 1974: The reversion of chromoplasts to chloroplasts in pumpkin fruits. Z. Pflanzenphysiol. **73**, 296–306.

DOUCE, R., 1974: Site of biosynthesis of galactolipids in spinach chloroplasts. Science **183**, 852–853.

ELLIS, R. J., 1976: Protein and nucleic acid synthesis by chloroplasts. In: The Intact Chloroplast (J. BARBER, Ed.). pp. 335–364. Elsevier, Amsterdam, New York, Oxford.

—, and M. R. HARTLEY, 1974: Nucleic acids of chloroplasts. In: Biochemistry of Nucleic Acids (K. Burton, Ed.). MTP International Review of Science. Biochemistry Ser. 1, Vol. **6**, pp. 323–348. Butterworths, London.

FALK, H., 1960: Magnoglobuli in Chloroplasten von *Ficus elastica* Roxb. Planta **55**, 525–532.

—, 1976: Chromoplasts of *Tropaelum majus* L.: Structure and development. Planta **128**, 15–22.

FALK, H. B. LIEDVOGEL, and P. SITTE, 1974: Circular DNA in isolated chromoplasts. Z. Naturforsch. **29c**, 541–544.
FREY-WYSSLING, A., 1959: Submicroscopic morphogenesis of plastids. Symp. Submicroscop. Cell Morphol. IX. Int. Bot. Congr. Montreal 1959. In: Recent Adv. Bot. Vol. **I**, pp. 737–740. University Press, Toronto, London 1961.
—, und E. KREUTZER, 1958a: Die submikroskopische Entwicklung der Chromoplasten in den Blüten von *Ranunculus repens*. Planta **51**, 104–114.
—, —, 1958b: The submicroscopic development of chromoplasts in the fruit of *Capsicum annuum* L. J. Ultrastruct. Res. **1**, 397–411.
—, and F. SCHWEGLER, 1965: Ultrastructure of the chromoplasts in the carrot root. J. Ultrastruct. Res. **13**, 543–559.
—, F. RUCH, und X. BERGER, 1955: Monotrope Plastiden-Metamorphose. Protoplasma **45**, 97–114.
GOODWIN, T. W., 1971: Biosynthesis. In: Carotenoids (O. ISLER, Ed.). pp. 577–636. Birkhäuser, Basel, Stuttgart.
—, 1973: Carotenoids. In: Phytochemistry (L. P. MILLER, Ed.). Vol. I, pp. 112–142. Van Nostrand, New York.
GRANICK, S., 1961: The chloroplasts: Inheritance, structure and function. In: The Cell (J. BRACHET and A. E. MIRSKY, Eds.). pp. 489–602 Academic Press, New York, London.
GREEN, P. B., 1964: Cinematic observations on the growth and division of chloroplasts in *Nitella*. Am. J. Bot. **51**, 334–342.
GREENWOOD, A. D., R. M. LEECH, and J. P. WILLIAMS; J. L. BAILEY, and A. G. WHYBORN, 1963: The osmiophilic globules of chloroplasts. I., II. Biochim. Biophys. Acta **78**, 148–174.
GRILLI, M., 1965a: Origine e sviluppo dei cromoplasti nei frutti di Zucca americana (*Cucurbita pepo* L. cv. Small Sugar). I. Origine dei cromoplasti da amiloplasti. Caryologia **18**, 409–433.
—, 1965b: —. II. Origine dei cromoplasti da cloroplasti e da proplastidi. Caryologia **18**, 435–459.
—, 1965c: Ultrastructure e stadi involutivi di alcuni di tipi di cromoplasti. G. Bot. Ital. **72**, 83–92.
GRÖNEGRESS, P., 1971: The greening of chromoplasts in *Daucus carota* L. Planta **98**, 274–278.
—, 1974: The structure of chromoplasts and their conversion to chloroplasts. J. Microscopie **19**, 183–192.
GUNNING, B. E. S., and M. W. STEER, 1975: Ultrastructure and the Biology of Plant Cells. E. Arnold, London.
HARRIS, W. M., 1967: Ultrastructural development of chromoplasts in three isogenic lines of *Lycopersicon esculentum* Mill., Pearson. Am. J. Bot. **54**, 642.
—, 1970: Chromoplasts of tomato fruits. III. The high-delta tomato. Bot. Gaz. **131**, 163–166.
—, and A. R. SPURR, 1969a: Chromoplasts of tomato fruits. I. Ultrastructure of low pigment and high-beta mutants. Carotene analysis. Am. J. Bot. **56**, 369–379.
—, —, 1969b: —. II. The red tomato. Am. J. Bot. **56**, 380–389.
HERRMANN, R. G., 1972: Do chromoplasts contain DNA? II. The isolation and characterization of DNA from chromoplasts, chloroplasts, mitochondria, and nuclei of *Narcissus*. Protoplasma **74**, 7–17.
HILL, H. M., S. K. CALDERWOOD, and L. J. ROGERS, 1971: Conversion of lycopene to β-carotene by plastids isolated from higher plants. Phytochemistry **10**, 2051–2058.
ISLER, O. (Ed.), 1971: Carotenoids, Birkhäuser, Basel, Stuttgart.
JUNGALWALA, F. B., and J. W. PORTER, 1967: Biosynthesis of phytoene from isopentenyl

and farnesyl pyrophosphates by a partially purified tomato enzyme system. Arch. Biochem. Biophys. **119**, 209–213.

KHUDAIRI, A. K., and O. P. ARBOLEDA, 1970: Phytochrome-mediated carotenoid biosynthesis and its influence by plant hormones. Physiol. Plant. **26**, 285–288.

KIRK, J. T. O., and B. E. JUNIPER, 1967: The ultrastructure of the chromoplasts of different colour varieties of *Capsicum*. In: Biochemistry of Chloroplasts (T. W. Goodwin, Ed.). Vol. **2**, pp. 691–701. Academic Press, London, New York.

—, and R. A. E. TILNEY-BASSETT, 1967: The Plastids. Freeman, London, San Francisco.

KLEINIG, H., and B. LIEDVOGEL, 1978: Fatty acid synthesis by isolated chromoplasts from the daffodil. [^{14}C]Acetate incorporation and distribution of labelled acids. Eur. J. Biochem. **83**, 499–505.

—, und M. WRISCHER, 1968: Die Feinstruktur von *Acetabularia*-Chloroplasten bei Sekundärcarotinoid-Bildung. Z. Pflanzenphysiol. **58**, 248–251.

KOLLMANN, R., H. KLEINIG, and I. DÖRR, 1969: Fine structure and pigments of plastids in *Orobanche*. Cytobiol. **1**, 125–128.

KOWALLIK, K. V., and R. G. HERRMANN, 1972: Do chromoplasts contain DNA? I. Electron-microscopic investigation of *Narcissus* chromoplasts. Protoplasma **74**, 1–6.

KUHN, H., 1970: Chemismus, Struktur und Entstehung der Carotinkriställchen in der Nebenkrone von *Narcissus poeticus* L. var. «La Riante». J. Ultrastruct. Res. **33**, 332–355.

—, A. FREY-WYSSLING, und P. STRICKLER, 1969: Über die Struktur pflanzlicher Karotinkriställchen. Experientia **25**, 972–973.

KUSHWAHA, S. C., C. SUBBARAYAN, D. E. BEELER, and J. W. PORTER, 1969: The conversion of lycopene-15,15′-^{3}H to cyclic carotenes by soluble extracts of higher plant plastids. J. Biol. Chem. **244**, 3635–3642.

KÜSTER, E., 1956: Die Pflanzenzelle. 3. Aufl. G. Fischer, Jena.

LABORDE, J. A., and A. R. SPURR, 1973: Chromoplast ultrastructure as affected by genes controlling grana retention and carotenoids in fruits of *Capsicum annuum*. Am. J. Bot. **60**, 736–744.

LAMARCK, J. P. B., 1778: Flore Française. Impr. royale, Paris.

LANCE-NOUGARÈDE, A., 1960: Developpement inframicroscopique des chromoplastes, au cours de l'ontogenèse des fleures ligulées de *Chrysanthemum segetum*. C. R. Acad. Sci. (Paris) **250**, 173–175.

LAVAL-MARTIN, D., 1974: La maturation du fruit de tomate cerise: Mise en évidence, par cryodécapage, de l'évolution des chloroplastes en deux types de chromoplastes. Protoplasma **82**, 33–59.

LEDBETTER, M. C., and K. R. PORTER, 1970: Introduction to the Fine Structure of Plant Cells. Springer, Berlin, Heidelberg, New York.

LEECH, R. M., and D. J. MURPHY, 1976: The cooperative function of chloroplasts in the biosynthesis of small molecules. In: The Intact Chloroplast (J. Barber, Ed.). pp. 365–401. Elsevier, Amsterdam, New York, Oxford.

LICHTENTHALER, H. K., 1969: Zur Synthese der lipophilen Plastidenchinone und Sekundärcarotinoide während der Chromoplastenentwicklung. Ber. Dt. Bot. Ges. **82**, 483–497.

—, 1970a: Die Lokalisation der Plastidenchinone und Carotinoide in den Chromoplasten der Petalen von *Sarothamnus scoparius* (L.) Wimm. ex Koch. Planta **90**, 142–152.

—, 1970b: Die Feinstruktur der Chromoplasten in plasmochromen Perigon-Blättern von *Tulipa*. Planta **93**, 143–151.

—, und B. SPREY, 1966: Über die osmiophilen globulären Lipideinschlüsse der Chloroplasten. Z. Naturforsch. **21b**, 690–697.

LIEDVOGEL, B., 1976: DNA content and ploidy of chromoplasts. Naturwissenschaften **63**, 248.
—, and H. KLEINIG, 1976: Galactolipid synthesis in chromoplast internal membranes of the daffodil. Planta **129**, 19–21.
—, —, 1977: Lipid metabolism in chromoplasts from the daffodil: glycosylation and acylation. Planta **133**, 249–253.
—, and P. SITTE, 1974: Lipids and proteins in lipid-rich chromoplast membranes. Naturwissenschaften **61**, 131.
—, —, and H. FALK, 1976: Chromoplasts in the daffodil: Fine structure and chemistry. Cytobiol. **12**, 155–174.
—, H. KLEINIG, J. A. THOMPSON, and H. FALK, 1978: Chromoplasts of *Tropaeolum majus* L.: Lipid synthesis in whole organelles and subfractions. Planta **141**, 303–309.
LINSBAUER, K. (Hrsg.), 1917: C. K. Schneiders illustriertes Handwörterbuch der Botanik. 2. Aufl. Engelmann, Leipzig.
LJUBEŠIĆ, N., 1968: Feinbau der Chloroplasten während der Vergilbung und Wiederergrünung der Blätter. Protoplasma **66**, 369–379.
—, 1972: Ultrastructural changes of plastids during the yellowing of the fruit of *Cucurbita pepo* var. pyriformis. Acta Bot. Croat. **31**, 47–53.
—, 1973: Transformations of plastids in white pumpkin fruits. Acta Bot. Croat. **32**, 59–62.
—, 1978: Chromoplasts in the petals of *Liriodendron tulipifera* L. Z. Pflanzenphysiol. **91**, 49–52.
MARZIANI, G., and M. ORSENIGO, 1965: Observations on the fine structure of the plastids in a «lycopenic» mutant of maize. Atti V. Congr. Ital. Microsc. Elletr., Bologna 1965. pp. 158–161.
MATIENKO, B. T., 1964: Organisation inframicroscopique des chromoplastes des Cucurbitacées et classification morphologique des chromoplastes. III. Eur. Reg. Conf. Electron Microscopy, Prague 1964. pp. 153–154.
—, S. M. SALINSKIJ, i V. K. SOLOVEJ, 1969: O vozmozhnosti prevrashchenija karotinoidoplastov (khromoplastov) v khloroplasty v epidermise ploda arbuza. In: Khloroplasty i Mitokhondrii (Voprosy Membranoj Biologii). Izd. Akad. Nauk SSSR. Moskva. pp. 190–198.
MATILE, P., 1975: The Lytic Compartment of Plant Cells. Springer, Wien, New York.
MAYER, F., und F.-C. CZYGAN, 1969: Änderungen der Ultrastrukturen in den Grünalgen *Ankistrodesmus braunii* und *Chlorella fusca* var. *rubescens* bei Stickstoffmangel. Planta (Berl.) **86**, 175–185.
MESQUITA, J. F., 1976: La différenciation des plastes dans les fleurs de *Narcissus* L. I. Modifications ultrastructurales et pigmentaires pendant la morphogénèse des chromoplastes chez *N. bulbocodium* L. Rev. Biol. (Lisbon) **10**, (4) 127–150.
MEYER, A., 1883a: Über Krystalloide der Trophoplasten und über die Chromoplasten der Angiospermen. Bot. Ztg. **41**, 489–498, 505–514, 525–531.
—, 1883b: Das Chlorophyllkorn in chemischer, morphologischer und biologischer Beziehung. A. Felix, Leipzig.
MOHR, W. P.. and M. STEIN, 1969: Fine structure of fruit development in tomato. Can. J. Plant Sci. **49**, 549–553.
MOLLENHAUER, H. H., and C. KOGUT, 1968: Chromoplast development in Daffodil. J. Microscopie **7**, 1045–1050.
NICHOLS, B. W., J. M. STUBBS, and A. T. JAMES, 1967: The lipid composition and ultrastructure of normal developing and degenerating chloroplasts. In: Symp. Biochem. Chloropl. (T. W. Goodwin, Ed.). Vol. **2**, pp. 677–690. Academic Press, London.

NIELSEN, O. F., 1974: Macromolecular physiology of plastids. XII. *Tigrina* mutants in barley: genetic, spectroscopic and structural characterization. Hereditas **76**, 269–304.

NOUGARÈDE, A., 1964: Evolution infrastructurale des chromoplastes au cours de l'ontogenèse des pétales chez le *Spartium junceum* L. (Papilionacées). C. R. Acad. Sci. (Paris) **258**, 683–685.

OSUMI, M., 1961: Electron microscopical studies on chromoplasts. I. The ultrastructure of chromoplasts in oranges. Bot. Mag. **74**, 165–168.

PURCELL, A. E., J. H. CARRA, and I. V. DEGRUY, 1963: The development of chromoplasts and carotenoids in colored grapefruit. Rio Grande Valley Hort. Soc. J. **17**, 123–127.

ROBERTS, E. A., and M. D. SOUTHWICK, 1948: Contribution of studies with the electron microscope to studies of the relationship of chromoplasts to carotene bodies and carotene bodies to vitamin A. Plant Physiol. **23**, 621–633.

ROSSO, S. W., 1967a: Studies on tomato fruit plastids. Am. J. Bot. **54**, 637.

—, 1967b: An ultrastructural study of the mature chromoplasts of the tangerine tomato (*Lycopersicon esculentum* var. «Golden Jubilee»). J. Ultrastruct. Res. **20**, 179–189.

—, 1968: The ultrastructure of chromoplast development in red tomatoes. J. Ultrastruct. Res. **25**, 307–322.

ROTHERT, W., 1912: Über Chromoplasten in vegetativen Organen. Bull. Acad. Sci. Cracovie, Sc. math. et nat. Sér. B, 189–336.

SALEMA, R., 1969: Amido – Estudo ultrastrutural da sua biogénese em plantas superiores. Brotéria, Sér. Ciênc. Nat. **38**, 1–126.

SAVELLI, R., 1937: Sur la distribution du carotinoide rouge dans les chlorochromoplastes. Protoplasma **29**, 601–607.

SAVIDGE, J. P., 1976: The angiosperm flower and related structures. In: Plant Structure, Function and Adaptation (M. A. Hall. Ed.). pp. 326–378. MacMillan, London.

SCHÄFER, E., 1971: Biochemische und strukturelle Veränderungen der Plastiden während der Umwandlung von Chloroplasten in Chromoplasten bei *Capsicum annuum*. Diplomarbeit, Darmstadt.

SCHIMPER, A. F. W., 1883: Über die Entwicklung der Chlorophyllkörner und Farbkörper. Bot. Ztg. **41**, 105–112, 121–131, 137–146, 153–163, 809–817.

—, 1885: Untersuchungen über die Chlorophyllkörper und die ihnen homologen Gebilde. Jahrb. Wiss. Bot. **16**, 1–247.

SCHMITZ, F., 1883: Die Chromatophoren der Algen. Verh. naturhist. Ver. preuss. Rheinlande u. Westf. Bonn, **40**, 1–180.

SCHNEPF, E., und F.-C. CZYGAN, 1966: Feinbau und Carotinoide von Chromoplasten im Spadix-Appendix von *Typhonium* und *Arum*. Z. Pflanzenphysiol. **54**, 345–355.

—, und W. NAGL, 1970: Über einige Strukturbesonderheiten der Suspensorzellen von *Phaseolus vulgaris*. Protoplasma **69**, 133–143.

SCHÜRHOFF, P. N., 1924: Die Plastiden. In: Handb. der Pflanzenanatomie (K. Linsbauer, Hrsg.) I. Abtg., 1. Teil. Borntraeger, Berlin.

SEYBOLD, A., 1954: Untersuchungen über den Farbwechsel von Blumenblättern, Früchten und Samenschalen. Sitzgsber. Heidelb. Akad. Wiss., Jahrg. 1953/54, 2. Abh., pp. 31–124. Springer, Berlin, Göttingen, Heidelberg.

SHARP, L. W., 1926: Introduction to Cytology. 2nd Ed. McGraw-Hill, New York.

SIMPSON, D. J., 1976: Aspects of plastid development. Ph. D. Thesis. Dept. Food Technol., Univ. of New South Wales. Kensington, Australia.

—, and T. H. LEE, 1974: Ultrastructure and formation of fibrils in chromoplasts. 8th Internat. Congr. Electron Microsc., Canberra. Vol. II, p. 626.

SIMPSON, D. J., and T. H. LEE, 1976: The fine structure and formation of fibrils of *Capsicum annuum* L. chromoplasts. Z. Pflanzenphysiol. **77**, 127–138.

—, —, 1977: Chromoplast ultrastructure of *Capsicum* carotenoid mutants. II. Effect of light and CPTA. Z. Pflanzenphysiol. **83**, 309–325.

—, M. R. BAQAR, and T. H. LEE, 1975a: Unusual ultrastructural features of the chloroplast-chromoplast transformation in *Solanum luteum* fruit. Aust. J. Plant Physiol. **2**, 235–245.

—, —, —, 1975b: Ultrastructure and carotenoid composition of chromoplasts of the sepals of *Strelitzia reginae* Aiton during floral development. Ann. Bot. **39**, 175–183.

—, —, —, 1977a: Chemical regulation of plastid development. III. Effect of light an CPTA on chromoplast ultrastructure and carotenoids of *Capsicum annuum*. Z. Pflanzenphysiol. **82**, 189–209.

—, —, —, 1977b: Chromoplast ultrastructure of *Capsicum* carotenoid mutants. I. Ultrastructure and carotenoid composition of a new mutant. Z. Pflanzenphysiol. **83**, 293–308.

—, —, —, 1977c: Fine structure and carotenoid composition of the fibrillar chromoplasts of *Asparagus officinalis* L. Ann. Bot. **41**, 1101–1108.

—, F. M. M. RAHMAN, K. A. BUCKLE, and T. H. LEE, 1974: Chemical regulation of plastid development. II. The effect of CPTA on the ultrastructure and carotenoid composition of chromoplasts of *Capsicum annuum* cultivars. Aust. J. Plant Physiol. **1**, 135–147.

SIREVÅG, R., and R. P. LEVINE, 1973: Transcription and translation for carotenoid synthesis in *Chlamydomonas reinhardtii*. Planta **111**, 73–84.

SITTE, P., 1974: Plastiden-Metamorphose und Chromoplasten bei *Chrysosplenium*. Z. Pflanzenphysiol. **73**, 243–265.

—, 1976: Jahresber. Sonderforschungsbereich «Molekulare Grundlagen der Entwicklung». pp. 134–135. Freiburg i. Br.

—, 1977a: Chromoplasten – bunte Objekte der modernen Zellbiologie. Biol. in unserer Zeit **7**, 65–74.

—, 1977b: Functional organization of biomembranes. In: Lipids and Lipid Polymers in Higher Plants (M. Tevini and H. K. Lichtenthaler, Eds.). pp. 1–28. Springer, Berlin, Heidelberg, New York.

—, 1978: Die lebende Zelle als System, Systemelement und Übersystem. Nova Acta Leopoldina **47**, Nr. 226, 195–216.

SMITH, L. L. W., and O. SMITH, 1931: Light and the carotenoid content of certain fruits and vegetables. Plant Physiol. **6**, 265–275.

SMITH, M., and R. D. BUTLER, 1971: Ultrastructural aspects of petal development in *Cucumis sativus* with particular reference to the chromoplasts. Protoplasma **73**, 1–13.

SPREY, B., 1970: Die Lokalisierung von Sekundärcarotinoiden von *Haematococcus pluvialis* Flotow em. Wille. Protoplasma **71**, 235–250.

SPURR, A. R., and W. M. HARRIS, 1968: Ultrastructure of chloroplasts and chromoplasts in *Capsicum annum*. I. Thylakoid membrane changes during fruit ripening. Am. J. Bot. **55**, 1210–1224.

STEFFEN, K., 1964a: Chromoplastenstudien. I. Der amöboide Chromoplastentyp. Planta **60**, 506–522.

—, 1964b: —. II. Die Ontogenese der Chromoplasten in den Beeren von *Convallaria maialis* L. Protoplasma **58**, 579–588.

—, und F. WALTER, 1955: Die submikroskopische Struktur der Chromoplasten. Naturwissenschaften **42**, 395–396.

STEFFEN, K., und F. WALTER, 1958: Die Chromoplasten von *Solanum capsicastrum* L. und ihre Genese. Planta **50**, 640–670.

STRASBURGER, E., 1884: Das botanische Practicum. G. Fischer, Jena.

STRAUB, O., 1976: Key to Carotenoids. Birkhäuser, Basel, Stuttgart.

STRAUS, W., 1953: Chromoplasts – development of crystalline forms, structure, state of pigments. Bot. Rev. **19**, 147–186.

STUBBE, W., 1971: Origin and continuity of plastids. In: Origin and Continuity of Cell Organelles (J. Reinert and H. Ursprung, Eds.). pp. 65–81. Springer, Berlin, Heidelberg, New York.

STUMPF, P. K., 1976: Membrane-bound enzymes in plant lipid metabolism. In: The Enzymes of Biological Membranes (A. Martonosi, Ed.). Vol. **2**, pp. 145–159. John Wiley, London, etc.

SUZUKI, S., 1974: Ultrastructural development of plastids in cherry peppers during fruit ripening. Bot. Mag. **87**, 165–178.

TEWARI, K. K., R. KOLODNER, N. M. CHU, and R. R. MEEKER, 1977: Structure of chloroplast DNA. In: Nucleic Acids and Protein Synthesis in Plants (L. Bogorad and J. H. Weil, Eds.). pp. 15–36. Plenum Press, New York, London.

THOMAS, G., and D. R. THRELFALL, 1974: Synthesis of polyprenyltoluquinols from homogentisate and polyprenyl pyrophosphates in particular fractions of *Euglena* and sugar beet. Biochem. J. **142**, 437–440.

THOMPSON, J. A., 1980: Isolation and characterization of DNA from different plastid types of *Tropaeolum majus*. Eur. J. Cell Biol. **21**, 37–42.

THOMSON, W. W., 1966: Ultrastructural development of chromoplasts in Valencia oranges. Bot. Gaz. **127**, 133–139.

–, 1969: Ultrastructural studies on the epicarp of ripening oranges. Proc. 1st Internat. Citrus Symp. Vol. **3**, 1163–1169.

–, and K. PLATT, 1973: Plastid ultrastructure in the barrel cactus, *Echinocactus acanthodes*. New Phytol. **72**, 791–797.

–, L. N. LEWIS, and C. W. COGGINS, 1967: The reversion of chromoplasts to chloroplasts in Valencia oranges. Cytologia **32**, 117–124.

TÔYAMA, S., 1967: Electron microscope studies on the morphogenesis of plastids. IV. Fine structure development of chromoplasts in the root of *Daucus carota* L. Sci. Rep., Tokyo Kyoiku Dagaiku, B, **12**, 245–253.

–, and K. FUNAZAKI, 1971: Electron microscope studies on the morphogenesis of plastids. V. Concerning one-dimensional metamorphosis of the plastids in *Cryptomeria* leaves. Bot. Mag. **84**, 123–136.

TRÉCUL, A., 1858: Des formations vesiculaires dans les cellules végétales. Ann. Sci. Nat., Bot. Sér. **4**, 10, 20–74, 127–163, 205–382.

TREFFRY, T., Y. BEN-SHAUL, and S. KLEIN, 1966: The origin, development and structure of carotene bodies in the cell of carrot root. 6th Internat. Congr. Electron Microsc. Kyoto. Vol. **2**, 337–338.

WALLES, B., and J. HUDÁK, 1975: Etioplast and chromoplast development in the lycopenic mutant of maize. J. submicr. Cytol. **7**, 325–334.

WILSON, E. B., 1896: The Cell in Development and Inheritance. MacMillan, New York, London.

WINKENBACH, F., H. FALK, B. LIEDVOGEL, and P. SITTE, 1976: Chromoplasts of *Tropaeolum majus* L.: Isolation and characterization of lipoprotein elements. Planta **128**, 23–28.

WRISCHER, M., 1972: Transformation of plastids in young carrot callus. Acta Bot. Croat. **31**, 41–46.

WUTTKE, H.-G., 1976a: Chromoplasts in *Rosa rugosa*: Development and chemical characterization of tubular elements. Z. Naturforsch. **31c**, 456–460.

—, 1976b: Circular DNA in chromoplasts of *Tulipa gesneriana*. Planta **132**, 317–319.

—, 1977: Untersuchungen zur Feinstruktur und Biochemie globulöser Chromoplasten. Thesis, Univ. of Freiburg i. Br.

Terpenoid, Especially Diterpenoid Pigments

Conrad Hans Eugster

Organisch-chemisches Institut der Universität Zürich

1. Summary

A review is presented on the occurrence and distribution in plants of terpenoid pigments (mono-, sesqui-, di-, tri- and tetraterpenes). The following chromophoric systems are known:

- linearly conjugated polyenes [only in carotenoids (tetraterpenes)];
- *p*-quinomethanes (« quinone methides »);
- fulvenes and azulenes;
- cyclic (cis)-α-diketones;
- acyl hydroquinones.

The largest group comprises carotenoids, the second largest being that of the diterpenoid pigments as a result of many new additions. Nearly all the structures of the latter group can be derived from the basic skeletons of abietane and totarane. They are classified as follows:

a) Tanshinones (10 compounds) – nor- or bisnor-abietanes with rings B and C as *o*- or *p*-naphthaquinone. They have been found exclusively in the rootbark of a few Labiatae.

b) Royleanones (15 compounds) – abietanes with quinonoid ring C. They are mostly found in the rootbark of Labiatae; a few cases of occurrence in other plant families are also known.

c) Coleones (ca. 75 compounds) – modified abietanes or totaranes comprising various chromophoric systems. They are present in special glands on the leaves and racemes of *Coleus-*, *Plectranthus-*, *Solenostemon-* and *Fuerstia*-species (Labiatae). Some of them are very labile in the pure isolated form. We assume that the cells containing them must be essentially free of any electrophiles and nucleophiles because they occur in the glands in astonishingly high concentration and can be kept unchanged over a long time. Many coleones are highly oxygenated and dehydrogenated and only a slight structural change is required to convert them into colorless aromatic compounds which have not yet been detected. In our opinion, the coleones represent a series of otherwise inaccessible intermediates of terpene catabolism. Dehydrogenation and oxygenation obviously starts with ring C or maybe in the side chains, goes on to ring B, and finally reaches ring A.

2. Introduction

Terpenoid pigments, in the comprehensive manner in which they are presented in this survey, should be understood as only those pigments whose biosynthesis involves mevalonic acid – either proven, or strongly suspected*. Prenylated polyketides, prephenic acid derivatives, etc., are not dealt with. Thus, the pigments treated here are always terpene skeletons which have been transformed into pigments by processes of dehydrogenation and oxygenation. From the chemical-structural standpoint there is no compelling reason to treat colored substances separately from colorless substances, merely because the former absorb in the visible region of the spectrum. This point will be made clear later, in examination of the quinone/hydroquinone pair since both compounds are related to one another by a relatively simple redox process. From the physiological standpoint, the division into colored and colorless compounds might be more significant; since the ability to oxidize (dehydrogenate and [or] oxygenate) compounds up to the pigment stage and to store the pigments in a high concentration is probably a characteristic property of the respective organism.

The ability to convert colorless precursors into pigments is particularly developed in plants in the case of tetraterpenes (i.e., carotenoids). Pigments in other terpene groups (mono-, sesqui-, di-, and triterpenes) have only been found in more recent time.

3. The More Important Chromophoric Systems

It is helpful to distinguish between the chromophoric systems responsible for selective light absorption, since such systems often show characteristic absorption curves which can be used to deduce partial structures. The most important of these selective chromophoric systems are:

- Linearly conjugated polyene systems. These can only occur in carotenoids. The transition from the UV to the visible range occurs at about seven conjugated double bonds. The absorption can be greatly altered by substituents and steric effects.
- Quinones, namely, *p*-benzoquinones(*1*), *p*-(*2*), and *o*-naphthaquinones(*3*), which in many cases contain an isopropyl group or the structural equivalent thereof (e.g., in a β-methylfuran).
- Extended quinones (*4*) in which the quinoid system extends over two or more rings.
- Fulvenes (*5*).
- Quinomethanes (*6*).
- Cyclic (*cis*)-α-diketones (*7*).
- Acyl hydroquinones (*8*).
- Azulenes (*9*).

The chromophores can also occur in combined form. They can be influenced greatly by steric and electronic effects. Compounds with especially strong absorption at long

* Proof of the biogenesis from mevalonic acid has only been verified by incorporation studies in a few cases. Placement into the class of terpenoids is often made without difficulty on the basis of the carbon skeleton or analysis of the accompanying substances in the same plant. Whenever this was not the case, the compound in question was not included in this discussion.

wavelengths are all-*trans*-polyenes of the carotenoid type (e.g., *48*), and *p*-quinomethanes (*6*), as well as certain *o*-quinones (*3*). The color of the compounds found to date extends from bright, green-tinted yellow to deep violet red (e.g., *36*)*.

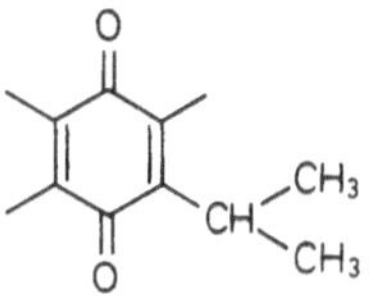

1

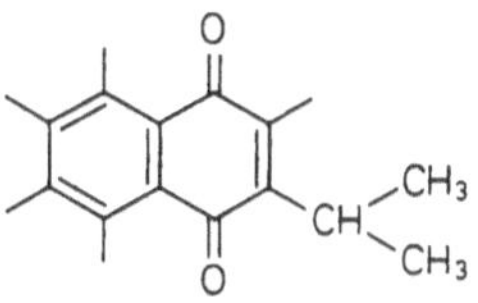

2

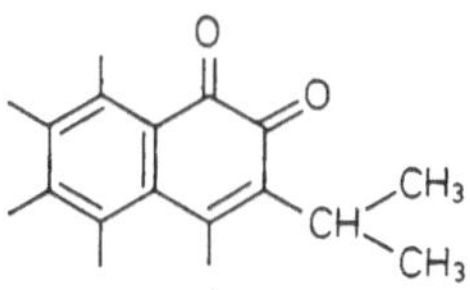

3

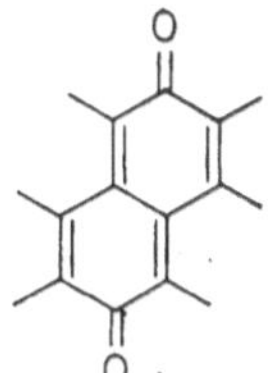

4

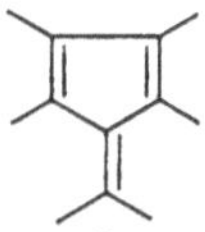

5

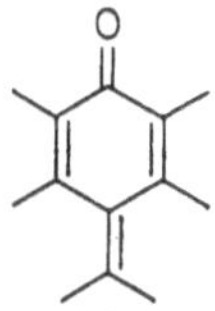

6

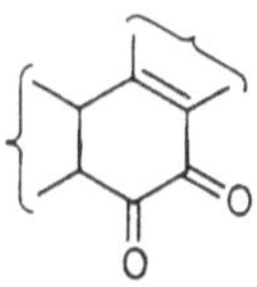

7

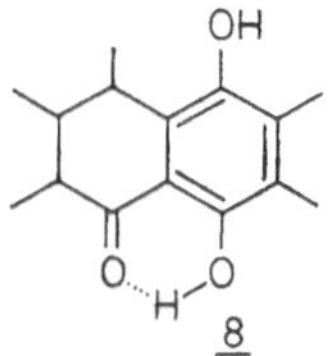

8

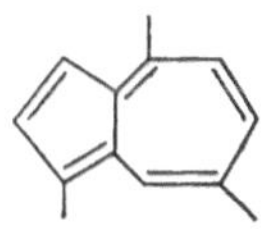

9

4. Known Terpene Structures**

4.1. Monoterpenes

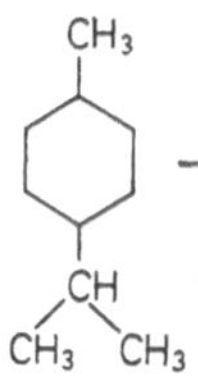

10

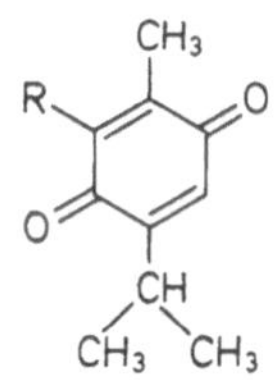

11

R = H, OH

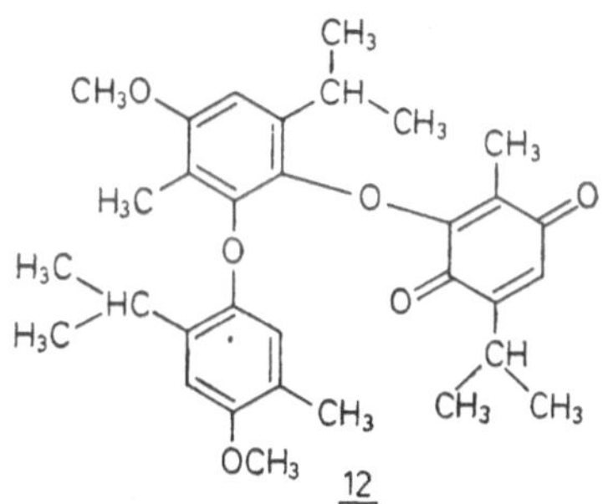

12

Few quinones have been found. All of them are derived from *p*-menthane (*10*), as, for example, the yellow thymoquinone (*11*, R=H) which occurs in essential oils of umbellifers and labiates, and in the seeds, leaves, and needles of ranunculaceans, composites, and conifers. This compound is perhaps produced in a few cases during the isolation by oxidation of the hydroquinone. The monohydroxy compound *11* (R = OH, red needles) has also been isolated (from the wood of *Juniperus chinensis*), but it is not yet certain whether this is a product of the host or its fungal parasite.

* A few of the tropolones from higher plants just reach the visible range of the spectrum (nootkatinol, α-thujaplicinol, utahin). They have not been included in this survey.

** Literature citations of compounds appearing in the comprehensive works of Thomson [1], Hegnauer [2], and Karrer [3] will not be given in the following text.

Moreover, in the heartwood of *Libocedrus decurrens* a red trimer of structure *12* has been found. It obviously has been formed from the monomer by oxidative phenol couplings.

4.2. Sesquiterpenes

A modified eremophilane skeleton is attributed to both yellow naphthaquinones maturone (*14*, R = OH) and maturinone (*14*, R = H). Its relation to the hypothetical precursor *13* is confirmed in this case by isolation of several accompanying substances which still carry the methyl, methylene, or aldehyde group*. On oxidation to the

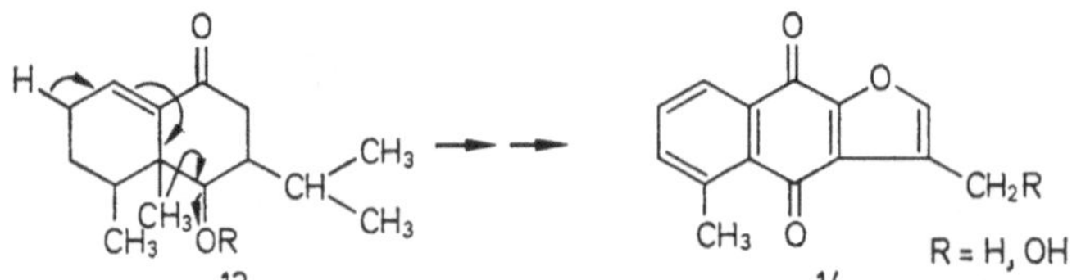

13 14

quinone this C atom is lost. The isopropyl group is also oxidized and transformed into a β-alkylated furan. Maturone and maturinone have been found in the roots of *Cacalia decomposita*, and recently sesquiterpenes with an unaltered eremophilane skeleton have been found in in *C. delphiniifolia* [4].

An unaltered eremophilane skeleton is found in the yellow warburgiadione (*15*) and in the bright yellow warburgin (*16*) from the heartwood of *Warburgia ugandensis* [5].

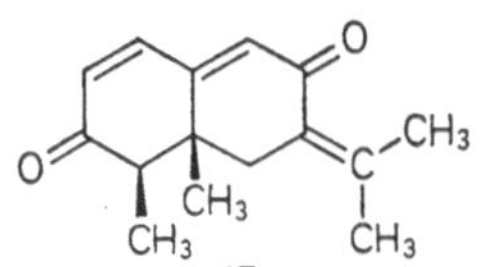

15

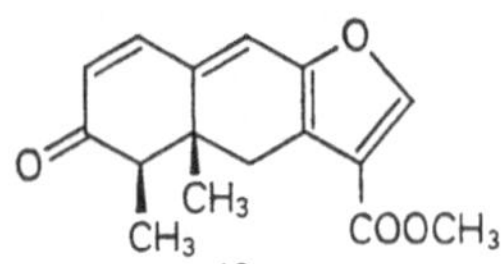

16

In the case of *15*, the absorption in the visible range is due to the presence of the conjugated diene-dione chromophore and in *16* to the extended conjugation between the carbonyl group and the substituted furan ring.

The rarely seen cuparane skeleton (*17*) is found in the yellow helicobasidin (*18*), a metabolite of *Helicobasidium mompa*. Similar compounds have also been isolated from cultures of *Coprinus lagopus* (lagopodin A, lagopodin B).

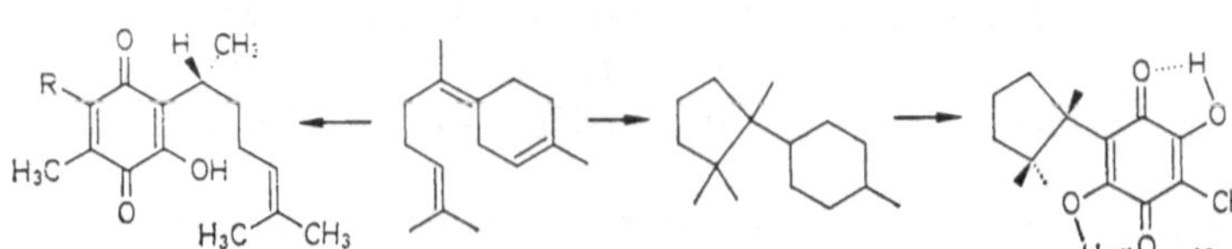

20 19 17 18

---

* See p. 151**

It has long been known that the mild laxative « Aurum vegetabile » prepared from the roots of Mexican composites contains quinones. The structures of perezone (*20*, R = H) and hydroxyperezone (*20*, R = OH), two orange red *p*-benzoquinones, were only established a few years ago. Biogenetically, *18* and *20* probably originate from a bisabolene (e.g., *19*).

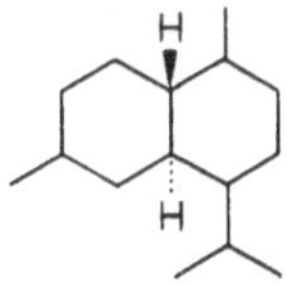

21

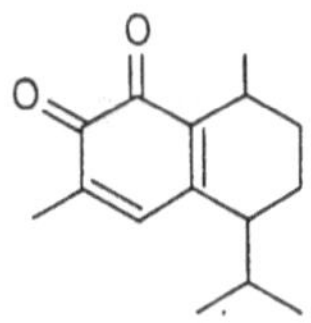

A
22

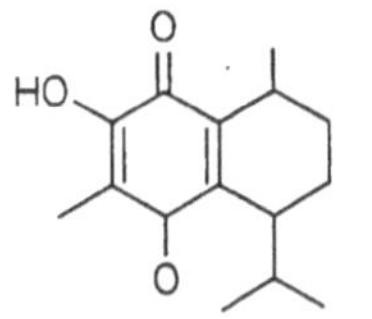

B
23

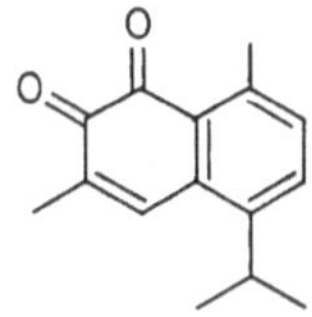

C
24

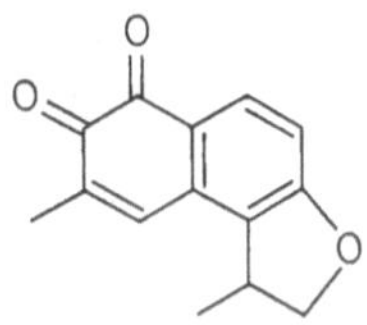

D
25

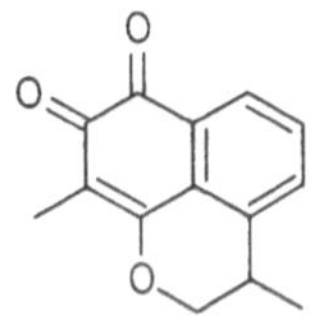

E
26

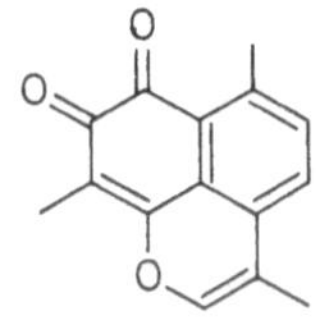

F
27

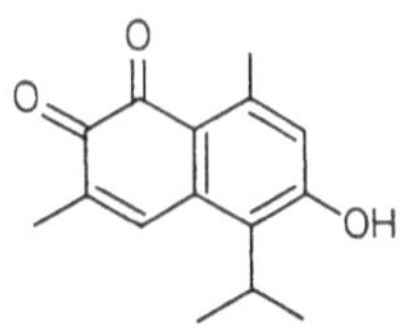

G
28

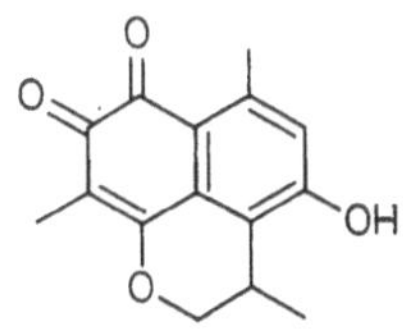

H
29

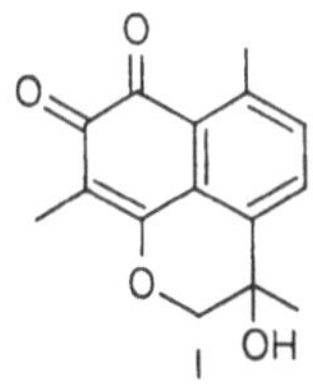

I
30

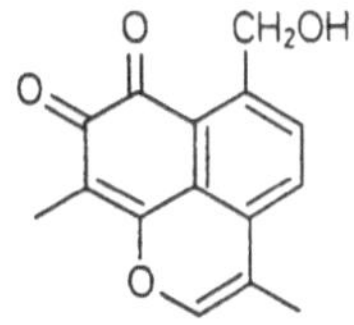

L
31

A series of colored compounds are derived from cadinane (*21*): both the mansonones from the heartwood of *Mansonia altissima* and *para*-hemigossypolone, and gossypol and the heliocides from *Bombax* and *Gossypium* species.

The structures of mansonone A, B, C, D, E, F, G, H, I, and L (*22* to *31*) show a yellow *p*-benzoquinone (*23*), orange-to-red *o*-naphthaquinones (*22*, *24* to *26*, *28* to *30*), and the violet *o*-quinones *27* and *31* (further conjugation with the oxaphenalene system). Mansonone C (*24*) is also found in the wood of numerous elm species.

The mansonones are an excellent example of the structural modifications which a parent compound in a biological system can undergo and at the same time be transformed into many new products with altered properties.

Gossypol (*32*), a highly unique yellow or brick red pigment, has been the subject of many investigations due to its toxicity and occurence in economically important plants. It is accumulated in special excretion cavities of the roots, stem, blossoms, fruit, and seeds of many malvaceans (*Gossypium*, *Cienfuegosia*, *Thespesia*, and *Kokia* species)*.

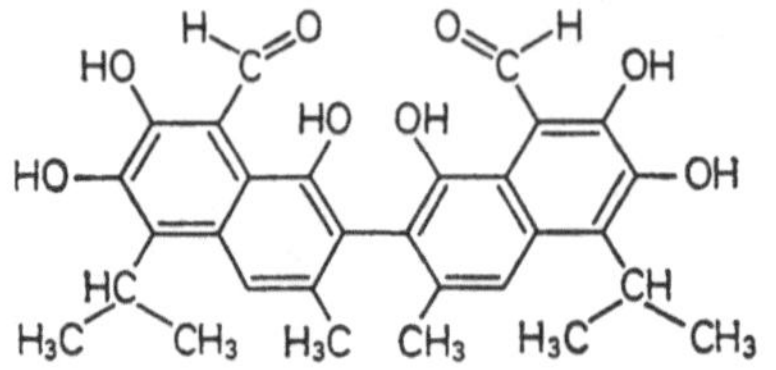

32

The color cannot be easily interpreted since the prevalent tautomeric form is not yet known. The structure involves a dimeric and fully aromatized cadinane skeleton presumably formed from suitable precursors by so-called phenol oxidation. In this process radicals are formed, the dimerization of which leads to a new C—C bond. Such precursors or derivatives thereof have been found recently in a search for natural inhibitors of the parasites *Heliothis virescens* and *H. zeae* in resistent *Gossypium* species. *Para*-hemigossypolone (*33*, R = H) and 6-O-methyl-*para*-hemigossypolone (*33*, R = CH_3) were isolated from the root bark of *Bombax malabaricum* [6]; and *33* (R = H) and the so-called heliocides H_1, H_2 (*34*), and H_3 (*35*) were isolated from *Gossypium hirsutum* (only from very young leaflets and flower buds) [7, 8]. The heliocides B_1, B_2, and B_3 whose structure is still unknown were provided by *G. barbadense* [8]. The color properties of *33*, *34*, and *35* are not given in [7] and [8]; presumably *33* is bright red and *34* and *35* are bright yellow.

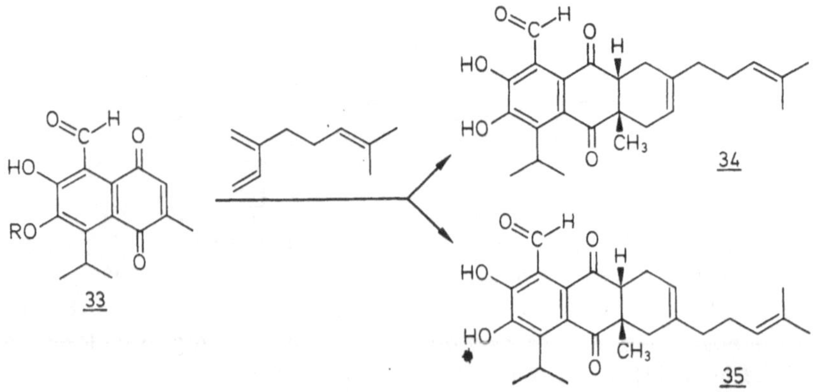

33 34 35

All of these compounds inhibit the growth of the cotton parasites mentioned above. The structures of *34* and *35* are remarkable because they are probably cycloaddition

* For a summary of the literature on gossypol, cf. [2].

products of *33* and myrcene (myrcene is present in a concentration of up to 8% in essential oils of the flower buds). Since this type of cycloaddition also occurs *in vitro* (whereby racemic *34* and *35* are formed), it presumably plays a role in the formation of the natural products as well. Generation of the above compounds during isolation cannot be ruled out, however. Unfortunately, the publications cited under [7] and [8] do not provide data on the possible optical activity of *34* and *35*.

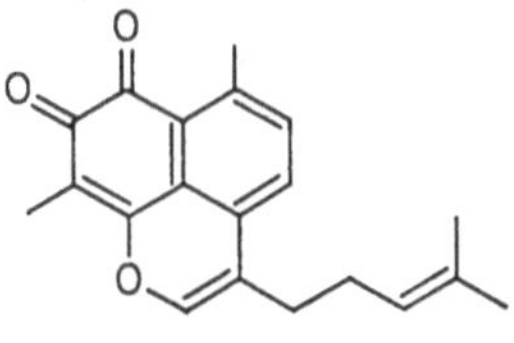

36

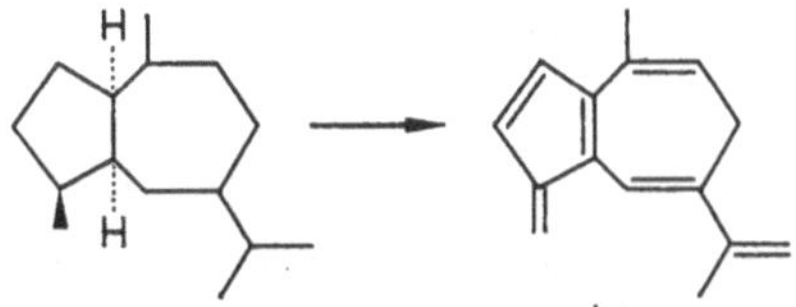

38 39

Biflorin (*36*) can be regarded as a prenylated cadinane derivative. It is found in leaves and blossoms of *Capraria biflora.* Like mansonone F and L it possesses an oxaphenalene system in addition to the *o*-quinone and is therefore violet red in color.

Azulene derivatives: The parent substance azulene (*9*) is a dark blue compound. The color can be strongly modified by substituents. Until recently all azulene derivates isolated from plants were referred to as artifacts which are easily formed from unstable precursors by elimination and oxidation processes during isolation. However, several azulenes have now been found which, according to all indications, are genuinely present in the respective plants. The azulenes *37a* and *37b* were isolated from the liverwort *Calypogeia trichomanes* [9] and also from *Pimpinella saxifraga* ssp. *nigra*, according to a Bulgarian paper cited in [9]. The azulene *37c* has been isolated from *Helichrysum bracteatum* [10]. It is supposed that these compounds are formed in the plant by loss of an isopropyl group from a still unknown sesquiterpene precursor.

37

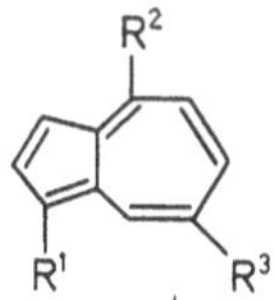

37 a $R^1 = R^2 = CH_3$, $R^3 = H$
b $R^1 = COOCH_3$, $R^2 = CH_3$, $R^3 = H$
c $R^1 = COOCH_3$, $R^2 = CHO$, $R^3 = H$
d $R^1 = CH_2OC(=O)-(CH_2)_{16}-CH_3$, $R^2 = CH_3$, $R^3 = C(=CH_2)CH_3$
e $R^1 = R^2 = CH_3$, $R^3 = C(CH_3)=CH_2$
f $R^1 = CH_2OH$, $R^2 = CH_3$, $R^3 = C(CH_3)=CH_2$
g $R^1 = CHO$, $R^2 = CH_3$, $R^3 = C(CH_3)=CH_2$
h $R^1 = CHO$, $R^2 = CH_3$, $R^3 = COCH_3$
i $R^1 = CHO$, $R^2 = CH_3$, $R^3 = CH(CH_3)CH_3$

Lactarius species of the sectio *Dapetes* are rich in azulenes and hydrogenated azulene precursors. The yellow orange milk of *Lactarius deliciosus* and its ease of oxidation to red, blue, violet, and even green pigments has long been the object of intensive investigation. The structures of the blue lactarazulene and the red violet lactaroviolin have been elucidated for a long time. They are known to be artifacts. Under careful conditions an orange red precursor (so-called lactarofulvene) has finally been isolated and characterized [11, 12] and found to have structure *39*. Its light absorption in the visible range is due to the conjugated fulvene chromophore. Biogenetically it is a guaiane derivative (*38*). Lately, azulene compounds have also been isolated from various *Lactarius* species and their structure has been determined, as in the case of the blue derivative *37 d* from *L. indigo*, the likewise blue compounds *37 e* and *37 f* (species not given), and the red azulenes *37 g*, *37 h* and *37 i* [13].

4.3. Diterpenes

All pigments isolated to date are derived from abietane (*51*) and totarane (*52*). See chapter 5.

4.4. Triterpenes

The pigments which have been identified are derivatives of the pentacyclic glutinane (*40*) and are found in the root bark of celastraceae. Pristimerin (*42*) is the methyl ester of celastrol (*41*) which is found in the North American *Celastrus scandens*, the Chinese *Tripterygium wilfordii*, and in a commercial insecticide prepared from the Chinese species. Celastrol has also been identified in the root bark of *Catha cassinoides* (*Celastraceae*) [14]. Pristimerin occurs in *Euonymus tingens*, *Catha cassinoides* [14], *Celastrus dispermus*, *Denhamia pittosporoides*, *Maytenus dispermus*, *M. chuchuhuasca* [15], *Pristimera indica*, *P. grahami*, and *Salacia macrosperma* [16]. Closely related to it are tingenone (*43*) and hydroxytingenone (*44*) from *E. tingens* [17] and *Salacia macrosperma* [16] and dispermoquinone (*45*) from *Maytenus dispermus* [19]. Tingenone has also been identified in *C. cassinoides* [14], *M. chuchuhuasca* [15], and *M. ilicifolia* [18].

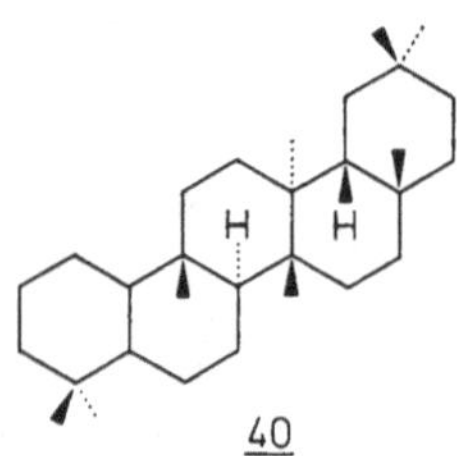

40

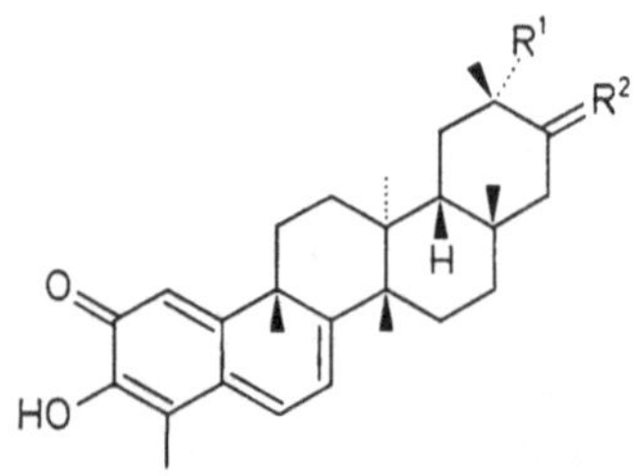

41 $R^1 = COOH$, $R^2 = H_2$
42 $R^1 = COOCH_3$, $R^2 = H_2$
43 $R^1 = H$, $R^2 = O$
44 $R^1 = OH$, $R^2 = O$

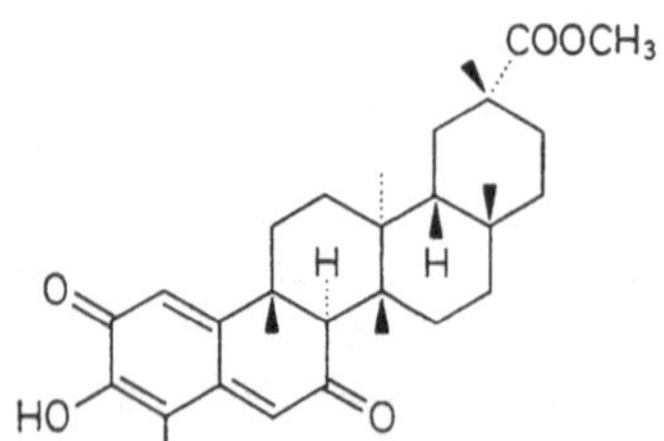

45

Iguesterin (*46*) [14] is a compound recently identified in *Catha cassinoides*. Structure *47* is attributed to a quinomethane from *Salacia macrosperma* [16]. The latter seems to replace an earlier bizarre structure postulated for « Saptarangiquinone A» from the same plant.

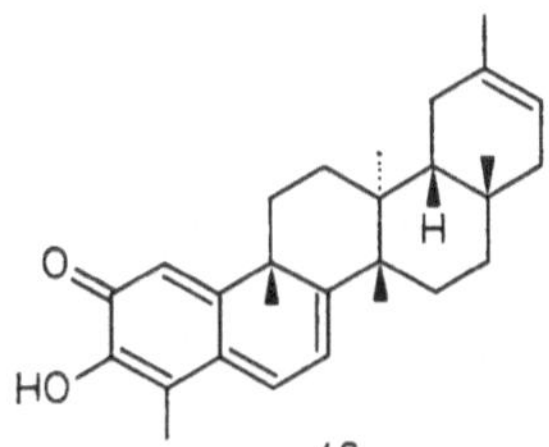

46

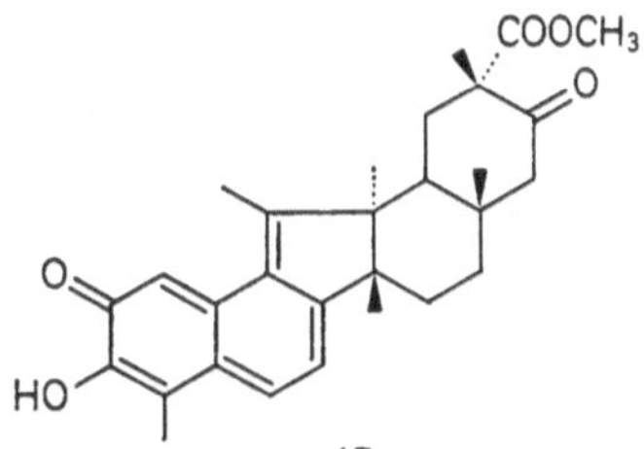

47

Pristimerin and tingenone crystallize in orange needles, celastrol in ruby red cubes, and hydroxytingenone in bright red crystals.

It seems as if other structurally related pigments are present in mother liquors of pristimerin, etc. Their separation presents great difficulty. Thus, a C(22)-hydroxytingenone was separated from the raw tingenone fraction (*43*) of an unspecified *Maytenus* species [20].

The pigments *41* to *44* are *p*-quinomethanes (« quinone methides »). Clarification of their structure lasted over several years due to the ready occurrence of rearrangements whose interpretation was not simple. Today so much spectral data is available that the structural analysis of analogs and homologs has become very easy, even with the smallest amounts of material.

Electron spectra of *p*-quinomethanes show a broad, usually structureless band at about 420–450 nm with ε-values above 10,000, which allows an easy differentiation from the $n \rightarrow \pi^*$ transition of simple or extended quinones (e.g., dispermoquinone (*45*)).

4.5. Tetraterpenes

These compounds are commonly known as carotenoids. They constitute by far the most extensive group of terpenoid pigments. To date approximately 400 different, naturally occuring carotenoids have been discovered [21, 22]. The formation of a long, conjugated, unsaturated polyene system is possible only with such compounds, e.g., lutein (*48*) with absolute configuration according to [23].

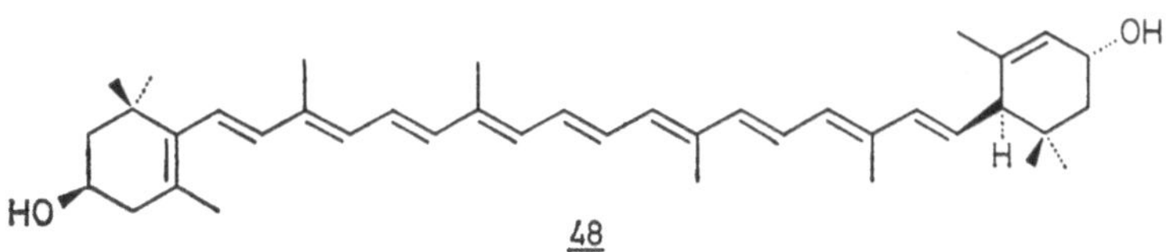

48

All carotenoids are derived from colorless C_{40} precursors which are formed by symmetrical tail-to-tail combination of geranyl-geranyl pyrophosphate [21, 24]. The intermediates of the stepwise dehydrogenation to colored polyenes are known by now. Polyenes with shorter chains, e.g., bixin (*49*, configuration according to [25])

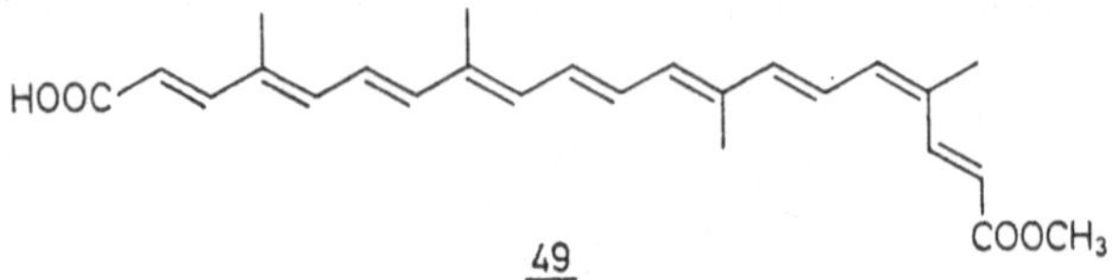

49

from the seeds of *Bixa orellana* and azafrin (*50*, absolute configuration according to

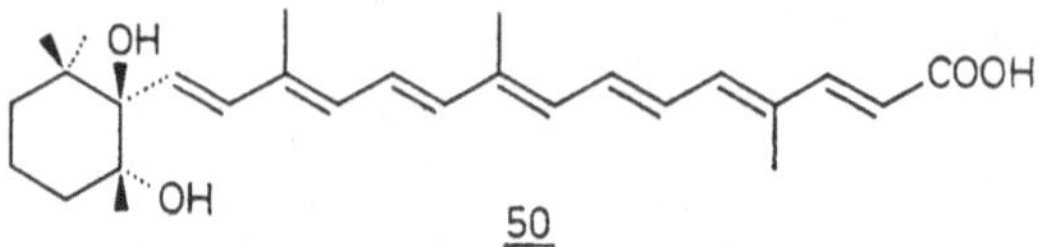

50

[26]) from the roots of *Escobedia* species are degradation products of C_{40} carotenoids*.

5. Diterpenoid Pigments

Due to the great number of these compounds – they have recently become the second largest group of terpenoid pigments – the diterpenoid pigments discovered to date will be treated in a separate chapter.

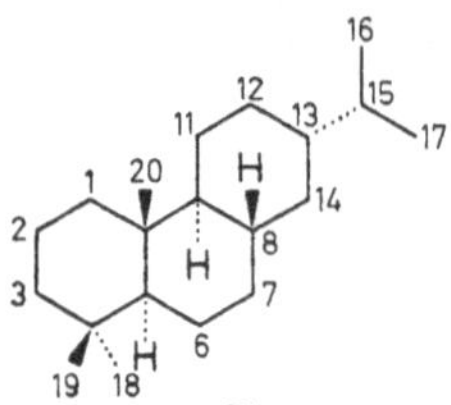

51

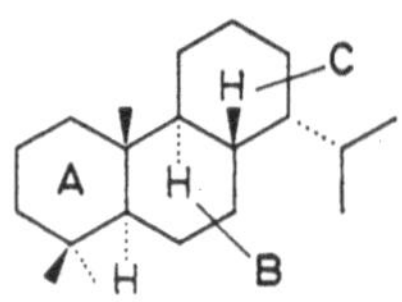

52

The carbon skeleton is abietane (*51*) in nearly every case; only occasionally totarane (*52*) is also found**.

For practical reasons the diterpenoid pigments are divided into three groups: tanshinones, royleanones, and coleones. The discovery of the first representatives of these pigments goes back to the search for physiologically active components in plants used in folk medicine. Tanshinones were thus found in the roots of *Salvia miltiorrhiza;* under the name « tan-shen » the roots are processed into an important

* A more detailed examination of the tetraterpenoid pigments is not in keeping with the goals of this survey.

** The totarane derivatives, e.g., totarol, are biogenetically linked with the corresponding abietane derivatives. The mechanism of isopropyl migration is not yet clarified. From the fact that all known totarane derivatives have an aromatized ring C carrying a hydroxyl group at C(13) it can be concluded that the rearrangement possibly occurs *via* an arene oxide.

drug with laxative and menstruation-promoting effects in China. In a similar manner royleanones were found in the roots of *Inula royleana* which is native to the West Himalayas and is considered to be toxic, insecticidic, and disinfectant. The first coleone was found in *Fuerstia africana*, a widely spread labiate in East Africa whose leaves were used by the Chagga tribe as a means of coloring butter, and as a galactagogue.

5.1. Tanshinones *

All ten quinones (*53* to *62*) have been found in the roots of *Salvia miltiorrhiza*. Each compound is a *nor*-abietane derivative from which the angular methyl group at C (10) has been removed during the course of dehydrogenation of ring B; one of the geminal methyl groups on C(4) is also eliminated whenever ring A has also been aromatized.

Only on miltirone (*60*) is the isopropyl group unchanged; in all other molecules it is incorporated into a β-methyldihydrofuran or a β-methylfuran. The most brightly colored compounds are the yellow *p*-benzoquinones *61* and *62*. The other compounds, above all the furan-substituted *o*-naphthaquinones, are red.

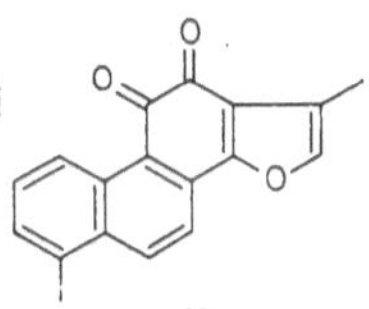
53
Tanshinone I

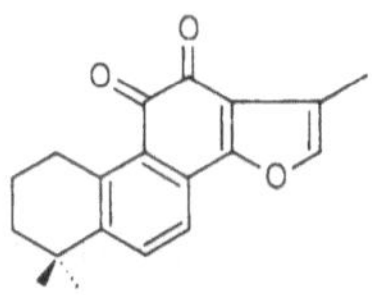
54
Tanshinone IIA

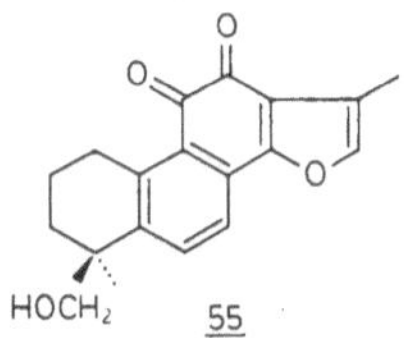
55
Tanshinone IIB

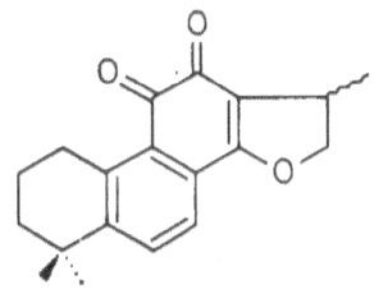
56
Kryptotanshinone

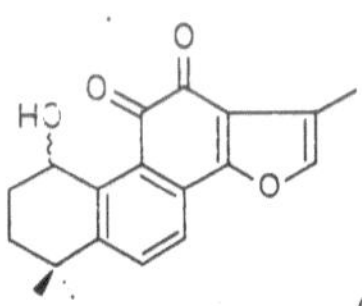
57
Hydroxytanshinone IIA

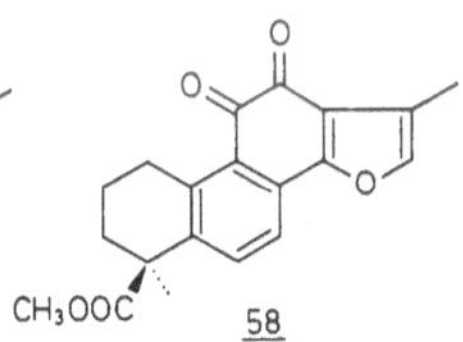
58
Tanshinone-methylester

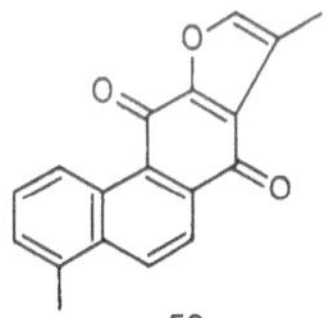
59
Isotanshinone I

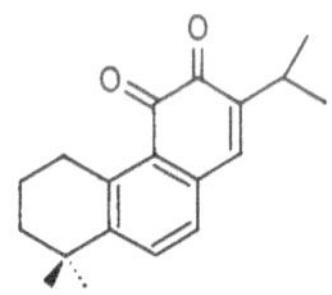
60
Miltirone

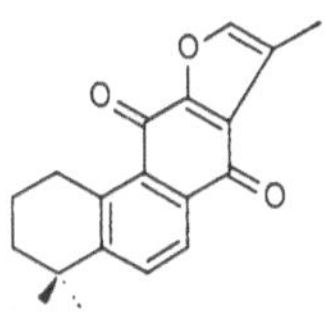
61
Isotanshinone II

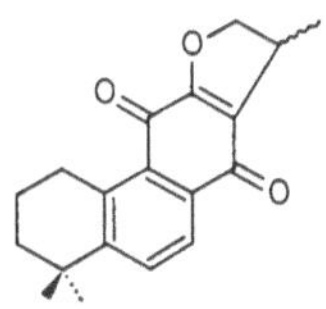
62
Isokryptotanshinone

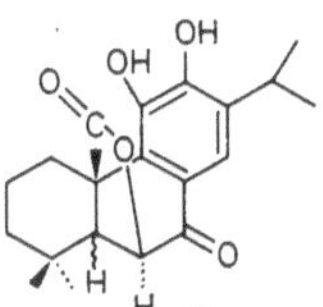
63

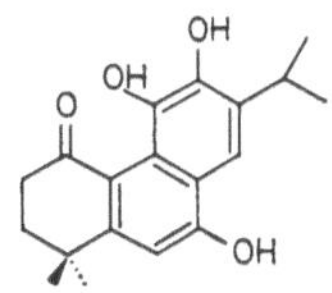
64

Refined chromatographic methods have permitted a substantial separation of the mixture of the pigments contained in the roots, and thus, similar to the mansonones, reasonable hypotheses on the course of the dehydrogenation steps can be made on

* For a review of the chemistry see [1] and [27].

the basis of the different structures of the substances found in the mixture. Obviously, dehydrogenation begins in ring C and then moves on to ring B. Dehydrogenation of ring A possibly starts with the oxidation of a geminal methyl group on C(4); whereupon the methyl group is removed either by retroaldol cleavage (or decarboxylation, cf., *115–119*). The further aromatization can be achieved by dehydrogenations or hydroxylations with subsequent elimination of water. Intermediates carrying oxidized methyl groups at C(10) have not yet been found in *S. miltiorrhiza;* however, cf., nemorone (*70*), galdosol (*63*) and arucatriol (*64*) from *S. canariensis* [28]. Compound *63* is reported to be red and *64* reddish yellow, which does not agree with the reported structures, but this could result from admixture of the corresponding *o*-quinones.

Oxidation of the C(10) methyl group can be seen – as previously mentioned – in galdosol (*63*). After elimination of water between C(5) and C(6), a vinylogous *β*-ketocarboxylic acid is formed. Its decarboxylation leads to the stable naphthol system. Arucatriol (*64*) is a further, oxidized product in this scheme.

All tanshinones are thus *nor*- and *bisnor*-diterpenoid quinones with dehydrogenated rings B and C; ring C is always quinonoid.

Extensive investigations on approx. 70 Russian *Salvia* species have recently shown that tanshinones occur in 31 of the investigated species*. Kryptotanshinone (*56*) was isolated from the roots of *Rosmarinus officinalis* a short time ago [31].

5.2. Royleanones

The compound which gave this group its name, royleanone (*65*), was originally isolated from the roots of a composite, *Inula royleana.* In the meantime this compound

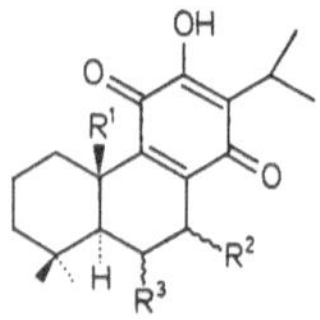

65 $R^1 = CH_3$, $R^2 = R^3 = H$
66 $R^1 = CH_3$, $R^2 = \beta\text{-OH}$, $R^3 = H$
67 $R^1 = CH_3$, $R^2 = \alpha\text{-OH}$, $R^3 = H$
68 $R^1 = CH_3$, $R^2 = \alpha\text{-OAc}$, $R^3 = H$
69 $R^1 = CHO$, $R^2 = \alpha\text{-OH}$, $R^3 = H$
70 $R^1 = CHO$, $R^2 = \alpha\text{-OAc}$, $R^3 = H$
71 $R^1 = CH_3$, $R^2 = O$, $R^3 = H$
72 $R^1 = CH_3$, $R^2 = \alpha\text{-OAc}$, $R^3 = \beta\text{-OH}$
73 $R^1 = CH_3$, $R^2 = \alpha\text{-OH}$, $R^3 = \beta\text{-OH}$
74 $R^1 = CH_3$, $R^2 = \alpha\text{-OCHO}$, $R^3 = \beta\text{-OH}$

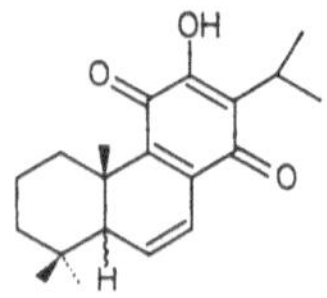

75

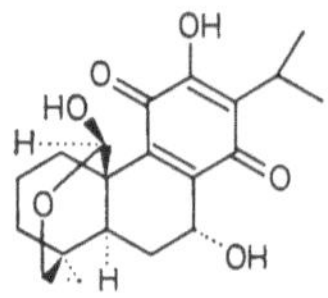

76

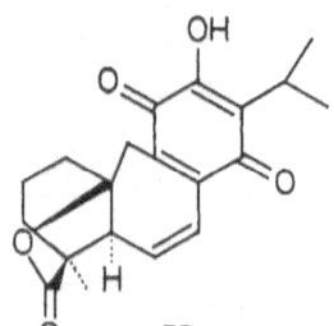

77

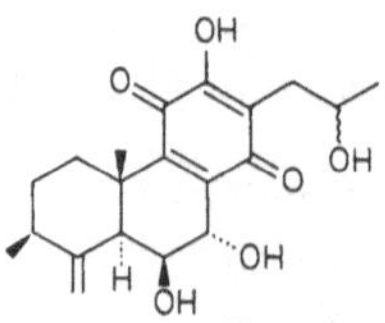

78

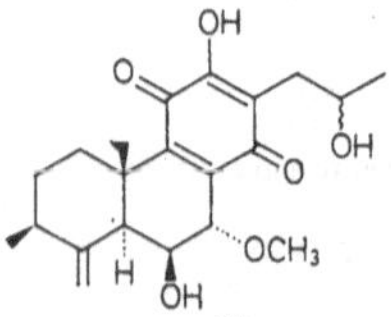

79

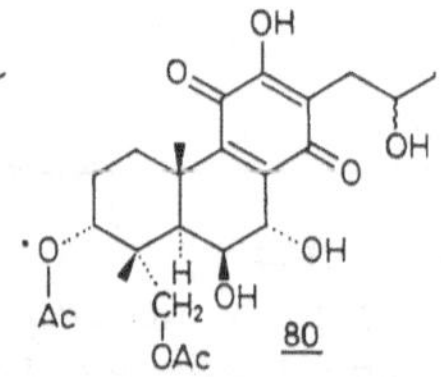

80

* In the publications available to the author, [29] and [30], no data is presented on the nature of the tanshinones.

has been identified in numerous other plants, and a series of structurally related compounds have been added to the list.

These include the following:*

65, yellow (or also orange) crystals; taxoquinone (*66*), orange needles; horminone (*67*), C(7) epimer of *66*, orange crystals; 7-O-acetylhorminone (*68*), yellow crystals. The aldehydes nemorone (*70*) and desacetylnemorone (*69*) are yellow. The orange red 7-keto compound (*71*) was recently isolated together with *65*, *66*, *67*, *75* and the golden yellow quinones *72* and *73* from leaf glands of a *Plectranthus* or *Coleus* species which has not yet been botanically classified [32]. The 7-oxoroyleanone (*71*) was also found in a reinvestigation of *Inula royleana* [33]. The compound *74*, which is closely related to *72*, occurs in the South African *Plectranthus myrianthus* together with *73* and several coleones (see coleone section) [34]. Dehydroroyleanone (*75*), a bright red quinone, occasionally accompanies royleanone *65* from which it cannot be easily separated. Like nemorone (*70*), the yellow conacytone (*76*) from the aerial parts of the Mexican *Salvia ballotaeflora* also possesses an aldehyde group at C(10) [35]. Since the C(19) methyl group has also been found oxidized to the alcohol stage, the hemiacetal group is easily formed. Structure *76* has been derived from an X-ray analysis; the absolute configuration given has not been proven so far, but is probable on the basis of biogenetic considerations.

A rearranged abietane skeleton is shown by the yellow icetexone (*77*) from *S. ballotaeflora* [33b, c; 36] as well as the greenish yellow compounds *78* and *79* and the orange red compound *80* from *Coleus coerulescens* [39q]. The possible biogenesis of such compounds will be discussed in the section on coleones.

Table 1 provides a summary of the occurrence of these compounds to date.

The royleanones are characterized structurally by the transformation of ring C of the abietane precursor into a hydroxy-*p*-benzoquinone. As a result they can be easily detected on the basis of color, spectra, acidity, color change to violet under NH_3 vapor, etc. As a rule, the isopropyl group in all known compounds is not oxidized; exceptions are the quinones *78*, *79*, and *80* whose unusual *n*-propyl side chain very probably stems from a coleone of the same type as *141*.

The occurrence of royleanones is not limited to certain plant organs, yet the majority of these compounds has been found in root barks (see Table 1). Their appearance in quite different plant families is remarkable; the labiates are the most represented, however. Recently both royleanones and coleones have been isolated from the same plant species: *Plectranthus myrianthus* [39o], *Coleus coerulescens* [39q].

Attention should also be called to the co-occurrence of taxoquinone (*66*) and horminone (*67*) [32]. Separation of the epimers is not an easy matter due to their very similar properties and could only be achieved by the use of modern chromatographic procedures [32]. As a result, it can be assumed that earlier samples of *66* and *67* were not free from epimers. This would mean that either the hydroxylation in the plant at C(7) does not occur stereospecifically (possibly not enzymatically), or that epimeri-

* For further reference see [1, 29–32].

Table 1: Occurrence of Royleanones *

Royleanone *65*:	*Taxodium distichum* (S), *Podocarpus ferrugineus* [37] (B), *Salvia nemorosa* (R), *S. officinalis* (R), *Plectranthus*** spp. (G), *Inula royleana* (R)
Taxoquinone (*66*):	*Taxodium distichum* (S), *Plectranthus*- (oder *Coleus*) sp. [32] (G)
Horminone (*67*):	*Horminum pyrenaicum* (A), *S. nemorosa, S. carduacea, S. deserta, S. officinalis, S. pachystachya, S. pratensis, S. verbenacea, Rosmarinus officinalis* (all R), *Plectranthus* (or *Coleus*) sp. [32] (G), *Hyptis fructicosa* [38] (R)
7-O-Acetylhorminone (*68*)	*Horminum pyrenaicum* (A), *Inula royleana, S. nemorosa, S. carduacea, S. deserta, S. officinalis, S. pachystachya, S. pratensis, S. verbenacea* (all R)
Desacetylnemorone (*69*):	*S. nemorosa* (R)
Nemorone (*70*):	*S. nemorosa* (R)
71, 72, 73, 74:	*Plectranthus* (or *Coleus*) sp. [32] (G), *71* from *Inula royleana* [33] (R), *73* and *74* from *P. myrianthus* [34] (G)
6,7-Didehydroroyleanone (*75*):	*Salvia officinalis* (R), *Rosmarinus officinalis* (R), *Inula royleana* (R), *Plectranthus* spp.**, *Plectranthus parviflorus*, var. *major* (G), *Plectranthus* (or *Coleus*) species [32] (G), *Podocarpus ferrugineus* [37] (B)
Conacytone (*76*) and Icetexone (*77*):	*Salvia ballotaeflora* [35] (A)
78, 79, 80:	*Coleus coerulescens* [39q] (G)

* Only definitively isolated compounds are considered. For qualitative investigations see [29, 30]; organs: R = root, B = bark, G = leaves with leaf glands, S = seeds, A = aerial parts.

** Various sources, cf. [1, 32].

zation occurs *via* an enolic *o*-quinomethane. *In vitro* attempts to epimerize C(7) have been unsuccessful to date, however.

Inula royleana contains *65, 68, 71* and *75* in addition to the diterpenoid alkaloid lycoctonin. Its absolute configuration at C(5) and C(10) is antipodal with respect to the quinones. This shows that different enantiomeric foldings of an acyclic precursor can be used for the formation of cyclic diterpenes in the plant.

5.3. Coleones

Assembled under this heading are mostly colored diterpenoid compounds stored in the glands of *Coleus, Plectranthus, Solenostemon* and *Fuerstia* species. The glands are usually located on the underside of the leaves; occasionally they are sufficiently accumulated along the edge such that a color effect can be seen with the naked eye, and they also appear on sepals and often also on blossoms*. Brief dipping a leaf into ether leads to quick extraction before chlorophyll is dissolved out of the inside of the leaf. Single pigments (e.g., coleone A) account for 1.5 to 1.9% of the dry weight of the leaf; the total amount of pigments accounts for more than 2%. The formation of the pigmented glands begins very early: cotyledons are usually already equipped with glands. Structurally related compounds have meanwhile been found in other plant species as well. Since they can also be classified as coleones they will be discussed in this section. The naming of the coleones corresponds to their order of discovery [39]. The discussion in the text follows the increasing complexity of structure, however.

By the end of 1977 over 70 various compounds were known which can be included under the coleones. Although almost all of these molecules can be derived from abietane (*51*), their structural diversity is so great that it is expedient to treat them as subgroups in the following discussion. The first group (under 5.3.1.) is formed by the *p*-quinomethanes (*p*-quinone methides) and extended quinones. The second group (5.3.2) comprises coleones with a hydroquinone ring C, and diketo or diosphenol groups in ring B as well. A third group (5.3.3) comprises the nearly colorless *spiro*-cyclopropyl-cyclohexenediones which can be derived biogenetically from the second group. The fourth group contains the coleones with an opened ring A (so-called *seco*-compounds, 5.3.4) and the recently discovered dimeric coleones (5.3.5). The simultaneous occurrence of coleones and royleanones has been encountered recently, as in *Plectranthus myrianthus* [39o], *Coleus coerulescens* [39q], and *P. lanuginosus* [40], and thus the long-suspected biogenetic connection between these groups is now documented experimentally. The simultaneous occurrence of royleanone (*65*) and taxoquinone (*66*) with the quinomethane *81*, the extended quinone *82*, and the colorless precursors *95* and *96* in the seeds of *Taxodium distichum* should also be mentioned in this connection [41].

5.3.1. p-Quinomethanes and extended quinones

The following compounds were known by the end of 1977:

Nilgherrone A and B (*184* and *185*) which are treated in the section on dimeric diterpenes (5.3.5) should also be included in the above mentioned list. The yellow compounds *81* and *82* were found in the seeds of *Taxodium distichum* in connection with the search for tumor inhibitors of plant origin [41]. The orange maytenoquinone *83* originates from the yellow root bark of *Maytenus dispermus* [42] where it occurs together with pristimerin (*42*) and dispermoquinone (*45*; see section 4.4 « Triterpenes »). It is the first and only compound with a totarane skeleton (*52*) found in this group to date. The compound 14-methoxytaxodione (*84*) isolated from the roots of *Hyptis fructicosa* (a labiate from northeast Brasil) in the form of orange needles is closely related to *81* and *82* [38]. The presence of a methoxy group in this compound is odd. Since the authors used methanol in the extraction, it has probably been formed from a

* For pictures of such glands see [39a (second paper), 39d, 39f].

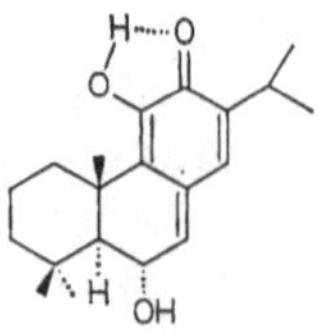

81
Taxodone
(316; 4.30)

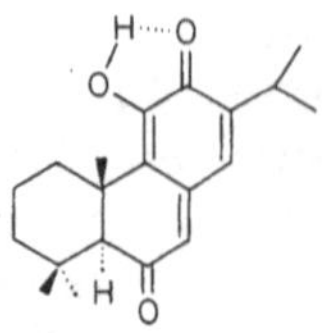

82
Taxodione
(400; 3.30)

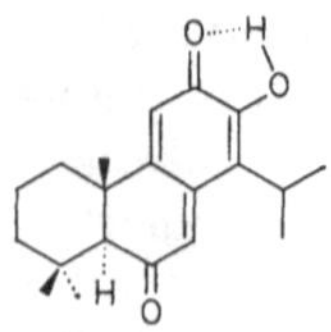

83
Maytenoquinone
(414; 3.33)

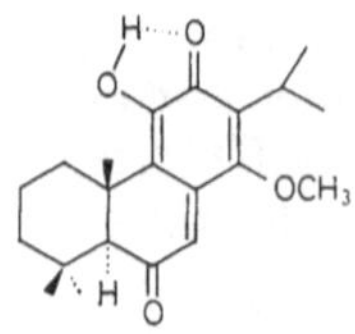

84
14-Methoxytaxodione
(420; 3.33)

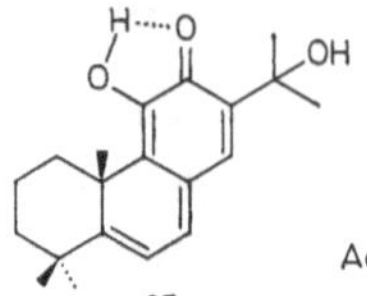

85
Fuerstione
(445; 4.10)

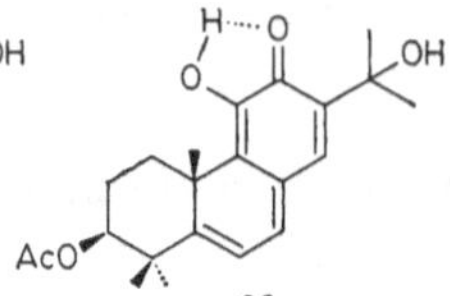

86
3β-Acetoxyfuerstione
(432; 3.95)

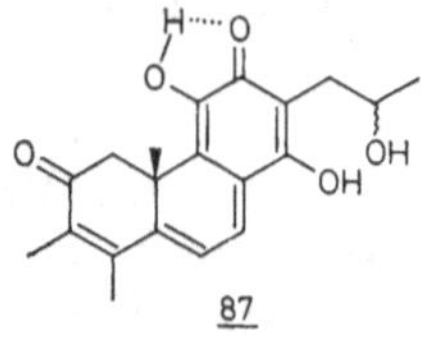

87
Coleone E
(435-440; 4.11)

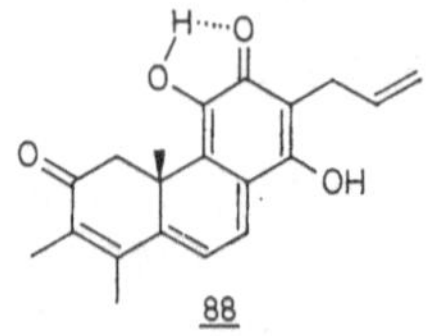

88
Coleone F
(436-440; 4.16)

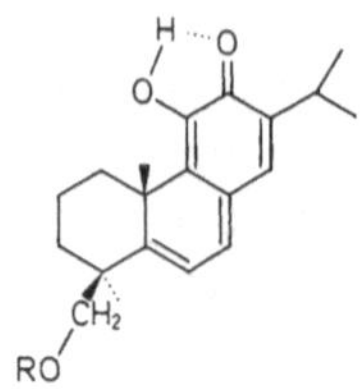

89 Parviflorone A, (420-425) R = 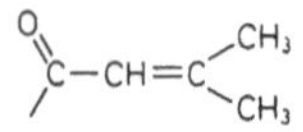

90 Parviflorone B, (420-425; 4.11) R = 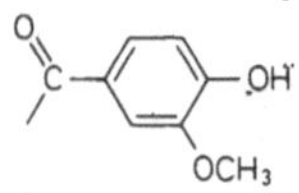

91 Parviflorone C, (420-425; 4.05) R = 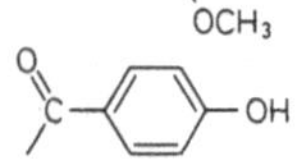

92 Parviflorone E, (415-425; 3.91) R = 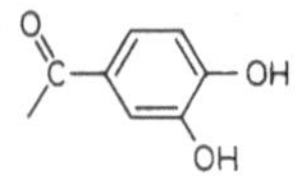

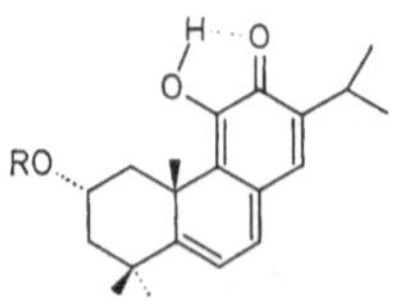

93 Parviflorone D, (420-425; 4.14) R = 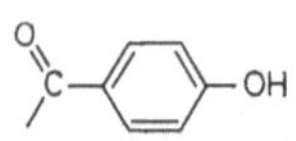

94 Parviflorone F, (415-425; 3.94) R = 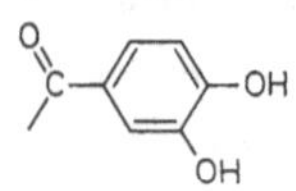

reactive precursor by addition of methanol and subsequent oxidation. Examples of more extensive dehydrogenation can be found in the quinomethanes *85*, *86*, and *89–94* and in the extended quinones *87* and *88*. Fuerstione (*85*) was the first representative of a quinomethane of the diterpene series to be found [39a]. This intense red and very unstable pigment is located in the glands of *Fuerstia africana* in very high concentration; up to 0.3% crystallized fuerstione can be isolated from the dried leaves. *Fuerstia* is very similar to the genus *Orthosiphon* and is widely distributed in the steppes of East Africa (West Kenya, Kilimanjaro). Its use in folk medicine has already been mentioned. Of the numerous characteristic reactions of fuerstione with acids, bases or phenols, only the very easily accomplished aromatization to stable

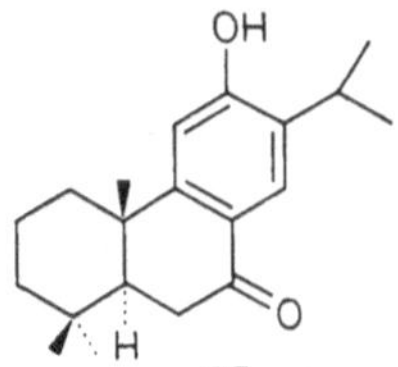

95

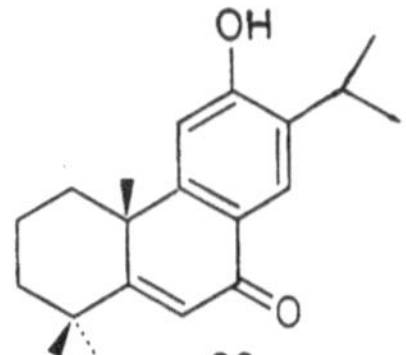

96

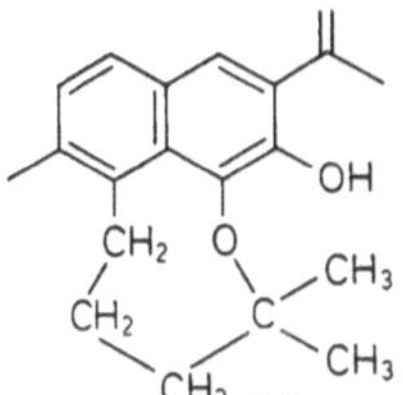

97

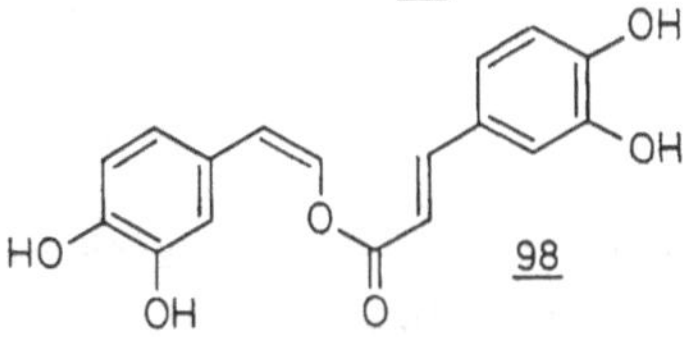

98

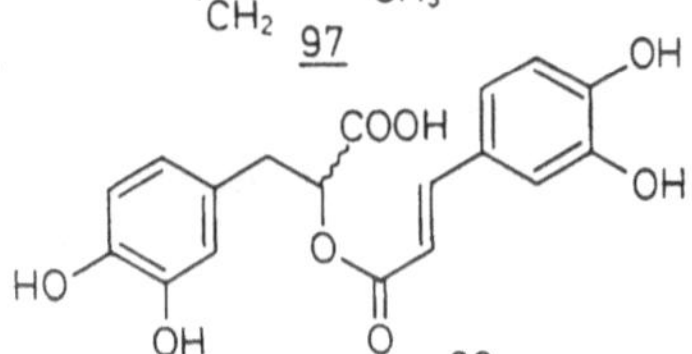

99

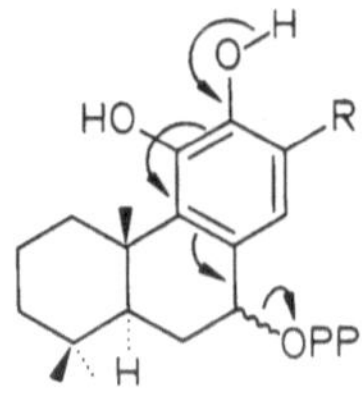

100

naphthalene derivatives, e.g., *97*, is mentioned here. This reaction takes place after migration of the quaternary methyl group from C(10) to C(5). Aromatization reactions will be referred to later in another connection. Fuerstione also occurs in the southern Indian *Plectranthus nilgherricus* together with the brilliant red 3β-acetoxyfuerstione (*86*) [39p]. Related to this compound are the parviflorones A to F (*89–94*) which were recently discovered in *Plectranthus parviflorus*. This plant is occasionally kept as an indoor evergreen. It is unusually rich in deep red pigmented glands arranged in a rosette shape. The very complex mixture of closely related and very labile *p*-quinomethanes contained therein was separated with great difficulty [39r]. The parviflorones possess the same chromophoric system as fuerstione (*85*) and 3β-acetoxyfuerstione (*86*). They do not show a hydroxyl function on the isopropyl group (C(15)). However, they are oxidized at C(2) or C(19) instead. Esterification of these hydroxyl functions with senecioic acid (parviflorone A), *p*-hydroxybenzoic acid (parviflorones C and D), protocatechuic acid (parviflorones E and F) and vanillic acid (parviflorone B) is a new feature. In these structures terpenes are united with prephenic acid derivatives! The caffeic acid ester (*98*) of dopaldehyde was already isolated from *Plectranthus caninus* earlier [44] ((Z)-enol-dopaldehyde-(E)-caffeate).

Since it is known that *Coleus blumei* forms rosmarinic acid (*99*) in callus cultures [45], it can be assumed that compound *98* is the product of an oxidative decarboxylation of *99*.

The exceptional instability, but not the structure, of fuerstione and the parviflorones is held in common with two further pigments: the deep red extended quinones coleone E (*87*) and coleone F (*88*). Both compounds were isolated from *P. kilimand-*

schari (Gürke) Agnew*, *P. barbatus* (Bentham) Agnew* and a *Coleus* species (Bally no. 10431). The plants used by us originated from the upland Kenyan bush, the Rift Valley, and the Taita Hills, respectively. Very different results were obtained by working with *Coleus* (*Plectranthus*) *barbatus* from Brasil [47] or India [48] (see later text). The long, conjugated system which now also includes the ring A of the abietane skeleton (*51*) is characteristic of the coleones E and F. To permit this dehydrogenation to occur, one of the geminal methyl groups on C(4) had to migrate. Such a migration can occur when position 3 is hydroxylated and the hydroxyl group is transformed into a good leaving group, e.g., by esterification with pyrophosphate (see section 5.3.3).

A very striking structural characteristic of the coleones E and F is the *n*-propyl side chain which was already mentioned in connection with the royleanones *78*, *79*, and *80*. Conceptions of the genesis of this unusual structural feature will be treated at the end of section 5.3.3.

Quinomethanes of the fuerstione (*85*) type show intensive and usually structureless absorption bands with the main absorption around 420–450 nm. The high molar extinction coefficients which can exceed 10,000 are especially characteristic. Such compounds thereby distinguish themselves characteristically from *o*- and *p*-benzoquinones. Further diagnostic characteristics are the carbonyl frequencies of unusually long wavelength in the IR spectra, as well as the NMR and mass spectra (detailed data in the corresponding papers under [39]). The special chemical reactivity of these compounds that leads to characteristic rearrangements should also be stressed at this point. The isolation and separation of such compounds requires a great deal of time and experience; although structure elucidation using modern spectroscopic techniques has become easier. Moreover, much data is available today for comparison. Advances in this field are thus primarily dependent on the application or development of suitable separation procedures. The extended quinones *82*, *83*, and *84* are much more stable than the quinomethanes; conventional chromatographic procedures suffice for these compounds.

When one examines the structures of the compounds isolated to date it soon becomes clear that dehydrogenation and functionalization begins in ring C. This is obriously connected to the biogenesis of the abietane structure which leads to a carbonium ion at C(13). Aromatization of ring C follows, and thereafter an oxygen function is introduced at the benzylic position (C(7)). This is already the case in taxoquinone (*66*), horminone (*67*), sugiol (*95*), and 5,6-didehydrosugiol (*96*). One easily then reaches the quinomethane series by an elimination reaction as is suggested in the postulated compound *100*. Further oxygenation at C(6) or elimination on a compound already hydroxylated at C(6) and C(7) (cf., compounds *72*, *73*, *74*, *78*, *79*, *80!*) gives taxodone (*81*). From taxodone, fuerstione (*85*) could be formed by oxidation and elimination. Apparently, the dehydrogenation of ring A follows that of ring B. To

* Attributed to the genus *Coleus* by both Gürke and Bentham and originally labelled so by us [39f, g]. Agnew recently assigned this species to the genus *Plectranthus* [46]. The same applies to other East African *Coleus* species.

achieve this, oxidations are necessary at C(2) and C(3) combined with elimination and rearrangement steps. Thus, the structures of the *p*-quinomethanes and the extended quinones already permit far-reaching conclusions about the course taken by the catabolic process on abietanoids.

5.3.2. Acylhydroquinones, C(6)/C(7)-diketones and -diosphenols

The colored diterpenes also include compounds which contain the chromophores named in the title of this section. For example, we find the acylhydroquinone chromophore *8* in the yellow inuroyleanol (*101*) from the roots of *Inula royleana* [33]* and in the bright-pink hyptol (*102*) from *Hyptis fructicosa* [43]**. Although the long-wave-

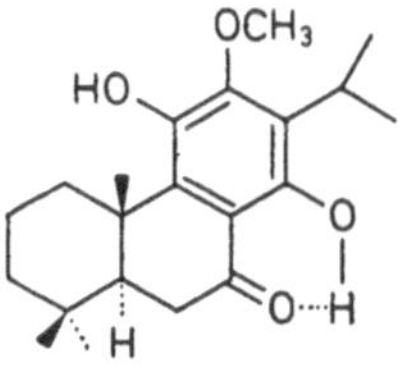

101 (370; 3,79)

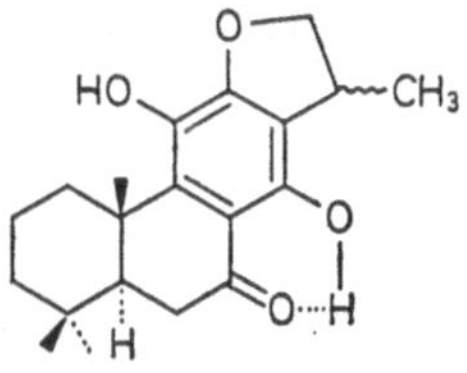

102 (356; 3,87)

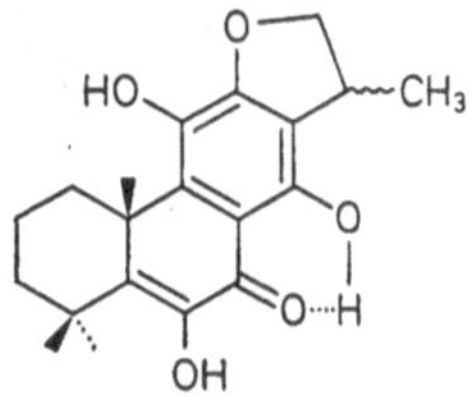

103 (385; 3,93)

length maximum of this compound is still in the UV range, the absorption reaches the visible range. When the hydroquinone group in ring C is lacking, the compounds are colorless because they are not acylhydroquinones: sugiol (*95*) has λ_{max} 284 (log ε_{max} 4.12) and 5,6-didehydrosugiol (*96*) has λ_{max} 316 (4.11). With an additional double bond between C(5) and C(6) the longest-wavelength absorption of the acylhydroquinones moves strongly towards the visible range. Such compounds have been discovered in recent years, e.g., lycoxanthol (*103*) and the coleones presented in Table 2. The chromophore is so characteristic that the spectral curve provides immediate information on the presence of a coleone C type. This helped a Canadian group, for example, to identify lycoxanthol (*103*), a yellow pigment from *Lycopodium lucidulum* [49]. A further diosphenol is the colorless compound (*122*, ($R^1{=}R^3{=}H$, $R^2{=}OH$) from the wood of *Juniperus rigida* [50] in which it was found together with xanthoperol (*121*, $R^1{=}R^3{=}H$, $R^2{=}OH$).

In the diterpenes with a diosphenol group at C(6)/C(7), further oxidations have been found to occur at C(2), C(3), C(16), C(17), and C(19). Compound (*122*) ($R^1 = R^3 = H$, $R^2 = OH$) is a ferruginol type and does not contain a hydroquinone group in ring C. The functionalization of C(16) or C(17) provides the possibility of ring closures with the aid of a suitable leaving group (e.g., pyrophosphate ester). Such ring closures have been performed *in vitro* on coleone C (*104*) [39d]; whereby a dihydrofuran

* Since methanol was used in the recrystallization, (*101*) could be an artifact.

** In the acylhydroquinones the strong hydrogen bond to the C(7) carbonyl group is the reason for the stability of the hydroquinone. The acylquinones of type *71* obtained by oxidation are very reactive and therefore rarely occur as natural products. The chemistry of the acylquinones has been investigated well; this appears to have escaped the authors of [43] who are surprised by the stability of hyptol.

Table 2: Acylhydroquinones with Diosphenol Grouping in Ring B

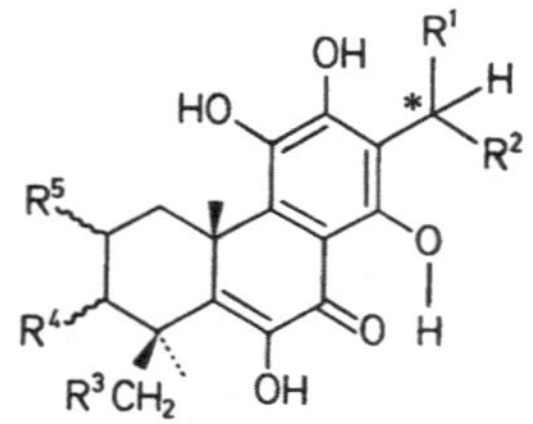

Coleones *104* to *114*

Name	Formula	Chirality at C(15) (*)	R^1	R^2	R^3	R^4	R^5	Color; λ_{max} ($\log \varepsilon_{max}$)	Isolated from	References
Coleone C	*104*	(R)	CH_2OH	CH_3	H	H	H	yel.; 398 (3,92)	a), b), c), d), k)	[39d] [39q] [39t] [39 l]
Coleone C′	*105*	(R)	CH_2OCOCH_3	CH_3	H	H	H	yel.; 381 (3.87)	b)	[39q]
Coleone H	*106*	(R)	CH_2OH	CH_3	H	β $OCOCH_3$	H	yel.; 398 (3.77)	d), e)	[39t] [39h]
Coleone L	*107*	?	CH_2OCOCH_3	CH_2OH	H	β $OCOCH_3$	H	yel.; 396 (3.91)	e)	[39m]
Coleone S	*108*	–	CH_3	CH_3	H	β OH	H	yel.; 385 (3.96)	f)	[39n]
Coleone U	*109*	–	CH_3	CH_3	H	H	H	yel.; 382 (3.91)	g)	[39o]
Coleone W	*110*	?	CH_2OCOCH_3	CH_2OH	H	H	H	yel.; 400 (3.65)	b), g), h)	[39q] [39o]

Coleone W′	*111*	–	CH_2OCOCH_3	CH_2OCOCH_3	H	H	H	yel.; 383 (3.91)	h), i)	[39u]
Mon-D	*112*	?	CH_2OH	CH_3	H	H	α OCH (=O)	yel.; 400 (3.90)	d)	[39t]
Lanu 15	*113*	–	H	$CH{=}CH_2$	OCOH	H	H	yel.; 384 (qual.)	c)	[40]
Lanu 16	*114*	–	H	$-CH(OH)-CH_3$	OCOH	H	H	yel.; 395 (qual.)	c)	[40]

a) *Coleus aquaticus*; b) *C. coerulescens*; c) *Plectranthus lanuginosus*; d) *Solenostemon monostachys*; e) *C. somaliensis*; f) *P. caninus*; g) *P. myrianthus*; h) *S. sylvaticus*; i) *C. garckeanus*; k) *P.* species according to W. Rauh Ma 7317.

ring is formed (*103*). The same compound (lycoxanthol) has then almost simultaneously been isolated as a natural product [49]. Diterpenes with such dihydrofuran rings or a corresponding furan ring are already known among the tanshinones (section 5.1, compounds *59*, *61*, and *62*). One may thus postulate compounds of the coleone C type as biogenetic precursors of the various tanshinones.

Similar to the case of coleones E (*87*) and F (*88*), in diosphenols of the coleone C type dehydrogenation can also involve ring A. The geminal substitution at C(4) which hinders the formation of the fully conjugated system is overcome in (*87*) and (*88*) by migration of a methyl group to C(3). The substituent can also be removed by retroaldol cleavage or decarboxylation:

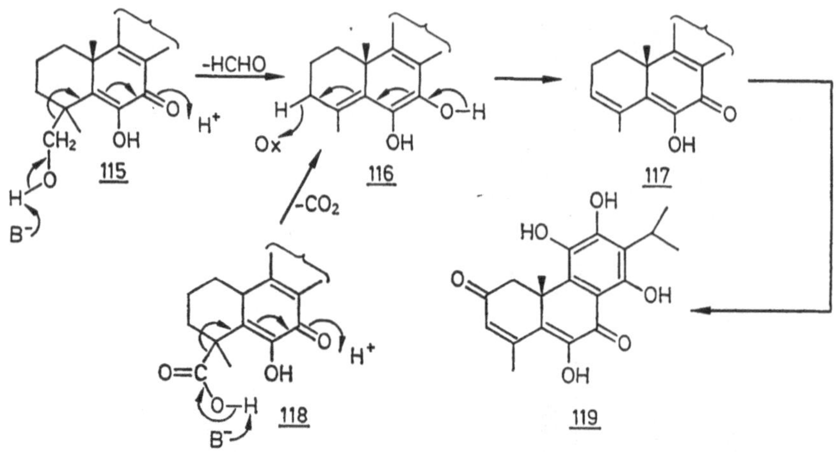

Compounds with oxidized C(4) methyl groups have been found, for example, among the coleones of *Plectranthus lanuginosus* (*113*, *114*; [40]); more highly oxidized stages (aldehyde, carboxylic acid at C(4)) have been found in coleone A (*176*) and edulone A (*183*). As vinylogous β-ketoaldehydes or β-ketocarboxylic acids such compounds should be easily transformed into *116* under suitable conditions. The path from *116* to *117* and *119* is obvious. The last of these compounds, coleone B (*119*), forms brilliant yellow orange crystals and is found in *Coleus igniarius*. The structure of *119* is very remarkable because its oxidation state substantially approaches the parent substance phenanthrene: the actual transition into the aromatic system is only hindered by the quaternary methyl group at C(10)! Nevertheless, the compound is more stable than one would expect from its structure. This can be attributed, above all, to the diene-dione system; it is also known that highly hydroxylated phenanthrenes convert relatively easily to the tautomeric keto form.

All diosphenols discussed up to this point can also occur in the tautomeric α-diketo form *120* or *121*.

A large number of such compounds have been discovered. The first of these was the yellow xanthoperol (*121*, $R^1 = R^3 = H$, $R^2 = OH$) from the wood of various *Juniperus*, *Podocarpus*, and *Cryptomeria* species (*Cryptomeria* species contain *121* ($R^1 = R^3 = H$, $R^2 = OH$) also in the bark and sap*). This compound shows rings

A and B *cis*-fused. Since this form is much more stable than the structure with *trans*-fused rings, it must be assumed that xanthoperol is a rearrangement product. Indeed, the *trans*-compound (*120*, $R^1 = R^3 = H$, $R^2 = OH$), the so-called praexanthoperol was later found in very small quantities in the wood of *Juniperus communis* [52].

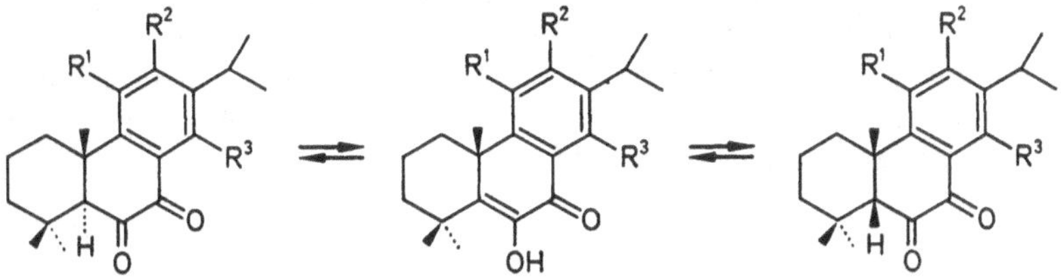

120 (341; 3.94) | 122 | 121 (352; 3.86)

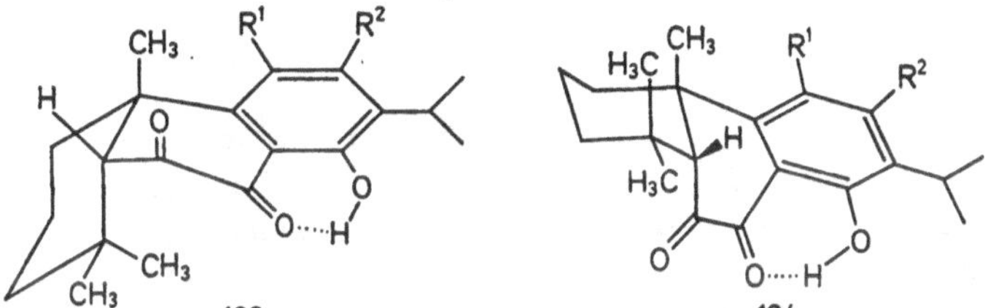

123 | 124

This compound is very unstable and converts easily to xanthoperol (*121*, $R^1 = R^3 = H$, $R^2 = OH$). As already mentioned, the compound *121* has been found together with its colorless diosphenol *122* ($R^1 = R^3 = H$, $R^2 = OH$) in *Juniperus rigida* [50]. The yellow color of the compounds *120* and *121* ($R^1 = R^3 = H$, $R^2 = OH$) can be attributed to the C(6)/C(7)-α-diketo group. An α-diketo group absorbs in the visible region when the carbonyl groups are forced into a cisoid conformation. This is the case with *120* and *121* ($R^1 = R^3 = H$, $R^2 = OH$). Depending on the size of the ring with the α-diketo group, the carbonyls are situated more or less planarly with respect to one another. Thus the X-ray analysis of *cis*-coleone D (*cis*-A/B-*125*) yields a torsion angle τ (O—C(6)—C(7)—O) = 10° [53]. The reciprocal rearrangements *120* ⇌ *122* ⇌ *121* have been studied in detail using the coleones C (*104*) and D (*125*) as examples [39 l]. The *cis*-A/B-C(6)/C(7)-diketones investigated to date occur solely in the steroidal conformation *123*; population of the likewise possible nonsteroidal conformation *124* is not detectable.

If the C(6)/C(7)-diketo chromophore is part of an acylhydroquinone, as in coleone D (*125*), then the light absorption of these compounds occurs at nearly as long wavelengths as that of quinones and extended quinones. For example, coleone D (*125*) has a $\lambda_{max}^{ethanol}$ of 421 nm (log ε 3.94). The position of the long-wavelength maximum

* Literature reference under [3], no. 3685/3686. In the reference the formulas of xanthoperol (*121*) and praexanthoperol (*120*) should be exchanged with one another; for further references see [37, 50, 51].

and the shape of the absorption curve is strongly influenced by the solvent used. The crystallized compounds are always orange red to bright red; see Table 3. Information on the compounds isolated from labiates to date is provided in this table; all compounds show *trans*-A/B-fused rings. In some cases the *cis*-A/B stereoisomers have been formed *in vitro* [39 l, m]; they show absorption bands of even longer wavelength.

Several biogenetic pathways from the diosphenols and diketones presented in section 5.3.2 to extended quinones and the *spiro*-compounds discussed in the following section can now be postulated. The first route has been demonstrated *in vitro*; a second route will be discussed in detail in 5.3.3.

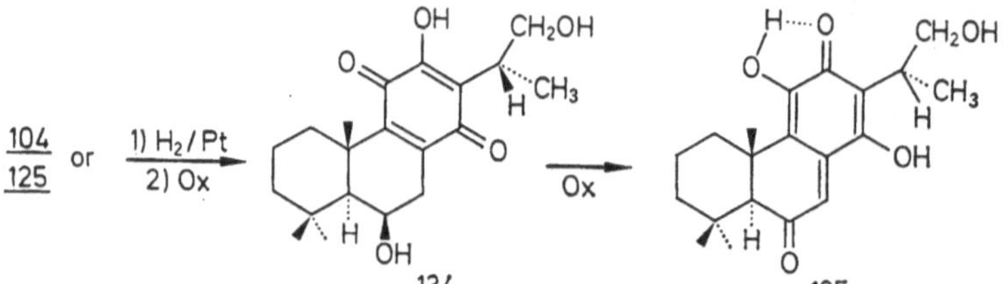

134 135

When coleone C (or coleone D) is catalytically reduced using Pt/acetic acid, 6β, 16-dihydroxyroyleanone (*134*) is formed by uptake of 2 equivalents of H_2 followed by air oxidation. Further oxidation with chromic acid results in formation of the extended quinone *135* which is closely related to taxodione (*82*) and 14-methoxytaxodione (*84*) [39 l, 40].

5.3.3. Coleones with spiro-cyclopropyl-cyclohexenedione structure

These compounds are generally only faintly yellow and actually need not be included in this discussion. Since they are formed from colored compounds and can themselves be transformed into pigments, they are mentioned here.

The most striking structural feature in the coleones E (*87*) and F (*88*) is the *n*-propyl side chain at C (13). Such a side chain has not yet been found in a diterpene. As an

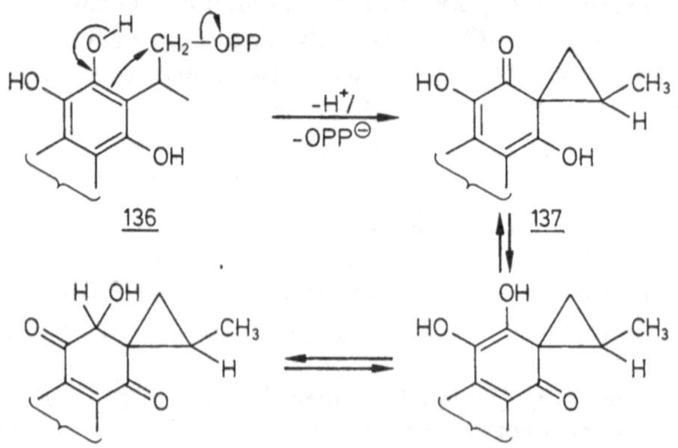

136 137 139 138

Table 3: Acylhydroquinones with C(6) keto-group

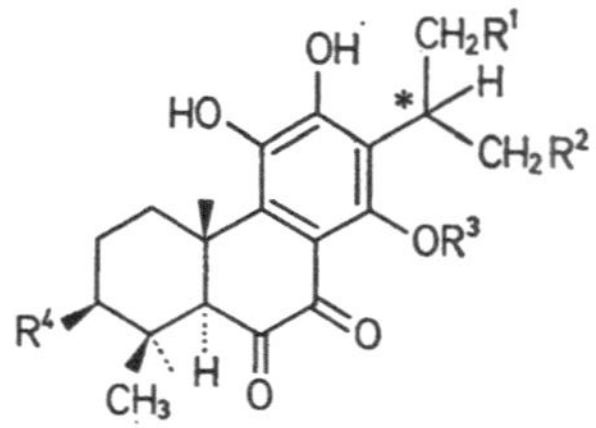

Coleones *125* to *133*

Name	Formula	Chirality at C(15) (*)	R^1	R^2	R^3	R^4	color; λ_{max}** ($\log \varepsilon_{max}$)	Isolated from	References
Coleone D	*125*	(R)	OH	H	H	H	orange red 421 (3.94)	a), b), c), k)	[39e] [39q] [40] [39l]
Coleone D′	*126*	(R)	$OCOCH_3$	H	H	H	bright red 385 (3.66)	b)	[39q]
Coleone I	*127*	(R)	OH	H	H	$OCOCH_3$	orange yellow 421 (3.83)	e), k)	[39h] [39l] [39m]
Coleone I′	*128*	(R)	OH	H	H	OCOH	orange yellow 421 (3.83)	k)	[39l] [39m]
Coleone K	*129*	?	$OCOCH_3$	OH	H	$OCOCH_3$	orange 395 (3.66)	e)	[39h] [39m]
Coleone T	*130*	–	H	H	H	OH	orange red 419 (3.66)	f)	[39n]
Coleone V	*131*	–	H	H	H	H	red 417 (3.89)	g), b)	[39o] [39q]
Coleone V′	*132*	–	H	H	COH	H	orange red 422 (3.54)	g)	[39o]
Coleone X	*133*	?	$OCOCH_3$	OH	H	H	orange 400 (3.64)	h), i)	[39u]

a) *Coleus aquaticus*; b) *C. coerulescens*; c) *Plectranthus lanugionsus*; d) *Solenostemon monostachys*; e) *C. somaliensis*; f) *P. caninus*; g) *P. myrianthus*; h) *S. sylvaticus*; i) *C. garckeanus*; k) *P.* species according to W. Rauh Ma 7317.

** Position of absorption maximum dependent on solvent. For solvent used see original publication.

explanation one can postulate a 1,2-methyl shift; the position of the double bond and the hydroxyl group then causes difficulties, however. Therefore, we postulated a *spiro*-cyclopropyl compound as an intermediate (e.g., *137*) [39f]. This hypothesis requires the presence of a monohydroxylated isopropyl group *ortho* on the phenol, which actually is the case in numerous compounds in Tables 2 and 3. With a suitable leaving group the aromatic ring can participate in the elimination, as suggested in *136*. The partialstructures *138* and *139* represent tautomers of *137*. From this conception, which is supported by already known results from mechanistical studies follow two important conclusions:

- Protonation of the diastereotopic enediol face in *138* must lead to two epimers.
- The *spiro*-ring closure can, depending on the chirality of the monohydroxylated isopropyl group and its conformation in the transition state, lead to four diastereoisomers of which there must be two pairs with quite different mechanistical behavior: *140* and its structural isomer (not designed) are shown by recent mechanistical studies [54]

to be extremely short-lived and subject to an immediate [1,5]-H-shift. Thus, *143* is formed with the correct position of the double bond as in coleone F (*88*); the other two compounds (*141* and its isomer) are not prone to this sigmatropic rearrangement due to steric reasons. Instead they can undergo an acid catalysed nucleophilic opening of the *spiro*-cyclopropane ring which leads to *142* with the correct position

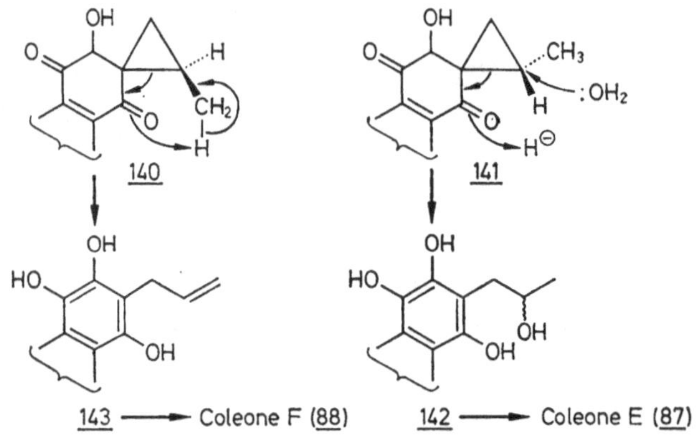

of the hydroxyl group in the product, as in coleone E (*87*). Examples of such cyclopropane ring openings will be given at the end of this section.

The structures *140* and *141* do not have long-wavelength-absorbing chromophores. It was a matter of great satisfaction to us as two groups [47, 39i] recently succeeded in identifying the postulated *spiro*-cyclopropyl cyclohexenediones of the *141* type and isolating them in crystalline form. Since then, more than 30 diterpenes of the *141* type have been isolated in rapid sequence, and their structures have been elucidated. A compilation of the structures found by the end of 1977 is provided in Table 4:

With mineral acid several *spiro*-cyclopropyl-cyclohexenediones produce stable purple colors which easily indicate the probable presence of these compounds in extracts.

Table 4: Coleones with spiro-Cyclopropyl-Cyclohexenedione Structure

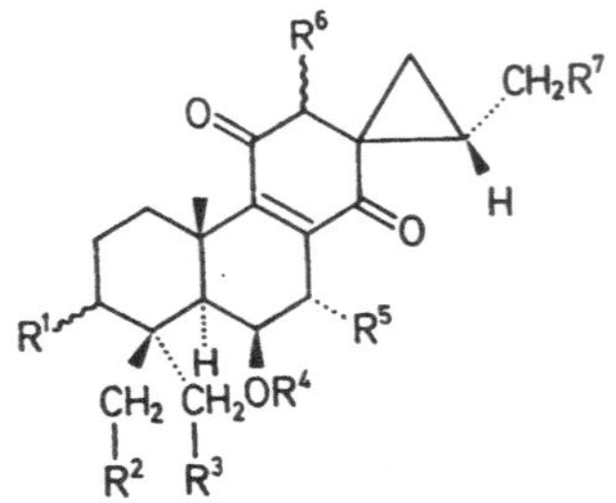

Compounds *144* to *159*

Name	Formula	R^1	R^2	R^3	R^4	R^5	R^6	R^7	Isolated from	References
Barbatusin	*144*	O=	H	H	$COCH_3$	OH	α $OCOCH_3$	H	a), b)	[47] [39k]
Coleone P	*145*	H	H	H	H	OH	β $OCOCH_3$	H	b)	[39k]
Coleone Q	*146*	H	H	H	H	OH	α $OCOCH_3$	H	b)	[39k]
Coleone R	*147*	α $OCOCH_3$	H	H	$COCH_3$	OH	α $OCOCH_3$	H	b)	[39k]
3β-Hydroxy-3-deoxo-barbatusin	*148*	β OH	H	H	$COCH_3$	OH	α $OCOCH_3$	H	a)	[47]
Coleone R'*	*149*	α $OCOCH_3$	H	H	$COCH_3$	OH	α OH	H	c)	[39q]
Coleone R''*	*150*	α $OCOCH_3$	H	H	H	OH	α OH	H	c)	[39q]
Coleone Y	*151*	α $OCOCH_3$	H	$OCOCH_3$	H	OH	α OH	H	c)	[39q]
Coleone Y'*	*152*	α OCOH	H	$OCOCH_3$	H	OH	α OH	H	c)	[39q]
Lanu 1*	*153*	H	OCOH	H	H	OCOH	β OH	H	d)	[40]
Lanu 8*	*154*	H	OCOH	H	H	OH	β OH	H	d)	[40]
CS-Ia*	*155*	H	$OCOCH_3$	H	$COCH_3$	OH	α $OCOCH_3$	H	e)	[39u]
CS-Id*	*156*	H	$OCOCH_3$	H	H	$OCOCH_3$	β OH	H	e)	[39u]
CS-IV*	*157*	H	$OCOCH_3$	H	H	$OCOCH_3$	α OH	H	e), h)	[39u]
Syl-A	*158*	H	H	H	$COCH_3$	OH	β OH	$OCOCH_3$	e)	[39u]
CS-VI*	*159*	β $OCOCH_3$	$OCOCH_3$	H	H	$OCOCH_3$	α OH	H	e)	[39u]

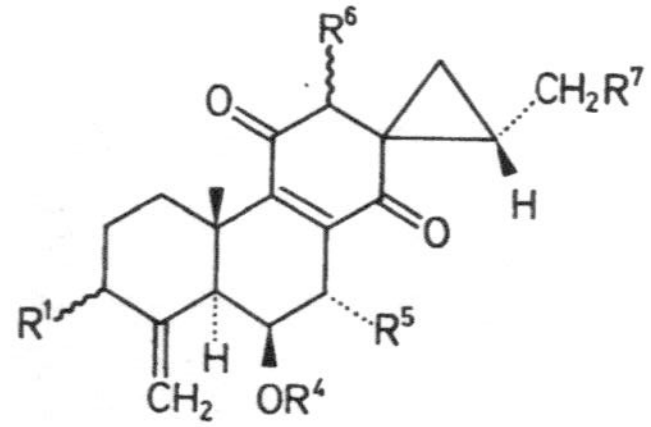

Compounds *160* to *169*

Name	Formula	R^1	R^2	R^3	R^4	R^5	R^6	R^7	Isolated from	References
Coleone G	*160*	α CH_3	–	–	H	$OCOCH_3$	α OH	H	f)	[39i]
Coleone J	*161*	α CH_3	–	–	H	OH	α OH	H	f)	[39i]
Syl-D*	*162*	$CH'_2=$	–	–	H	$OCOCH_3$	α OH	H	e)	[39u]
Syl-B*	*163*	βCH_3 + α OAc	–	–	H	H	α OH	H	e)	[39u]
Mon-A*	*164*	$CH_2=$	–	–	H	$OCOCH_3$	β $OCOCH_3$	H	g)	[39u]
Mon-B*	*165*	$CH_2=$	–	–	H	OCOH	β $OCOCH_3$	H	g)	[39u]
Mon-C*	*166*	$CH_2=$	–	–	H	$OCOCH_3$	β OCOH + β $OCOCH_3$	$OCOCH_3$	g)	[39u]
Coleone N	*167*	β CH_3	–	–	H	$OCOCH_3$	α $OCOCH_3$	H	b)	[39k]
Coleone N′	*168*	β CH_3	–	–	H	$OCOCH_3$	α OH	H	c)	[39q]
Coleone N″	*169*	β CH_3	–	–	H	OH	α OH	H	c)	[39q]

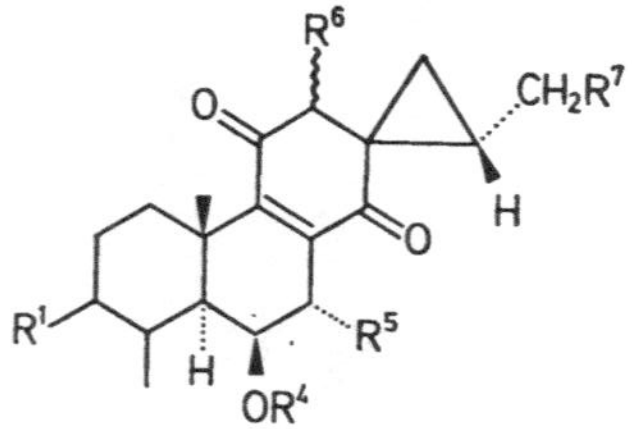

Compounds *170* to *172*

Name	Formula	R^1	R^2	R^3	R^4	R^5	R^6	R^7	Isolated from	References
Coleone O	*170*	CH_3	–	–	H	$OCOCH_3$	α OH	H	c), e), f), h)	[39k] [39q] [39u]
Coleone M	*171*	CH_3	–	–	H	$OCOCH_3$	α $OCOCH_3$	H	b)	[39k]
Syl-C*	*172*	CH_2OCOCH_3	–	–	H	$OCOCH_3$	α OH	H	e)	[39u]

a) *Coleus barbatus* (from Brasil); b) *Plecthrantus caninus*; c) *C. coerulescens*; d) *P. lanuginosus*; e) *Solenostemon sylvaticus*; f) *C. somaliensis*; g) *S. monostachys*; h) *P. garckeanus*. * Provisional Name

If one examines the oxidation pattern found to date in *spiro*-coleones (see Table 4), a distinct similarity is found with that of the acylhydroquinones. There is a significant difference, however, in the oxidation states of C(6) and C(7): to date there are no *spiro*-coleones with diosphenol or diketo functions in ring B. On the basis of the similar oxidation pattern, we conclude that the transition from *136* to *139* is a late metabolic step, and the oxidation of C(6) and C(7) to the diketone probably occurs in an independent pathway. The shift of a methyl group from C(4) to C(3), which was already mentioned in connection with a few royleanones (section 5.2) and fewer *p-quinomethanes* (section 5.3.1), is also found in numerous *spiro*-coleones. We would like to point out that the C(3) methyl group in the coleones G (*160*) and J (*161*) has α-configuration, and in the coleones N (*167*), N′ (*168*), and N″ (*169*) β-configuration, however. This can only point to the existence of a stereospecific methyl shift. The following proposal is made with the mechanistically plausible assumption that the shift is initiated by elimination of an oxygen function from C(3):

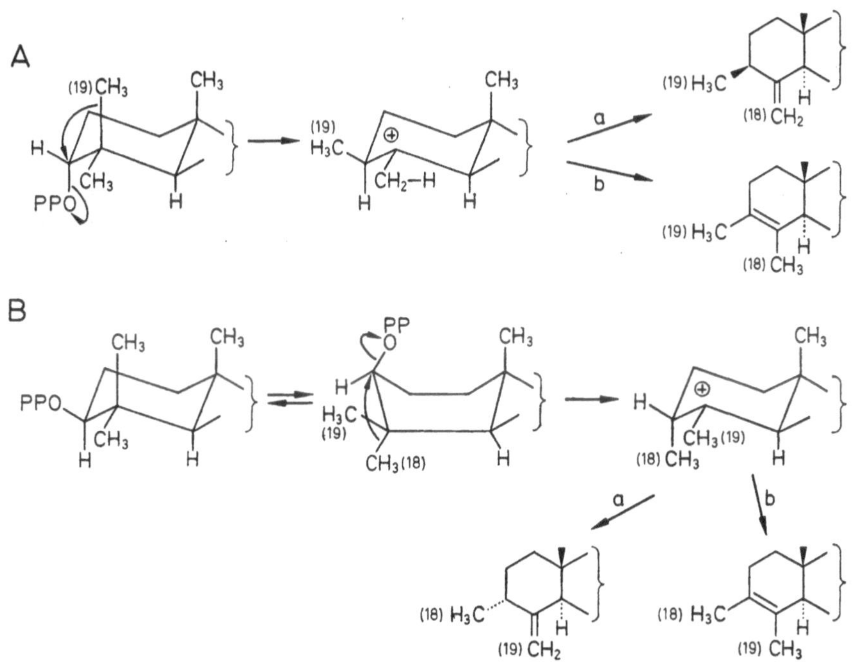

According to the reaction sequence A, the elimination of the C(3)-α-OH (e.g., as a pyrophosphate) induces migration of the axial CH_3(19) to C(3) with retention of β-configuration. Depending on the fate of the resulting carbonium ion, compounds with an exocyclic methylene group (route a) or with an endocyclic double bond (route b) are formed. Both types have been found. In the sequence B an analogous elimination of the C(3)-β-OH *via* the boat conformation leads to migration of the CH_3(18) to C(3) with α-configuration. These conceptions are supported by *in vitro* experiments on triterpenes (for literature references see [39k]) and, furthermore, by the fact that *Coleus somaliensis* contains the C(3)-β-oxygenated coleones H (*106*), I (*127*), and K (*129*) together with coleones G (*160*) and J (*161*) which both possess a C(3)-α-CH_3 group. Furthermore, coleone R (*147*) which contains a C(3)-α-acetoxy group occurs in *Plectranthus caninus* together with coleone N (*167*) which contains C(3)-β-CH_3. An analogous situation is found among the coleones of *C. coerulescens* [39q].

Worthy of special note are the compounds with two neighboring exocyclic methylene groups which form a cisoid butadiene system, such as syl-D (*162*) from *Solenostemon sylvaticus* and mon-A (*164*), mon-B (*165*) and mon-C (*166*) from *S. monostachys*. To our knowledge, such a structural type is new among natural terpenoid substances. The listed compounds are very unstable and easily undergo further reactions. It is possible that they have been formed in the plant by the following route:

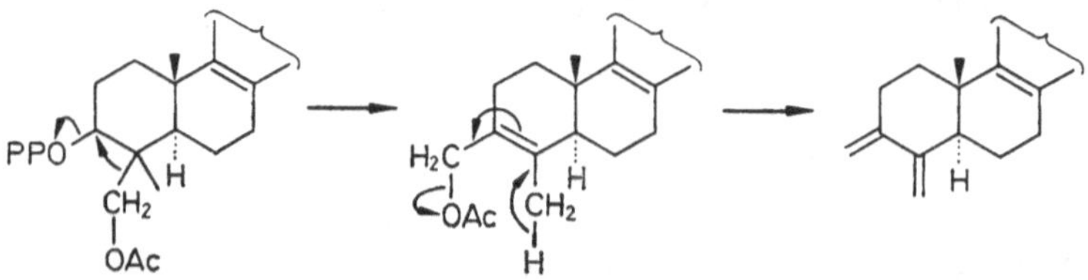

Two of the isolated compounds, syl-A (*158*) and mon-C (*166*), show two cyclopropylcarbinol groups. Alternatively to a subsequent hydroxylation, formation of the *spiro*-cyclopropane from the double oxygenated isopropyl group could be postulated here:

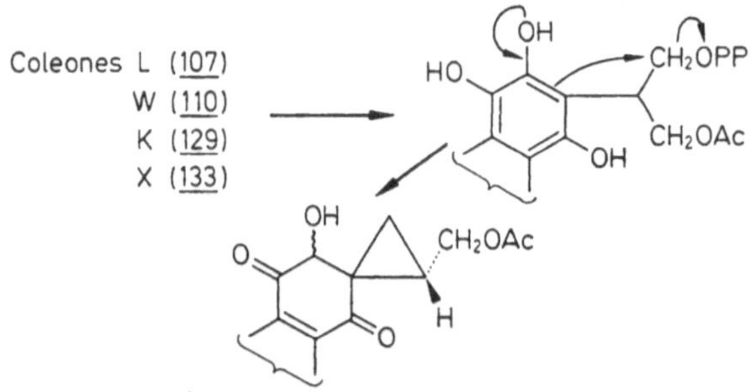

e.g., 158, 166

Before discussing some specific rearrangements of *spiro*-coleones, cyclobutatusin (*173*) found in *Coleus barbatus* should be mentioned:

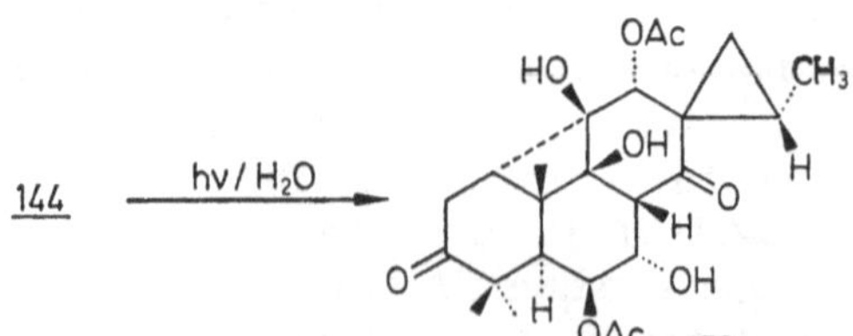

173

This pentacyclic diterpene was probably produced from barbatusin (*144*) by a photochemical process followed by water addition to the enone system. Similar cyclobutane ring closures between C(1) and C(11) have been recently observed *in vitro* on royleanones [55].

A remarkable transition from *spiro*-coleones into a royleanone or a *p*-quinomethane was brought about *in vitro* in a methanol-water solution containing a trace of base [39q]:

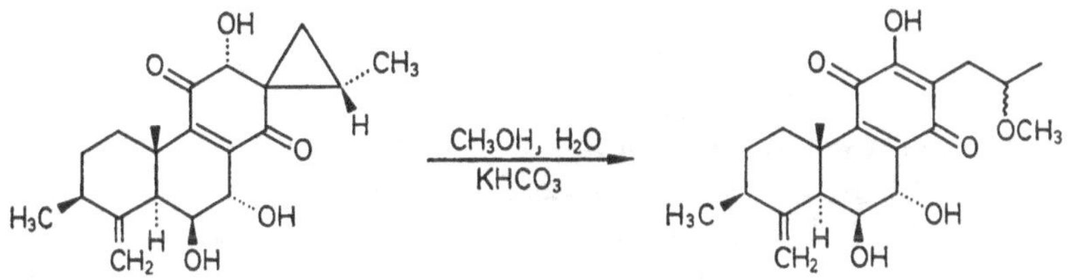

169 174

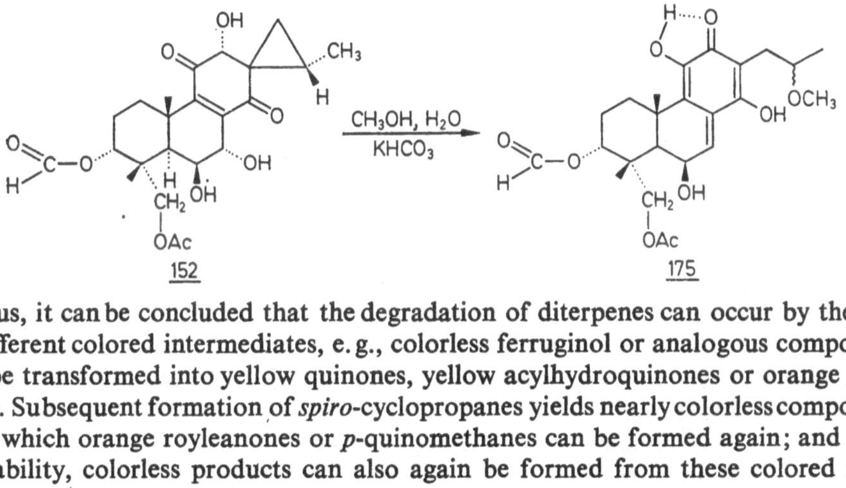

152 175

Thus, it can be concluded that the degradation of diterpenes can occur by the way of different colored intermediates, e.g., colorless ferruginol or analogous compounds can be transformed into yellow quinones, yellow acylhydroquinones or orange diketones. Subsequent formation of *spiro*-cyclopropanes yields nearly colorless compounds from which orange royleanones or *p*-quinomethanes can be formed again; and in all probability, colorless products can also again be formed from these colored intermediates in the course of further degradation.

5.3.4. Secocoleones

These are compounds with oxidatively cleaved ring bonds. To date coleone A (*176*) and edulone A (*183*) are known. As mentioned earlier, the first of these compounds occurs together with coleone B (*119*) in considerable amounts in the leaves of *Coleus igniarius* (Lake Dschalla, Kilimandjaro). It crystallizes in beautiful, dark red crystals and is quite stable, but undergoes numerous noteworthy reactions [39b]. The accumulation of functional groups is very unusual for a naphthaquinone derivative, and structure elucidation required a substantial amount of chemical-analytical work [39b]. The structural link to abietane in particular becames clear only after tedious determination of the relative positions of all functional groups.

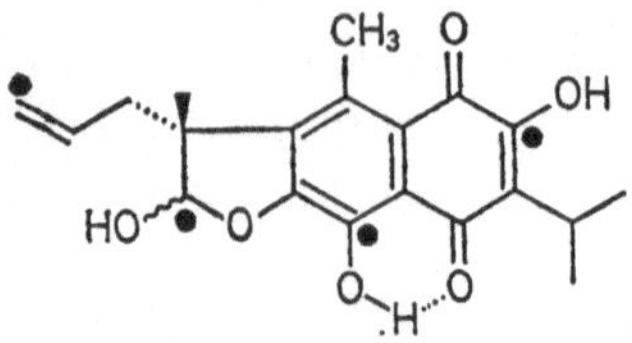

176

This link was proven by incorporation experiments with 2-^{14}C-mevalonate applied to branches of *C. igniarius* and chemical degradation of the labelled coleones A and B thereby obtained. Incorporation occurred in *176* – as shown by degradation – at C(1), C(7), C(12), and C(18) in accordance with its biogenesis from abietane (labelled atoms in coleone A, cf., *176*) [56]. It is thus clear that a ring cleavage between C(10) and C(1) is involved and could occur as follows: *177* → *178* → *176*.

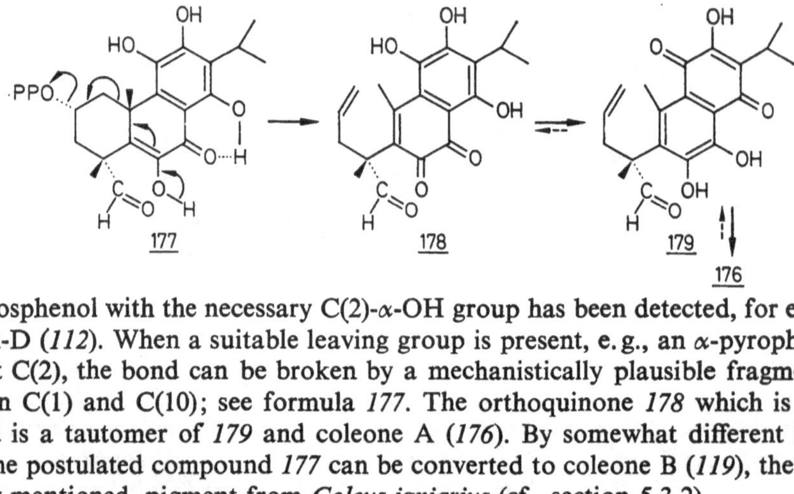

177 178 179 176

A diosphenol with the necessary C(2)-α-OH group has been detected, for example, in mon-D (*112*). When a suitable leaving group is present, e.g., an α-pyrophosphate ester at C(2), the bond can be broken by a mechanistically plausible fragmentation between C(1) and C(10); see formula *177*. The orthoquinone *178* which is thereby formed is a tautomer of *179* and coleone A (*176*). By somewhat different reaction steps the postulated compound *177* can be converted to coleone B (*119*), the second, already mentioned, pigment from *Coleus igniarius* (cf., section 5.3.2).

Edulone A, the second *seco*-coleone discovered to date, is a red orange pigment from *Coleus edulis* (origin: Kenya). Its relatively high acidity and strong reducing power are noticeable: Tollens and Tillman's reagents are promptly reduced. This property may be attributed to an enediol group. The very unusual absorption spectrum is different from those of all previously discovered coleones; it is closest to a 6-acylisocoumarin (*180*) type. Structure *183* was derived for edulone A from spectral data [39s]; the absolute configuration is not yete definite.

This unusual structure can be derived from a coleone, assuming an oxidative cleavage of the C(4)-C(5) bond as suggested in *181*. Formation of a lactone with C(12)-OH and ring closure between C(3) and C(11) must follow. Similar ring closures have been observed *in vitro* with royleanones in strong sulfuric acid [57].

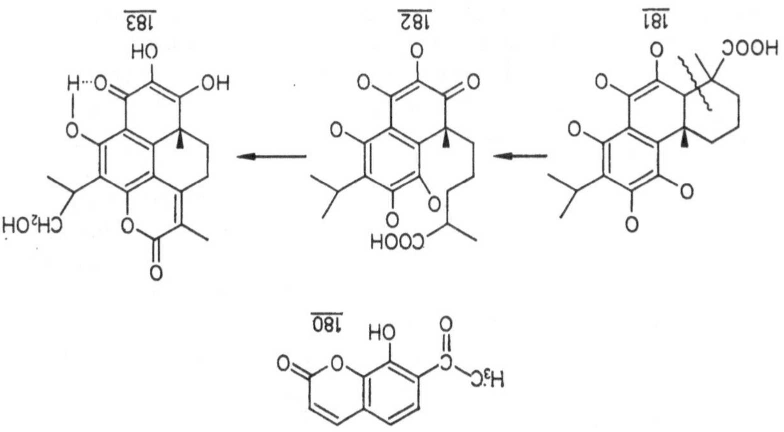

5.3.5. Dimeric Diterpenes

A survey of the structures of the coleones described in sections 5.3.1, 5.3.2 and 5.3.3 allows the conclusion that some of these highly reactive molecules easily enter into coupling and addition reactions with other molecules. Such reaction products have only been observed recently, presumably because the coleones in the glands are well protected and hardly come into contact with other molecules, with the exception of molecules similar to themselves, i.e., molecules with similar reactivity and polarity.

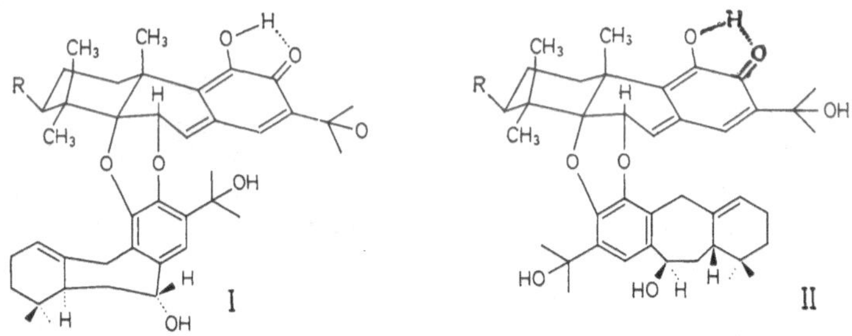

184 R = H, 185 R = OAc (I) 184 R = H, 185 R = OAc (II)

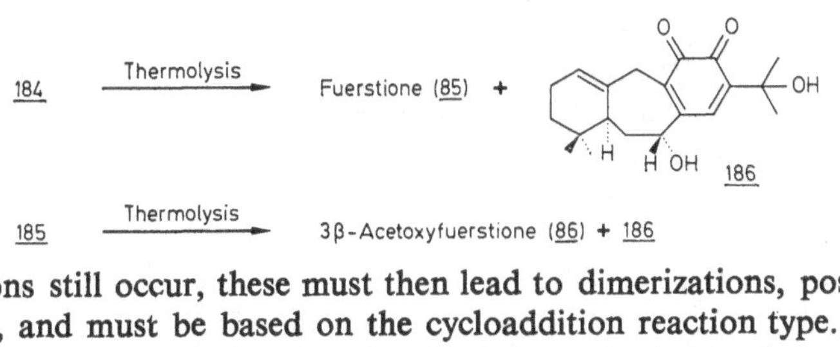

Should reactions still occur, these must then lead to dimerizations, possibly even oligomerizations, and must be based on the cycloaddition reaction type. Naturally, dimerizations by way of so-called « phenol oxidation » and esterifications can also be expected. Dimeric diterpenes, that is, those which are C-C and C-O linked were hardly heard of previously, with the exception of the colorless, C-C linked podototarin from *Podocarpus totara* [58] and the likewise colorless maytenone from *Maytenus dispermus* [59]. The first colored dimeric diterpenes which were recently discovered as glandular pigments from the South Indian *Plectranthus nilgherricus* are the yellow compounds nilgherrone A (*184*) and nilgherrone B (*185*) [39p]. Neither of the structural possibilities I or II can be decided on to date. The compounds can be thermally cleaved and the cleavage products have proven to be fuerstione (*85*) in the case of nilgherrone A, and 3β-acetoxyfuerstione (*86*) in the case of nilgherrone B. As previously mentioned, both of these compounds can also be isolated in free form from *P. nilgherricus* (section 5.3.1). The component common to both dimers is the *ortho*-quinone *186*, in which the ring B of the abietane skeleton has been expanded to a seven-membered ring. Thus, in the dimerization step a *p*-quinomethane undergoes reaction with an *ortho*-quinone in the sense of a so-called heterodiene synthesis, whereby a benzodioxane ring is formed:

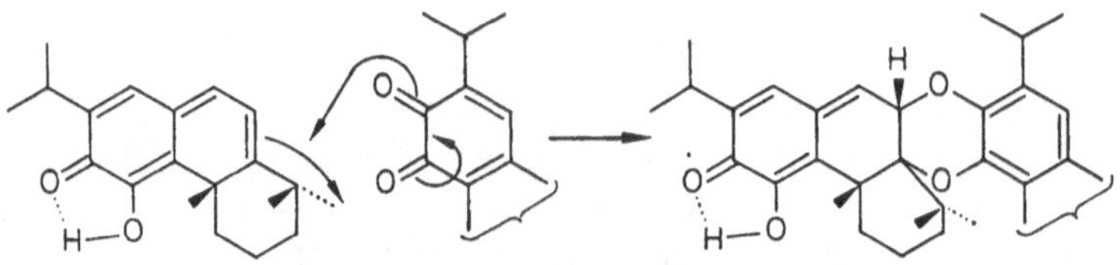

A further, this time red, dimeric diterpene was recently detected in *P. myrianthus* [39 o]. Its structure is not yet established with certainty; the two molecule halves are connected by a five-membered ketal function.

6. Summary and Conclusions

I have attempted to provide a survey of the structures and occurrence of terpenoid pigments in plants. In the process, isolation procedures and the argumentation of the determination of structures have not been mentioned since they are probably of slight nterest to the botanist. The chemist, however, is the one responsible for these questioins, whereas he is incapable of answering the subsequent questions regarding the significance to the plant of the isolated materials. However, in all the years during which we raised *Coleus*, *Plectranthus*, *Fuerstia*, and *Solenostemon* species, it was always noticeable that the plants never suffered from snail damage, insect attack, and fungal diseases. It is thus possible that repellents, antifeedants, and substances with other protective functions are contained in the glandular substances. Only the relatively high antibiotic effect of coleone A and a few *p*-quinomethanes have been experimentally proven, however.

This survey shows that the structurally most uniform group of the terpenoid pigments stems from the diterpenes (not considering the carotenoids). All isolated compounds belong biogenetically to the abietane type, including totarol. One may therefore construct an approximate sequence of structural modification or degradation on the basis of the wealth of isolated compounds, especially when one also considers the related colorless compounds which were not included in this survey. General experenice shows that more highly dehydrogenated and oxygenated compounds stem from later biogenetic steps. The postulated sequence would be more convincing, of course, if it could be supported by compounds from the same plant. The tanshinones, for example, could be helpful here. They all have very similar oxidation levels, however. An exception is provided by the *Plectranthus* from Abessinia whose royleanones offer a clearly wider range of oxidation levels. Otherwise, plants can apparantly only store one or two compounds with a similar oxidation level (dehydrogenation and oxygenation) at high concentration. The individual plant thus provides a sort of momentary picture from a movie whose message only becomes clear after correct sequential ordering of as many short pictures as possible. Even today relatively few labiates have been thoroughly investigated for terpenoid pigments. We may therefore assume that many more findings may result from investigation of *Coleus, Plectranthus, Salvia,* and related species, and thus that the message of the movie will become clearer.

The most unstable compounds originate from leaf glands. It has already been mentioned that their survival is dependent on the fulfillment of very special requirements by the accompanying substances, e.g., pH, absence of electrophiles and/or nucleophiles.

Structural modification begins by oxidation of ring C. This can be understood from the scheme of the biogenesis of the abietane skeleton: the formation of the isopropyl group leads to a reactive intermediate that is presumably the start for dehydrogenations and oxygenations. Many abietane derivatives therefore possess double bonds and hydroxyl groups in ring C or are phenolic; a marked tendency towards dehydrogenation and oxygenation on rings B and A is then noticeable. The driving force for this structural change is unknown. Some compounds are very close to the colorless aromatic state, i.e., the transformation of the colored diterpene into a colorless, stable aromatic system is only blocked, for instance, by the presence of a single quaternary methyl group. An example is fuerstione (*85*). The stable aromatic state can be easily reached by a rearrangement or an oxidative elimination. Such substances have been produced from coleones *in vitro* (see the original papers), but have not yet been identified in the plants. However, there are indications for their occurrence. For example, on thin-layer chromatograms of many extracts one finds colorless spots which fluoresce strongly under UV light.

We interpret the coleones (and also the royleanones) *as intermediates of terpene catabolism.* In surveying the structures discovered to date one can only be impressed by the diversity of possibilities which are available to plants for transforming or degrading primary products with terpene structure. Several of the molecules which have been found are so highly oxidized and functionalized, that further degradation steps would presumably lead to a collapse of the C_{20} skeleton into several smaller fragments. Such fragments and their connection to the original molecules have not yet been encountered. We suspect that a few individual coleones are subject to an active metabolism; in any case we have found a spectrum of substances in the same plant in June/July which was different from the spectrum in September/October.

The concept of « plant phenolics » is used nearly exclusively today for derivatives of the shikimate and polyacetate pathway; although terpenoid phenols, such as ferruginol, sugiol, totarol, etc., are typical phenols. The numerous newly found royleanones and coleones show that this concept must now also be extended to include compounds of pure terpenoid origin.

In regard to my own publications which have been mentioned in this survey, they have only been made possible by the collaboration of numerous undergraduates, graduate students and postdoctoral co-workers (see list of references). I would like to especially mention the successful collaboration over many years of Dr. Peter Rüedi. Financial support was provided by the Swiss National Science Foundation and the Kanton Zürich. For aid in the acquisition of plant material we would like to thank Messrs. P.R.O. Bally (Nairobi), G. A. Classen (Nairobi), P. J. Greenway (Nairobi), A. D. Q. Agnew (Aberystwyth), J. P. Brenan (Kew), J. B. Harborne (Reading), J. D. Supthut (Zürich), and numerous botanical gardens for shipment of seeds.

References

[1] R. H. Thomson, 1971: Naturally occurring quinones, Academic Press, New York.
[2] R. Hegnauer, 1973: Chemotaxonomie der Pflanzen, Birkhäuser, Basel.
[3] W. Karrer, 1958: Konstitution und Vorkommen der organischen Pflanzenstoffe, Birkhäuser, Basel; W. Karrer, E. Cherbuliez and C. H. Eugster, 1977: Ergänzungsband 1, Birkhäuser, Basel.
[4] K. Naya, K. Takai, M. Nakanishi and K. Omura, 1977: Chemistry Letters (Tokyo), 1179.
[5] C. J. W. Brooks and G. H. Draffan, 1969: Tetrahedron **25**, 2865.
[6] V. Seshadri, A. K. Batta and S. Rangaswami, 1971: Current Sci. **40**, 630; ref. Chem. Abstr. **76**, 83512 (1972).
[7] J. R. Gray, T. J. Mabry, A. A. Bell, R. D. Stipanovic and M. J. Lukefahr, 1976: Chem. Commun., 109.
[8] R. D. Stipanovic, A. A. Bell, D. H. O'Brien and M. J. Lukefahr, 1977: Tetrahedron Letters, 567; *iidem*, 1978: Phytochem. **17**, 151.
[9] D. Meuche and S. Huneck, 1966: Chem. Ber. **99**, 2669; 1969: **102**, 2493.
[10] F. Bohlmann and Ch. Zdero, 1973: Chem. Ber. **106**, 1337.
[11] K. Vokáč, Z. Samek, V. Herout and F. Šorm, 1970: Coll. czech. chem. Commun. **35**, 1296.
[12] D. J. Bertelli and J. H. Crabtree, 1968: Tetrahedron **24**, 2079.
[13] U. Weiss, 1977: Abstracts Sec. Int. Mycol. Congress University South Florida, Tampa, S. 727 and private communication of 5/10/77.
[14] A. G. González, C. G. Francisco, R. Freire, R. Hernández, J. A. Salazar and E. Suárez, 1975: Phytochem. **14**, 1067.
[15] P. Martinod, A. Paredes, F. Delle Monache and G. B. Marini-Bettòlo, 1976: Phytochem. **15**, 562.
[16] G. C. S. Reddy, K. N. N. Ayengar and S. Rangaswami, 1976: Indian J. Chemistry **14 B**, 131; ref. Chem. Abstr. **85**, 63197 (1976).
[17] P. M. Brown, M. Moir, R. H. Thomson, T. J. King, V. Krishnamoorthy and T. R. Seshadri, 1973: J. chem. Soc. Perkin I, 2721; see also K. Nakanishi, V. P. Gullo,

I. Miura, T. R. Govindachari and N. Viswanathan, 1973: J. Amer. chem. Soc. **95**, 6473, and F. Delle Monache, G. B. Marini-Bettòlo et al., 1973: J. chem. Soc. Perkin I, 2725.

[18] F. Delle Monache, G. B. Marini-Bettòlo, O. Gonçalves de Lima, I. L. D'Albuquerque and J. S. de Barros Côelho, 1972: Gazzetta chim. ital. **102**, 317.

[19] J. D. Martín, 1973: Tetrahedron **29**, 2997.

[20] F. Delle Monache, M. Pomponi, G. B. Marini-Bettòlo and O. Gonçalves de Lima, 1974: An. Quim. **70**, 1040; ref. Chem. Abstr. **83**, 131782 (1975).

[21] O. Isler, H. Gutmann and U. Solms, 1971: Carotenoids, Birkhäuser, Basel.

[22] O. Straub, 1976: Key to Carotenoids, Birkhäuser, Basel.

[23] R. Buchecker, P. Hamm and C. H. Eugster, 1974: Helv. chim. Acta **57**, 631.

[24] B. D. Davies, this volume p. 31 ff.

[25] G. Pattenden, J. E. Way and B. C. L. Weedon, 1970: J. chem. Soc. C, 235.

[26] W. Eschenmoser and C. H. Eugster, 1975: Helv. chim. Acta **58**, 1722; U. Vögeli, W. Eschenmoser and C. H. Eugster, 1975: Helv. chim. Acta **58**, 2044.

[27] Y. Inouye and H. Kakisawa, 1969: Bull. chem. Soc. Japan **42**, 3318; T. Hayashi, H. Kakisawa, H.-Y. Hsu and Y. P. Chen, 1970: Chem. Commun., Chem. Soc. London, 299.

[28] A. G. Gonzalez, B. M. Fraga, J. G. Luis and A. G. Ravelo, 1975: An. Quim. **71**, 701; ref. Chem. Abstr. **85**, 2505 (1976).

[29] A. S. Romanova, G. F. Pribylova, A. V. Patudin, E. S. Leskova, D. A. Pakaln and A. I. Baňkovskii, 1972: Khim. Prirod. Soedin., 237.

[30] A. S. Romanova, A. V. Patudin, G. F. Vlasova and A. I. Baňkovskii, 1973: Ref.: Chem. Abstr. **79**, 15793.

[31] C. H. Brieskorn and L. Buchberger, 1973: Planta medica **24**, 190.

[32] M. Hensch, P. Rüedi and C. H. Eugster, 1975: Helv. chim. Acta **58**, 1921.

[33] S. V. Bhat, P. S. Kalyanaraman, H. Kohl, N. J. de Souza and H.-W. Fehlhaber, 1975: Tetrahedron **31**, 1001.

[34] T. Miyase, P. Rüedi and C. H. Eugster, 1977: Helv. chim. Acta **60**, 2770.

[35] a) Z. Taira and W. H. Watson, 1976: Acta cryst. **B 32**, 2149; b) W. H. Watson, Z. Taira, X. A. Domínguez, H. González, M. Gutiérrez and R. Aragón, 1976: Tetrahedron Letters, 2501. c) X. A. Domínguez, H. González, R. Aragón, M. Gutiérrez, J. S. Marroquín and W. Watson, 1976: Planta medica **30**, 237.

[36] Z. Taira, W. H. Watson and X. A. Domínguez, 1976: J. chem. Soc. Perkin II, 1728.

[37] E. Wenkert, J. de Paiva Campello, J. D. McChesney and D. J. Watts, 1974: Phytochem. **13**, 2545.

[38] F. Marletti, F. Delle Monache, G. B. Marini-Bettólo, M. Do Carmo, M. De Aranjo, M. De Salete Barros Cavalcanti, I. L. D'Albuquerque and O. Gonçalves de Lima, 1976: Gazzetta chim. ital. **106**, 119.

[39] a) Fuerstione: P. Karrer and C. H. Eugster, 1951: Z. Naturforsch. **6**b, 276; 1952: Helv. chim. Acta **35**, 1139; J. S. Scarpa, M. Ribi and C. H. Eugster, 1966: *ibid* **49**, 858; D. Karanatsios, J. S. Scarpa and C. H. Eugster, 1966: *ibid.* **49**, 1151. b) Coleone A: C. H. Eugster, H.-P. Küng, H. Kühnis and P. Karrer, 1963: *ibid.* **46**, 530; D. Karanatsios and C. H. Eugster, 1965: *ibid.* **48**, 471. c) Coleone B: M. Ribi, A. Chang Sin-Ren, H.-P. Küng and C. H. Eugster, 1969: *ibid.* **52**, 1685. d) Coleone C: P. Rüedi and C. H. Eugster, 1971: *ibid.* **54**, 1606. e) Coleone D: P. Rüedi and C. H. Eugster, 1972: *ibid.* **55**, 1736. f) Coleone E: P. Rüedi and C. H. Eugster, 1972: *ibid.* **55**, 1994. g) Coleone F: P. Rüedi and C. H. Eugster, 1973: *ibid.* **56**, 1129. h)

Coleones H, I, K: M. Moir, P. Rüedi and C. H. Eugster, 1973: *ibid.* **56**, 2534. i) Coleones G, J: M. Moir, P. Rüedi and C. H. Eugster, 1973: *ibid.* **56**, 2539. k) Coleones M, N, O, P, Q, R: S. Arihara, P. Rüedi and C. H. Eugster, 1975: *ibid.* **58**, 343. l) Coleones C, D, I, I′: P. Rüedi and C. H. Eugster, 1975: *ibid.* **58**, 1899. m) Coleone L: P. Rüedi and C. H. Eugster, 1977: *ibid.* **60**, 1233. n) Coleones S, T: S. Arihara, P. Rüedi and C. H. Eugster, 1977: *ibid.* **60**, 1443. o) Coleones U, V, W: 14-O-Formyl-coleone-V: T. Miyase, P. Rüedi and C. H. Eugster, 1977: *ibid.* **60**, 2770. p) 3β-Acetoxyfuerstione, Nilgherrone A, Nilgherrone B: T. Miyase, P. Rüedi and C. H. Eugster (1977): *ibid.* **60**, 2789. q) 16-O-Acetylcoleone C, 16-O-Acetylcoleone D, Coleone Y and others: K. Grob, P. Rüedi and C. H. Eugster, 1978: *ibid.* **61**, 871. r) Parviflorones A, B, C, D, E, F: P. Rüedi and C. H. Eugster, 1978: *ibid.* **61**, 709. s) Edulone A: G. Buchbauer, P. Rüedi and C. H. Eugster, 1978: *ibid.* **61**, 1969. t) Coleones C, H and spirocoleones with butadiene structure: T. Miyase, P. Rüedi and C. H. Eugster, 1977: Chem. Commun., 859; Helv. chim. Acta **62**, in preparation, u) Coleones W, W′, X and spirocoleones: T. Miyase, P. Rüedi and C. H. Eugster, 1977: Chem. Commun., 859; Helv. chim. Acta **62**, in preparation; N'Tanda Kabengele, P. Rüedi and C. H. Eugster, 1979: *ibid.* **62**, 2374.

[40] M. Schmid, 1978: unpublished results.

[41] S. M. Kupchan, A. Karim and C. Marcks, 1968: J. Amer. chem. Soc. **90**, 5923; 1969: *iid.*, J. org. Chem. **34**, 3912.

[42] J. D. Martín, 1973: Tetrahedron **29**, 2553.

[43] F. Delle Monache, F. Marletti, G. Marini-Bettólo, J. F. De Mello and I. L. D'Albuquerque, 1977: Gazzetta chim. ital. **107**, 319.

[44] S. Arihara, P. Rüedi and C. H. Eugster, 1975: Helv. chim. Acta **58**, 447.

[45] M. H. Zenk, H. El-Shagi and B. Ulbrich, 1977: Naturwiss. **64**, 585.

[46] A. D. Q. Agnew, 1974: Upland Kenya Wild Flowers, Oxford University Press, Nairobi.

[47] A. H.-J. Wang, I. C. Paul, R. Zelnik, K. Mizuta and D. Lavie, 1973: J. Amer. chem. Soc. **95**, 598; A. H.-J. Wang, I. C. Paul, R. Zelnik, D. Lavie and E. C. Levy, 1974: *ibid.* **96**, 580; R. Zelnik, D. Lavie, E. C. Levy, A. H.-J. Wang and I. C. Paul, 1977: Tetrahedron **33**, 1457.

[48] J. S. Tandon, 1977: private communication.

[49] R. H. Burnell, L. Mo and M. Moinas, 1972: Phytochem. **11**, 2815.

[50] T. Yanagawa and Y. Hirose, 1971: Ref. Chem. Abstr. **75**, 148545.

[51] K. Hata and M. Sogo, 1959: Chem. Abstr. **53**, 22925; K. Doi and T. Shibuya, 1972: Phytochem. **11**, 1175.

[52] J. B.-son Bredenberg and J. Gripenberg, 1956: Acta chem. scand. **10**, 1511; J. B.-son Bredenberg, 1960: *ibid.* **14**, 385.

[53] H. P. Weber, T. J. Petcher, P. Rüedi and C. H. Eugster, 1976: Helv. chim. Acta **59**, 1221.

[54] H.-J. Hansen, 1971: In: Thyagaraian, Mechanisms of molecular migrations, Vol. 3, S. 177, Wiley-Interscience, New York.

[55] E. O. Edwards and P. T. Ho, 1978: Canad. J. Chemistry **56**, 733.

[56] M. Ribi, Di Tullio, C. H. Eugster and D. Arigoni, 1971: Unpublished results.

[57] M. Hensch, C. H. Eugster, and H. P. Weber, 1975: Helv. chim. Acta **58**, 1934.

[58] R. C. Cambie, W. R. J. Simpson and L. D. Colebrook, 1962: Chemistry and Industry, 1757.

[59] A. W. Johnson, T. J. King and R. J. Martin, 1961: J. chem. Soc., 4420.

Recent developments in flavonoid biosynthesis*

H. Grisebach

Lehrstuhl für Biochemie der Pflanzen,
Biologisches Institut II der Universität Freiburg i. Br.

1. Introduction

Early tracer studies led to a basic knowledge of the building units of flavonoids and to an understanding of some details of their biosynthesis. The biogenetic relationships among flavonoids which followed from these *in-vivo* studies are shown in Fig. 1.

The work of the last few years has now concentrated on the enzymology of this biosynthetic pathway and on its regulation. A correlation between genetic make up and the biochemistry of the flavonoid pathway is also now emerging through work with mutants. Plant cell suspension cultures have played a major role in some of these developments.

The biosynthesis of flavonoids has recently been covered in two extensive reviews (Hahlbrock and Grisebach, 1974; Wong, 1976). In this chapter I shall therefore concentrate the discussion on some selected topics which are at present at the center of research.

2. Flavanone synthase**

They key enzyme for the formation of the flavonoid skeleton is flavanone synthase. It catalyses the condensation of 4-coumaroyl-CoA with 3 molecules of malonyl-CoA to form the flavanone naringenin (Fig. 2). After a long search for this enzyme it was finally discovered in 1972 by Kreuzaler and Hahlbrock in cell suspension cultures of parsley (*Petroselinum hortense*) and in very young parsley leaves (Kreuzaler and Hahlbrock 1972). In retrospect it is clear why earlier attempts to find this enzyme met with failure. The enzyme is strongly inhibited by coenzyme A ($K_i \approx 4 \times 10^{-5}$ M) (Kreuzaler and Hahlbrock, 1975). Since 4-coumaroyl-CoA was not available in earlier investigations the incubation mixture contained 4-coumaric acid, ATP and larger amounts of CoASH, as well as a coenzyme A ligase from beef liver and labelled malo-

* Our own work described in this review was supported by the Deutsche Forschungsgemeinschaft (SFB 46) and by Fonds der Chemischen Industrie.

** See Addendum.

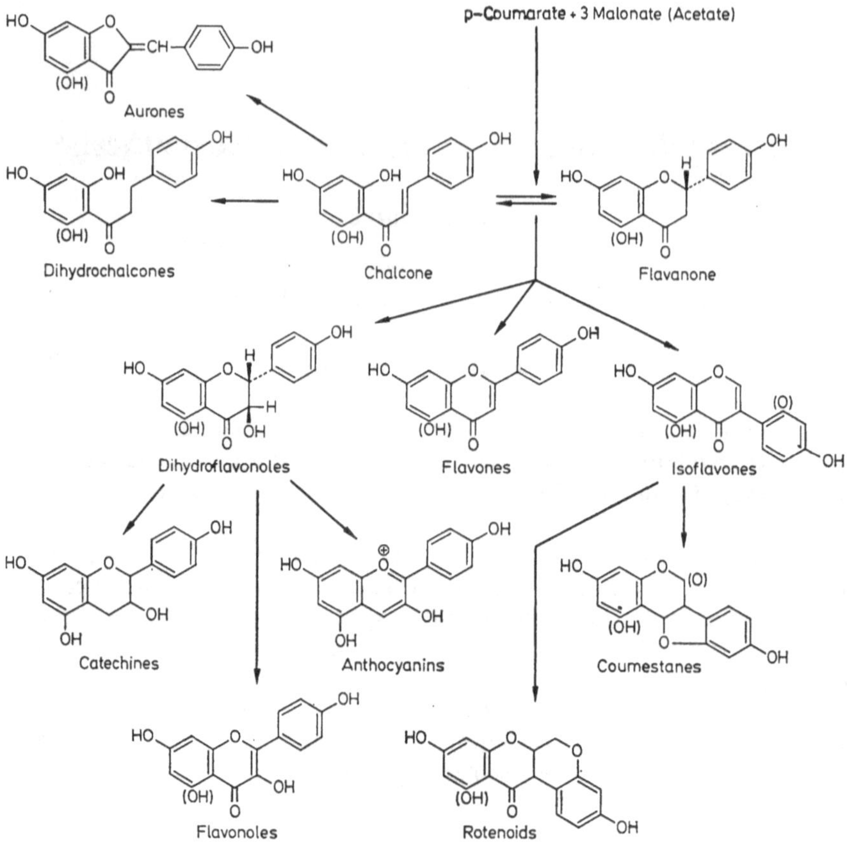

Fig. 1: Biogenetic relationships of flavonoids as deduced from tracer experiments *in vivo*

nyl-CoA. The concentration of CoASH present in the incubation would have inhibited naringenin formation. Only when the CoA ester of 4-coumaric acid itself was used as substrate was naringenin formed.

Flavanone synthase has in the meantime also been found in cell cultures of *Haplopappus gracilis* (Saleh et al., 1978), in red cabbage seedlings (G. Hrazdina and Creasy, 1979), in elicitor induced cotyledons of soybean (U. Zähringer et al., 1977, 1978), and in anthers of *Tulipa co.* »Apeldoorn» (Sütfeld and Wiermann, 1976).

It had formerly been assumed that the first product of the condensation reaction shown in Fig. 2 would be the tetrahydroxychalcone which could then be cyclized by the action of chalcone isomerase (EC 5.5.1.6) to the flavanone naringenin (Grisebach, 1962). However with a partially purified synthase from parsley cell cultures it was found that the product is the flavanone and not the chalcone (Kreuzaler and Hahlbrock, 1975). The synthase which was free of chalcone isomerase was incubated with 4-coumaroyl-CoA and [2-^{14}C]-malonyl-CoA in the presence or

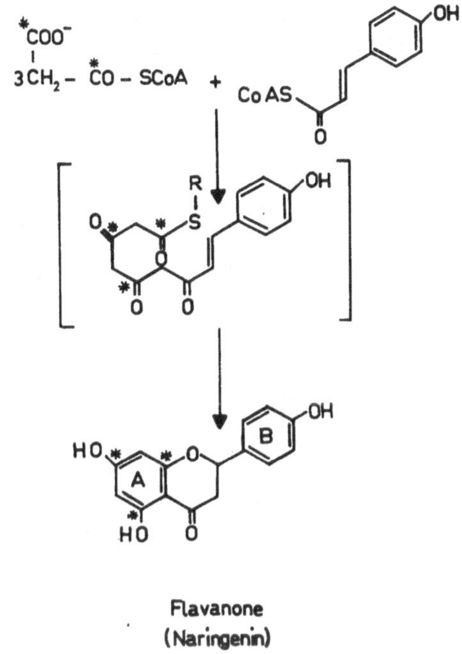

Fig. 2: Flavanone synthase reaction. R could be the coenzyme A residue or enzyme

absence of the unlabelled chalcone. The conditions allowed reisolation of the chalcone, despite its rapid non-enzymic isomerization. With incubation times between 0.5 and 6 min, radioactivity was only detected in naringenin and bis-noryangonin (see below) but in no case in the chalcone. The lowest amount of chalcone detectable would have been about 0.1 to 1% of the radioactivity isolated as naringenin. These results leave no doubt that with the synthase from parsley cell cultures flavanone is the first product released from the enzyme. It has, however, not yet been shown that the naringenin which is produced is the natural (–) 2S isomer.

In the synthase reaction with 4-coumaroyl-CoA three labeled products in addition to naringenin were observed on the chromatograms. By comparison with synthetic reference compounds the products were identified as p-hydroxybenzalacetone (Fig. 3; I), 4-hydroxy-5, 6-dihydro-6(4-hydroxyphenyl)2-pyrone (II) (Hrazdina et al., 1976) and bis-noryangonin (III) (Kreuzaler and Hahlbrock, 1975). These products obviously originate by condensation of 4-coumaroyl-CoA with only one or two molecules of malonyl-CoA, respectively. The release of these products from the enzyme prior to the formation of flavanone is influenced by thiol reagents, the ionic strength of the buffer and by the nature of the cinnamoyl-CoA « starter molecule » (see below). While the occurrence of styrylpyrones and the dihydropyrone derivates in some higher plant species is known (Mors et al., 1961) these compounds could not be detected in the parsley cell cultures. These products are therefore artifacts originating from the premature release of enzyme-bound intermediates under the *in vitro* incubation conditions.

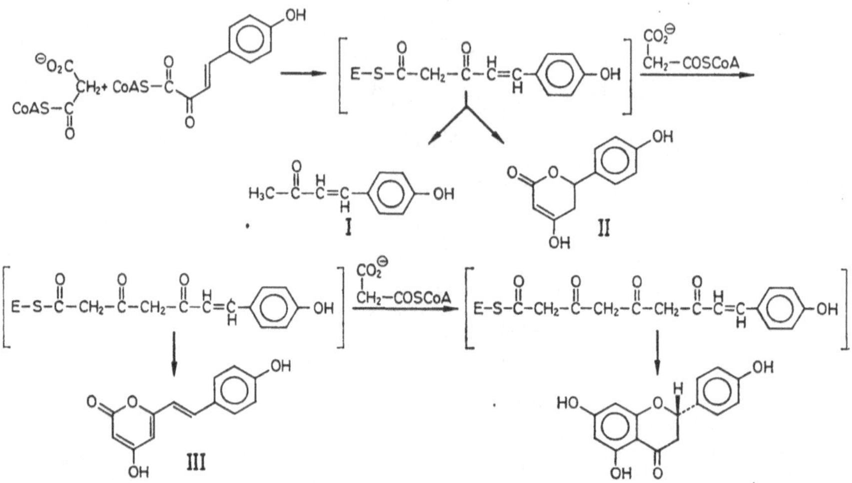

Fig. 3: Structures of products which are formed in the synthase reaction under *in vitro* conditions

The substrate specificity of flavanone synthase was investigated with the purified enzyme from parsley cell cultures (Hrazdina et al., 1976) and *Haplopappus gracilis* cell cultures (Saleh et al., 1978). At the pH-optimum for naringenin formation (pH 8) only a very small amount of eriodictyol (5,7,3′,4′-tetrahydroxyflavanone) was formed with caffeoyl-CoA as substrate, besides large amounts of the corresponding release products mentioned above. With feruloyl-CoA no detectable amount of the expected flavanone homoeriodictyol was detectable at pH 8. Again the corresponding release products were formed at considerably higher rates under these conditions than in the presence of 4-coumaroyl-CoA. These results led us to conclude that naringenin is the only natural product of the synthase reaction. In contrast, it appeared that with the crude synthase from *Tulipa* a substantial amount of eriodictyol was formed from caffeoyl-CoA, although quantitative data was not available (Sütfeld and Wiermann, 1976). It was then discovered that the pH optimum for eriodictyol formation with caffeoyl-CoA is not 8, but is between pH 6.5 and 7 (Fig. 4) (Saleh et al., 1978). The results shown in Fig. 4 were obtained with the *H. gracilis* enzyme. An essentially similar pH curve was found for the parsley synthase. With feruloyl-CoA a low yield of a radioactive product was formed at lower pH values (6–6.5) which had the same Rf-value in 4 different solvent systems as homoeriodictyol. But at all pH levels the amount of release products predominated when feruloyl-CoA was used as substrate.

According to these results one can conclude that both 4-coumaroyl-CoA and caffeoyl-CoA are physiological substrates for the synthase from parsley and *H. gra-*

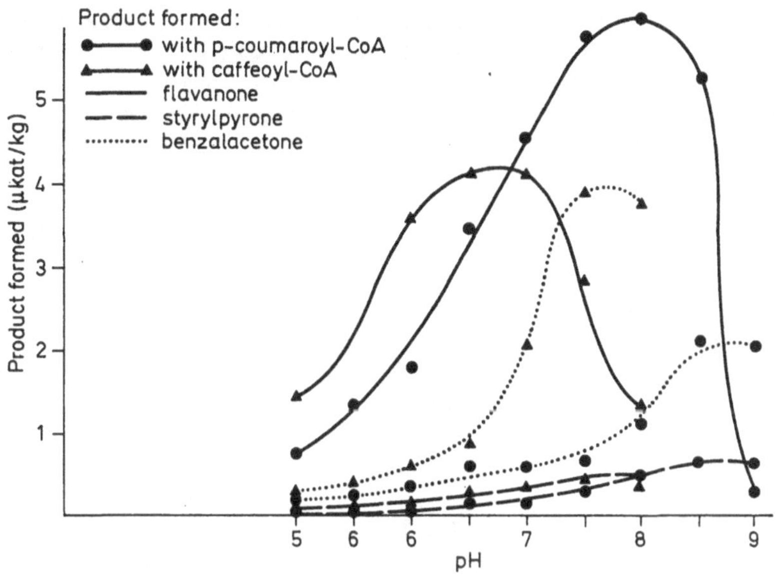

Fig. 4: pH-Optima for flavanone synthase with 4-coumaroyl-CoA and caffeoyl-CoA as substrates

cilis cell cultures. Besides concentration of the substrates, the pH at the site of synthase reaction could play a role in determining how much naringenin or eriodictyol is formed. In contrast, feruloyl-CoA does not seem to be a physiological substrate. This is in agreement with the fact that specific methyltransferases for flavonoids exist which catalyse methylation in the 3′ position (see section 5). Km values for 4-coumaroyl-CoA and malonyl-CoA are 1.6 and 35 µM, respectively, for the parsley synthase. For caffeoyl-CoA a Michaelis constant is not yet known. In kinetic measurements difficulties arise due to strong product inhibition and because several products are formed.

With an improved purification procedure Kreuzaler and Hahlbrock obtained a homogeneous enzyme from parsley cell cultures (unpublished results). The synthase has a molecular weight of about 80,000 and is composed of two apparently identical subunits of Mw 40,000. No pantothenic acid could be found to be present in the enzyme. Furthermore, an amino acid analysis showed no β-alanin and cysteamin. Flavanone synthase therefore does not have a phosphopantetheine prosthetic group which is present in both fatty acid synthase and 6-methylsalicylic acid synthase (Dimroth et al., 1970).

These results are surprising because they seem to disprove the hypothesis that the mechanism of flavanone synthase closely resembles that of the condensation steps involved in the syntheses of fatty acids and 6-methylsalicylic acid (Kreuzaler and Hahlbrock, 1975). On the other hand, the inhibition of flavanone synthase by the antibiotic cerulenin (2S, 3R 2,3-epoxy-4-oxo-7,10-dodecadienoylamide) indicates a close relationship with the condensation step in fatty acid synthesis. Cerulenin inhibits both reactions of β-ketoacyl-acyl carrier protein synthetase: fatty acyl transacylase and malonyl-acyl carrier protein decarboxylase (D' Agnolo et al. 1973).

So far no indication of covalently-bound intermediates has been found. At present it seems that the condensation reaction occurs with the CoA-esters. However, further work is necessary to clarify the mechanism of this interesting enzyme.

3. The role of chalcone isomerase**

Chalcone isomerase (EC 5.5.1.6) catalyses the stereospecific isomerisation of chalcone to (–) (2S) flavanone (Fig. 5). Despite intensive studies on this enzyme its role in flavonoid biosyntheis is not clear (Hahlbrock and Grisebach, 1974). The enzyme from soybeans has recently been purified to apparent homogeneity (Boland and Wong, 1975). With the aid of this enzyme the equilibrium constant of the isomerisation of isoliquiritigenin (4,2',4'-trihydroxychalcone) to (–) liquiritigenin (7,4'-dihydroxyflavanone) at pH 7.6 was shown to be 37 in favour of the flavanone. Attempts to determine the equilibrium constant for the reaction 4,2',4',6'-tetrahydroxychalcone→ naringenin, were unsuccessful owing to the fact that the reaction proceeds so strongly toward formation of the flavanone. The reverse reaction could not be demonstrated even with the use of very high concentrations of naringenin. These results indicate that with tetrahydroxychalcone an equilibrium greater than 1,000:1 exists in favour of flavanone. However, if the chalcone is removed *in vivo* as an intermediate in the biosynthsis of other flavonoids, it is conceivable that naringenin could be converted to the tetrahydroxychalcone by the isomerase.

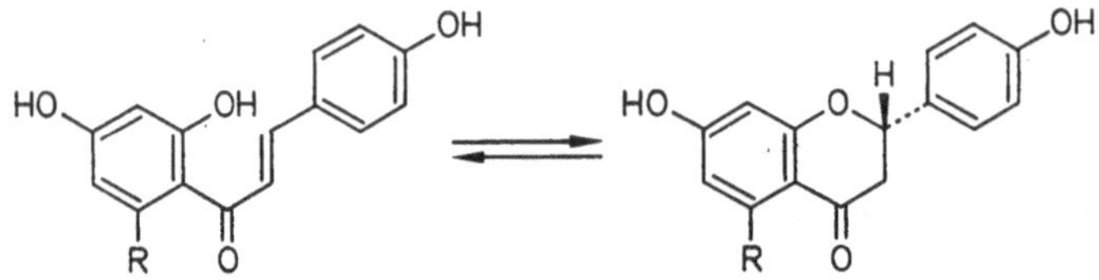

Fig. 5: Chalcone isomerase reaction. R = H or OH depending on the substrate specificity of the enzyme

In assessing the role of the isomerase for flavonoid biosynthesis two major questions arise: a) is flavanone the product of the synthase from all plants and b) do further conversions to other flavonoids take place with the flavanone or chalcone as substrate.

Concerning a), the direct formation of flavanone with the synthase from parsley cell cultures has been discussed. In contrast, results of investigations on the genetic control of chalcone isomeras in *Callistephus chinensis* can be rationalized best with the assumption that 4,2',4',6'-tetrahydroxychalcone, and not naringenin, is the pro-

** See Addendum.

duct of the synthase in this plant (Kuhn et al., 1977). In a recessive genotype (chch) of *C. chinensis* which is blocked in anthocyan synthesis, tetrahydroxychalcone-2′-glucoside (isosalipurpuroside) is accumulated in the flowers. It was shown that chalcone isomerase is absent in this mutant whereas the wildtype which synthesizes anthocyanins and flavonol glycosides contains an isomerase specific for tetrahydroxychalcone. The accumulation of isosalipurpuroside should therefore be due to lack of the isomerase (Fig. 6). The Ch gene could be directly responsible for the synthesis of the isomerase.

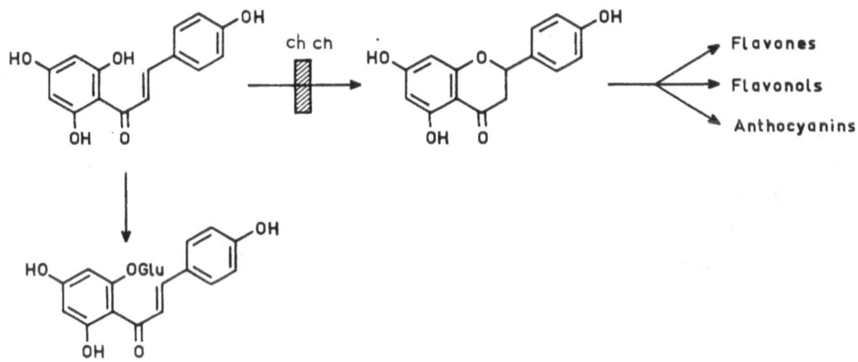

Fig. 6: Accumulation of tetrahydroxychalcone-2′-glucoside in blocked mutant of *Callistephus chinensis*

In a similar case it was reported that tetrahydroxychalcone is accumulated in pollen of an inbred line of *Petunia hybrida* which is homozygous recessive for a gene called W (De Vlaming and Kho, 1976). Furthermore, chalcones accumulate during the development of anthers in *Tulipa co.* «Apeldoorn» (Quast and Wiermann, 1973) and maximum chalcone isomerase activity was measured at a point when the concentration of chalcones was decreasing rapidly and the concentration of flavonols was increasing (Wiermann, 1972).

Final proof that chalcone is indeed the first product of the synthase reaction in the above mentioned plants awaits isolation of the enzyme from these sources and careful analysis of the products.

With respect to b), there is only one example at the enzymatic level, in which a definite decision could be reached as to whether chalcone or flavanone is the substrate for further reactions: this is the oxidation of the flavanone to flavone. Cell-free extracts from very young primary leaves of parsley plants catalyse the oxidation of naringenin to apigenin and of 7,4′-dihydroxyflavanone to 7,4′-dihydroxyflavone (Sutter et al., 1975). No oxidation to the flavone took place with 4,2′,4′-trihydroxychalcone as substrate. However, when the chalcone was preincubated with chalcone isomerase, subsequent incubation with the parsley extract showed conversion to the flavone. These experiments together with other controls not mentioned here prove unequivocally that the flavanone is the substrate for the oxidation to flavone.

Support for the role of flavanone in further conversion to other flavonoids comes also from work with the mutants of *Callistephus* mentioned earlier. In genotypes which lack chalcone isomerase only the chalcone glucoside accumulates. In genotypes with high isomerase activity no chalcone is present but flavones, flavonols and anthocyanins now appear (Kuhn et al. 1977).

It has been suggested that peroxidases may take part in the conversion of chalcones to other flavonoids (Rathmeel and Bendall 1972), and Wong has recently carefully analyzed the peroxidase-catalysed oxidation of 4,2′,4′-trihydroxychalcone (Wong and Wilson 1976). But it is by no means proven, that these reactions have physiological significance.

Further information on the role of chalcone isomerase in flavonoid biosynthesis can be obtained from studies on the localization of this enzyme in different cell types. In primary leaves of *Avena sativa*, which accumulate C-glycosylflavones, chalcone isomerase was found to be localized almost entirely in the mesophyll and not in the epidermis, whereas phenylalanine ammonia lyase was about equally distributed in these tissues in young leaves (Weissenböck and Sachs, 1977). Since flavone accumulation was found to be nearly the same in lower and upper epidermis and mesophyll, chalcone isomerase seems not to be necessary for flavone biosynthesis in the epidermis.

4. Biosynthesis of anthocyanins

Compared with the rapid progress in the enzymology of the flavonoid glycoside pathway in parsley cell cultures (Hahlbrock and Grisebach, 1974), the elucidation of the pathway to anthocyanins is rather slow. A cell-free system for anthocyanin biosynthesis is still unknown. Nevertheless, work with labeled precursors together with work on flower mutants does allow the formulation of a biosynthetic pathway via dihydroflavonols which is consistent with all experimental facts. Furthermore glycosyltransferases for anthocyanins have recently been characterized.

The high incorporation of tritium-labelled dihydrokaempferol and dihydroquercetin into cyanidin with low dilution has been demonstrated in buckwheat seedlings and cell cultures of *Haplopappus gracilis* (Fritsch et al., 1971). Dimethylsulfoxide-treated cells from a cell culture of *H. gracilis* and a microsomal preparation from these cell cultures were capable of converting naringenin to eriodictyol and dihydrokaempferol (Fritsch and Grisebach, 1975). A further product was tentatively identified as dihydroquercetin. The requirement for NADPH and O_2 by the microsomal system characterizes it as a microsomal mixed-function oxidase (H. Fritsch, unpublished). But attempts to convert [^{3}H]-dihydroquercetin to cyanidin with cell-free extracts of *H. gracilis* cell cultures have so far been unsuccessful. We have proposed a scheme for anthocyanin biosynthesis in which conversion of naringenin to dihydroquercetin and possibly also its conversion to cyanidin take place at a membrane (Fritsch and Grisebach, 1975) (Fig. 7). Glucosylation of cyanidin then occurs by action of a glucosyltransferase which may be located in the vacuole or at the inner side of the tonoplast; the hydrophilic cyanin could then be excreted into the aqueous milieu of the vacuole.

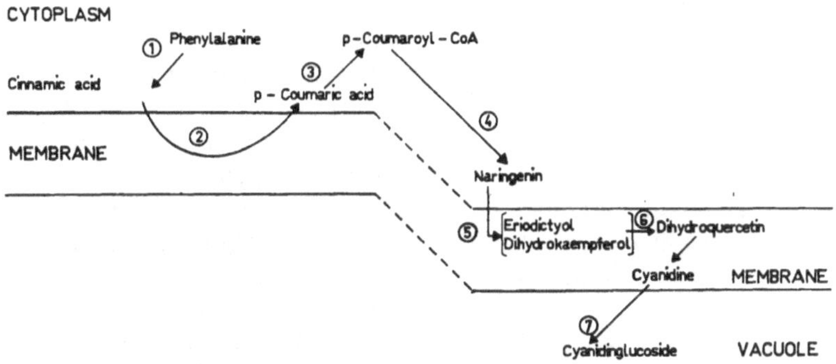

Fig. 7: Hypothesis for biosynthesis of anthocyanin on membrane. 1: Phenylalanine ammonia-lyase; 2: cinnamate 4-hydroxylase; 3: 4-coumarate: CoA ligase; 4: flavanone synthase; 5: 6 hydroxylases for 3 and 3′ position; 7: cyanidin 3-glucosyltransferase

Enzymes, which catalyse the glucosylation of cyanidin at the 3-position using UDP-D-glucose as glucosyldonor have recently been isolated and purified for the first time from young red cabbage (*Brassica oleracea*) seedlings (Saleh et al., 1976b) and cell cultures of *H. gracilis* (Saleh et al., 1976a). The properties of the enzymes from both sources are very similar. The reaction was inhibited by low concentrations of the reaction product, cyanidin-3-glucoside. In addition to cyanidin the enzyme could also glucosylate other anthocyanins and flavonols but not dihydroquercetin.

Recent studies on subcellular localization of flavanone synthase, chalcone isomerase and UDP-glucose: 3-0-glucosyltransferase in petals of *Hippeastrum* and *Tulipa* showed that these enzymes are predominantly localized in the cytosol fraction (Hrazdina et al., 1978). Vacuoles which where obtained by fractionation of protoplasts did not contain activity of these 3 enzymes. This finding does not support the hypothesis mentioned above, that glucosylation of cyanidin could take place in the vacuole or on the inner side of the tonoplast.

Further strong support for the role of dihydroflavonols as intermediates in anthocyanin biosynthesis has recently come from work with genetically defined acyanic lines of *Matthiola incana* (G. Forkmann, 1977), *Petunia hybrida* (Kho et al., 1975; Kho et al., 1977), and *Antirrhinum majus* (Harrison and Stickland, 1974; Stickland and Harrison, 1974). In principle, a complementation technique was used. Dihydroflavonols or their glycosides which accumulated in white flowering mutants were supplied to acceptor mutants in which the synthesis of dihydroflavonols was blocked, and this lead to anthocyanin synthesis in the acceptor mutant. In flower buds of the white flowering mutant W 19 of *Petunia hybrida* dihydroquercetin-7-glucoside, dihydroquercetin-4′-glucoside, dihydroquercetin, and dihydrokaempferol-7-glucoside accumulated. When dihydroquercetin was supplied to *in vitro* cultured corollas of the white flowering mutant W 18, cyanidin 3-glucoside was formed as the major pigment together with smaller amounts of cyanidin 3-di-glucoside and delphinidin 3-glucoside. These anthocyanins are the same as those detected in flowers of a red flowering genotype, R 3, of *Petunia hybrida.*

Similar results were obtained with acyanic genotypes of *Matthiola incana* and *A. majus.* In lines of the genotype ff of Matthiola, administration of dihydrokaempferol, dihydroquercetin, naringenin or the corresponding chalcone, and tetrahydroxychalcone-2′-glucoside permitted synthesis of anthocyanins. In contrast, hydroxycinnamic acids were ineffective. This indicates that the genetic block in these lines lies before naringenin and could concern flavanone synthase. Lines with the genotype ee and gg did not produce anthocyanins with the precursors mentioned above. This supports the assumption that the genes e and g block the pathway after dihydroflavonol synthesis but before anthocyanin formation. The probable position of the genetic blocks in Mathiola is shown in Fig. 8 (Forkmann, 1977). In genotype ff dihydroflavonols were clearly superior to the other precursors with respect to anthocyanin formation (Fig. 9).

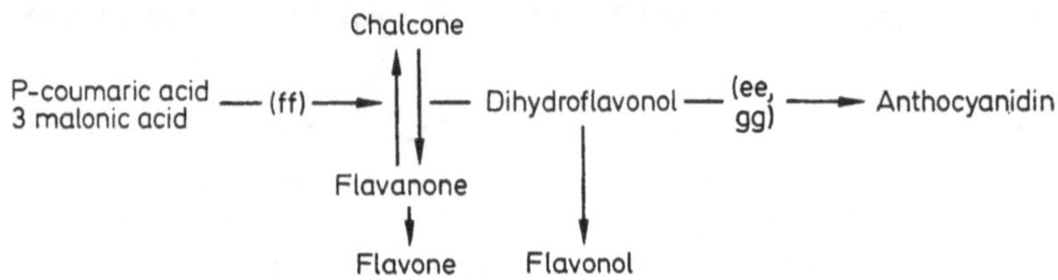

Fig. 8: Interrelationship between some flavonoid classes and the probable position of the genetic blocks in *Matthiola* (Forkmann, 1977)

Additional evidence for dihydroflavonols as intermediates in anthocyanin biosynthesis comes from experiments with inhibitors of phenylalanine ammonia-lyase. The most powerful inhibitor of PAL is L-α-aminooxy-β-phenylpropionic acid or its N-benzyloxycarbonyl derivative (Amrhein et al., 1976; Amrhein and Gödeke, 1977) which inhibits this enzyme competitively with a Ki of 1.4×10^{-9} M. This compound is also a potent inhibitor of cyanidin synthesis in buckwheat hypocotyls, in buds of *Ipomea purpurea* and *Catharanthus roseus*, and in red cabbage seedlings (Amrhein, 1978). Inhibition of cyanidin synthesis can be fully or partially reversed by precursors of the B-ring of cyanidin(*trans*-cinnamate, 4-coumarate) and by naringenin, dihydrokaempferol or dihydroquercetin. In contrast, nonreversion was observed with apigenin, kaempferol, quercetin and dihydrofisetin (Amrhein, 1978).

In conclusion, the results of tracer and inhibition experiments and the work with mutants are all consistent with the view that dihydroflavonols are intermediates in anthocyanin biosynthesis. Dihydroflavonols themselves can be formed from flavanones with a microsomal mixed-function oxidase. The conversion of dihydroquercetin into cyanidin-3-glucoside requires dominant alleles of two genes (An 1 and An 2) (Kho et al., 1977). There is no evidence for glucosylation of dihydroqercetin at posi-

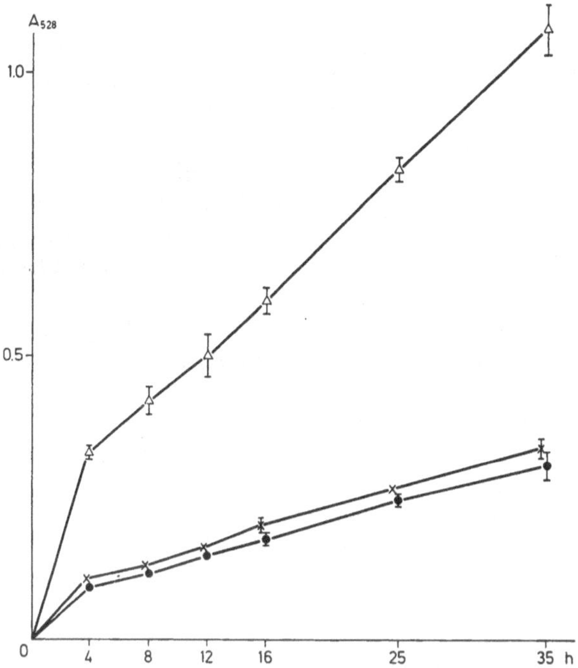

Fig. 9: Anthocyanin formation in line 18 after imbibition of different precursors (mean values of 5 experiments). △——△ dihydroquercetin, ×——× naringenin, ●——● tetrahydroxychalcone (Forkmann, 1977)

tion 3. This is in agreement with the specificity of the cyanidin-3-0-glucosyltransferase mentioned above and with the conclusion that glycosylation is the last step in the biosynthesis of cyanidin-3-O-glucoside.

5. Flavonoid Methyltransferases

In the section on flavanone synthase it was mentioned that feruloyl-CoA is apparently not a physiological substrate for this enzyme from parsley and *H. gracilis* cell cultures. O-Methyl groups in ring B seem therefore to be formed at the flavonoid level. Strong support for this assumption is the occurrence of O-methyltransferases directly related to flavone glycoside biosynthesis in cell cultures of parsley (Ebel et al., 1972) and soybean (Poulton et al. 1976a; Poulton et al. 1977).

The enzyme from parsley cell cultures catalyses the transfer of the methyl group of S-adenosyl-L-methionine to the meta position of *o*-dihydric phenols. Luteolin (5,7,3',4'-tetrahydroxyflavone) and its 7-O-glucoside were the best substrates for the enzyme with apparent Michaelis constants of $4.6 \cdot 10^{-5}$ and $3.1 \cdot 10^{-5}$ M, respectively (Ebel et al., 1972). Caffeic acid was a much poorer substrate for the enzyme (Km = $1.6 \cdot 10^{-3}$ M). The enzyme has now been classified as luteolin methyltransferase (EC 2.1.1.42). It requires Mg^{2+} for activity and is not inhibited by thiol reagents.

In soybean (*Glycine max* L. var Mandarin) cell cultures two distinct methyltransferases exist. One enzyme is specific for methylation of substituted cinnamic acids, such as caffeic acid and 5-hydroxyferulic acid (CMT) (Poulton et al., 1976b), whereas the other methylates only flavonoid substrates such as luteolin and quercetin (FMT) (Poulton et al., 1977). Both enzymes have been separated and partially purified. The flavonoid methyltransferase from soybean cell cultures is closely related in its properties to the luteolin methyltransferase from parsley cell cultures. The soybean FMT also requires Mg^{2+} and shows a preference for flavonoid substrates. Luteolin and quercetin had the highest V/Km values. Within the group of flavonoid compounds tested, flavones and flavonols were more efficient substrates than flavanones and flavanols. These properties, which were also noted for the parsley FMT suggest that the *in vivo* methylation of flavonoid compounds occurs at the flavone/flavonol level of oxidation. As with the parsley enzyme, both luteolin and its 7-O-glucoside were good substrates for the soybean enzyme. The possible significance of the two methyltransferases in the phenylpropanoid metabolism of the soybean cell culture is illustrated in Fig. 10.

Methyltransferases for anthocyanin substrates have not yet been characterized.

A specific methyltransferase for methylation of isoflavones at the 4′-hydroxyl group has been isolated and purified 360-fold from cell suspension cultures of chick pea (*Cicer arietinum* L.). The enzyme was also detected in seedlings of this plant (Wengenmeyer et al. 1974).

This methyltransferase does not require Mg^{2+} and is highly specific for methylation of 4′hydroxyisoflavones. Intermediates in the biosynthetic pathway to isoflavones, e.g., 4-hydroxycinnamic acids and 5,7,4′-trihydroxyflavanone, were not methylated. Neither were other flavonoid substrates. These results make it very likely that methylation is the last step in the biosynthesis of 4′-methoxyisoflavones.

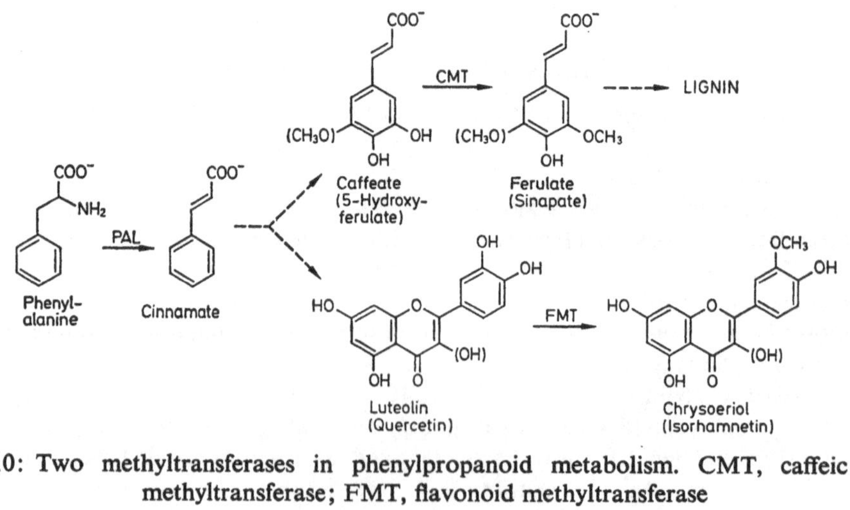

Fig. 10: Two methyltransferases in phenylpropanoid metabolism. CMT, caffeic acid methyltransferase; FMT, flavonoid methyltransferase

6. Regulation of flavonoid biosynthesis

Most of the studies done in Freiburg on regulation of flavonoid biosynthesis were carried out with plant cell cultures. While the results obtained with these cell cultures do not necessarily represent the situation in the plant, they do, however, reveal important facts about possible regulatory control mechanisms for this pathway.

Initially it was demonstrated that large, concomitant changes in the activities of a number of enzymes involved in the formation of flavonoid glycosides (Kreuzaler

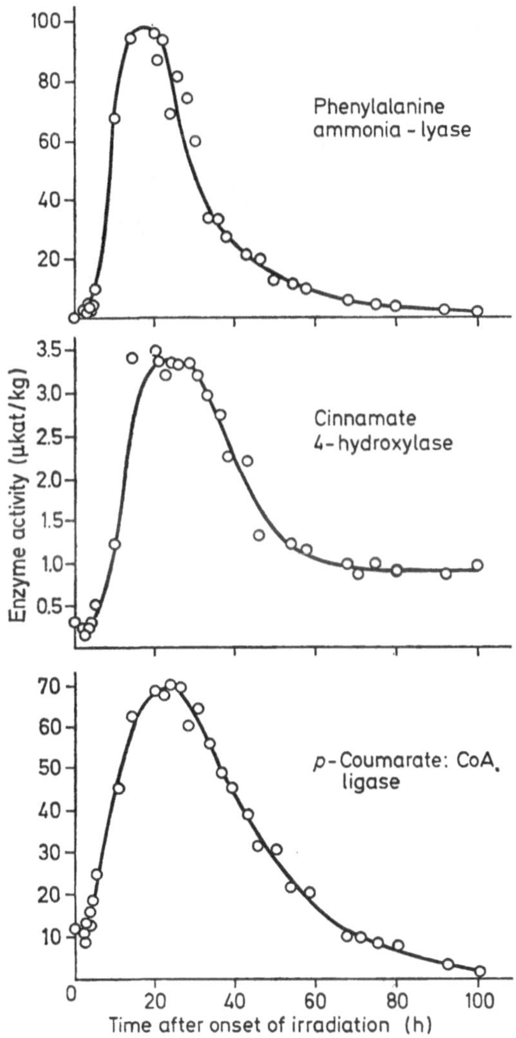

Fig. 11: Time courses of the light-induced changes in the enzyme ectivities of group I (HAHLBROCK et al., 1976)

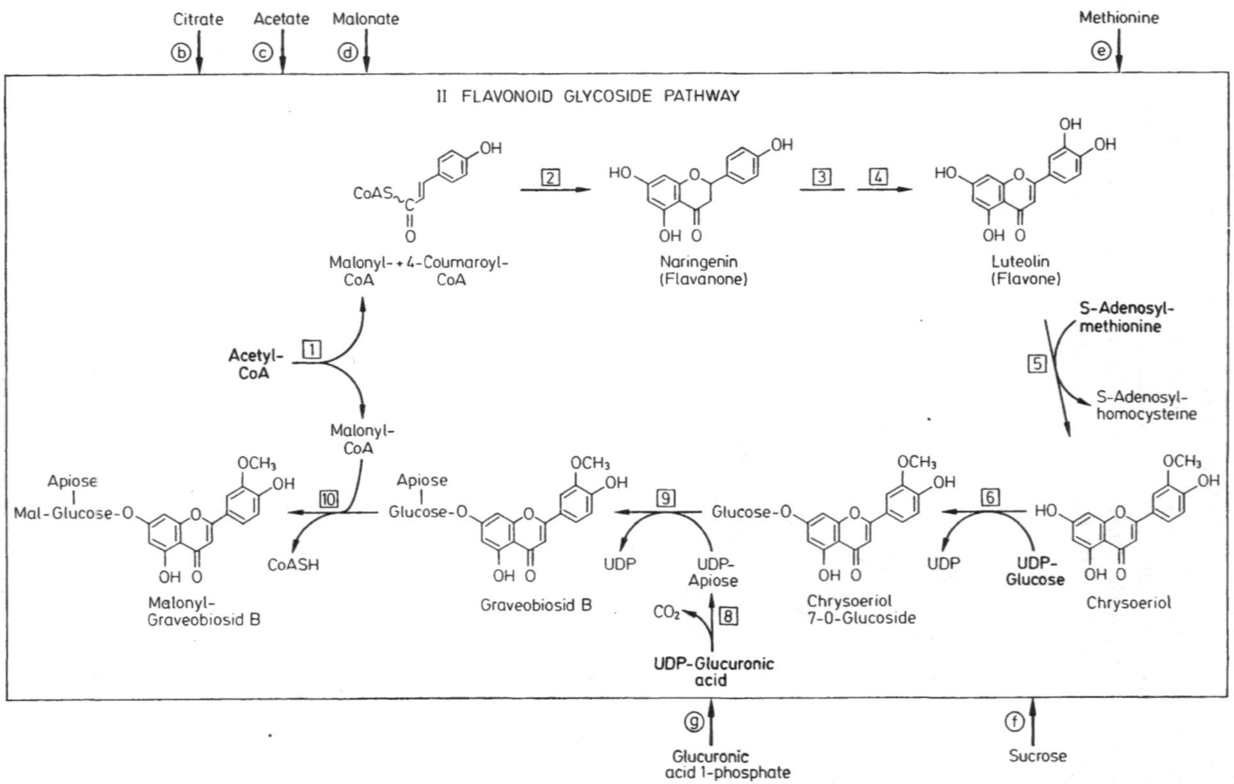

Fig. 12: Reactions of the flavonoid glycoside pathway. 1: acetyl-CoA carboxylase; 2: flavanone synthase; 3 and 4: enzymes involved in as yet not well characterized oxidative steps; 5: methyltransferase; 6: 7-O-glucosyltransferase; 8: UDP-apiose synthase; 9: apiosyltransferase; 10: malonyltransferase. The reactions leading to intermediary metabolites, which are utilized as substrates for the flavonoid glycoside pathway (shaded areas), are indicated by arrows outside the frame (Ebel and Hahlbrock, 1977)

and Hahlbrock, 1973) occur during the development of cotyledons and leaves in parsley seedlings (Hahlbrock et al., 1971b). Subsequently it was shown that cell cultures of parsley responded specifically to irradiation with ultraviolet light by accumulating a number of flavone and flavonol glycosides, concomitant with large and rapid increases, and subsequent decreases, in the enzyme activities related to their synthesis. Detailed examination of the light-induced activity changes showed that the enzymes can be divided into two groups, each of which is specific for a series of closely related reactions (Hahlbrock et al., 1971a, 1976). Group I consists of a sequence of three enzymes involved in the conversion of L-phenylalanine to 4-coumaroyl-CoA (Fig. 11). Group II included the enzymes catalyzing the formation of the flavone glycoside, malonylapiin (Hahlbrock, 1972), from 4-coumaroyl-CoA (Ebel and Hahlbrock, 1977). The reactions catalyzed by the seven enzymes of group II for which the activity changes have until now been determined are shown in Fig. 12.

Table 1: Characteristic Data for Induction of Enzymes of Flavonoid Biosynthesis in Parsley Cell Cultures

Group	Enzyme	Lag period	Peak position	Apparent half-life
		h		
I	Phenylalanine ammonia-lyase	2–2.5	17	10
	Cinnamate 4-hydroxylase	2–2.5	22	15
	4-Coumarate: CoA ligase	2–2.5	23	17
II	Flavanone synthase	4	37	37
	7-O-Glucosyltransferase	4	26	45
	UDP-Apiose synthase	4	27	71

The three enzymes of group I show a lag period of 2–2.5 h, a peak position between 17 and 23 h and an apparent half-life of 10–17 h (Table 1). This group can also be induced independently of group II in the absence of light by dilution of the parsley cells into fresh medium (Hahlbrock and Wellmann, 1973) or water (Hahlbrock and Schröder, 1975). In contrast, the enzymes of the flavonoid glycoside pathway (group II) have the following properties: (a) light-induced increases in enzyme activity occuring after lag periods of about 4 h; (b) highest activities reached within about 25–40 h after onset of induction; and (c) apparent half-lives of enzyme activity during the subsequent periods of decline of 30–70 h (Table 1). As an example, Fig. 13 shows the activity changes for the first enzyme of each group, phenylalanine ammonia-lyase and flavanone synthase. Depending on the growth stage of the cell cultures used for the induction experiment, the lag periods and peak positions can vary within the range mentioned above.

Reactions 5 and 6 on Fig. 12 involve a 3′-O-methyltransferase and a 7-O-glucosyltransferase, both of which are, on the basis of their substrate specificities, involved in the formation of flavone and flavonol glycosides. In contrast, the 3-O-glucosyltrans-

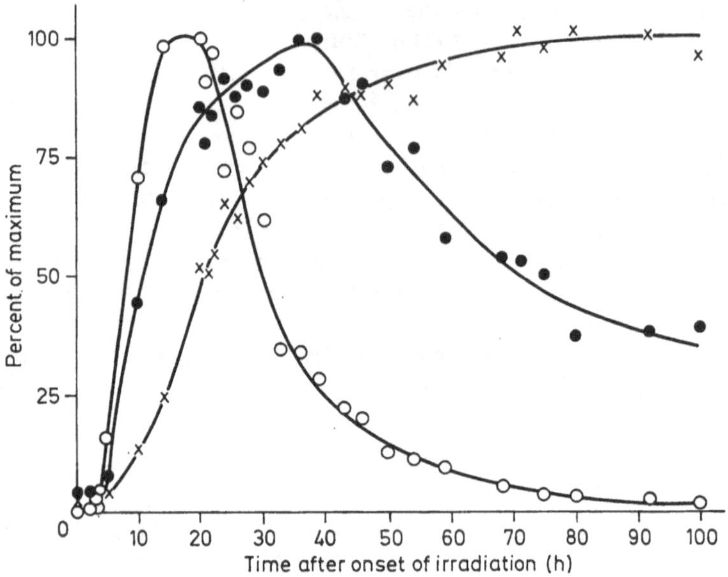

Fig. 13: Time courses of light-induced changes in the activities of phenylalanine ammonia-lyase (○) and flavanone synthase (●), and in the accumulation rate of flavonoid glycosides (×) in 10-day-old parsley cell cultures

ferase is specific for the synthesis of flavonol glycosides (Sutter and Grisebach, 1973). The pattern of light-induced activity of this enzyme clearly showed that it belongs to group II and is a strong indication that the enzymes of both the flavone and flavonol glycoside pathways are members of the same group.

Somewhat surprising was the fact that acetyl-CoA carboxylase is induced concomitantly with the enzymes of group II. This was not only shown by the typical time course of light-induced changes of this enzyme but also by the absence of an effect on enzyme activity upon diluting the cells. As already discussed above, the dilution effect is typical for enzymes of group I.

Acetyl-CoA carboxylase provides the malonyl-CoA for the flavanone synthase reaction. In addition, in parsley cell cultures malonyl-CoA is used for a malonyl-transfer reaction to graveobiosid B (Fig. 12). It is not yet known whether a single acetyl-CoA carboxylase in parsley cells provides malonyl-CoA for both fatty acid and flavonoid biosynthesis or whether two enzymes exist for the different pathways.

The inducing effect of the light treatment is strictly confined to the enzymes of flavonoid glycoside biosynthesis, according to all results obtained so far. The following twelve enzymes from various other metabolic pathways showed no significant changes upon irradiation of the parsley cell cultures: acetate: CoA ligase, UDP-xylose synthase, homocysteine methyltransferase, glutamate dehydrogenase, glucose-6-phosphate dehydrogenase, cholinephosphotransferase, phenylalanine: 2-oxoglutarate aminotransferase, malonate: CoA ligase, sucrose synthase, UDP glucuronate pyrophosphorylase, methionine adenosyltransferase, and succinate:CoA ligase. Most of these enzymes are assumed to be indirectly related to flavanoid glycoside formation by supplying the various primary metabolites which are utilized as substrates for this pathway. The reactions catalyzed by these enzymes are indicated by bold arrows outside the frame in Fig. 12. They participate in the synthesis of phenylalanine, acetyl-CoA, S-adenosylmethionine, UDP-glucose, and glucuronic acid.

It must be emphasized that light induction of the enzymes mentioned is not a general phenomenon in plant cell cultures. In soybean (*Soja hispida*) cell cultures light is not required for induction of enzymes of phenylpropanoid metabolism. The enzymes of group I are induced during growth of the cell cultures in the dark. The induction occurs at the same stage of growth in which inducibility of the enzymes by light in the parsley cell cultures is observed. A sharp peak in the enzyme activities is observed when nitrate is completely exhausted from the medium (Ebel et al., 1974). Although nothing is known at present about the endogenous mechanism of this induction, it is interesting to note that the amount of ethylene produced by the cells is also significantly increased at this growth stage of the soybean cultures (Larne and Gamborg, 1971).

The question of whether a relationship exists between enzyme activities and product accumulation in the parsley cell cultures was also investigated (Hahlbrock et al., 1976). The open circles in Fig. 14 represent the total concentration of the flavone and flavonol glycosides throughout the 100 h period of continuous irradiation of the cells. 187 h after onset of induction the absorbance at 340 nm had only slightly decreased, suggesting a low rate of degradation, if any, of the flavonoid products. Fig. 14 also shows the integrations of the curves obtained for the changes of phenylalanine ammonia-lyase, 4-coumarate:CoA ligase and flavanone synthase activities (Fig. 13). A very close correspondence between the shapes of the curves obtained for the integrated phenylalanine ammonia-lyase activities and the rates of accumulation of the flavonoid glycosides can be seen. No similar direct relationship is apparent for the changes in the rates of product accumulation and in the 4-coumarate:CoA ligase or flavanone synthase activities. From these data, it can be concluded that phenylalanine ammonia-lyase, and not an enzyme of the flavonoid pathway proper, is rate limiting for the formation of the flavonoid glycosides under the conditions used. The striking coincidence of the two curves further indicates that an inhibition of the phenylalanine ammonia-lyase by the flavonoid compounds cannot play a major role in the regulation of this enzyme activity *in vivo* in the cell cultures.

In contrast to these results, Margna has recently critically discussed the possible control functions of phenylalanine ammonia-lyase and concludes from a survey of experimental results that substrate (phenylalanine) supply rather than enzymic activity is the most likely limiting factor in controlling phenylpropanoid accumulation (Margna, 1977). However, correlations between product accumulation and phenylalanine ammonia-lyase activity have to be interpreted with great caution. There is always the possibility that more than one enzyme and product pool exists in the plant and that a correlation between enzyme activity and product accumulation can only be expected when the interrelated pools are compared. A good example of such a case

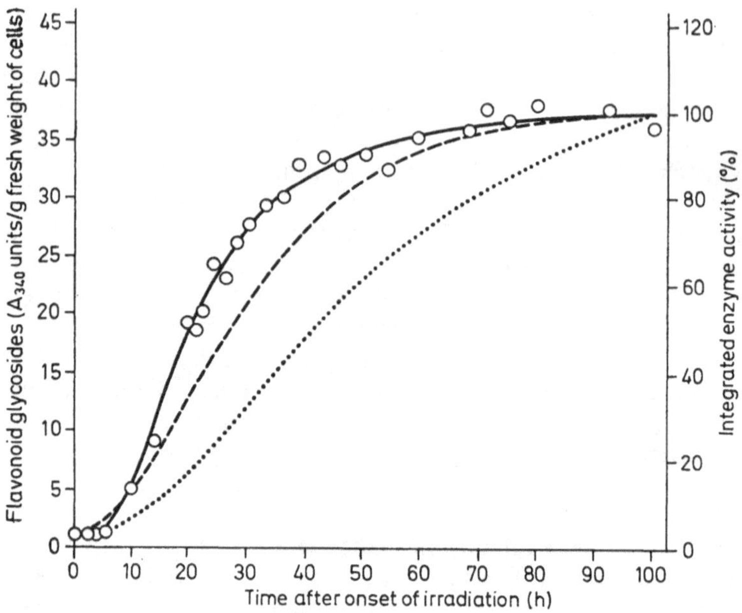

Fig. 14: Time course of the light-induced changes in the total concentration of flavonoid glycosides (○) in parsley cell cultures. The three theoretically-derived curves represent the integrations of the activities of phenylalanine ammonia-lyase (———), 4-coumarate:CoA ligase (– – –) and flavanone synthase (· · · · ·). Compare Fig. 13

is the following investigation on the control of enzyme activity and flavonoid synthesis in cotyledons of mustard seedlings (*Sinapis alba*) (Wellmann, 1974). It was shown earlier that in the cotyledons anthocyanin synthesis and the activity of phenylalanine ammonialyase and chalcone isomerase are under control of phytochrome. However, there did not seem to be a quantitative correlation between anthocyanin accumulation and the activity changes of these enzymes after induction with far red light. These results were obtained with whole cotyledons. A closer examination of the occurrence of flavonoids in the cotyledons then showed that flavonols are predominantly localized in the upper epidermis and anthocyanin in the lower epidermis. After removal of the epidermal layers neither flavonoids nor the two enzymes related to their synthesis could be detected in the remaining tissue. It is therefore evident that a meaningful correlation can only be made between flavonols in the upper, and anthocyanins in the lower epidermis, on the one hand, and the enzymes localized in each respective cell layer on the other hand. Indeed, when the anthocyanin and quercetin content of the separated epidermi and the activity of phenylalanine ammonia-lyase and chalcone isomerase was determined in these tissues a good correlation between PAL activity and anthocyanin accumulation was found (Fig. 15). Without doubt many more such examples can be found. Only when these criteria are taken into consideration is it

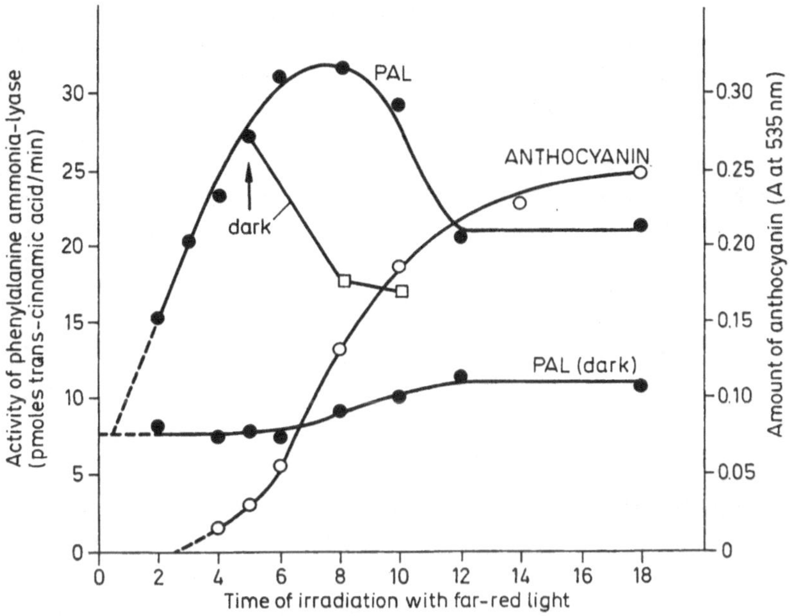

Fig. 15: Correlation between anthocyanin accumulation in lower epiderms of *Sinapsis alba* seedlings and time course of light-induced changes in activity of phenylalanine ammonia-lyase

possible to decide whether substrate supply as suggested by Margna could be a controlling factor in flavonoid and, more generally, phenylpropanoid accumulation.

In addition to the work discussed above, Hahlbrock, Schröder and coworkers have carefully analyzed the molecular basis for the light-induced changes in phenylalanine ammonia-lyase activity in recent years. This work has not been discussed here. It should only be mentioned that the rise in enzyme activity after induction is due to *de novo* synthesis, and furthermore, that a strict correlation exists between enzyme synthesis and synthesis of the messenger RNA for the PAL subunits. This work could ultimately lead to an understanding of how light functions as a trigger for flavonoid biosynthesis.

Addendum

Flavanon Synthase

With the homogeneous enzyme from parsley cell cultures (Kreuzaler et al., 1979) the nature of the condensation product(s) was reinvestigated. Incubation of synthase with [2-^{14}C]malonyl-CoA and 4-coumaryl-CoA at pH 6.8 (instead of pH 8.0 used in earlier work) in the presence of 1.4 mM 2-mercaptoethanol resulted in the predominant formation of labelled naringenin *chalcone* (Fig. 5, R = OH) (W. Heller and K. Hahlbrock, Arch. Biochem. Biophys., **200**, 617–619 (1980)). The time course of the rates of chalcone and flavanone production proved that flavanone formation is only due to chemical cyclization. The results obtained prove that 4.2'.4'.6'-tetrahydroxychalcone and not flavanon (naringenin) is the product of the synthase reaction. Logically the enzyme should therefore be named *chalcone synthase.*

The fact that chalcone is the first product is in agreement with the results obtained with flower mutants discussed above. It now also defines a clear role to the chalcone isomerase. This enzyme is necessary for stereospecific cyclization to the 2-S-naringenin which can serve as substrate for conversion into other flavonoids.

Chalcone synthase also catalyzes CO_2 exchange and decarboxylation of malonyl-CoA both in the presence and in the absence of 4-coumaryl-CoA (F. Kreuzaler, R. J. Light, and K. Hahlbrock, FEBS Letters **94**, 175–178 (1978)). Decarboxylation of the malonate residue is also catalyzed by fatty acid synthases from various sources and by 6-methylsalicylic acid synthase. CO_2 exchange has been observed with several fatty acid synthases.

Incorporation of [1,2-^{13}C] acetate into pisatin in *Pisum sativum* proved that cyclization of ring A in the chalcone synthase reaction occurs in a clockwise fashion (A. Stoessl and J. B. Stothers, Z. Naturforsch. **34c**, 87–89 (1979)). The results also established that removal of the oxygen function in 5-deoxyflavonoids occurs before cyclization of the A-ring.

References

D'Agnolo, G., Rosenfeld, I. S., Awaya, J., Omura, S., and Vagelos, P. R., 1973: Inhibition of Fatty Acid Synthesis by the Antibiotic Cerulenin. Specific Inactivation of β-Ketoacyl-acyl Carrier Protein Synthetase, Biochim. Biophys. Acta **326**, 155–166.

Amrhein, N., 1979: Biosynthesis of Cyanidin in Buck wheat Hypocotyls, Phytochemistry **18**, 585–589.

Amrhein, N., and Gödeke, K. H., 1977: α-Aminooxy-β-Phenylpropionic Acid – A Potent Inhibitor of L-Phenalalanine Ammonia-Lyase in vitro and in vivo, Plant Science Letters **8**, 313–317.

Amrhein, N., Gödeke, K. H., and Kefeli, V. I., 1976: The Estimation of relative intracellular Phenylalanine Ammonia-Lyase (PAL)-Activities and the Modulation in vivo and in vitro by competitive Inhibitors, Ber. Deutsch. Bot. Ges. **89**, 247–259.

Boland, M. J., and Wong, E., 1975: Purification and Kinetic Properties of Chalcone-Flavanone Isomerase from Soy Bean, Eur. J. Biochem. **50**, 383–389.

Dimroth, P., Walter, H., and Lynen, F., 1970: Biosynthese von 6-Methylsalicylsäure, Eur. J. Biochem. **13**, 98–110.

Ebel, J., and Hahlbrock, K., 1977: Enzymes of Flavone and Flavonol-Glycoside Biosynthesis. Coordinated and Selective Induction in Cell-Suspension Cultures of *Petroselinum hortense*, Eur. J. Biochem. **75**, 201–209.

Ebel, J., Hahlbrock, K., and Grisebach, H., 1972: Purification and properties of an O-dihydricphenol meta-O-methyltransferase from cell suspension cultures of parsley and its relation to flavonoid biosynthesis, Biochim. Biophys. Acta **269**, 313–326.

EBEL, J., SCHALLER-HEKELER, B., KNOBLOCH, K. H., WELLMANN, E., GRISEBACH, H., and HAHLBROCK, K., 1974: Coordinated changes in Enzyme Activities of Phenylpropanoid Metabolism during Growth of Soybean Cell Suspension Cultures, Biochim. Biophys. Acta **362**, 417–424.

FORKMANN, G., 1977: Precursors and Genetic Control of Anthocyanin Synthesis in *Matthiola incana* R. Br., Planta **137**, 159–163.

FRITSCH, H., and GRISEBACH, H., 1975: Biosynthesis of Cyanidin in Cell Cultures of *Haplopappus gracilis*, Phytochemistry **14**, 2437–2442.

FRITSCH, H., HAHLBROCK, K., and GRISEBACH, H., 1971: Biosynthese von Cyanidin in Zellsuspensionskulturen von *Haplopappus gracilis*, Z. Naturforsch. **26**b, 581–585.

GRISEBACH, H., 1962: Die Biosynthese der Flavonoide, Planta Medica **10**, 385–397.

HAHLBROCK, K., 1972: Malonyl coenzyme A: flavone glycoside malonyltransferase from illuminated cell suspension cultures of parsley, FEBS Letters **28**, 65–68.

HAHLBROCK, K., and GRISEBACH, H. 1974: Biosynthesis of Flavonoids. In «The Flavonoids» (Harborne, J. B., Mabry, T. J., and Mabry, H.) Chapman and Hall, London 1975, p. 866–915.

HAHLBROCK, K., and SCHRÖDER, J., 1975: Specific Effect on Enzyme Activities upon Dilution of *Petroselinum hortense* cell cultures into water, Arch. Biochem. Biophys. **171**, 500–506.

HAHLBROCK, K., and WELLMANN, E., 1973: Light-independent induction of enzymes related to phenylpropanoid metabolism in cell suspension cultures from parsley, Biochim. Biophys. Acta **304**, 702–706.

HAHLBROCK, K., EBEL, J., ORTMANN, R., SUTTER, A., WELLMANN, E., and GRISEBACH, H., 1971a: Regulation of enzyme activities related to the biosynthesis of flavone glycosides in cell suspension cultures of parsley (*Petroselinum hortense*), Biochim. Biophys. Acta **244**, 7–15.

HAHLBROCK, K., KNOBLOCH, K. H., KREUZALER, F., POTTS, J. R. M., WELLMANN, E., 1976: Coordinated Induction and Subsequent Activity Changes of Two Groups of Metabolically Interrelated Enzymes. Light-Induced Synthesis of Flavonoid Glycosides in Cell Suspension Cultures of *Petroselinum hortense*, Eur. J. Biochem. **61**, 199–206.

HAHLBROCK, K., SUTTER, A., WELLMANN, E., ORTMANN, R., and GRISEBACH, H., 1971b: Relationship between organ development and activity of enzymes involved in flavone glycoside biosynthesis in young parsley plants, Phytochemistry **10**, 109–116.

HARRISON, B. J., and STICKLAND, R. G., 1974: Precursors and Genetic Control of Pigmentation. 2. Genotype Analysis of Pigment Controlling Genes in Acyanic Phenotypes in *Antirrhinum Majus*, Heredity **33**, 112–115.

HRAZDINA, G., KREUZALER, F., HAHLBROCK, K., and GRISEBACH, H., 1976: Substrate Specificity of Flavanone Synthase from Cell Suspension Cultures and Structure of Release Products in Vitro, Arch. Biochem. Biophys. **175**, 392–399.

HRAZDINA, G., WAGNER, G. J., and SIEGELMAN, H. W., 1978: Subcellular Localization of Enzymes of Anthocyanin Biosynthesis in Protoplasts, Phytochemistry **17**, 53–56.

HRAZDINA, G., and CREASY, L. L., 1979: Light Induced Changes in Anthocyanin Concentration, Activity of Phenylalanine Ammonia-Lyase and Flavanone Synthase and some of their Properties in *Brassica Oleracea*, Phytochemistry **18**, 581–584.

KHO, F. F., BENNINK, G. J. H., and WIERING, H., 1975: Anthocyanin Synthesis in a White Flowering Mutant of *Petunia hybrida* by a Complementation Technique, Planta **127**, 271–279.

KHO, K. F. F., BOLSMAN-LOUWEN, A. C., VNIK, J. C., and BENNINK, G. J. H., 1977: Anthocyanin Synthesis in a White Flowering Mutant of *Petunia hybrida*. II. Accumulation of

Dihydroflavonol Intermediates in White Flowering Mutants. Uptake of Intermediates in Isolated Corollas and Conversion into Anthocyanins, Planta **135**, 109–118.

KREUZALER, F., and HAHLBROCK, K., 1972: Enzymatic synthesis of aromatic compounds in higher plants: formation of naringenin (5.7.4′-trihydroxyflavanone) from p-coumaroyl coenzyme A and malonyl coenzyme A, FEBS Letters **28**, 69–72.

KREUZALER, F., and HAHLBROCK, K., 1973: Flavonoid glycosides from illuminated cell suspension cultures of *Petroselinum hortense*, Phystochemistry **12**, 1149–1152.

KREUZALER, F., and HAHLBROCK, K., 1975a: Enzymic Synthesis of an Aromatic Ring from Acetate Units. Partial Purification and Some Properties of Flavanone Synthase from Cell-Suspension Cultures of *Petroselinum hortense*, Eur. J. Biochem. **56**, 205–213.

KREUZALER, F., and HAHLBROCK, K., 1975b: Enzymatic Synthesis of Aromatic Compounds in Higher Plants. Formation of Bisnoryangonin (4-Hydroxy-6 [4-hydroxystyryl] 2-pyrone) from p-coumaroyl-CoA and malonyl-CoA, Arch. Biochem. Biophys. **168**, 84–90.

KUHN, B., FORKMANN, G., and SEYFFERT, W., 1978: Genetische Kontrolle der Chalkon-Flavanon Isomerase Aktivität in *Callistephus chinensis*, Planta **138**, 199–203.

LARNE, T. A. G., and GAMBORG, O. L., 1971: Ethylene Production by Plant Cell Cultures. Variations in Production During Growth Cycle and in Different Plant Species, Plant Physiol. **48**, 394–398.

MARGNA, U., 1977: Control at the level of Substrate Supply – An Alternative in the Regulation of Phenylpropanoid Accumulation in Plant Cells, Phytochemistry **16**, 419–426.

MORS, W. B., MAGELHEAS, M. T., and GOTTLIEB, O. R., 1961: Naturally Occurring Aromatic Derivativs of Monocyclic α-Pyrones, Progr. Chem. Natural. Prod. **20**, 131–164.

POULTON, J., GRISEBACH, H., EBEL, J., SCHALLER-HEKELER, B., and HAHLBROCK, K., 1976a: Two Distinct S-Adenosyl-L-Methionine: 3,4 Dihydric Phenol 3-O-Methyltransferases of Phenylpropanoid Metabolism in Soybean Cell Suspension Cultures, Arch. Biochem. Biophys. **173**, 301–305.

POULTON, J., HAHLBROCK, K., and GRISEBACH, H., 1976b: Enzymic Synthesis of Lignin Precursors. Purification and Properties of the S-Adenosyl-L-Methionine: Caffeic Acid 3-O-Methyltransgerase from Soybean Cell Suspension Cultures. Arch. Biochem. Biophys. **176**, 449–456.

POULTON, J., HAHLBROCK, K., and GRISEBACH, H., 1977: O-Methylation of Flavonoid Substrates by a Partially purified Enzyme from Soybean Cell Suspension Cultures, Arch. Biochem. Biophys. **180**, 543–549.

QUAST, L., and WIERMANN, R., 1973: Über das Vorkommen verschiedener substituierter Chalkone während der Mikrosporogenese bei Tulipa, Experienta **29**, 1165–1166.

RATHMEEL, W. G., and BENDALL, D. S., 1972: The Peroxidase-Catalyzed Oxidation of a Chalcone and its Possible Physiological Significance, Biochem. J. **127**, 125–132.

SALEH, N. A. M., FRITSCH, H., WITKOP, P., and GRISEBACH, H., 1976a: UDP-Glucose: Cyanidin 3-O-Glucosyltransferase from Cell Cultures of *Haplopappus gracilis*, Planta **133**, 41–45.

SALEH, N. A. M., FRITSCH, H., KREUZALER, F., and GRISEBACH, H., 1978: Flavanone Synthase from Cell Suspension Cultures of *Haplopappus gracilis* and Comparison with the Synthase from Parsley, Phytochemistry **17**, 183–186.

SALEH, N. A. M., POULTON, J. E., and GRISEBACH, H., 1976b: UDP-Glucose: Cyanidin 3-O-Glucosyltransferase from Red Cabbage Seedlings, Phytochemistry **15**, 1865–1868.

STICKLAND, R. G., and HARRISON, B. J., 1974: Precursors and Genetic Control of Pigmentation. L. Induced Biosynthesis of Pelargonidin, Cyanidin and Delphinidin in Antirrhinum Majus, Heredity **33**, 108–112.

SUTTER, A., and GRISEBACH, H., 1973: UDP-glucose: flavonol 3-O-glucosyltransferase from cell suspension cultures of parsley, Biochim. Biophys. Acta **309**, 289–295.

SUTTER, A., POULTON, J., and GRISEBACH, H., 1975: Oxidation of Flavanone with Cell-Free Extracts from Young Parsley Leaves, Arch. Biochem. Biophys. **170**, 547–556.

SÜTFELD, R., and WIERMANN, R., 1976: The Formation of Flavanones from Hydroxy-cinnamoyl-CoA Thiol Esters and Malonyl-CoA by Enzyme Extracts from Anthers, Z. Physiol. Chem. **79**, 467–472.

DE VLAMING, P., and KHO, K. F. F., 1976: 4,2',4',6'-Tetrahydroxychalcone in Pollen of *Petunia Hybrida*, Phytochemistry **15**, 348–349.

WEISSENBÖCK, G., and SACHS, G., 1977: On the Localization of Enzymes Related to Flavonoid Metabolism in Sections and Tissues of Oat Primary Leaves, Planta **137**, 49–52.

WELLMANN, E., 1974: Gewebsspezifische Kontrolle von Enzymen des Flavonoidstoffwechsels durch Phytochrom in Kotyledonen des Senfkeimlings (*Sinapis alba* L.), Ber. Deutsch. Bot. Ges. **87**, 275–279.

WENGENMAYER, H., EBEL, J., and GRISEBACH, H., 1974: Purification and Properties of a S-Adenosylmethionine: Isoflavone 4'-O-Methyltransferase from Cell Suspension Cultures of *Cicer arietinum* L., Eur. J. Biochem. **50**, 135–143.

WIERMANN, R., 1972: Aktivität der Chalkon-Flavanon Isomerase und Akkumulation von phenylpropanoiden Verbindungen in Antheren, Planta **102**, 55–60.

WONG, E., 1976: Biosynthesis of Flavonoids in: Chemistry and Biochemistry of Plant Pigments (T. W. Goodwin, ed.) Vol. 1, 464–526, Academic Press London, New York, San Francisco.

WONG, E., and WILSON, J. M., 1976: Products of the Peroxidase-catalyzed Oxidation of 4.2'.4'-Trihydroxychalcone, Phytochemistry **15**, 1325–1332.

ZÄHRINGER, U., EBEL, J., KREUZALER, F., and GRISEBACH, H., 1977: Biosynthesis of the Elicitor-Induced Phytoalexin, Glyceollin in Soybean (*Glycine max*), Hoppe-Seyler's Z. Physiol. Chem. **358**, 1303.

ZÄHRINGER, U., EBEL, J., and GRISEBACH, H., 1978: Induction of Phytoalexin Synthesis in Soybean. Elicitor-Induced Increase in Enzyme Activities of Flavonoid Biosynthesis and Incorporation of Mevalonate into Glyceollin, Arch. Biochem. Biophys. **188**, 450–455.

Degradation of Flavonoids and Isoflavonoids

Wolfgang Barz

Lehrstuhl für Biochemie der Pflanzen, Westfälische Wilhelms-Universität
Münster/Germany

1. Introduction

This article will describe some aspects of flavonoid and isoflavonoid metabolism in higher plants in order to document some principles of the degradation of aromatic and heterocyclic plant constituents in the producing plants. This area of research represents a rather new field of plant biochemistry because until recently secondary plant constituents were regarded as metabolically inactive end products which are permanently stored in specific tissues, cells or vacuoles (Reznik, 1960; Zenk, 1967). With respect to synthesis, turnover and degradation of plant constituents however, recent studies, seem to prove that the dynamics of secondary metabolism is comparable to what is known from primary metabolism. Even those compounds which appear to be solely accumulated during certain phases of plant growth and development should only be regarded as being in a transitory, metabolically inactive pool. Various regulatory parameters such as enzyme induction can transfer such compounds into a metabolically active pool and thus into further turnover and degradation processes (cf., Zenk, 1967). Quite often, especially in actively growing and developping plant tissues, secondary metabolism is characterized by simultaneous synthesis and degradation. Therefore, the stationary concentrations of secondary plant products in living tissue result from the ratio of the rate of synthesis and rate of degradation (Barz and Hösel, 1975). In view of the wide range of possible metabolic rates and the often great number of closely related secondary constituents in one plant, each case must be individually analyzed in order to determine the actual extent and velocity of turnover and degradation (Barz, 1977a; Strack and Reznik, 1976). Any generalizations, however, must still be avoided due to the great variety of as yet unknown factors influencing secondary metabolism in plants. Furthermore, some experimental data show that synthesis and degradation of secondary compounds may be regulated independently from each other. Thus alterations in endogenous and exogenous regulatory factors may preferentially influence only one or the other of these two metabolic branches in either the same or opposing directions (Barz et al., 1971; Barz, 1977b; Ellis, 1974).

2. General Aspects of Flavonol and Isoflavone Catabolism

Germination of seeds, growth of buds and development of seedlings constitute very active metabolic processes often accompanied by an intense flavonoid metabolism. Thus, according to Harborne (table 1, no. 1) and Reznik and associates (nos. 3 to 5) various characteristic flavonol and flavone glycosides completely disappear from seeds or buds upon further growth. The complete disappearance of the « youth anthocyanin » from *Catalpa ovata* leaves (Zenk, 1967) (table 1, no. 2) can also best be explained by assuming catabolic processes. Various other examples of this kind point to the importance of degradative reactions involved in polyphenol metabolism.

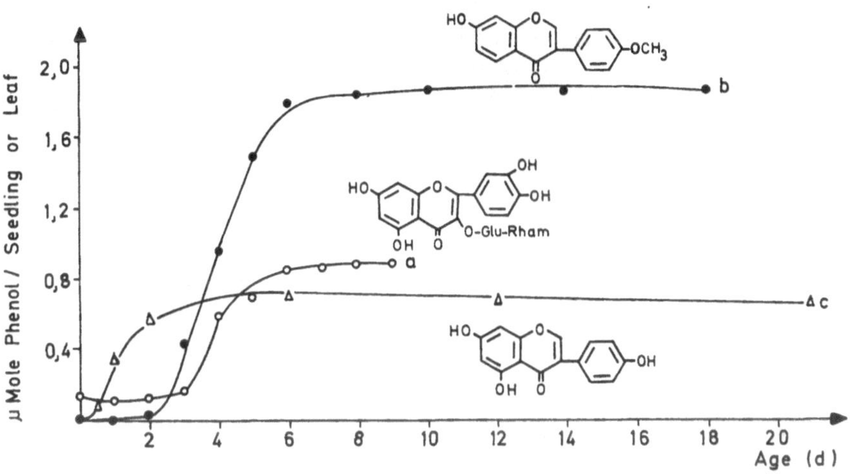

Fig. 1: Accumulation curves of a) rutin in buckwheat seedlings (Amrhein and Zenk, 1970), b) formononetin in seedlings of *Cicer arietinum* L. (Barz and Adamek, 1970), c) genistein in leaves of *Trifolium subterraneum* (Rossiter and Beck, 1967a, b)

The accumulation curves (figure 1) of the flavonol rutin in buckwheat seedlings, of the isoflavone genistein in *Trifolium subterranum* and of the isoflavone formononetin in *Cicer arietinum* demonstrate a characteristic phenomenon of the metabolism of secondary plant constituents. Such curves show that the isolatable amounts of compound per individual plant finally remain constant over longer phases of intense plant growth. These constant yields can be explained either by cessation of synthesis with undisturbed accumulation of secondary compounds or by permanent turnover at a rate equal to that of synthesis. With the aid of pulse-labelling experiments which allow measurements of the dynamics of flavonoid metabolism, the second alternative was proven (cf. Barz, 1969; Barz and Hösel, 1971). Such sigmoid accumulation curves with a pronounced plateau may thus be assumed to describe systems characterized by identical intensities of both synthesis and turnover of the constitutive compounds. However, it must again be mentioned that in some cases no turnover of secondary compounds could be found due to complete cessation of synthesis (Barz, 1977a; Barz and Hösel, 1979). Pulse-labelling experiments have also been helpful in determining the great variations in the flow of material through the pool of a secondary compound. Such variations were mainly caused by alterations of external conditions (i.e., light/darkness) (Barz, 1969; Barz and Hösel, 1975). Experiments of this type led to the conclusion that synthesis and degradation of a compound are possibly unlinked metabolic routes. Figure 2 summarizes our results of numerous pulse-labelling experiments with the flavonol kaempferol and the isoflavone formononetin in the garbanzo bean (*Cicer arietinum* L.). It is shown that the changes in flavonol and isoflavone content obtained during a light-darkness-light treatment of the experimental plants result predominantly from changes in the rate of turnover. The importance of

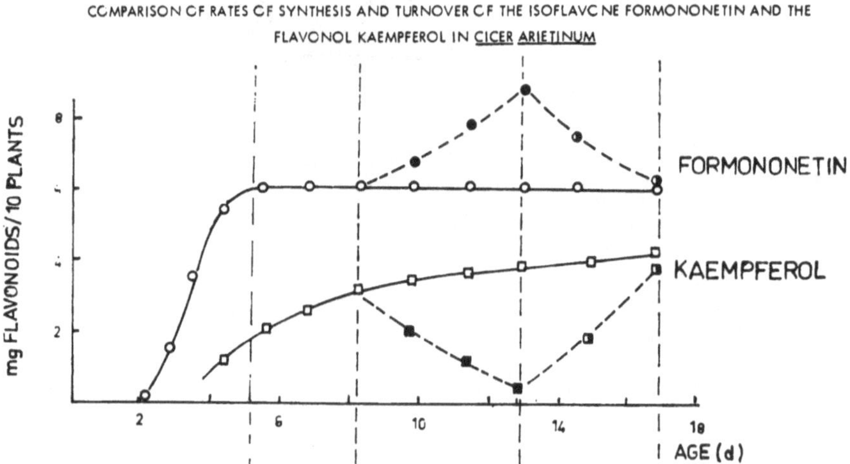

Compound investigated	white light	dark period	white light
Formononetin	synthesis equals turnover	synthesis unaltered turnover decreases	synthesis normal turnover increases
Kaempferol	synthesis greater than turnover	synthesis unaltered turnover increases	synthesis prevails because turnover decreases

Fig. 2: Accumulation curves of the isoflavone formononetin and the flavonol kaempferol in *Cicer arietinum* and comparison of rates of synthesis and turnover depending on illumination and darkness

catabolic routes for regulating the level of phenolic plant constituents is thus documented. The same explanation may possibly be used for the repeatedly observed diurnal changes in the flavonoid content of plants (Nöll, 1955; Tissut and Egger, 1972). Similarly, substantial seasonal variations in the flavonoid level of various plants

(Ahlgrim, 1956; Staude and Reznik, 1973, Tissut, 1968) constitute examples for the dynamics of secondary metabolism based on changing rates of synthesis and degradation.

3. Catabolism of Flavonoids

Various analytical studies (cf., table 1), pulse-labelling experiments, application of radioactive flavonoids to plants and plant cell cultures, as well as enzymatic studies have provided unequivocal evidence for degradative reactions of flavonoids and iso-

Table 1: Examples of turnover and degradation of flavonoids in actively growing plants

1) Several kaempferol-glycosides are only found in the seeds, but no longer in adult plants of *Solanum tuberosum* (Harborne, 1962).
2) Anthocyanin is completely degraded when young red leaves of *Catalpa ovata* grow to mature green leaves (Zenk, 1967).
3) Kaempferol-, quercetin- and apigenin-glycosides completely disappear upon germination of *Acer pseudoplatanus* seeds (Rumpenhorst, 1968).
4) Within 10 days after germination of 2 *potatoe* cultivars, kaempferol 3-diglucoside-7-rhamnoside and kaempferol 3-triglucoside-7-rhamnoside completely disappear (Reznik, cited after Strack, 1973).
5) During growth of buds of *Corylus avellana*, kaempferol-3-glucosyl-(p-coumarate) completely disappears (Staude and Reznik, 1973).

flavonoids. Table 2 lists the hitherto known examples and demonstrates that concrete evidence for turnover and degradation is available for essentially all classes of flavonoids and isoflavonoids (Barz and Hösel, 1975). Elucidation of flavonoid catabolism in plants has turned out to be very difficult (cf. Barz et al., 1970). This mainly due to the complex tissue differentiation of higher plants combined with a rather reduced capability for long distance transport of secondary compounds, and due to the low concentrations of catabolites, to very pronounced polymerisation reactions of phenols, and to the ever-present microbial contamination of plant tissues (cf. Barz 1975, Barz and Hösel, 1979). Such difficulties have partly been overcome by the recent introduction of plant cell suspension cultures for investigations on catabolic pathways of secondary plant constituents (Barz, 1975; Barz 1977a, b; Ellis, 1974). The main advantages of plant cell suspension cultures are:

a) unequivocal absence of microorganisms.

b) a more homogenous cell material which can partially be directed towards specific metabolic processes by means of culture conditions and application of phytohormones (Zenk et al., 1975).

c) lack of complex tissue differentiation leading to a highly improved rate of substrate uptake.

d) higher concentrations of catabolites.

e) facilitated isolation procedures of enzymes.

Table 2: Flavonoids and isoflavonoids for which turnover and degradation in plants has been shown (cf., Barz and Hösel, 1975; Barz, 1977a)

Chalcones	2',4,4'-Trihydroxychalcone 2',4,4',6'-Tetrahydroxychalcone 2',4'-Dimethoxy-4-methoxychalcone
Dihydrochalcone	Phloretin
Aurone	Hispidol
Flavanones	Naringenin 7,4'-Dihydroxyflavanone Eriodyctiol
Dihydroflavonols	Dihydrokaempferol, Dihydroquercetin, Garbanzol
Flavonols	Galangin, Kaempferol, Quercetin, Isorhamnetin, Myrizetin
Flavones	Apigenin, Luteolin, 7,4'-Dihydroxyflavone, Vitexin, Isovitexin
Anthocyanidins	Delphinidin, Petunidin, Cyanidin, Malvidin
Catechines	Epicatechin, Epigallocatechin, Epicatechingallate, Epigallocatechin-gallate
Isoflavones	Daidzein, Dihydrodaidzein, Formononetin, Biochanin A, Genistein, 7-Hydroxy-2',4',5'-trimethoxyisoflavone
Coumestane	Coumestrol
Rotenoids	Rotenone, Amorphigenin, Amorphin

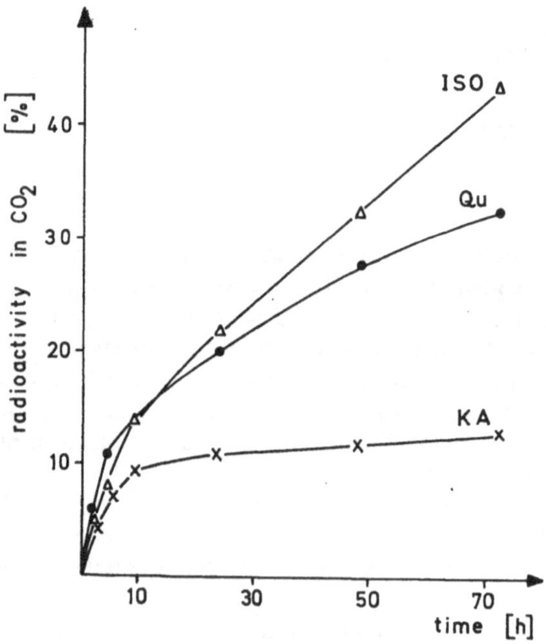

Fig. 3: Formation of $^{14}CO_2$ after application of (U$-^{14}$C)-kaempferol (KA), (U$-^{14}$C)-quercetin (Qu) and (U$-^{14}$C)-isorhamnetin (Iso) to soybean cell suspension cultures (cf. Hösel et al., 1972)

Restrictions in the use of plant cell cultures may result from the observation that in some cases cell cultures no longer produce the characteristic secondary compounds (Teuscher, 1973) or that enzyme patterns have been obtained which differ qualitatively or quantitatively from what is found in intact plants (Hösel et al., 1977). It therefore seems appropriate to interpret results obtained in cell cultures with the understanding that the regulatory parameters may be different. Still, cell suspension cultures have been ideal systems for demonstrating the great capacity of plant cells for the degradation of aromatic structures.

As an example of the complete and rapid catabolism of flavonoids figure 3 shows the formation of $^{14}CO_2$ in soybean cell suspension cultures measured upon the application of the ^{14}C-labelled flavonols kaempferol, quercetin and isorhamnetin. In addition to the high $^{14}CO_2$ values, analyses of the nutrient medium and of the cell extracts proved that the applied substrates had been completely degraded (Hösel et al., 1972). Similar studies using ^{14}C-labelled kaempferol 3-glucoside and kaempferol 7-glucoside demonstrated (figure 4) that the substrates are rapidly deglucosilated and that the rate of glucose oxidation from the 3-glucoside was almost identical to that

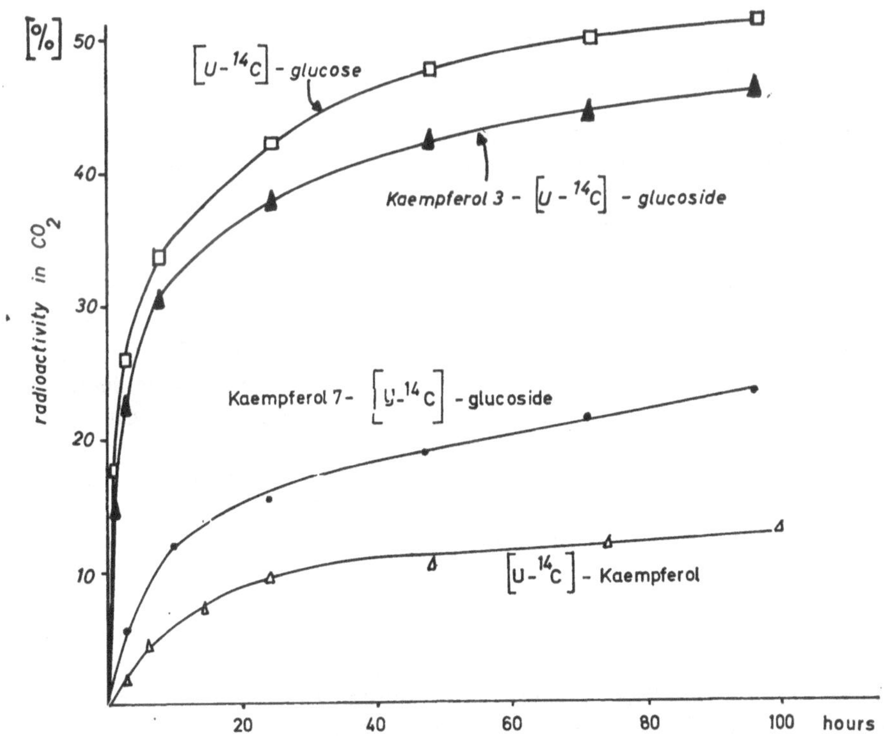

Fig. 4: $^{14}CO_2$ production from labelled kaempferol, kaempferolglucosides and glucose demonstrating the efficient degradation of glucosides in soybean cell suspension cultures (Muhle et al., 1976)

of free glucose itself (Hösel et al., 1977). It seems remarkable that the oxidation of glucose from position 3 was consistently two or three times more rapid than from position 7 (Muhle et al., 1976).

Investigations of this type have been carried out with numerous substrates and they all led to the conclusion that the capacity of plant cells for the degradation of aromatic and heterocyclic compounds seems to be more pronounced than previously expected. Some recent data on the degradation of flavonoids and isoflavones obtained by use of plant cell suspension cultures or isolated enzymes are summarized in figure 5. In the case of chalcones and flavanones, specifically ^{14}C labelled substrates were used to demonstrate that A-rings with both the resorcinol, and phloroglucinol, type of substitution pattern are well catabolized. Such data are of importance in view of the great number of natural products showing this ring structure.

A remarkable feature of the presently known degradative pathways of chalcones, aurones, flavanones and flavonols is the fact that the B-ring of these flavonoids is converted to the equivalent substituted benzoic acids. These benzoic acids can be metabolized in turn by hydroxilation, O-demethylation, ring fission and oxidative decarboxylation (Berlin et al., 1971; Harms et al., 1972; Barz, 1977a; Berlin and Barz, 1975b). Therefore, benzoic acids can be regarded as intermediates in flavonoid degradation and they are funnelled either into further catabolism or oxidative polymerization reactions (Berlin and Barz, 1975a).

Recent results however, have revealed that flavanones can also be degraded to chromones by removal of ring B as a simple phenol (Janistyn and Stocker, 1976, Patzlaff and Barz, 1978). These reactions, shown in figure 6, are catalyzed by peroxidases and require a 4'-hydroxy group to proceed. The available data on peroxidative flavanone degradation demonstrates that such compounds can be attacked at either ring A or carbon atom 2 (Barz and Hösel, 1978).

Until now, catabolites of isoflavones could not be isolated regardless of whether isoflavone aglycones (Berlin et al. 1974) or glycosides (Hösel et al. 1977) were used as substrates. Investigations with appropriately ^{14}C-labelled isoflavones, however, indicated that the whole carbon skeleton of these compounds became degraded.

2,3-Dihydroxyflavanones are intermediates in flavonol degradation (Hösel and Barz, 1972; Hösel et al., 1972). The reaction is catalyzed by peroxidases in the presence of H_2O_2 (Hösel et al., 1973) and requires free hydroxyl substituents in positions 3 and 4' (Frey, 1977). All peroxidase isoenzymes isolated from *Cicer arietinum* cell cultures and *Sinapis alba* seedlings were shown to possess this property without any quantitative difference in specificity (Frey, 1973). A set of flavonol-oxidases which have so far only been isolated from *Mentha piperita* Mitcham plants and cell cultures also catalyze the conversion of quercetin to 2,3,5,7,3',4'-hexahydroxyflavanone in the presence of oxygen only (Barz, 1977a; Frey, 1977). In contrast to peroxidases, this latter group of flavonol-oxidases has no porphyrin prostetic group.

In addition to the various known aspects of peroxidases, especially in connection with the well-studied function as an IAA-oxidase (Galston, 1969), our investigations on the degradation of flavonols, flavanones (Berlin et al., 1973; Patzlaff, 1974), aurones (Teufel, 1974) and benzoic acids (Berlin and Barz, 1975a, b) may possibly

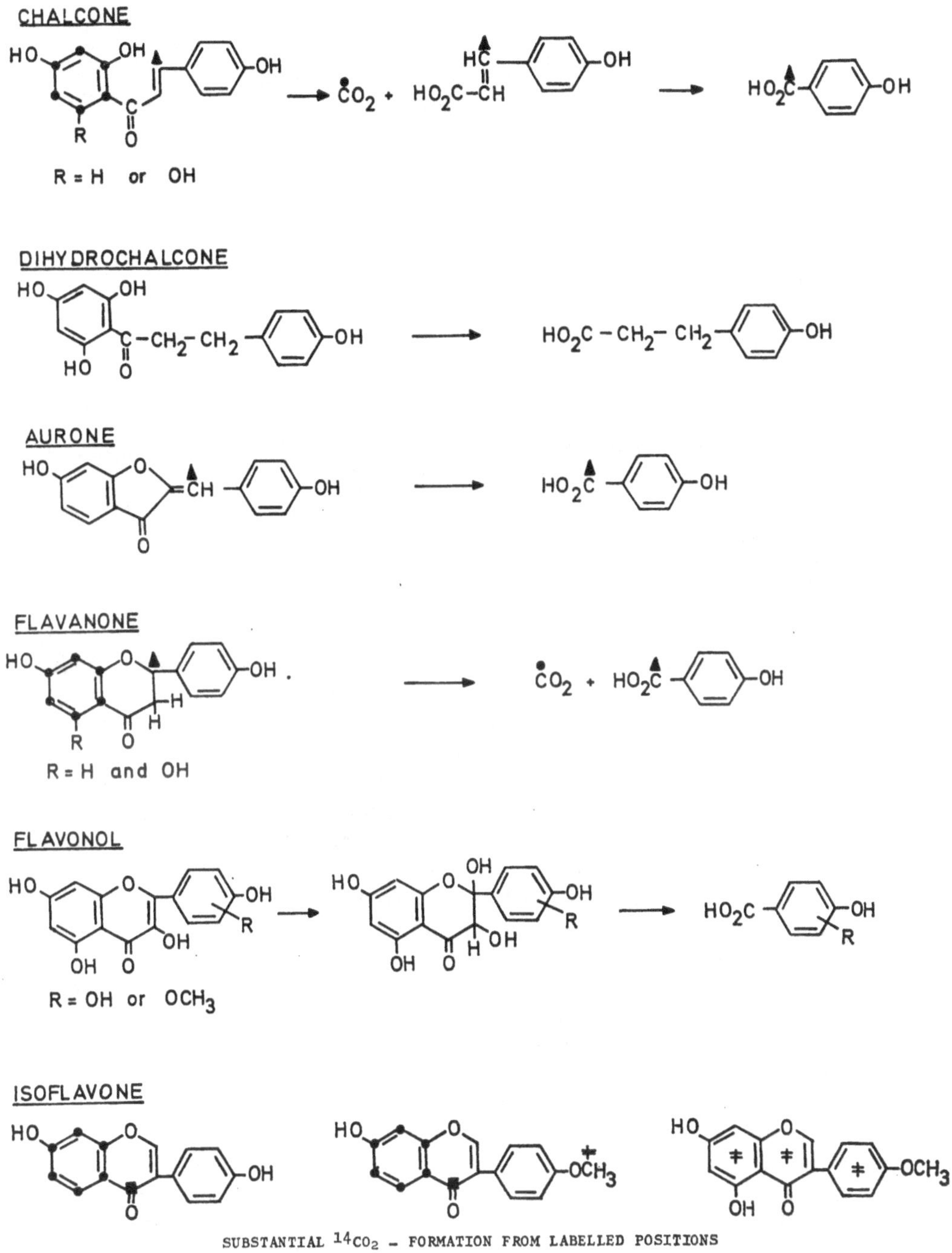

Fig. 5: Degradative pathways of flavonoids and isoflavones (cf., Barz and Hösel, 1975)

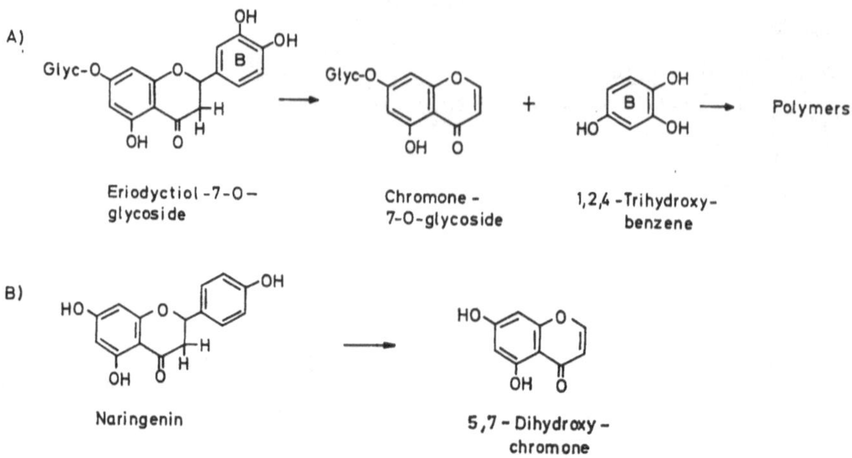

Fig. 6: Peroxidative degradation of flavanones to chromones (A) Janistyn and Stocker, 1976; (B) Patzlaff and Barz, 1978

lead to a new functional role of these enzymes which is associated with the destruction of polyphenols.

Studies with isolated enzymes and cell suspension cultures have led to the postulate that peroxidases may participate in flavonol degradation. It is not known, however, whether peroxidases carry out flavonol degradation under in vivo conditions in an intact plant. It is therefore of interest to discuss a correlation between peroxidases and the occurrence of flavonols resulting from studies by Wiermann (1973). These investigations deal with the accumulation of the flavonols kaempferol, quercetin and isorhamnetin in anthers of *Tulipa*. During differentiation of anthers ten developmental stages can be differentiated during which the various intermediates and the corresponding enzymes of flavonoid biosynthesis follow each other over a long period of time. Thus, phenylalanine ammonia lyase is followed by transcinnamic acid, the 2′,3,4,4′,6′-pentahydroxychalcone is followed by chalcone-flavanone isomerase and, finally, the flavonol pigments appear in the last stage. It can therefore be concluded that the appearance of an enzyme with its substrate is correlated by time and site. In the early stages of anther differentiation high activities of peroxidase occur. Prior to and during the final accumulation of flavonols peroxidase activity falls to a very low level. One may therefore conclude that flavonols can only accumulate to a significant extent in this in vivo system when peroxidase activity has strongly decreased.

Elucidation of polyphenol catabolic pathways in plants has been severely hampered by polymerization reactions which convert the applied substrates to ethanol-insoluble, lignin-like products (Dietermann, 1969; Berlin and Barz, 1971; Harms et al., 1972; Barz, 1977a). These polymers are not lignin, however, (Berlin and Barz, 1971). Rather, they stem from reactions of many different polyphenols, are catalyzed by phenol oxidases (Butt, 1972) or peroxidases (Berlin and Barz, 1975b), and are formed via ortho- or parachinoid structures. These polymerization reactions require specific substitution patterns as documented by our studies on isoflavones in cell suspension cultures. In figure 7 the investigated isoflavones are categorized as good and poor substrates for polymerization reactions. The well-polymerized isoflavones either carry

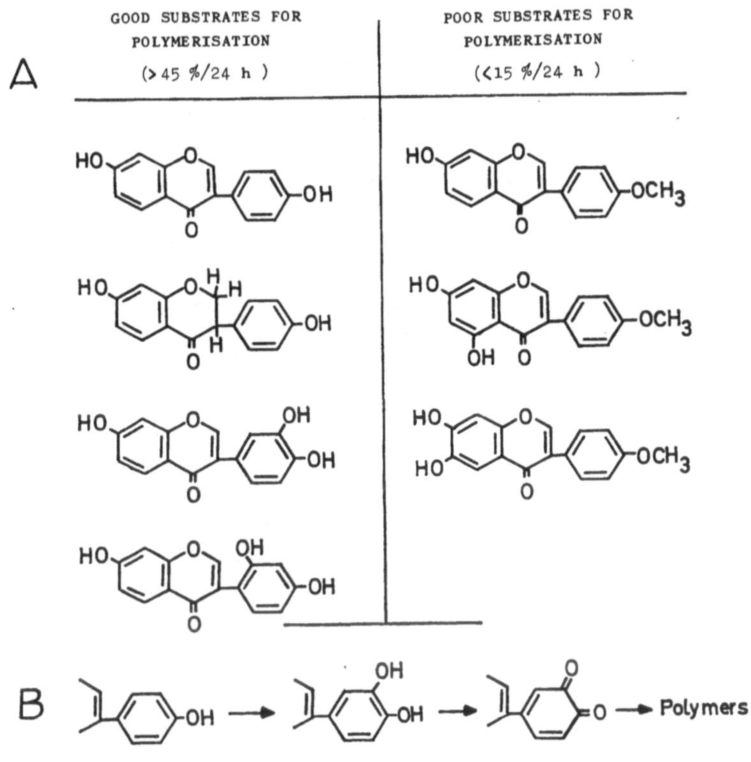

Fig. 7: Correlation of polymer formation in cell suspension cultures of mung bean with substitution pattern of isoflavones: a) structures of investigated substrates, b) mechanism of phenoloxidase-catalyzed polymer formation

a 4′-hydroxy group which allows a second hydroxilation in position 3′, or they already possess a 3′,4′-dihydroxy grouping. According to the reaction sequence in the lower part of figure 7 the polymerization can be visualized to proceed via an ortho-quinoid structure in ring B of the isoflavones. A methyl ether substituent in the 4′-position inhibits polymerization and even an ortho-dihydroxy grouping in ring A in positions 6 and 7 does not represent the substitution pattern required for polymerization by phenoloxidase and/or peroxidase. The feeding experiments summarized in figure 7 could also be corroborated by studies with enzyme preparations from mung bean cell suspension cultures (Berlin et al., 1974). Those isoflavones in figure 7 which were not predominantly polymerized in the cell cultures were alternatively funnelled into catabolic pathways or conjugation reactions with carbohydrates (Berlin and Barz, 1971). According to feeding experiments with various substrates (Barz and Hösel, 1975) it can be concluded that these metabolic alternatives may always occur simultaneously, but to a very different extent.

4. Conclusions

Numerous studies in recent years have provided evidence that flavonoids and isoflavones are metabolically active compounds. Apart from the well-known interconversions of flavonoids (Grisebach and Barz, 1969; Hahlbrock and Grisebach, 1975), the metabolism of these polyphenols is characterized by either total degradation or polymerization reactions. These alternatives are based mainly on the availability of enzymes in the living tissue or on the substitution pattern of the compounds under investigation. Both possibilities may result from the obvious lack of an excretory system in higher plants. Even those compounds stored in vacuoles for a certain period of time may be refunnelled into active metabolism (Barz, 1977a, b). As soon as plant constituents reside in a metabolically active pool, degradation or polymerization reactions proceed fully, and thus only low levels of catabolites seem to be present. The metabolism of secondary plant constituents can be summarized in a dynamic, multiregulated metabolic grid (figure 8). The ratio of degradation to polymerization must be understood quantitatively because the importance of each pathway will vary according to the chemical structure of the substrates and the available

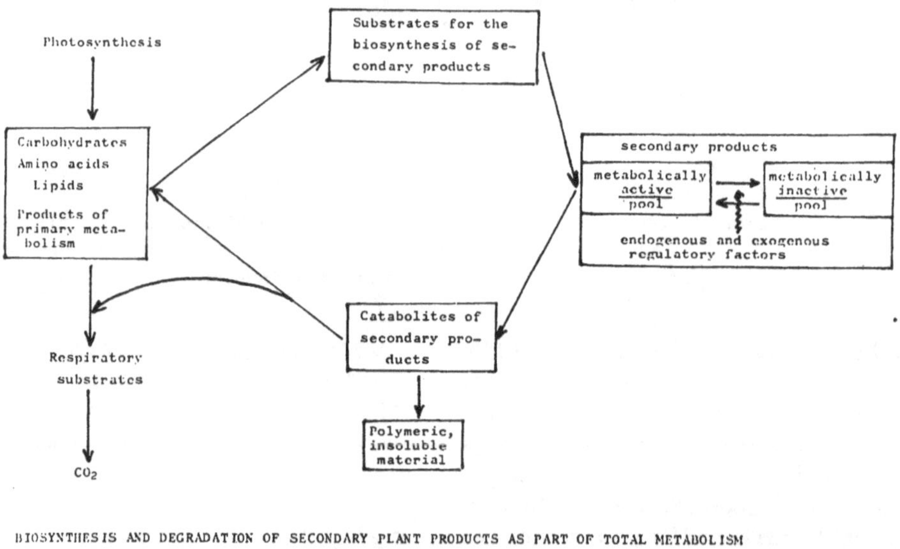

Fig. 8: Metabolic grid of secondary plant products depicting metabolically active and inactive pools

enzyme pattern. Differentiation between a metabolically active and a metabolically inactive pool of secondary plant products results from the biochemical laws of metabolism and various experimental data on the transient inactivation of plant constituents (cf., Barz and Hösel, 1975).

I express my best thanks to my co-workers for their cooperation. Our investigations have been generously supported by the Deutsche Forschungsgemeinschaft and Fonds der Chemischen Industrie.

References

AHLGRIM, E.-D., 1956: Beiträge zur Frage der Biogenese sekundärer Stoffwechselprodukte dargestellt an *Mentha piperita* und an *Fagopyrum*-Arten. Planta **47**, 255–298.

AMRHEIN, N., und M. H. ZENK, 1970: Aktivität der Phenylalanin-Ammonium-Lyase (PAL) und Akkumulation von phenylpropanoiden Verbindungen während der Keimung von Buchweizen (*Fagopyrum esculentum* Moench) im Dunkeln. Z. Pflanzenphysiol. **63**, 384–388.

BARZ, W., 1969: Über den Umsatz von Isoflavonen und Cumöstanen in *Cicer arietinum* L. und *Phaseolus aureus* Roxb., Z. Naturforschg. **24b**, 234–239.

BARZ, W., 1975: Abbau von aromatischen und heterocyclischen Pflanzeninhaltsstoffen durch Zellsuspensionskulturen. Planta medica, Supplement 117–133.

BARZ, W., 1977a: Degradation of polyphenols in plants and plant cell suspension cultures. Physiol. Vég. **15**, 261–277.

BARZ, W., 1977b: Catabolism of Endogenous and Exogenous Compounds by Plant Cell Cultures. In: W. Barz, E. Reinhard, M. H. Zenk (eds.), Plant Tissue Culture and its Biotechnological Application. Springer-Verlag, Berlin, Heidelberg, New York, pp. 153–171.

BARZ, W., and CH. ADAMEK, 1970: Über den Einfluß von Licht auf den Umsatz der Isoflavone Formononetin und Biochanin A in *Cicer arietinum* L. Planta (Berlin) **90**, 191–202.

BARZ, W., CH. ADAMEK and J. BERLIN, 1970: The Degradation of Formononetin and Daidzein in *Cicer arietinum* L. and *Phaseolus aureus*. Phytochemistry **9**, 1735–1744.

BARZ, W., and W. HÖSEL, 1971: Über den Umsatz von Flavonoiden und Isoflavonen in *Cicer arietinum*. Phytochemistry **10**, 335–341.

BARZ, W., W. HÖSEL and CH. ADAMEK, 1971: Über den Einfluß von Licht auf den Umsatz von Flavonolen und Isoflavonen in *Cicer arietinum*. Phytochemistry **10**, 343–349.

BARZ, W., and W. HÖSEL, 1975: Flavonoid Metabolism. In: J. B. Harborne, T. J. Mabry and H. Mabry (eds.), The Flavonoids. Chapman and Hall, Ltd, chapter 17.

BARZ, W., and W. HÖSEL, 1979: Metabolism and Degradation of phenolic compounds in Plants. Recent Adv. Phytochem., **12** 339–369.

BERLIN, J., and W. BARZ, 1971: Stoffwechsel von Isoflavonen und Cumöstanen in Zell- und Callussuspensionskulturen von *Phaseolus aureus* Roxb. Planta (Berlin) **98**, 300–314.

BERLIN, J., W. BARZ, H. HARMS and K. HAIDER, 1971: Degradation of phenolic compounds in plant cell cultures. FEBS-Letters **16**, 141–146.

BERLIN, J., and W. BARZ, 1975a: Stoffwechsel von Benzoesäuren in pflanzlichen Zellsuspensionskulturen. Planta Medica, Supplement 134–139.

BERLIN, J., and W. BARZ, 1975b: Oxidative Decarboxylation of para-Hydroxy-benzoic Acids by Peroxidases under in vivo and in vitro Conditions. Z. Naturforschg. **30c**, 650–658.

BERLIN, J., M. PATZLAFF, F. MOHR, E. TEUFEL and W. BARZ, 1973: Über den Abbau von

Flavanonen, Auronen und Benzoesäuren mit Enzymen aus pflanzlichen Zellsuspensionskulturen. Hoppe-Seyler's Z. Physiol. Chem. **354**, 1170.

BERLIN, J., P. KISS, D. MÜLLER-ENOCH, D. GIERSE, W. BARZ and B. JANISTYN, 1974: Über den Abbau von Chalkonen und Isoflavonen in pflanzlichen Zellsuspensionskulturen. Z. Naturforschg. **29c**, 374–383.

BUTT, V. S., 1972: Polyphenol Oxidases. Hoppe-Seyler's Z. Physiol. Chem. **353**, 131–132.

DIETERMANN, L. J., S. H. WENDER, W. CHORNEY and K. SKOK, 1969: Incorporation of ^{14}C-Quercetin into an isopropyl alcohol- and acid-insoluble fraction of tobacco. Phytochemistry **8**, 2321–2323.

ELLIS, B. E., 1974: Degradation of Aromatic Compounds in Plants. Lloydia **37**, 168–184.

FREY, G., 1973: Diplomarbeit Universität Freiburg/Brsg.

FREY, G., 1977: Untersuchungen zum enzymatischen Abbau von Flavonolaglyka in höheren Pflanzen. Dissertation, Universität Münster/Westf.

GALSTON, A. W., 1969: Flavonoids and photomorphogenesis in peas. In: J. B. Harborne and T. Swain (eds.), Perspectives in Phytochemistry, 193–204. Academic Press, London–New York.

GRISEBACH, H., and W. BARZ, 1969: Biochemie der Flavonoide. Naturwissenschaften **56**, 538–544.

HAHLBROCK, K., and H. GRISEBACH, 1975: Biosynthesis of Flavonoids. In: J. B. Harborne, T. J. Mabry and H. Mabry (eds.), The Flavonoids. Chapman and Hall, Ltd., chapter 16.

HARBORNE, J. B., 1962: The flavonol glycosides of wild and cultivated potatoes. Biochem. J. **84**, 100–104.

HARMS, H., K. HAIDER, J. BERLIN, P. KISS and W. BARZ, 1972: Über O-Demethylierung und Decarboxylierung von Benzoesäuren in pflanzlichen Zellsuspensionskulturen. Planta (Berlin) **105**, 342–351.

HÖSEL, W., and W. BARZ, 1972: Enzymatic Transformation of flavonols with a cell-free preparation from *Cicer arietinum* L. Biochim. Biophys. Acta **261**, 294–303.

HÖSEL, W., P. D., SHAW, G. FREY and W. BARZ, 1973: Abbau von Flavonolen in höheren Pflanzen. Hoppe-Seyler's Z. Physiol. Chem. **354**, 1203.

HÖSEL, W., P. D. SHAW and W. BARZ, 1972: Über den Abbau von Flavonolen in pflanzlichen Zellsuspensionskulturen. Z. Naturforschg. **276**, 946–954.

HÖSEL, W., G. BURMEISTER, P. KREYSING and E. SURHOLT, 1977: Enzymological Aspects of flavonoid Catabolism in Plant Cell Cultures. In: W. Barz, E. Reinhard, M. H. Zenk (eds.), Plant Tissue Cultures and its Bio-technological Application. Springer-Verlag, Berlin, Heidelberg, New York, pp. 172–177.

JANISTYN, B., and M. STOCKER, 1976: Enzymatischer Abbau von (Ring B–U–^{14}C)-5,7,3′,4′-Tetrahydroxyflavanon-7-O-glucosid zu 5,7-Dihydroxychromon-7-O-glucosid und (Ring–U–^{14}C)-1,2,4-Trihydroxybenzol mit einem zellfreien Extrakt von *Mentha longifolia.* Z. Naturforschg. **31c**, 408–410.

MUHLE, E., W. HÖSEL and W. BARZ, 1976: Catabolism of flavonol glucosides in plant cell suspension. cultures. Phytochemistry **15**, 1669–1672.

NÖLL, G., 1955: Untersuchungen über den Einfluß verschiedener Faktoren auf den Flavonolgehalt von *Fagopyrum esculentum* L. Moench und *Fagopyrum tartaricum* L. Gaertner. Pharmazie **10**, 609–615 und 679–691.

PATZLAFF, M., 1974: Diplomarbeit, Universität Münster.

PATZLAFF, M., and W. BARZ, 1978: Peroxidatic Degradation of Flavanones. Z. Naturforschg. **33c**, 675–684.

REZNIK, H., 1960: Erg. Biol. **23**, 14.

ROSSITER, R. C., and A. B. BECK, 1967a: Physiological and ecological studies on the oestrogenic isoflavones in subterranean clover (*T. subterraneum* L.) III. Effects of Light. Austral. J. Agric. Res. **18**, 23–37.

ROSSITER, R. C., and A. B. BECK, 1967b: Physiological and ecological studies on the oestrogenic isoflavones in subterraneum clover (*T. subterraneum* L.). V. Autogenic changes. Austral. J. Agric. Res. **18**, 561–573.

RUMPENHORST, H., 1968: Dissertation Universität Münster.

STAUDE, M., und H. REZNIK, 1973: Das Flavonoidmuster der Winterknospen und Laubblätter von *Corylus avellana* L. Z. Pflanzenphysiol. **68**, 346–356.

STRACK, D., 1973: Dissertation Universität Köln.

STRACK, D., and H. REZNIK, 1976: Die Dynamik von Flavonolglycosiden während der Keimlingsentwicklung von *Cucurbita maxima* Duchesne. Z. Pflanzenphysiol. **79**, 95–108.

TEUFEL, E., 1974: Diplomarbeit, Universität Freiburg/Brsg.

TEUSCHER, E., 1973: Probleme der Produktion sekundärer Pflanzenstoffe mit Hilfe von Zellkulturen. Pharmazie **28**, 6–18.

TISSUT, M., 1968: Etude de cycle annuel des dérivés phénoliques de la feuille de hêtre, *Fagus sylvatica* L. Physiol. Veg. **6**, 351–365.

TISSUT, M., et K. EGGER, 1972: Les glycosides flavoniques foliaires des quelques arbes, au cours de cycle vegetatif. Phytochemistry **11**, 631–634.

WIERMANN, R., 1973: Über die Beziehungen zwischen flavonolaufbauenden Enzymen, einem flavonolumwandelnden Enzym und der Akkumulation phenylpropanoider Verbindungen während der Antherenentwicklung. Planta (Berlin) **110**, 353–360.

ZENK, M. H., 1967: Biochemie und Physiologie sekundärer Pflanzenstoffe. Ber. Deutsch. Bot. Ges. **80**, 573–591.

ZENK, M. H., H. EL-SHAGI and U. SCHULTE, 1975: Anthraquinone production by cell suspension cultures of *Morinda citrifolia* Planta Medica, Supplement 79–101.

Flavonoids as Chemotaxonomic Markers in Plants

Tony Swain

Department of Biology, Boston University
Boston, Massachusetts U.S.A.

This Chapter is dedicated to Holger Erdtman, E. C. Bate-Smith and Jeffrey Harborne, without whose percipience and perseverance it could not have been written.

1. Introduction

The classification of living organisms has traditionally been made on the basis of direct morphological and anatomical observations of extant and fossil materials. The more recent applications of genetic, especially cytological, data and the use of refined electron microscopical techniques has extended the evidence which has been employed (Kubitzki, 1977a, b), but the overall approach still appears to be almost mystical in character. For example, Heywood (1974) stated that, « in constructing classifications [of plants] we do not start off with an array of pieces of information or characters and then set about to seek correlations between these so as to produce groups. On the contrary, classes or groups are first perceived mentally – and then our impression can be analyzed and given precision by extracting characters from it ». Nevertheless, the taxonomic decisions which are made on the basis of this approach are generally valid, although usually based on a very limited number of individual examples of the organisms under study. There is normally a lack of samples covering the geographical range or temporal and other important factors which affect species diversity. As a result, typological procedures are an essential part of present day taxonomy, but they are recognized as such and taxonomic descriptions of a given species, genus or family are not meant to have the same precision as the name or formula of a chemical compound.

If we note these differences between taxonomic and chemical procedures we should be better able to appreciate how to make use of comparative chemical information to elucidate systematic problems. Unfortunately, there is a tendency for most chemists and biochemists to assume that the data they produce, because of its « apparent » precision, has a higher order of value in plant classification than the morphological judgements made by traditional taxomists. One should remember, however, that the majority of chemical data, including that on flavonoids, is based on even more limited sampling than that used in traditional taxonomy. Much is obtained from a single organ from an individual member of a plant species gathered without thought for possible populational variation and there are few cases where the effects of geographical range, season, climate, edaphic or other factors on the chemical composition of plants have been examined for *taxonomic* purposes. On the other hand, where adequate chemical data is available our knowledge of the biosynthetic interrelationships between the individual components often allows us to discern associations between plant taxa which cannot always be determined using morphological criteria alone (Swain, 1974a, b; Aparecida et al., 1978). Furthermore, we are now becoming aware of the ecological importance of plant chemicals which allows us to comment on wider systematic problems of the interactions between plants and all other organisms (Levin, 1976; Swain, 1977) and thus begin to understand the reason for the enormous chemical diversity of the living world.

In this chapter, we will examine the distribution of flavonoid and related compounds in plants which throw some light on their evolutionary and taxonomic relationships.

2. The Plant Kingdom

The majority of present-day biologists subscribe to the classification of living organisms into five kingdoms: Monera (prokaryotes, including bacteria and blue-green algae); Protista (unicellular and colonial eukaryotes); and the more complex eukaryotes Fungi; Animalia; and Plantae (Whittaker and Margulis, 1978). The boundaries between the Protista and the last three kingdoms of eukaryotes are still subject to argument. In the case of the algae, for example, Whittaker and Margulis place only the red and brown algae in the Plantae, putting all others, including the green algae, which are generally recognized as progenitors of the land plants, in the Protista.

The Plantae proper include two major divisions: (i) the Bryophyta (mosses and liverworts) which have no vascular system and whose first fossil record is about 380 million years ago; and (ii) the Tracheophyta or vascular plants. The sub-divisions of the latter in order of appearance in the fossil record (and of increasing morphological advancement) are the Psilopsida (400 m.y.a.), Lycopsida and Sphenopsida (380 m.y.a.) and the Pteropsida which are divided into the classes Filicinae (ferns, 375 m.y.a.), Gymnospermae (370 m.y.a.) and Angiospermae (135 m.y.a.) (Swain, 1974).

The evolution of the angiosperms is still a matter of controversy (Sporne, 1974; Kubitzki, 1977b). While I do not subscribe to a polyphyletic origin of the flowering plants, it is obvious that two distinct chemical lines have emerged, one dependent mainly on flavonoids and the other on alkaloids for their chemical defenses against herbivores and phytopathogens (Swain, 1975). Chemically, the taxa in the first group have retained (and also elaborated) flavonoid diversity from more primitive taxa (ferns and gymnosperms), but at the same time have evolved many morphological and anatomical features which are adjudged as advanced. Conversely, taxa in the second group have many morphological features which are agreed to be primitive, but it should be recognized that their chemistry is unique in that they accumulate

« aromatic » alkaloids heretofore unknown in the plant kingdom. It seems unlikely, therefore, that these latter are the progenitors of all angiosperms as believed by many taxonomists (Sporne 1974; Rezende et al., 1976; Kubitzki, 1967; Bate-Smith, 1974).

3. Flavonoids as Taxonomic and Evolutionary Markers in Plants

The overall distribution of many different classes of chemical compounds have been used effectively as criteria in determining evolutionary and taxonomic relationships in plants. Obviously, a comparison of the sequences of nucleotides in homologous nucleic acids in different plant taxa are likely eventually to yield the most cogent information on evolutionary processes. But such information is at present extremely limited and, because much of the DNA in higher organisms plays an ill-understood complex control function or may be redundant, it will be some time before we can appreciate its significance (Ayala, 1977). Nevertheless, the data so far available allows us to postulate the probable evolution of prokaryotes and to propose that chloroplasts, and probably mitochondria, in eukaryotes originated by symbiosis of prokaryotes into a protoeukaryotic cell (Margulis, 1979).

Similarly, the determination of amino acid sequence in homologous proteins has thrown much light on the evolution of various plant and animal taxa, but again the data is too sparse from within the plant kingdom to enable us to examine relationships with any more precision than is given from the admitted inadequate fossil record (Ayala, 1977).

Information from the distribution of so-called micromolecules (secondary compounds) i.e., with MW <1,000) is at present just as informative for land plants as that from macromolecules because so much more is available (Swain, 1974b). The advent of rapid chromatographic and electrophoretic separatory techniques (paper, TLC, GLC, etc.) and identification of compounds by a combination of UV, IR, NMR and mass spectra has almost tripled our knowledge of such compounds each decade since the introduction and increased sophistication of these methods in the 1950's. Our knowledge of the variation in structure and distribution of the lipids, various types of terpenoids, alkaloids, non-protein amino acids, flavonoids and other phenolics, and a range of other secondary compounds has expanded enormously and has enabled us to indicate taxonomic relationships which were problematical before this revolution began (Bendz & Santesson, 1974; Smith, P., 1976).

Of all the various classes of compounds which have been examined, the flavonoids and related phenolic compounds have proved so far to be the most useful. This is because of their universal distribution in most aerial parts of land plants, usually varying in different organs (flowers, leaves, bark, wood etc.); their ease of separation and accurate identification using simple chromotographic (paper, TLC) and spectroscopic (UV) techniques; and the number of structural variations (skeletal, glycosidation, hydroxylation, etc). (Harborne, 1977; Harborne et al., 1975). Even prior to the advent of modern techniques (1930's), flavonoids were the compounds of choice in early experiments on chemical genetics (Harborne, 1967). Now that their biosynthetic pathways are accurately known we can postulate approximate sequences in their acquisition by various taxa in plants and thus can use them to determine both evolutionary and taxonomic relations.

A number of workers have stressed the importance of comparing the taxonomic value of flavonoids and other compounds in relation to their biosynthetic complexity (Swain, 1974a). Thus, if a plant species possesses the ability to accumulate the relatively rare 6-methoxyflavonols (e.g., patuletin, *1*), then it must contain the enzyme

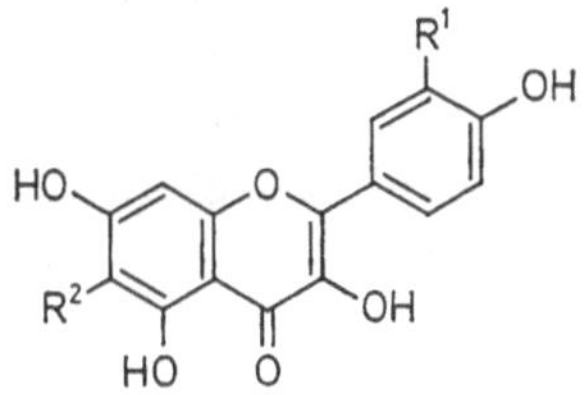

1 $R^1 = H$, $R^2 = OCH_3$
2 $R^1 = R^2 = H$
3 $R^1 = OH$, $R^2 = H$

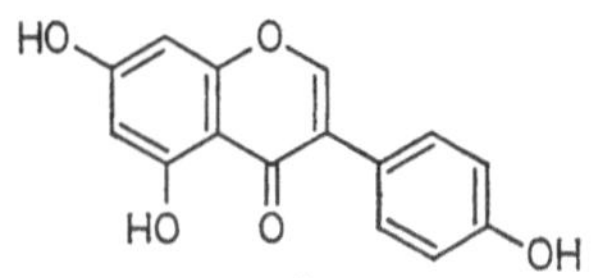

4

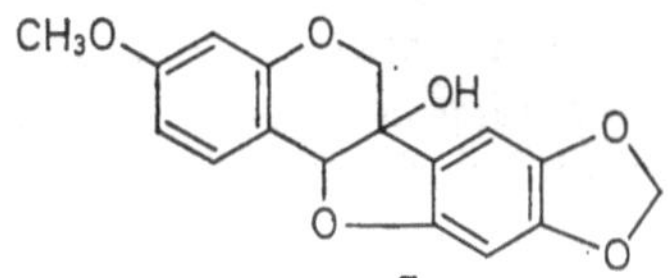

5

systems (and regulatory functions) necessary for the introduction of a hydroxyl at the 6-C atom of the flavonoid (whether done at the flavanone or later level) and its subsequent *O*-methylation. In general the greater the number of biosynthetic steps involved to give rise to a given compound, the less likely is it to occur; but of course, there are «target» molecules which one can regard as the norm within any given skeletal group. Thus, in the flavonols, kaempferol (*2*) can be regarded as the target, quercetin (*3*) requiring one extra biosynthetic step (the introduction of 3′-OH) and patuletin 3 extra steps (introduction of 3′-OH and 6-OH and methylation of the latter).

The most comprehensive approach in analysis of this sort has been that given by

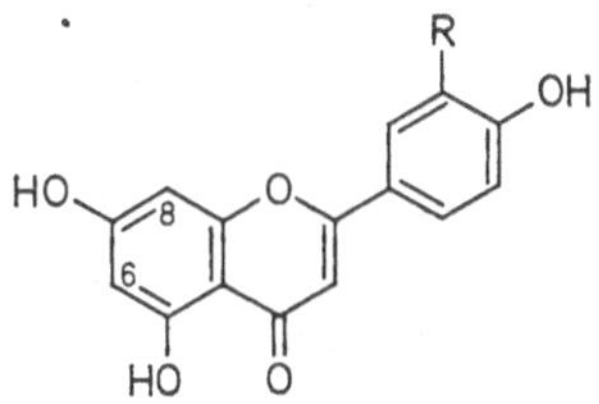

6 R = H
7 R = OH

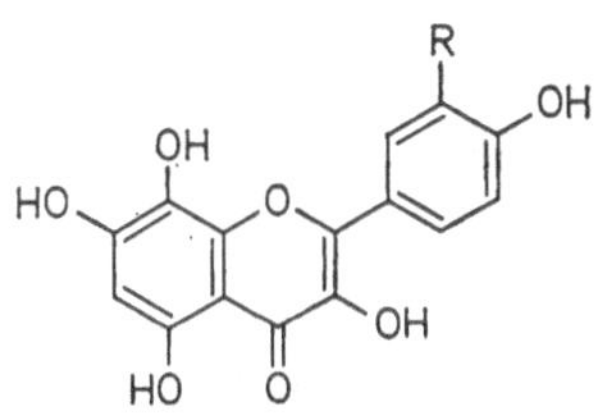

8 R = H
9 R = OH

Gottlieb and his co-workers (Rezende and Gottlieb, 1973; Rezende et al., 1975; Cagnin et al., 1976; Aparecida et al., 1978). Their treatment of the isoflavonoids (isoflavones, pterocarpanoids, rotenoids, 3-aryl coumarins, angolensins) is masterly in this respect (Aparecida et al., 1978). They calculated the relative probability of occurence (RPO) for each *skeletal* variation by dividing the number of assigned reaction steps from the basic isoflavone (given the value 1) by the number of known derivatives in that variant reported. Thus, for the isoflavones (*4*) as such the RPO = 1/152 (0.0066) whereas for 6a-hydroxypterocarpans (*5*) RPO = 4 (reaction steps)/5 = 0.80. A similar treatment of hydroxy-substitution patterns (5, 7, 4′ = 1, up to 7, 8, 2′, 5′ = 6 reaction steps involving introduction or elimination of -OH functions) gave a series of substitutional RPO's. Combination of the skeletal and substitutional RPO's into a two-dimensional matrix now allows one to use the isoflavonoid data in a meaningful systematic manner. As Aparecida et al. (1978) point out, it is only within the Leguminosae at present that there is a sufficient diversity of isoflavonoids and species analyzed to justify the application of RPO data for systematic purposes. In this family, several useful associations result in relation to the evolution of the legumes, and give some support for changes in taxonomic relationships proposed recently by Polhill (quoted in Aparecida et al., 1978). The Brazilian authors also showed that even with the paucity of information on other isoflavonoid-containing families the application of their approach led to useful ideas about the relative advancement of the families concerned. My only comment is that we cannot as yet decide the importance of the different RPO's (e.g., A v B ring substitution) and there are undoubtedly cases where evolutionary advancement is caused by *lost* (c.f., Harborne 1977) as well as gained biosynthetic elaborations. Furthermore, a simpler treatment of the data of flavonoid distribution (e.g., Bate-Smith's 1962 code for presence or absence of proanthocyanidins (a, a_0) or B-ring trihydroxy derivatives (b, b_0) and one might compare also the treatment of data in Bate-Smith et al. 1975, and Smith et al. 1977) gives equally meaningfull results.

4. The Overall Distribution of Flavonoids in the Plant Kingdom

Flavonoids are universally present in all species of land plants (Swain, 1975). They also occur in two species of the relatively advanced green algal family, the Characcae, whose ancestors are believed to have been the progenitors of land plants over 400 million years ago. The flavonoids of the Characeae are all *C*-glycosylflavones. *Nitella hookeri* contains 6,8-di-*C*-glycosides of both apigenin and luteolin (*6*, *7*) and similar compounds were found in *Chara* species (Swain, 1975).

The most primitive class of vascular plants, the Psilopsida, contain biflavones and traces of their *O*-glycosides (Cooper-Driver, 1977), which also occur in the Sellaginellales of the almost equally ancient Lycopsida (Voirin and Jay, 1978). The other two orders in this class, the Lycopodiales and Isoetales, both contain only simple flavone-*O*-glycosides. The Sphenopsida, horsetails, which according to the fossil record were almost co-eval with the lycopods, show a unique advancement in flavonoid chemistry in that they not only contain simple flavonols (*2*, *3*) (which indicates the ability to introduce the 3-hydroxyl group into the flavanone precursor), but also the corresponding 8-hydroxy derivatives of these compounds (herbacetin, *8*, and gossypetin, *9*) and the oligomeric proanthocyanidins, procyanidin, *10* and prodelphinidin (*11*), (Swain, 1975). Thus, almost all the biosynthetic features which characterize the most advanced members of the plant kingdom probably evolved before the end of the Devonian period.

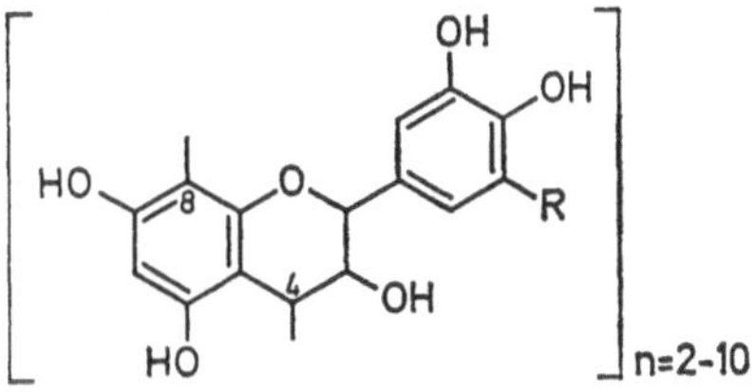

10 R=H
11 R=OH

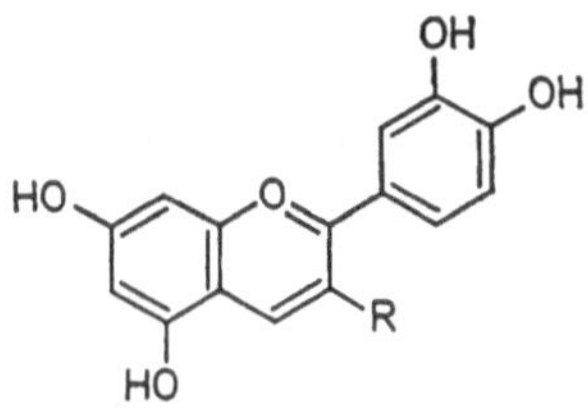

12 R=H
13 R=OH

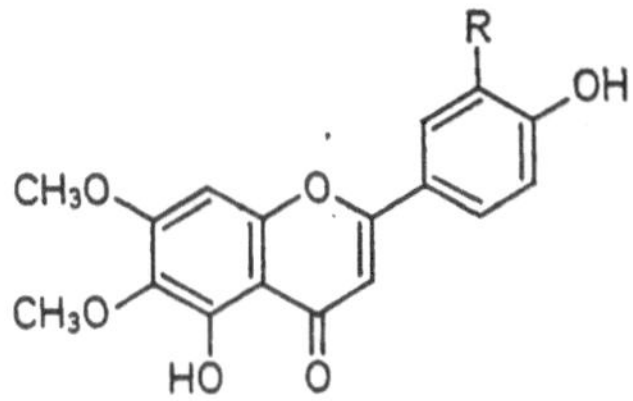

14 R=OH
15 R=H

The ferns, Filicinae, show fewer new excursions into novel flavonoid structures than the Sphenopsida. Notable is the almost universal presence of the common flavonols (*2*, *3*) and proanthocyanidins (*10*, *11*) (except in the more recent Ophioglossides) and their ability to produce 3-deoxyanthocyanins (*12*), both *C*- and *O*-methylflavonoids, and to accumulate chalkones and dihydrochalkones in the farina on the underside of the fronds (Wollenweber, 1977). Similarly the gymnosperms show few new biosynthetic classes of flavonoids, although the anthocyanidins (*13*) proper have been isolated from the cones, seeds and leaves of several species of Cupressaceae, Podocarpaceae and Pinaceae. Characteristic compounds, besides the common flavonols and procyanidins are the biflavones (present in all taxa except the Pinaceae and Gnetales) and the *C*-methylflavones. However, B-rings without hydroxyl substituents are prevelant.

As might be expected, the full flowering of flavonoid diversity is found only in the angiosperms. New ring systems include the isoflavones neoflavones, and aurones, but the diversity of hydroxylation (at C-8, C-6, C-2′ etc.); *C*- and *O*-alkylation (including prenylation); and arylation, glycosylation, and *O*-acylation (of both the flavonoid and sugar hydroxyls by organic and inorganic radicals) far outweighs those of other plant classes. Since the distribution of these diverse features are restricted to certain flowering plant taxa, it is not surprising that the flavonoids have been the major group of compounds used for systematic studies in this class.

The non-vascular land plants, the mosses and liverworts (Bryophyta), in spite of their ancient fossil record and the presumption that they are the progenitors of all vascular plants, are as advanced as the Sphenodsida in their ability to produce a wide diversity of flavonoid structures. Thus, flavonols, B-ring trihydroxy flavones, and so on, have been isolated from liverworts and mosses (Markham and Porter, 1978). It would appear, therefore, that extant bryophytes represent either a morphologically reduced offshoot of vascular plant evolution or have advanced in parallel from a more ancient, now extinct, class.

The diversity and evolutionary importance of flavonoids in the angiosperms has been recently discussed in detail by Harborne (1972, 1975, 1977). The main evolutionary features found in advanced taxa (Sporne, 1974) are: (a) loss of proanthocyanidins (*10, 11*) (e.g., 100% in Dilleniaceae, 0% in Labiatae); (b) loss of biflavones; (c) replacement of flavonols (*2, 3*) by flavones (*6, 7*); (d) change from hydroxylation at C-8 to that at C-6; (e) increase in *O*-methylation; (f) increase in complex glycosidation including sugar acylation; (g) increase in complexity of rarer flavonoids, (isoflavones, chalkones, etc). Not all these features are equivalent in an evolutionary sense. For example, those given in (a) to (c) as mentioned above, are ancient plant features occurring almost universally in ferns and gymnosperms, whereas (d) to (f) are almost uniquely found in the angiosperms themselves. Thus, 8-hydroxylation, occurring rarely in non-flowering plants, is generally a feature of angiosperm families showing medium advancement (average advancement index, 46%, Sporne, 1974) whereas 6-hydroxylation occurs in families which are at the height of evolutionary change (average index = 85%) (Harborne, 1977). However, Harborne did not use increases in biosynthetic complexity to « score » these parameters and this may well be necessary if his observations are to be used to the full (Aparecide et al., 1978).

From the evolutionary point of view, the data outlined above points to a relatively rapid evolution of basic flavonoid structures during the Devonian period (400 m.y.a.) followed by a long period of relative stagnation until the rise of the angiosperms in the mid-Cretaceous (130 m.y.a.). Such changes might be expected if the original rise of the flavonoid diversity was due to their importance in protecting plants against light and the more primitive herbivores and phytopathogens, the post-Cretaceous changes in the flavonoid complement in flowering plants presumably reflecting their close association with the co-evolving diverse pollinators (birds, insects) and herbivores (reptiles, mammals, insects). Overall, however, we can see that during the course of evolution, flavonoid diversity has depended first on an increase in biosynthetic steps required to produce novel compounds (e.g., isoflavonoids) and second on the ability of plants to apply certain such steps to give complexity to the more primitive flavonoids (e.g., flavones). Both are part of the evolutionary process.

5. Flavonoids as Systematic Markers in Higher Plants

The first real investigation on the use of flavonoids as systematic markers in higher plants was carried out in Sweden on *Pinus* species by Holger Erdtman and his colleagues starting in the 1940's. To be sure, Robert Robinson had given some leads a decade or so earlier, but the Stockholm group, who began by using the classical methods of organic chemistry, showed a remarkable dichotomy between the Haploxylon and Diploxylon subgenera. The former were all shown to contain varying amounts of flavones (chrysin, tectochrysin) and 6-*C*-Me-derivatives of chrysin and its flavonone and flavonol congeners which were absent from the latter (Erdtman, 1956).

The major breakthrough in flavonoid chemotaxonomy, however, was that of Bate-Smith in the early 1950's who in a series of elegantly simple papers surveyed the distribution of the flavonols (kaempferol, quercetin and myricetin) the proanthoacyanidins (procyanidin and prodelphinidin) and the cinnamic acids (p-coumaric, caffeic, ferulic and sinapic) by paper chromatography in the hydrolyzed leaves (no prior extraction was carried out) of over 1 500 species from nearly one half the families of flowering plants. He demonstrated for the first time how common these constituents were and indicated their association with certain orders and families (Bate-Smith, 1962, 1968).

Somewhat later, Alston and his co-workers (1962), Reznik (1956) and Geissman and Harborne (1956), using techniques first demonstrated in an examination of glycosidic flavonoid petal pigments by Nordstrom and Swain (1953), started to look at the usefulness of flavonoids as systematic markers at the family and generic level. The untimely death of Ralph Alston in 1966 deprived chemotaxonomy of a world leader, and his contribution to *Baptisia* hybridization studies is a model which has not been surpassed. Fortunately, the torch he laid down has been well carried by Harborne, Mabry, and many others including Bate-Smith himself who has, in the last few years, greatly illuminated our knowledge of proanthocyanin and ellagitannin distribution in plants (Bate-Smith, 1974), as well as relationships between flavonoids and evolutionary advancement in several families such as the Cunoniaceae (Bate-Smith, 1977).

However, the clear leader here is certainly Harborne, who with his co-workers, has published over 100 papers, chapters and books on the chemotaxonomy of flavonoids. We may look first, therefore, at some of his many perceptive contributions to the field, (Harborne, 1975).

One of the most striking of his earlier demonstrations of the supportive nature of flavonoid distribution in plants to their taxonomy was in the Gesneriaceae, where it was shown that there was an absolute correlation between the taxonomic arrangement and the possession of 3-deoxyanthocyanins in the flowers (Harborne, 1966). These pigments were shown to be restricted to the New-World (American) isocotylous sub-family, Gesnerioidae and absent from the other sub-family, the Cyrtandroidae which are restricted to the Old-World (Africa, etc) and whose flowers contain only normal anthocyanins. This mode of distribution completely supported the revised treatment of the family by Burtt (1962) and aided in the placement of certain species, as well as demonstrating that flower color is associated with pollinator preferences (Harborne 1972).

Harborne and his co-workers have since published multifarious other studies at the family level (Umbelliferae, Ericaceae, Graminae, Palmae, Polemoniaceae, Juncacae and so on). Much of the data his group has accumulated in these studies have been summarized and synthetized into thought-provoking reviews on the way in which flavonoid distribution illuminates angiosperm evolution and on the role of flavonoids in plants (Harborne, 1972, 1975, 1966, 1977).

Harborne's use of flavonoid data in systematic studies is well illustrated by work on the Polemoniaceae (the phlox family). This family of 18 genera and over 300 species is chiefly found in North America and is traditionally associated with the Boraginaceae, Convolvulaceae and Solanaceae (Stebbins 1974, Cronquist 1968, Takhtajan 1969). Early data on the flavonoid distribution within the family was in agreement with this grouping (Harborne 1967). Species in this group of families are marked by commonly having flavonols, delphinidin glycosides, in the flowers and rarely possessing unusual or methylated flavonoids. A more thorough survey started in 1971 (Smith et al., 1971, 1977; Dean et al., 1978 and Harborne and Smith 1978) and showed that while a few genera of the Polemoniaceae did indeed have the characteristics outlined above, the majority did not fit into this rubric. For example, 10 of the genera were shown to contain the rare 6-methoxyflavonols, patuletin (*1*), eupatolitin (*14*) and eupalitin (*15*); *C*-glycosyl flavones (*6*, *7*) were found to be universal in *Phlox*, *Loeselia*, *Leptodactylon*, *Linanthus* and present in three others; pelargonidin and cyanidin (*13*) glycosides were found in 11 genera. This data suggests that the Polemoniaceae are more closely linked with other families in the Asteridae such as Verbenaceae, Scrophulariaceae, Bignoniaceae and so on.

Within the family, the flavonol data support the distinction of *Lianthus* from *Leptodactylon*, the latter probably being misplaced within the family since they appear more like members of the tribe Polemonieae than the Gilieae in which they are put: the flavonoids in the desert annual *Langloisia* divide them into two distinct groups as suggested by Brand (1907), while the treatment of the *Gilia* complex by Grant (1959) is generally supported.

Harborne and his co-workers analyzed the flavonoid data on the Polemoniaceae in terms of an evolutionary advancement (see earlier p. 230) giving « low numerical scores » for biosynthetic features in the flavonoids which are correlated with primitive features in angiosperms (c.f., Harborne, 1977). On this basis, the *absence* of the common flavonols, kaempferol, quercetin and myricetin and the glycosylflavones are given a « score » of 1; the *presence* of the *O*-methylated querectin derivatives also count as 1, while the methylated flavones and patuletin (1) are given the value 2 and the rarer eupatolitin (*14*) and eupalitin (*15*) score 3. On this basis, it was found that the tropical genera have low overall scores (1–5), confirming Grant's suggestion (1959) that they are the most primitive, while the *Ipomopsis* and *Eriastrum* score consistently high (10–12). The data also indicates that the Polemonieae are much less advanced than the Gilieae.

The distribution of anthocyanidins in the flowers of the Polemoniaceae (Harborne and Smith 1978) is almost entirely consistent with the pollination ecology of the species concerned (Grant and Grant, 1955). Thus, hummingbird pollinated species which, as in the Gesneriaceae, have mainly orange red flowers generally contain pelargonidin derivatives, while bee species were usually blue and contained delphinidin glycosides. The lepidopteran species contained cyanidin or mixtures of this pigment with del-

Table 1: Recent Examples of the Application of Flavonoid Distribution to Systematic Problems in Plants

Taxon	Level of Systematic Study	Reference
Bryophyta		
Hymenophyton	Species	Markham et al. 1976
Marchantia berteroana	Sexual development	Markham et al. 1978
Psilophyta		
Psilotum nudum	Evolution	Cooper-Driver 1977
Lycopsida		
Lycopodium	Generic and evolutionary	Voirin and Jay 1978a
Lycopsida	Phylogenetic	Voirin and Jay 1978b
Ferns		
Adiantum	Sub-generic, geographical and evolutionary	Cooper-Driver and Swain 1976
Asplenium	Hybridization	Harborne et al. 1973
Pityrogramma triangularis	Chemical races	Smith et al. 1971
Polypodiaceae	Generic	Wollenweber 1978
Gumnosperms		
Podocarpaceae	Generic	Lowry 1972
Dacrydrium sp.	Hybridization	Quinn and Rattenbury 1972
Angiosperms		
General	Phytogeography	Mabry 1973
Bidens	Floral UV patterns	Scogin and Zakar 1976
Briza	Polyploidy	Murray and Williams 1973
Bromeliaceae	Sub family	Williams 1978
Caprifoliaceae	Generic and evolutionary	Bohm and Glennie 1971
Compositae	All aspects	Heywood et al. 1977
Coreopsis mutica	Hybridization	Crawford 1972
Cyperaceae	Phytogeography and evolution	Williams and Harborne 1977
Dahlia	Species and phyto-geography	Giannasi 1975
Ericaceae	Tribal	Harborne and Williams 1973
Fouquieriaceae	Species	Scogin 1977
Gibasis schiedeana	Genetic and evolution	Martinez and Swain 1977
Gramineae	Tribal and related families	Harborne and Williams 1976
Hazardia	Species	Clark and Mabry 1978
Iley	Sectional	Santamour 1973
Liliaceae	Order and evolution	Williams 1975
Phlox spp	Geographic and polyploidy	Levy 1976 Levy and Fuji 1978
Ribes	Evolution	Bate-Smith 1976
Sarracenia	Ecological and sectional	Romeo et al. 1977
Valerianella	Phytogeography and sectional	Greger and Ernet 1973
Viburnum and *Hydrangea*	Evolutionary	Bate-Smith 1978
Umbelliferae	Tribal and evolutionary	Crowden et al. 1969 Harborne and King 1976
Zygophyllaceae	Sub-family	Saleh and El-Hadidi 1977

phinidin. It is of interest, however, that the distribution of glycosidic types, including acylated glycosides (always by *p*-coumaric acid), correlated with the systematic position of the species rather than the pollinating vector.

The application of the distribution of flavonoid compounds in plants to systematic problems has grown apace in the last ten tears. About 100 papers a year are now being published which deal with some aspect or other of flavonoid taxonomy, many, however, at the elementary level of merely recording differences in the appearance of «spots» on two-dimentional chromatograms of related taxa. Nevertheless, the more detailed investigations, in which the majority (or all) flavonoid compounds are identified in a representative sample of a given plant species, have thrown light on systematic problems at all taxonomic levels, albeit there is still a need to apply biosynthetic criteria to their interpretation. A detailed discussion of this data would take a book in itself and a selection of the more recent results are, therefore, summarized in Table 1.

6. Conclusions

This brief outline will, I hope, have drawn attention to the current importance of flavonoids as chemotaxonomic markers in plants. Nevertheless, it sould be made clear that because of restrictions in space I have not been able to highlight the many important contributions made by the now numerous excellent practitioners in the field. The examples given in Table 1 only cover a few of these and I recommend that the reader follow up the references given to see the breadth of interest in the subject. I would equally urge that the importance of other chemical classes (acetylenes, alkaloids, non-protein amino acids, nucleic acids, polyketides, proteins, terpenoids, etc.) as systematic markers be clearly noted for I feel strongly that the more information we have on the distribution of all classes of defined chemical components in plants the closer we shall come to an understanding of the importance of evolutionary and ecological change in determining the present variation in the plant kingdom.

Acknowledgement: I am grateful to Mrs. Shirley Manditch for careful preparation of this chapter.

References

ALSTON, R. E., and TURNER, B. L., 1963: *Biochemical Systematics*, Prentice Hall, New Jersey.

APARECIDE, M., CAGNIN, M., and GOTTLIEB, O. R., 1978: Biochem. Syst. Ecol., **6**.

AYALA, F. J. (ed.), 1977: *Molecular Evolution*. Sinauer, Sunderland, Mass.

BATE-SMITH, E. C., 1962: J. Linn. Soc. Bot. **58**, 95.

BATE-SMITH, E. C., 1974: In: Chemistry in Botanical Classification, (G. Bends and J. Santesson eds.), Academic Press, New York.

BATE-SMITH, E. C., 1977: Biochem. Syst. Ecol., 95.

BATE-SMITH, E. C., 1976: Biochem. Syst. Ecol. **4**, 13.

BATE-SMITH, E. C., 1978: Phytochem. **17**, 267.

BATE-SMITH, E. C., FERGUSON, I. K., HUTSON, K., JENSEN, S. R., NIELSON, B. J., and SWAIN, T., 1975: Biochem. Syst. Ecol. **3**, 79.

BATE-SMITH, E. C., 1968: J. Linn. Soc. Bot. **60**, 325.

BENDZ, G., and SANTESSON, J., 1974: *Chemistry in Botanical Classification*, Academic Press, London.
BIERNER, M. W., DENNIS, M., and WOFFORD, B. E., 2977: Biochem. Syst. Ecol. **5**, 23.
BOHM, B. A., and GLENNIE, C. W., 1971: Canad. J. Bot. **49**, 1799.
BRAND, A., 1907: Das Pflanzenreich: Band IV, p. 112. (Engler, A., and Diels, L. eds.). Englemann, Weinheim.
BURTT, B. L., 1962: Notes, R. Bot. Gdn. Edinb. **14**, 205.
CLARK, W. D., and MABRY, T. J., 1978: Biochem. Syst. Ecol. **6**, 19.
COOPER-DRIVER, G. C., 1977: Science **198**, 1260.
COOPER-DRIVER, G. A., and SWAIN, T., 1976: Bot. J. Linn. Soc. **74**, 1.
CRAWFORD, D. J., 1972: Taxon **21**, 27.
CROWDEN, R. K., HARBORNE, J. B., and HEYWOOD, V. M., 1969: Phytochemistry, **8**, 1996.
CAGNIN, M. A. H., GOMES, C. M. R., GOTTLIEB, O. R., MARX, C. M., ROCHA, A. I., DA SILVA, M. M., and TEMPERINII, J. A., 1977: In: Kubitzki, K., 1977b.
CRONQUIST, A., 1968: *The evolution and classification of flowering plants*, Houghton Mifflin, Boston.
DEAN, F. M., COSTA, A. S., HARBORNE, J. B., and SMITH, D. M., 1978: Phytochemistry **17**, 505.
ERDTMAN, H., 1956: In: *Perspectives in Organic Chemistry*. (A. R. Todd, ed.) p. 430. Interscience, New York.
GEISSMAN, T. A., and HARBORNE, J. B., 1956: J. Amer. Chem. Soc. **78**, 832.
GIANNASI, D. E., 1975: Mem. New York Bot. Gdn. **26**, 1.
GRANT, V., 1959: *Natural history of thė Phlox family*. Nijhoft, The Hague, Netherlands.
GRANT, V., and GRANT, K., 1965: *Flower pollination in the Phlox family*. Columbia Univ. Press, New York.
GREGER, H., and ERNET, D., 1973: Phytochemistry **12**, 1693.
HARBORNE, J. B., 1966: Phytochemistry **5**, 589.
HARBORNE, J. B., 1967: *Comparative biochemistry of the flavonoids*. Academic Press, London.
HARBORNE, J. B., 1972: Rec. Adv. Phytochem. **4**, 107.
HARBORNE, J. B., and WILLIAMS, C. A., 1973: Bot. J. Linn. Soc. **66**, 37.
HARBORNE, J. B., 1975: In: *The Flavonoids*. (J. B. Harborne, T. J. Mabry and H. Mabry eds). p. 1060. Chapman and Hall, London.
HARBORNE, J. B., 1976: In: *Chemistry and Biochemistry of Plant Pigments*, 2nd Edn. (T. W. Goodwin ed.) p. 736.
HARBORNE, J. B., 1977: Biochem. Syst. Ecol. **5**, 7.
HARBORNE, J. B., MABRY, T. J., and MABRY, H., 1975: *The Flavonoids*, Chapman and Hall, London.
HARBORNE, J. B., and KING, L., 1976: Biochem. Syst. Ecol. **4**, 111.
HARBORNE, J. B., and SMITH, D. M., 1978: Biochem. Syst. Ecol. **6**, 127.
HARBORNE, J. B., WILLIAMS, C. A., and SMITH, D. M., 1973: Biochem. Syst. **1**, 51.
HARBORNE, J. B., and WILLIAMS, C. A., 1976: Biochem. Syst. Ecol. **4**, 267.
HEYWOOD, V. H., HARBORNE, J. B., and TURNER, B. L., 1977: *The Biology and Chemistry of the Compositae*. 2 Vols. Academic Press, London.
KUBITZKI, K., 1967: Ber. Deutsch. Bot. Ges. **80**, 757.
KUBITZKI, K., 1977a: Forschr. der Botanik **39**, 192.
KUBITZKI, K., 1977b: *Flowering Plants: Evolution and Classification of Higher Categories*. Springer, Vienna.
LEVIN, D., 1976: Ann. Rev. Ecol. Syst. **7**, 121.

LEVY, M., 1976: Biochem. Syst. Ecol. **4**, 249.
LEVY, M., and FUJI, K., 1978: Biochem. Syst. Ecol. **6**, 117.
LEVY, M., and LEVIN, D., 1975: Evolution **29**, 487.
LOWRY, J. B., 1972: Phytochemistry **11**, 3267.
MABRY, T. J., 1973: In: *Chemistry in Evolution and Systematic*, (T. Swain ed.) p. 316.
MARGULIS, L., 1979: Evolution of Cells. Harvard Univ. Press., Cambridge, Mass.
MARKHAM, K. R., and PORTER, L. J., 1978: Progress in Phýtochem. **5** (in press), Pergamon Press, Oxford.
MARKHAM, K. R., MOORE, N. A., and PORTER, L. J., 1978: Phytochemistry **17**, 911.
MARKHAM, K. R., PORTER, L. J., CAMPBELL, E. O., CHOPIN, J., and BOUILLANT, M., 1976: Phytochemistry **15**, 1517.
MARTINEZ, M. A., and SWAIN, T., 1977: Biochem. Syst. Ecol. **5**, 37.
MURRAY, B. G., and WILLIAMS, C. A., 1973: Nature **243**, 87.
NORDSTROM, C. G., and SWAIN, T., 1953: J. Chem. Soc. 2764.
QUINN, C. J., and RATTENBURY, J. A., 1972: New Zealand J. Bot. **10**, 427.
REZENDE, C. M., and GOTTLIEB, O. R., and MARX, M. C., 1975: Biochem. Syst. Ecol. **3**, 63.
REZENDE, C. M., and GOTTLIEB, O. R., 1973: Biochem. Syst. **1**, 111.
REZNIK, H., 1956: Sbr. heidelb. Akad. Wiss. 125.
ROMEO, J. T., BACON, J. D., and MABRY, T. J., 1977: Biochem. Syst. Ecol. **5**, 117.
SALEH, N. A. M., and EL-HADIDI, M. N., 1977: Biochem. Syst. Ecol. **5**, 121.
SANTAMOUR, F. S., 1973: Phytochemistry **12**, 611.
SCOGIN, R., 1977: Biochem. Syst. Ecol. **5**, 265.
SCOGIN, R., and ZAKAR, K., 1976: Biochem. Syst. Ecol. **4**, 165.
SMITH, D. M., CRAIG, S. P., and SANTAROSA, J., 1971: Amer. J. Bot. **58**, 292.
SMITH, D. M., GLENNIE, C. W., and HARBORNE, J. B., 1971: Phytochemistry **10**, 3115.
SMITH, D. M., GLENNIE, C. W., HARBORNE, J. B., and WILLIAMS, C. A., 1977: Biochem. Syst. Ecol. **5**, 107.
SMITH, P., 1976: *The Chemotaxonomy of Plants*, Elsevier, New York.
SPORNE, K. R., 1974: *The Morphology of Angiosperms*. Hutchinson, London.
STEBBINS, G. L., 1974: *Flowering Plants – evolution above the species* level. Harvard Univ. Press., Cambridge, Mass.
SWAIN, T. (ed.), 1966: *Comparative Phytochemistry*. Academic Press, New York.
SWAIN, T., 1974a: In: *Comprehensive Biochemistry*, (M. Florkin and E. H. Stotz, eds.). Vol. 29 A, pp. 125. Elsevier, Amsterdam.
SWAIN, T., 1974b: In: *Chemsitry in Botanical Classification*, (G. Bendz and J. Santesson eds.). pp. 81–92 Academic Press, New York.
SWAIN, T., 1975: In: *The Flavonoids*, (J. B. Harborne, T. J. Mabry and H. Mabry eds.) pp. 1096.
SWAIN, T., 1977: *Ann. rev. plant physiol.* **28**, 479.
TAKHTAJAN, A., 1969: *Flowering plants: origin and dispersal.* Oliver and Boyd, Edinburgh.
VOIRIN, B., and JAY, M., 1978: Biochem. Syst. Ecol. **6**, 99.
VOIRIN, B., and JAY, M., 1978b: Biochem. Syst. Ecol. **6**, 99.
WILLIAMS, C. A., 1975: Biochem. Syst. Ecol. **4**, 229.
WILLIAMS, C. A., 1978: Phytochemistry **17**, 729.
WILLIAMS, C. A., and HARBORNE, J. B., 1977: Biochem. Syst. Ecol. **5**, 45.
WHITTAKER, H., and MARGULIS, L., 1978: Biosystems **16**, 1.
WOLLENWEBER, E., 1978: Amer. Fern. J. **68**, 13.

Chlorophyll Biosynthesis. Enzymes and Regulation of Enzyme Activities

Hansjörg Schneider

Botanisches Institut der Universität Köln

1. Introduction

As far as can be deduced from experiments simulating the primaeval atmosphere of the earth, porphyrins were formed in addition to nucleotides, aminoacids and carbohydrates under the conditions of early geological ages (e.g. see Simonescu et al. 1978). Since porphyrins were present and proved to be unique substances for harvesting and transferring energy, the direction of evolution was determined by them from early beginnings. Now descendents of porphyrins are found in a wide range of positions regulating fundamental biological processes. In these respects porphyrins are quite different from most of the other pigments in plants.

The present chapter treats of chlorophylls, their biosynthesis and of the sequence and the properties of proteins forming chlorophyll step by step. The initial steps are identical with steps in the biosynthesis of other porphyrins. The location and the sites of synthesis of chlorophyll forming enzymes will be discussed as well as mechanisms regulating chlorophyll synthesis. The chapter begins with a survey on the distribution and the structures of chlorophylls and on intermediates in the chlorophyll biosynthetic pathway.

Abbreviations

ALA	δ-aminolaevulinic acid
ALAD	δ-aminolaevulinic acid dehydratase
ALAE	δ-aminolaevulinic acid synthesizing enzyme(s)
ALAS	δ-aminolaevulinic acid synthetase
BChl	bacteriochlorophyll
ChChl	Chlorobium chlorophyll
Chl	chlorophyll
Chlase	chlorophyllase
Chlid	chlorophyllide
Copro	coproporphyrin
Coprogen	coproporphyrinogen
EDTA	(ethylenedinitrilo)tetraacetic acid·
GSH	glutathione
LA	laevulinic acid
Mg-Proto	magensium protoporphyrin
Mg-Proto-ME	magnesium protoporphyrin monomethyl ester
PBG	porphobilinogen

PBGase	porphobilinogenase
PChl	protochlorophyll
Pchlid	protochlorophyllide
Proto	protoporphyrin
Protogen	protoporphyrinogen
Uro	uroporphyrin
Urogen	uroporphyrinogen

2. The Distribution of Chlorophylls, their Structures, and Intermediates in the Chlorophyll Biosynthetic Pathway

2.1. Distribution

Chl *a* is the principal chlorophyllous pigment in all classes of algae and in higher plants. *Cyanophyta* and *Xanthophyta* only contain this chlorophyll. In addition higher plants, *Chlorophyta* and *Euglenophyta* possess Chl *b*. PChl accumulates in higher plants and *Euglenophyta* if these organisms are kept in the dark. Chl *c* was isolated from *Phaeophyta*, *Chrysophyta* and *Pyrrophyta* and Chl *d* was sporadically detected in *Rhodophyta*.

Photosynthetic bacteria contain two types of Chls, BChl and ChChl. BChl *a* or *b* was found in *Thio*- and *Athiorhodaceae* and $ChChl_{650}$ or $ChChl_{660}$ in *Chlorobacteriaceae* which may also contain BChls.

2.2. Structures

Most forms of Chl are discernible by different spectroscopic properties; however, their structures are also well known. Stimulated by the interest in the compounds catalysing unique reactions, intensive analytical and preparative work has been performed since the beginning of this century (for summaries see Smith (ed.) 1975).

Chl *b* differs from Chl *a* in the formyl group on position 3 replacing the 3-methyl group of Chl *a*. PChl as compared with Chl *a* has an additional double bond on the ring D. Lacking phytylation, a 7-acrylic acid side chain instead of a 7-propionic acid side chain and a double bond on ring D, like PChl, distinguishes Chl *c* from Chl *a*. One form of Chl *c* (Chl c_2) possesses a 4-vinyl group instead of the usual 4-ethyl group (e.g. see Budzikiewicz and Taraz 1971). Chl *d* is derived from Chl *a* by an oxidized substituent on position 2. The 2-vinyl group of Chl *a* is replaced by a formyl group (Holt 1961).

A great variety of Chls has been found in photosynthetic bacteria.

BChls are characterized by two reduced rings, the rings D and B, and by a 2-acetyl group. BChl *b* differs from BChl *a* in a 4-ethylidene group (Schaer et al. 1974). BChls may be esterified with all-trans-geranylgeraniol (Katz et al. 1972b).

Up to now 18 forms of ChChls are known (see Kenner et al 1976). These ChChls all have a 2-hydroxyethyl group and lack a substituent on position 10, but are different with respect to the substituents on the carbons 4 and 5 and on the δ-carbon. The series absorbing at 650 nm ($ChChl_{650}$ = BChl *d*) comprises 6 forms with hydrogen on the δ-carbon. This position is occupied by a methyl or an ethyl group in the forms absorbing at 660 nm ($ChChl_{660}$ = BChl *c*). Methyl and ethyl groups are the substituents of the carbon 5 of both forms whereas the position 4 is occupied by either an ethyl, a n-propyl or an i-butyl group.

Beside these ChChl forms another series with at least three members was isolated by Brockmann (1976). This ChChl is named BChl *e* and possesses a 3-formyl group. The esterifying alcohol of ChChls is farnesol.

Alcohols other than phytol have also been detected in PChl and during the biosynthesis of Chl *a* (see 3.11).

As recently shown, the mixtures of Chls extracted from bacteria were even more complex than indicated by earlier findings. For example, different BChlides were found to be esterified with stearylalcohol, cis-9-hexadecen-1-ol, or 4-undecyl-2-furan-methanol (Gloe and Risch 1978, Caple et al. 1978, Risch et al. 1979, Kemmer et al. 1979). Chls lacking Mg, i.e. pheophytins, were recognized to play a role in the photosynthetic reaction centers of bacteria (see Dutton 1976). Formulas of Chls are given in figure 1.

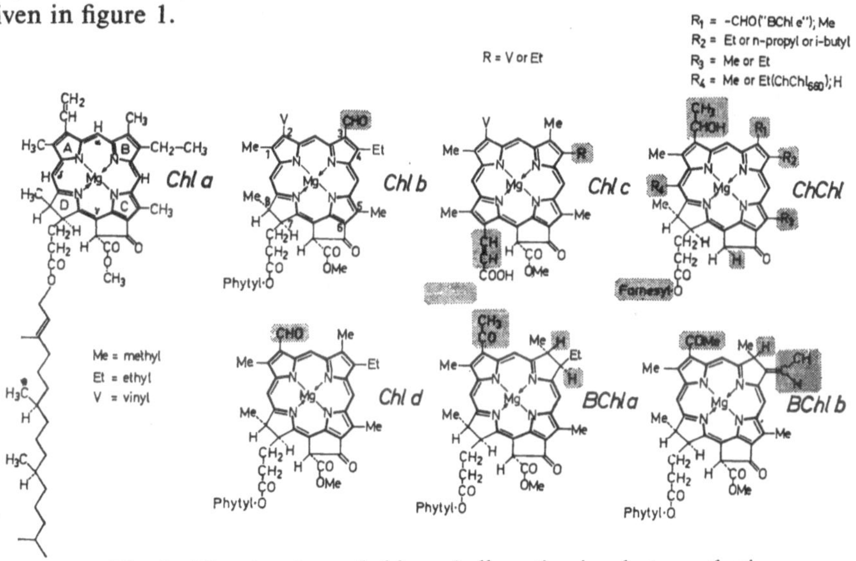

Fig. 1: The structure of chlorophylls active in photosynthesis

2.3. Intermediates

Intermediates in Chl biosynthesis have mostly been isolated from mutants or plants fed with ALA and blocked by inhibitors (e.g. see Granick 1948a, b, 1950, 1961, 1967; Jones 1963a–c; Ellsworth and Aronoff 1968, 1969; Duggan and Gassman 1974; Ryberg and Sundquist 1976 or Wong 1978a, b, see also Mascia 1978) but they can also be detected in cell-free systems (see Rebeiz and Castelfranco 1973 or Cohen et al. 1977, Smith and Rebeiz 1979). The sequence of steps in the Chl biosynthetic pathway arises when the intermediates are ordered according to their increasing complexity (e.g. see Granick 1967) (compare figure 2). The order of some steps is still discussed (see parts 3.7.–3.9.).

The biosynthetic pathways for Chls other than Chl *a* may be formally derived from the general pathway by adding or changing a few steps at the end of the general pathway. For example, the extra methylation present in the side groups of BChls has been shown to be methionine derived (Kenner et al. 1978).

Although Chl degradation is a common periodocally returning phenomenon there are only speculations about biochemical reactions involved in this process (but see chapter xv of this book: R. Kufner: The Biological Degradation of Chlorophyll in Scenescent Tissues).

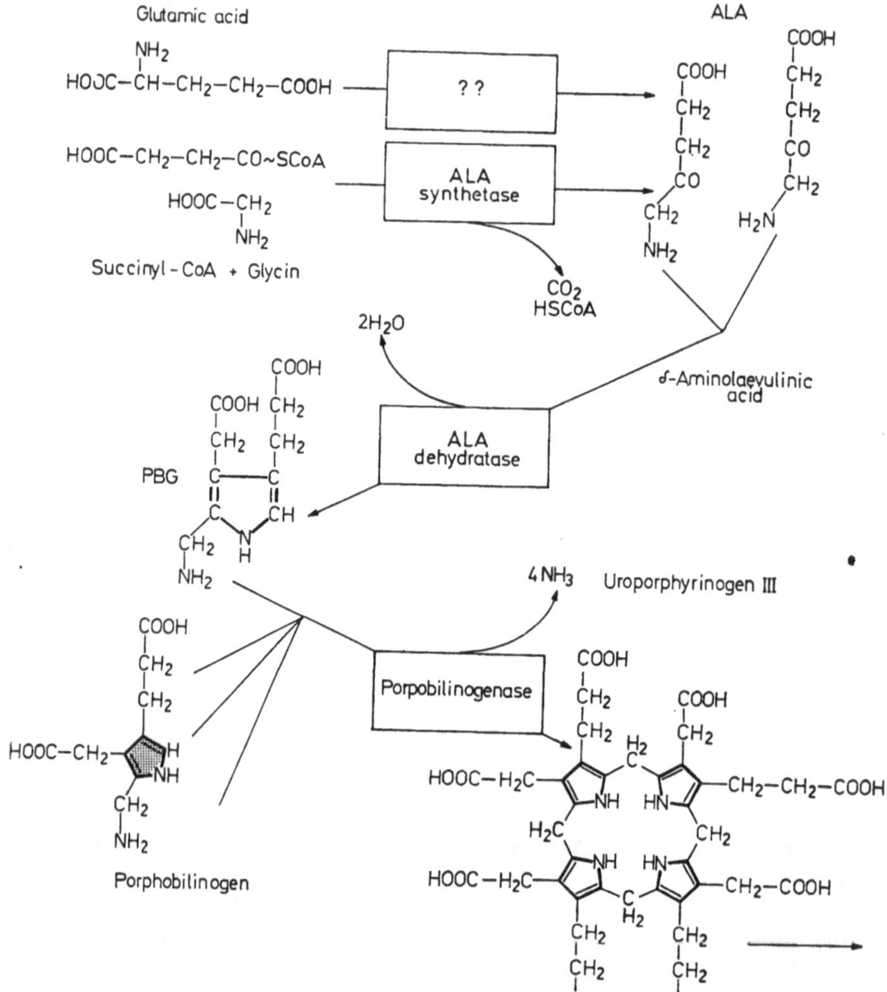

Fig. 2a

Fig. 2: Intermediates and enzymatic steps in the biosynthesis of chlorophyll *a*. (The sequence of some steps is yet discussed)

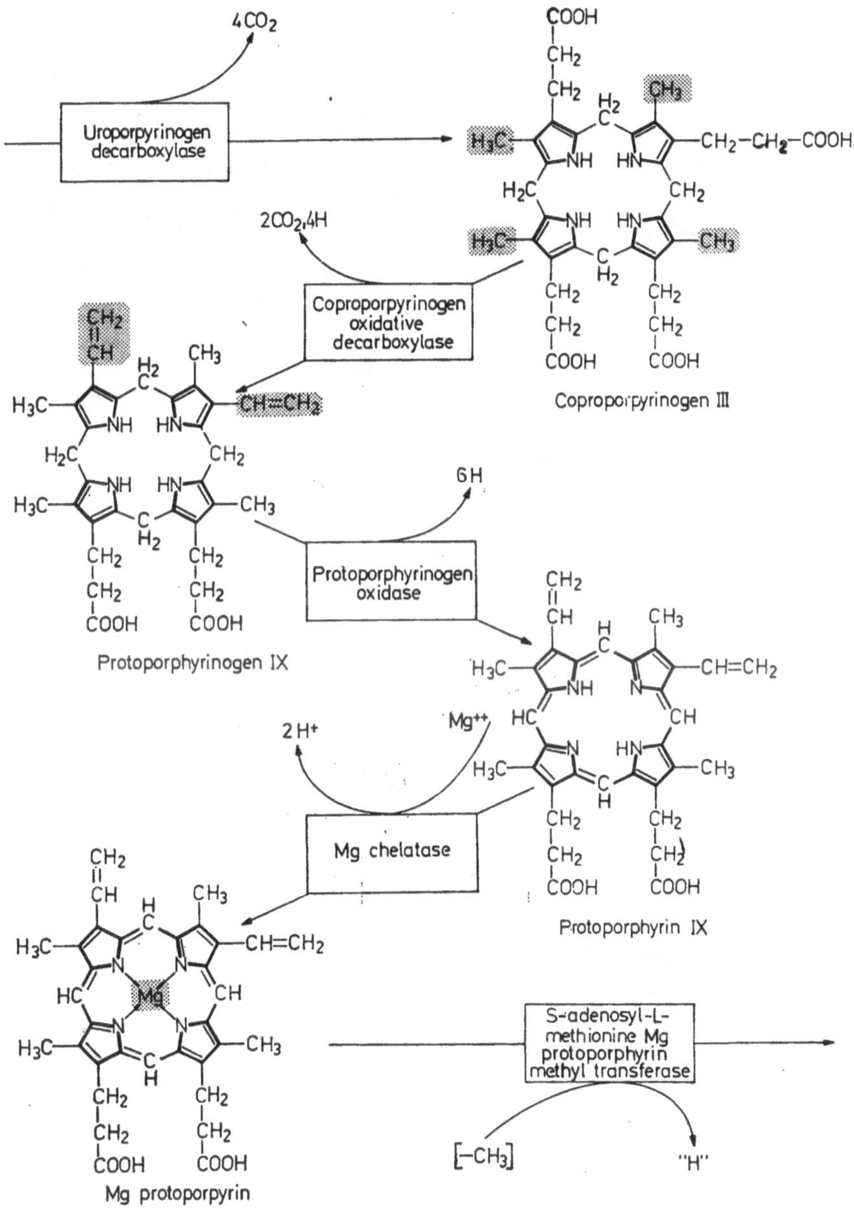
4CO2
Uroporpyrinogen decarboxylase
2CO2,4H
Coproporpyrinogen oxidative decarboxylase
Coproporpyrinogen III
Protoporphyrinogen IX
6H
Protoporphyrinogen oxidase
Protoporphyrin IX
2H+
Mg++
Mg chelatase
Mg protoporpyrin
S-adenosyl-L-methionine Mg protoporphyrin methyl transferase
[-CH3]
"H"

Fig. 2b

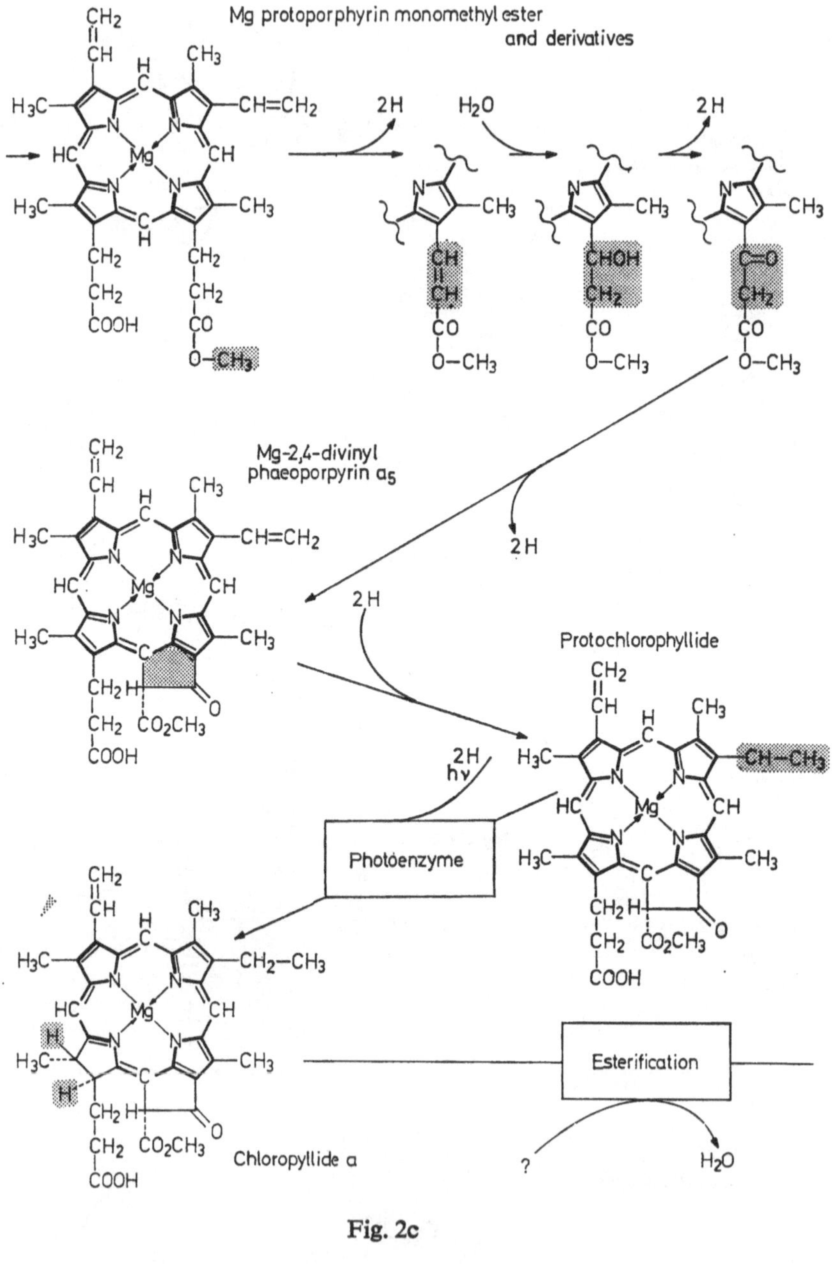
Mg protoporphyrin monomethyl ester
and derivatives
2H
H2O
2H
Mg-2,4-divinyl
phaeoporpyrin a5
2H
2H
Protochlorophyllide
2H
hν
Photoenzyme
Esterification
Chloropyllide a
?
H2O

Fig. 2c

3. Enzymes in the Biosynthetic Pathway of Chlorophyll

3.1. δ-Aminolaevulinic Acid Synthetase (E. C. 2.3.1.37) and other Enzyme Systems Synthesizing δ-Aminolaevulinic Acid

ALA is the first compound specific to the porphyrin and chlorophyll biosynthetic pathway. In animals (Gibson et al. 1958), yeast (Porra et al. 1972), bacteria (Tait 1972b) and photosynthetic bacteria (Kikuchi et al. 1958) ALA is synthesized from glycin and succinyl CoA by the action of ALAS. Pyridoxalphosphate is required as a cofactor.

Purification procedures resulting in highly purified preparations of ALAS were described for enzymes from various sources (e.g. Aoki et al. 1971 (rabbit reticulocytes), Tait 1973 (*Micrococcus denitrificans*), Tuboi et al. 1970a, Fanica-Gaigner and Clément-Métral 1973a, and Nandi and Shemin 1977 (*Rhodopseudomonas spheroides*)). All these enzymes share in common a pH optimum of about 7.5. ALAS from *Rhodopseudomonas spheroides* has a molecular weight of 80,000 to 85,000 as determined by sedimentation equilibrium (Nandi and Shemin 1977) and is composed of two subunits which have a molecular weight of about 40,000 to 45,000 each. Previous molecular weight determinations revealed values of about 60,000 (Shemin 1970, Warnick and Burnham 1971), 60,000 to 80,000 (Tuboi et al. 1970a), and about 100,000 (Tuboi and Hayasaka 1971, Fanica-Gaignier and Clément-Métral 1973b). The molecular weight of ALAS from *Micrococcus* (Tait 1973) was found to be in the same range of molecular weights in contrast to animal enzymes which were found to have much higher molecular weights (Aoki et al. 1971, Hayashi et al. 1970).

Different forms of ALAS were isolated from cells of *Rhodopseudomonas spheroides* (Tuboi et al. 1970a). Hayasaka and Tuboi (1974) presented evidence for two low activity and two high activity forms and Sandy et al. (1975) separated at least three forms, two with low activity and one with high activity. The low activity forms were convertible to high activity forms by cystine and an endogenous activating enzyme (Hayasaka and Tuboi 1974). Sandy et al. (1975) found that the activating agent was cystine trisulphide, synthesized by cystathionase from cystine (de Xifra et al. 1976) and other trisulphides as glutathione trisulphide and tri- and polytrisulphone disulphonates and hybrids of these compounds. The mechanism of activation is not clear.

Fanica-Gaigner and Clément-Métral (1973b, 1973c) described two different forms of ALAS of *Rhodopseudomonas*, one form associated with the chromophores, the other one present in the cytoplasm. Both forms had the same molecular weight but different numbers of HS-groups. 8 HS-groups were found at the chromatophore enzyme, 2 HS-groups at the cytoplasmic enzyme (Clément-Métral and Fanica-Gaignier 1975). Two low activity forms, each with one HS-group less were obtained by growing the cells in mediums containing higher amounts of iron. The chromophore enzyme was found to be inducible by light (Fanica-Gaignier and Clément-Métral 1973c), however, a more complicated pattern for enzyme induction was reported by Tuboi et al. (1970b). Besides iron, ATP was found to be inhibitory (Fanica-Gaignier and Clément-Métral 1973a, 1973b, Gajdos and Gajdos-Török 1973). Other well known

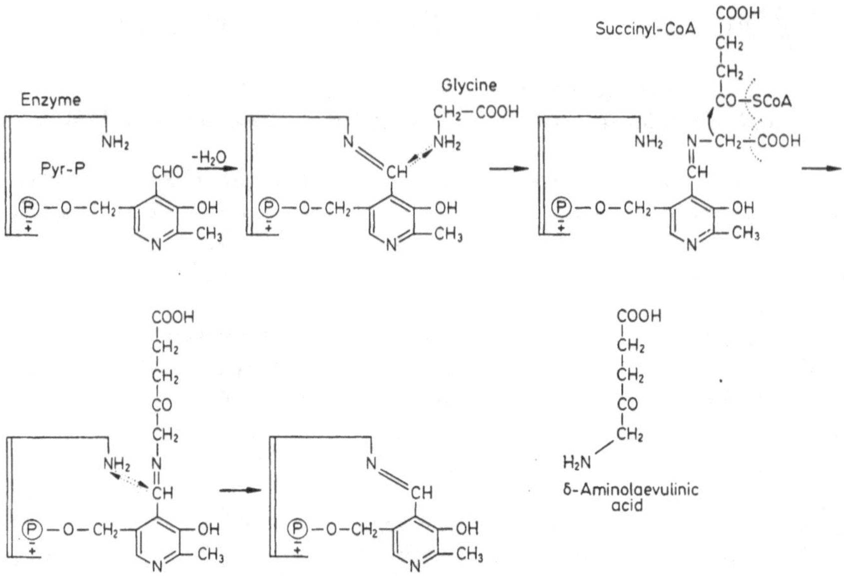

Fig. 3: The mechanism of δ-aminolaevulinic acid formation (Shemin 1970)

inhibitors of ALAS from *Rhodopseudomonas spheroides* are haem and Proto (Lascelles 1960, Burnham 1970, Porra et al. 1972).

A scheme for the reaction mechanism for ALAS was elaborated by Shemin (1970). The cofactor pyridoxalphosphate is bound to the enzyme protein by a covalent linkage, an iminogroup and by an ionic bond between the phosphate group of the cofactor and a positive group on the enzyme. By transimination glycine combines with pyridoxal phosphate. The carbanion on C_2 of glycine, stabilized by the arising SCHIFF base, makes a nucleophilic attack onto C_1 of succinyl CoA. CoA and CO_2 are eliminated and after transimination back to the SCHIFF base with the amino group of the enzyme ALA is released. ALA may either occur free or concurrently transiminate a SCHIFF base with the ensuing enzyme ALAD (compare fig. 3).

Lysine is the binding site for pyridoxal phosphate in the formation of ALA by ALAS from *Rhodopseudomonas spheroides* (Nandi 1978a). Based on new results Nandi (1978b) worked out another model for the reaction mechanism of ALAS. This model is shown in figure 4.

In contrast to the multitude of knowledge about ALAS from the photosynthetic bacterium *Rhodopseudomonas spheroides* there is a striking lack of information about ALAS from higher plants. There are only a few reports on ALAS in crude homogenates. Obviously attempts to detect ALAS in higher plants failed, despite the special interest in this enzyme which was assumed to play a main regulatory role in Chl biosynthesis (see part 5). Adding pyridoxal phosphate, ATP and a succinyl CoA regenerating system, glycine, Mg^{2+} and glutathione to extracts of soybean, de Xifra et al. (1971) and de Barreiro (1975) achieved formation of ALA, similar findings were published by Ramaswamy and Nair (1973) from a system prepared from

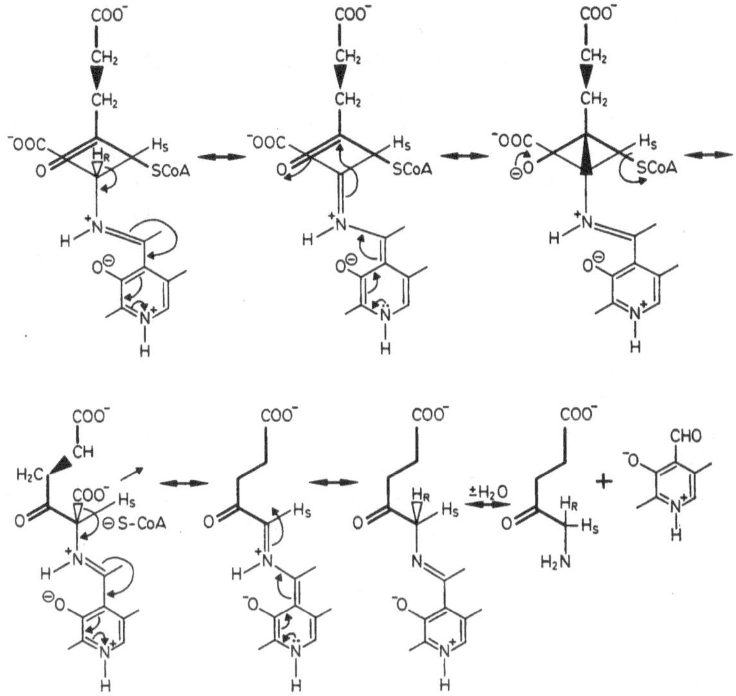

Fig. 4: Outline of chemical mechanism of ALA synthase from succinyl-CoA and glycine stabilized by Schiff base formation with pyridoxal phosphate. *Deep black* in the figure is behind the plane. The decarboxylation step occurs with retention of configuration, while the initial attack on succinyl-CoA occurs on the opposite that of from which the H_K proton of glycine is removed. This direction is presumably directed by the electrophilic nature of the carbonyl carbon of succinyl-CoA. Note also the rotation of the α-carbon of glycine that accompanies elimination of CoA so that the carbonyl group will be the out-of-plane position required for decarboxylation (Nandi 1978b)

greening potato peels. To a certain extent the potato peel extract was also working without supplementation of a succinyl CoA regenerating system.

Despite these spare reports on ALAS of higher plants *in vitro*, ALA synthesis *in vivo* can easily be followed. ALAD, the enzyme converting ALA to PBG, is competitively inhibited by laevulinic acid (Nandi and Shemin 1968b) or by hydroxylaevulinic acid (Yamasaki and Moriyama 1971). Blocking ALAD with LA, the quantities of ALA which accumulate before this block, can be taken as a measure for the capacity of any ALA synthesizing enzyme, if the results are interpreted with precaution. This method was successfully applied to get information about regulating processes in the porphyrin and Chl biosynthetic pathway (Beale 1970, 1971, Ho and Lascelles 1971, Harel and Klein 1972, Schneider 1973a, b, 1976, Porra and Grimme 1974, Masoner and Kasemir 1976, Klein et al. 1975, Fluhr et al. 1975, see part 5).

Beale and Castelfranco (1973, 1974b) used the possibility to accumulate ALA to scrutinize the biosynthetic pathway leading to the synthesis of ALA. In combination with LA these authors incubated greening cucumber cotyledons with labeled substrates tentatively presumed to be precursors of ALA. Glycine and succinate proved to be very poor substrates. In addition label from C_1 and C_2 of glycine was incorporated into ALA at about equal amounts. If ALA were formed by ALAS, C_1 should not be in the ALA. As much as 30 times more label than with glycine was incorporated into ALA when metabolically closely related five-carbon compounds, such as glutamate, glutamine or α-ketoglutarate were used. Both C_1 and C_3-C_4-labelled glutamate were equally able to label ALA.

By chemical degradation of ALA formed in greening barley leaves, Beale et al. (1975) showed that [1-^{14}C]-glutamate labelled the C_5 fragment of degraded ALA while [3,4-^{14}C]-glutamate labelled the C_1-C_4 fragment. These results made clear that the intact carbon skeleton of glutamate was incorporated into ALA which was formed in plants. Confirming results were published by Castelfranco and Jones (1975) and Weinstein and Castelfranco (1977). The latter found Proto synthesis from glutamate in a cell-free chloroplast preparation. Results from respective experiments done with turkey blood reconfirmed the glycine-succinyl CoA pathway in animals. (Beale et al. 1975). Glycine and succinate also seem to be ALA precursors in Chlorella (Porra and Grimme 1974).

In a mutant of *Scenedesmus obliquus* both pathways were active (Klein and Senger 1978). From experiments with 2.2'-bipyridyl-treated barley leaves it was concluded (Hendry and Stobart 1978b) that the classical pathway for ALA synthesis was operative in the dark while the glutamate consuming pathway was of importance in the light.

Beale (1976) considered a series of hypothetical pathways leading from glutamate to ALA. Intermediates could be: glutamate 5-semialdehyde → pyrroline-5-carboxylic acid → 3-pyrrolidone-5-carboxylic acid, or hydantoin 5-propionic acid → imidazole 5-propionic acid, or glutamate 1-semialdehyde, or 2-ketoglutarate → 4,5-dioxovaleric acid, or 2-hydroxyglutarate → 4,5-dioxovaleric acid. Transamination of 4,5-dioxovaleric acid was previously reported to occur in extracts of photosynthetic bacteria (Neuberger and Turner 1963) and *Chlorella* (Gassman et al. 1968).

From maize leaves Lohr and Friedmann (1975) partially purified a NADH-dependent enzyme reducing the 1-carboxyl group of α-ketoglutarate and an alanine-dependent transaminase. ALA synthesis via 4.5-dioxovaleric acid was also demonstrated in greening barley leaves (Kannangara and Gough 1978) and in acellular preparations of *Euglena* (Salvador 1978). The enzymic activities were demonstrated by the use of labeled substrates.

The new interesting findings at present do not quite rule out the possible existence of different ALA forming pathways in plants which may be active depending on species and developmental stage (see also part 5). However, it seems that the pathway starting with glutamate or α-ketoglurarate is the predominant one.

Table 1: Properties of δ-aminolaevulinic acid dehydratase

Source	Purification	molecular weight	pH Optimum	K_m (ALA)	necessary ions	authors
Rhodopseudomonas	70	–	7.8–8.0	–	K	Burnham and Lascelles (1963)
Triticum	16	–	7.2–7.8	$1 \cdot 10^{-3}$	Mg, Mn	Nandi and Waygood (1967)
Rhodopseudomonas	390	250,000	8.5	$7 \cdot 10^{-4}$	K	Nandi et al. (1968)
Nicotiana	80	–	7.4	$6.3 \cdot 10^{-4}$	Mg	Shetty and Miller (1969)
Glycine	40	280,000	8.4–8.8	$3.5 \cdot 10^{-4}$	–	Tigier et al. (1970)
Spinacia	290	324,000 ±14,000	8.2	$2.4 \cdot 10^{-4}$	Mg (Mn)	Schneider (1970 b)
Phaseolus	–	–	7.5–8.0	$3.0 \cdot 10^{-4}$	–	Sluiters-Scholten et al. (1973)
Raphanus	290	282,000	8.0	$3.9 \cdot 10^{-4}$	Mg	Shibata and Ochiai (1977)
Chlorella	737	316,000	8.5	$5 \cdot 10^{-4}$	Mg, Mn	Tamai et al. (1979)

In the meantime a highly efficient purification of ALAD from spinach was achieved by means of monoclonal antibodies (W. Liedgens, R. Grützmann, and Hj. A. W. Schneider). The subunit molecular weight was found to be 50,000 daltons. The use of monoclonal antibodies had additional advantages: PBG was synthesized in a continuous process by ALAD bound to these antibodies and an evolutionary tree based on the antigenic characters of ALADs from 24 different species could be constructed. These highly specific monoclonal antibodies (see Köhler and Milstein, Nature 256, 495 (1975)), which are directed against single determinants of a protein, can be prepared without the prior availability of the pure enzyme and allow to precipitate selectively the enzyme from crude extracts.

3.2. δ-Aminolaevulinic Acid Dehydratase (E. C. 4.2.1.24) and the Formation of Porphobilinogen

As shown in the scheme for the biosynthesis of Chls ALAD catalyzes the condensation of two molecules of ALA to give one molecule of PBG. The properties of ALAD are much better known than those of the preceding ALA synthesizing enzymes. After the removal of inhibitory substances from the cell sap ALAD activity can easily be detected in plant extracts. The enzyme was purified several times from plant sources. A survey on reported properties is given in table 1.

Mg^{2+} ions are essential for the action of ALAD from higher plants (Nandi and Waygood 1967, Shetty and Miller 1969, Schneider 1970b) although this requirement was not always reported (Tigier et al. 1970). In dialysed enzyme preparations activity was also partly restored by Mn^{2+}. Heavy metal ions were inhibitory (Schneider 1970b). Enzymes from heterotropic bacteria were also reported to be dependent on Mg^{2+} ions (Ho and Lascelles 1971, Yamasaki and Moriyama 1971), whereas the enzyme from the fungus *Ustilago* seems to be a copper enzyme (Komai and Neilands 1969). Although it was assumed that animal ALADs were active without metal ions, it turned out that ALAD from bovine liver is a Zn enzyme with 4 to 6 atoms Zn per molecule (Shemin 1975). The enzyme of *Rhodopseudomonas spheroides* requires K^+ (Mg^{2+}, Na^+) ions for full activity (Burnham and Lascelles 1963). K^+ ions control enzyme activity by an allosteric action (Nandi et al. 1968a, b, van Heyningen and Shemin 1971). With exception of ALAD of *Rhodopseudomonas spheroides* the enzyme from all other sources seems to be inhibited by metal chelators. SH-group activating reagents like cystein, mercaptoethanol or Cleland's reagens have to be added to all enzymes to maintain activity in preparations.

Haemin and Proto were found to be inhibitory in experiments with ALAD from *Rhodopseudomonas spheroides* (Burnham and Lascelles 1963). Inhibition by these substances and the requirement for metal ions might play a role in the regulation of Chl synthesis via ALAD activity

The molecular weight of ALAD is relatively high. By density gradient centrifugation a molecular weight of $324{,}000 \pm 14{,}000$ was found for the spinach enzyme (Schneider 1970b). A molecular weight of similar magnitude was reported from *Chlorella regularis* (Tamai et al. 1979). ALAD from soybean tissue showed an apparent molecular weight of 280,000 Daltons (Tigier et al. 1970) and ALAD purified from radish cotyledons revealed 282,000 Daltons (Shibata and Ochiai 1977). Six subunits were found to compose ALAD from *Rhosopseudomonas spheroides* (van Heyningen and Shemin 1971). This enzyme had a molecular weight of about 240,000 (Nandi and Shemin 1968a, Nandi et al. 1968). In the presence of K^+ ions aggregates were formed with molecular weights of 480,000 and 720,000. In comparison ALAD isolated from bovine liver had 8 subunits with a molecular weight of about 35,000 each. Two very active SH-groups in a total of 6 to 7 per subunit were found (Shemin 1975). Besides the functional group lysine (Nandi 1978c), histidine is also assumed to play a catalytic role (Tsukamoto et al. 1975).

The role of SH-groups for the substrate binding of beef liver ALAD was investi-

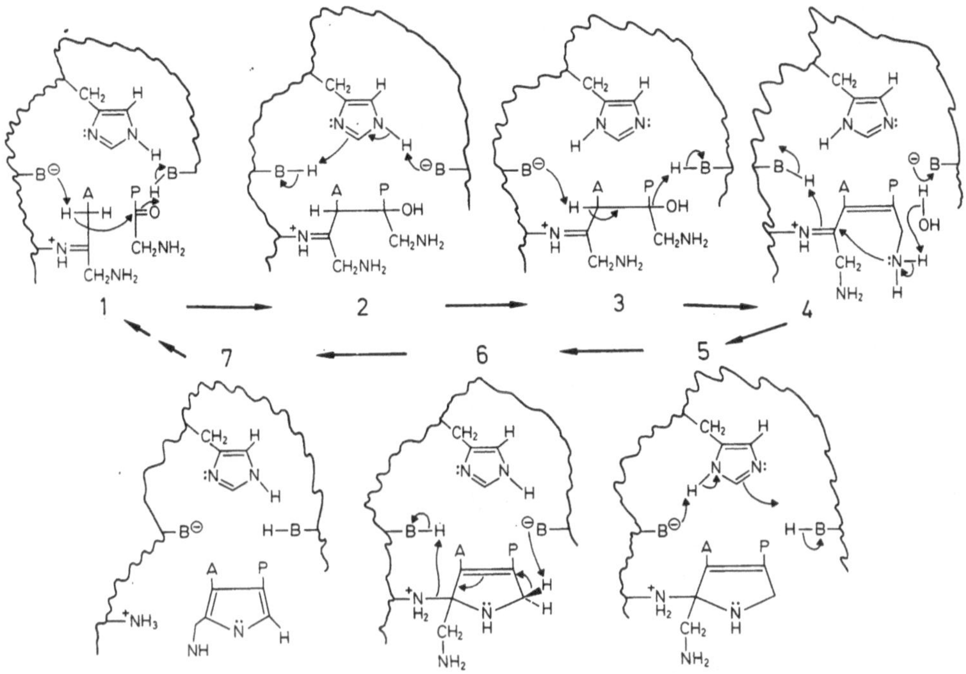

Fig. 5: Proposed mechanism of PBG synthesis from 2 molecules of δ-aminolevulinic acid as catalyzed by the enzyme.
A *site*, acetic acid side chain; P *site*, propionic acid side chain.

The mechanism involves the initial deprotonation, by B^- (possibly a thiolate anion) of the δ-aminolevulinic acid molecule at the A site and the protonation of the carbonyl group by BH (possibly thiol) at the P site (Fig. 12, Step 1). The starting B and BH status is regenerated by mediation of the imidazole bridge (Step 2). This is followed by a second protonation/deprotonation sequence with the elimination of water (Step 3). The subsequent transamination steps together with the mediation of the imidazole bridge system complete PBG formation and regenerate the initial state of the enzyme (Steps 4 to 7). (Barnard et al. 1977)

gated by Barnard et al. (1977). The SH-groups were assumed to be essential in the acid/base catalysis required for the enzymic protonation/deprotonation sequences in the synthesis of PBG. The elaborated mechanism of action (Fig. 5) also includes the participation of a histidinyl residue (Tsukamoto et al. 1975).

The pH for maximum activity of ALAD from photosynthetic organisms was found to center about pH 8. Generally speaking the pH optima were lower for animal enzymes (e.g. see Wilson et al. 1972) and higher for microbial enzymes (de Barreiro 1967, Ho and Lascelles 1971, Yamasaki and Moriyama 1971). In greening tobacco callus Kaul and Sabharwal (1974), however, found a pH optimum of 9.5 for ALAD.

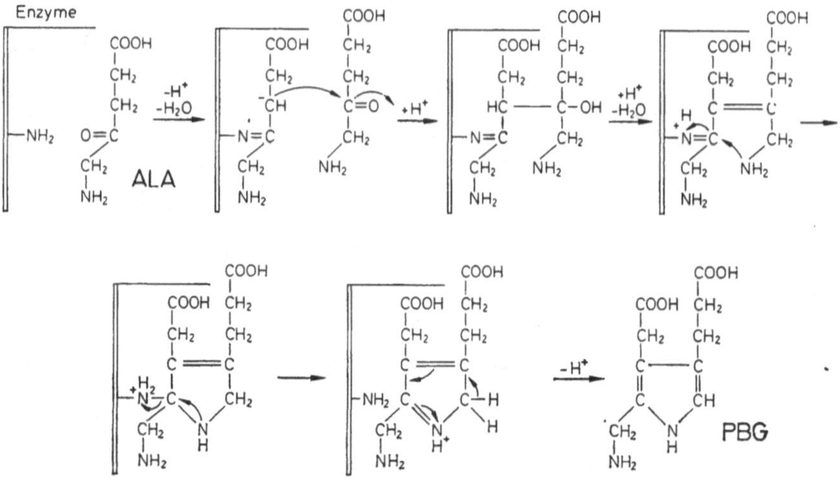

Fig. 6: The mechanism of porphobilinogen formation (Shemin 1970)

The finding that ALAD is competively inhibited by laevulinic acid (Nandi et al. 1968b) or hydroxy laevulinic acid (Yamasaki and Moriyama 1970, 1971) has been very useful in studying biosynthetic pathways forming ALA and in studying regulative mechanisms involved in ALA synthesis. In living material ALA accumulates due to the inhibited ALAD (see 3.1).

The reaction mechanism of ALAD was elaborated by Shemin (1970). According to Shemin one molecule of ALA gets bond to the enzyme protein by a SCHIFF base. By this reaction the elements of water are eliminated from the carbonyl carbon of ALA and the amino group of a lysine residue. The free carboxyl group of ALA also seems to be essential for fixing the substrate. A carbanion essential for the condensation of ALA with a second molecule of ALA can readily be stabilized by the SCHIFF base and attacks the carbonyl carbon atom of the second molecule. PBG is formed, following elimination of H_2O and transimination (Fig. 6). Additions to and changes of this scheme were published by Barnard et al. (1977) and Batlle and Stella (1978). Studies with ALAD from bovine liver and some other enzymes of the porphyrin biosynthetic pathway immobilized on Sepharose were performed by Batlle et al. (1978a, b).

The idea that the carbanion could be stabilized by a metal ion (see Granick 1967) is supported by the Mg^{2+} requirement of plant enzymes.

3.3. The Formation of Uroporphyrinogen III. Porphobilinogenase, i.e. Porphobilinogen Deaminase (E.C. 4.3.1.8) and Uroporphyrinogen III Cosynthetase

By the action of PBGase, i.e. by the synergetic action of PBG deaminase and Urogen III cosynthetase, one molecule of Urogen III is formed from four molecules of PBG. If only PBG deaminase is present Urogen I arises (Bogorad 1958a), a compound synthesized *in vivo* under pathological conditions (for literature see e.g. Doss (ed.) 1976). No conversion at all is observed if Urogen III cosynthetase is incubated with PBG but without PBG deaminase (Bogorad 1958b). Urogen I proved to be not a substrate for Urogen III cosynthetase, either.

The Urogen I isomer is characterized by regular alternating acetic acid and propionic acid side chains. In isomer III the sequence of side chains is changed in ring D. By means of specific labelled PBG Battersby et al. (1976b) have shown that the PBG unit which forms ring D is built in with an intramolecular rearrangement of the carbon atom in the aminomethyl side chain forming the δ-bridge. Isomers I, II, III and IV and their respective derivatives which are used to determine specific properties of the enzymes following in the prophyrin and chlorophyll biosynthetic pathway are available by chemical synthesis (see Smith (ed.) 1975). None of a great variety of synthetic analogues of PBG was metabolized by the Urogen forming enzymes. Many of them proved to be inhibitory (Frydman and Feinstein 1974, Frydman et al. 1976).

PBG deaminase and Urogen III cosynthetase were first isolated by Bogorad to study and to partly purify PBG deaminase from spinach; Urogen III cosynthetase, which proved to be more labile than the deaminase, was destroyed by heat treatment (Bogorad 1958a).

Urogen III cosynthetase was available free from PBG deaminase activity from wheat germ (Bogorad 1958b).

Stevens and Frydman (1968) and Frydman and Frydman (1970) used wheat germ to separate both activities from one source. Reaction mixtures were buffered with tris-HCl pH 8.2 or phosphate pH 7.4. The respective enzymes isolated from human erythrocytes showed properties similar to the properties of wheat germ enzymes (Frydman and Feinstein 1974). The cosynthetase was more labile than the deaminase in several respects. Stevens et al. (1968) reported a great excess of cosynthetase as compared with the deaminase. However, no molecular weights were submitted. Llambias and Batlle (1971) separating and purifying PBG deaminase and Urogen III cosynthetase from cultured soybean cells found by gelfiltration molecular weights of about 40,000 and 210,000 for the deaminase and the cosynthetase, respectively. Surprisingly the PBGase complex had only a molecular weight of about 22,000. The pH optimum for both enzymes was 7.2.

Deaminases from wheat germ and spinach both had a molecular weight of 40,000 and the synthetase from wheat germ a molecular weight of 62,000 (Higuchi and Bogorad 1975). These molecular weight determinations were done by means of gel filtration, sucrose density centrifugation and SDS acrylamid electrophoresis. The reaction mixtures were buffered to give a pH of 8.2. Gel filtration and electrophoresis (Davis and Neuberger 1973) and density gradient centrifugation (Jordan and Shemin 1973) of the deaminase from *Rhodopseudomonas spheroides* resulted in molecular weights of 36,000 and 39,000. The pH optimum of this enzyme was 7.6.

A molecular weight of only 25,000 was found for deaminase from human erythrocytes (Frydman and Feinstein 1974). Cow liver deaminase appears to exist as a single polypeptide chain with a molecular weight of about 40,000 (Sancovich et al. 1976). SH-groups seem to be essential for activity.

Experiments with the inhibitor butanedione infere that an arginine residue is required for the activity of deaminase isolated from wheat germ (Pollack and Russell 1978).

After the investigations of the properties of the separated enzymes, the association of the two enzymes, i.e. the stability of the PBGase complex was studied by different methods and with different results. Higuchi and Bogorad (1975) found that PBG deaminase sedimented more rapidly in the presence of Urogen III cosynthetase and PBG, but not if the experiment was run without PBG. On the other hand Frydman and Feinstein (1974) immobilizing the deaminase on sepharose found the cosynthetase associated with the deaminase when the cosynthetase was filtered through the system. Urogen III was formed by the immobilized complex after addition of PBG.

Although 20 to 30 models (for literature see e.g. Frydman et al. 1976) for the co-operative action of the PBG deaminase and the Urogen III cosynthetase have been elaborated, the real mode of action is not yet clear (but see below!). Intermediates between PBG and Urogen III (or I) are not liberated into the reaction medium, except under special conditions.

Open chain oligopyrroles accumulated when the reaction was run in the presence of ammonium or high concentrations of PBG (Pluscec and Bogorad 1970, Radmer and Bogorad 1972). Similar results were contributed by Davis and Neuberger (1973), using hydroxylamine, ammonium and methoxy-amine. They found that the inhibitory amines were incorporated into the oligopyrroles and were released by non-enzymic cyclisation.

Compounds assumed to be tripyrroles according to molecular weight determinations by gel filtration accumulated in a system from soybean cells (Llambias and Batlle 1970, Stella et al. 1971). However, it should be proven that pyrroles are really separated according to their molecular weights. The elution volume of some pyrroles was found to be dependent on other criteria than molecular weight (Schneider 1973 d).

Some of the accumulated products can be assumed to be intermediates because they were incorporated into tetrapyrroles. Llambias et al. (1970 b) reported on the incorporation of the products formed in soybean preparations. Pluscec and Bogorad (1970) found a dipyrrole – formally derived from a head-to-tail reaction of two PBG units – to be a substrate for the formation of Urogen I, but only when incubated with PBG. In addition Frydman et al. (1973), incubating a dipyrrole – formally derived from a head-to-head reaction of two PBGs – with PBGase, achieved incorporation of this dipyrrole into Urogen III. The externally added dipyrroles, however, competed very unfavourably with the intermediates bound to the enzymes. Dipyrroles with structures other than the mentioned ones were no substrates. In neither case tetrapyrroles were formed by dimerisation of dipyrroles. Exogenous tripyrroles were found to be even less substrates than dipyrroles. Only a tripyrrole – synthesized by a regular head-to-tail condensation of PBGs – was incorporated to a minor extend into Urogen I (Frydman et al. 1974).

From all these findings it appears that the reaction proceeds by a stepwise condensa-

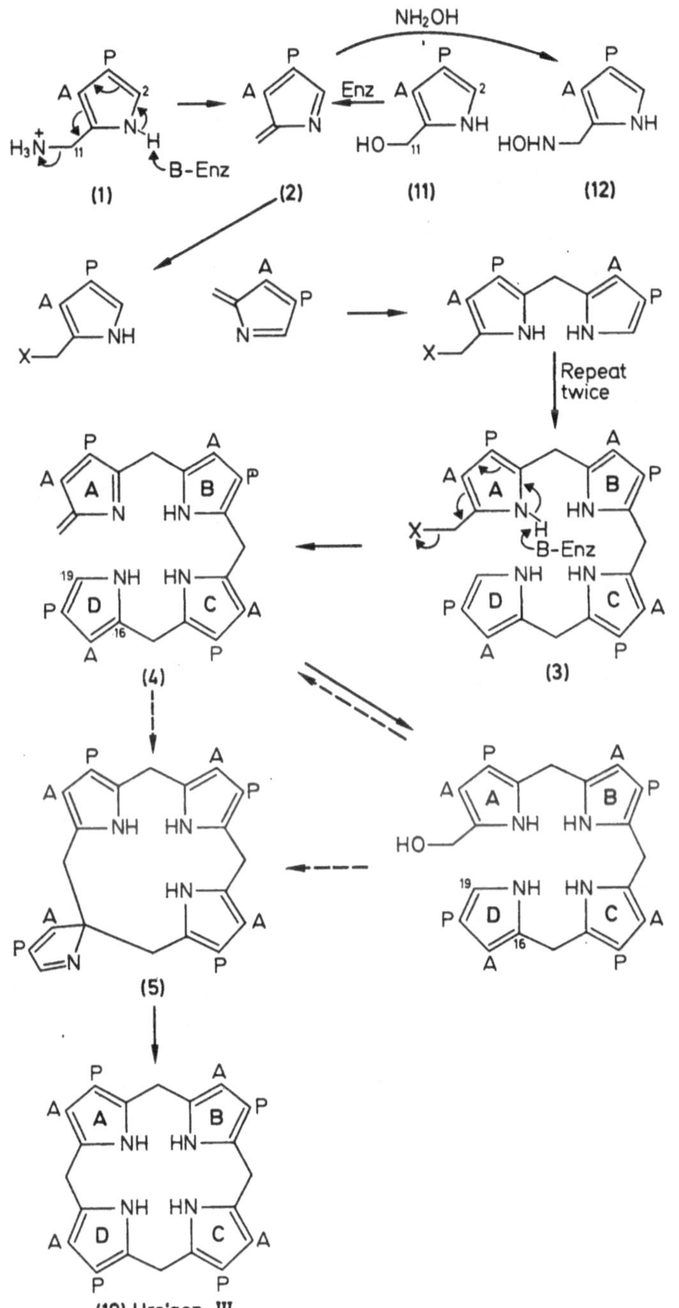

Fig. 7: Scheme of the formation of Uroporphyrinogen from porphobilinogen *via* pre-uro-porphyrinogen (Battersby et al. 1979)

tion of the four PBG units. Already the first step separates Urogen III from Urogen I synthesis by a head-to-head condensation instead of the head-to-tail reaction. Urogen III cosynthetase acts as a specifier protein of PBG deaminase (Frydman et al. 1976). It was guessed that PBG deaminase has at least two active sites, one site for fixation on the protein and condensation and another one for cyclisation (Frydman and Frydman 1970). Frydman et al. (1976) assumed that the porphobilinogenase complex might be a sort of methyl transferase. The head-to-head condensation is thought to take place by displacement of an activated methyl group arising from the aminomethyl side chain of PBG. « The ‹ active methyl › group of the first unit is transferred and remains either covalently to the dipyrrylmethane, or in a partial equilibrium with the protein. In the next steps the third and forth units may attack (A) either through the α-free position, or (B) through the α-occupied position, producing a constant transfer of the initial ‹ active methyl › group ». Further experiments will be necessary to test this model.

But scarcely had these latter lines been written, when a competitive hypothesis became more and more probable. Battersby et al. (1978, 1979a) and Burton et al. (1979) presented evidence that the biosynthesis of Urogen III occurred by head-to-tail assembly of four PBG units followed by intramolecular rearrangement. It was shown that there was a noncyclic, soluble intermediate, pre-Urogen, which accumulated in PBG/PBG deaminase assays up to 43 percent and which was subsequently converted into Urogen III by Urogen III cosynthetase at a rate of 87 percent (Jordan et al. 1979). The assembly of PBG units was found to start from ring A with sequential addition of rings B, C, and D. It is ring D which is rearranged (Battersby et al. 1979b). The sequence of steps according to these findings is shown in figure 7.

3.4. The Synthesis of Coproporphyrinogen III. Uroporphyrinogen Decarboxylase

Urogen decarboxylase catalyses the elimination of the carboxyl groups from each of the acetic acid side chains of Urogen. The product is Coprogen with four methyl groups in place of the acetic acid groups of Urogen. Evidence that porphyrinogens were the intermediates of porphyrin biosynthesis had come from experiments in the midfifties (Bogorad 1955, Neve et al. 1956, Mauzerall and Granick 1958).

Activity of Urogen decarboxylase was proven in extracts of algae and higher plants which accumulated various porphyrins after addition of ALA (Bogorad and Granick 1953, Bogorad 1958 a—c, Carell and Kahn 1964, Rebeiz et al. 1970, 1972, Battersby et al. 1976c).

Chen and Miller (1974) have presented a survey on Urogen decarboxylase activity in different plant species. They purified the enzyme from tobacco leaves about 70 fold by means of ammonium sulphate precipitation and calcium phosphate gel absorption. The pH optimum was 6.5. Thiol reagents, heavy metal ions, chloride and fluoride inhibited enzyme action. Urogen III was found to be a better substrate than Urogen I. Chen and Miller (1974) reported the enzyme to be mainly located in the cytosol. This finding, however, may be due to broken chloroplasts during chloroplast preparation (see part 4).

First information about the mode of action of Urogen decarboxylase came from Mauzerall and Granick (1958). They found that each of the four possible isomers of Urogen was metabolized. The reaction rate was highest with the biologically important isomer III, the reaction rate decreased in the order of the isomers III > IV > II > I. In the reaction mixture intermediates with seven, six or five carboxyl groups were detected.

Large amounts of the heptacarboxylic intermediate accumulated also in preparations of avian erythrocytes (Garcia et al. 1973) purified according to Tomio et al. (1970). These authors achieved a 220fold enzyme purification. The reaction rate of the first decarboxylation surpassed the following ones. Decarboxylation was inhibited by the substrates. Na^+ was also inhibitory; EDTA and GSH showed enhancing effects. Recently, Jackson et al. (1976b) reported that the acetic acid groups are decarboxylated in a sequential clockwise fashion, starting with that on the D ring and followed by those on the A, B, and C rings. Corresponding results were obtained by Battersby et al. (1976c). A synthetic heptacarboxylic porphyrin with one methyl group at the D ring was shown to be identical with the oxidized natural occurring heptacarboxylic porphyrinogen biosynthesized by avian erythrocyte preparations.

3.5. The Formation of Protoporphyrinogen IX. Coproporphyrinogenase (E.C. 1.3.3.3)

Coprogenase (i.e. Coproporphyrinogen oxidative decarboxylase) converts Coprogen into Protogen. The propionic acid side chains in positions 2 and 4 (on ring A and B) are converted to vinyl groups. Molecular oxygen is essential in this reaction.

Animal enzyme preparations converted isomeres III and IV at similar overall rates into the corresponding protoporphyrins (Al-Hazimi et al. 1976). In contrast previous communications had reported a much slower reaction rate for isomer IV than for isomer III (Porra and Falk 1964) or no conversion of isomer IV at all (Sano and Granick 1961). Isomers I and II were never found to be substrates of Coprogenase. By reason of this finding and of the observation that mesoporphyrinogen-VI (a porphyrinogen with the same arrangement of side chains in the rings A and B but only with methyl and ethyl groups on the rings C and D) was also metabolized, it was concluded that the minimal structural requirement for a substrate of Coprogenase was the sequence of side chains: methyl, methyl-propionate, methyl. The vicinal propionic acid groups on rings C and D are not essential for the substrates of this enzyme (Jackson et al. 1976a).

It was also found that a pentacarboxylate porphyrinogen III, the immediate precursor of Coprogen III, was also metabolized (Elder and Evans 1978).

The reaction sequence begins with the conversion of the propionic acid group in position 2. Coprogen with one propionic acid side chain in position 2 already converted to a vinyl group was converted faster by *Euglena* preparations than those with a 4-vinyl group (Calvaleiro et al. 1973). During incubation with animal enzyme preparations a tricarboxylic intermediate accumulated in appreciable amounts. By synthesis and nuclear mass resonance spectrometry this substance was characterized as harderiporphyrinogen (Coprogen with a 2-vinyl group) (Al-Hazimi et al. 1976, Games et al. 1976). The sequence and the velocity of the two decarboxylation reactions has recently been confirmed by experiments with rat liver enzyme and Coprogen labeled in the propionic acid side chains (Elder et al. 1978).

Coprogenase was isolated from tobacco and purified about 70 fold by Hsu and Miller (1970). Enzyme activity was mainly found in a mitochondrial fraction rather than in chloroplasts. In contrast to the animal enzyme from rat mitochondria (Batlle et al. 1965) the tobacco enzyme was inhibited by chelating agents (o-phenanthroline, EDTA) and activated by Fe^{2+} and to a lesser extent by Co^{2+} and Mn^{2+}; Fe^{3+} and Cu^{2+} were inhibitory. No effects of chelators or Fe^{2+} were reported of the enzyme from the sulphur bacterium *Chromatium* (Mori and Sano 1972). Plant enzyme as animal enzyme preparations show maximum activity at about pH 7.4.

The mechanism for the conversion of the propionic side chains to vinyl groups is yet under discussion. Conventional β-oxidation of the propionic acid side chains was excluded by Sano and Granick (1961). Porphyrinogens expected to have structures of intermediates in such a reaction were not converted. Therefore Sano and Granick (1961) proposed a mechanism comprising removal of a hybrid ion from the β-C of the propionic acid side chain and subsequent decarboxylation.

Later Sano (1966) found that 2,4-bis (β-hydroxy-propionic acid) deuteroporphyrinogen IX was converted to protogen. French et al. (1970) isolating a substance from meconium different from coproporphyrinogen by an acrylic acid side chain in positions 2 and 4 discussed these substances as possible intermediates. Results, however, obtained by Battersby et al. (1972) (Zaman et al. 1972) disfavour the latter substances as intermediates. Deuterium label in the α-methyl group of the propionic acid side chain is not lost during conversion to a vinyl group; that should happen if acrylic acid intermediates were existent. Jackson et al. (1974) pointed out that a reaction *via* an oxygenase reaction, removal of H_2O and decarboxylation as postulated by Sano (1966) is the most probable one. Zaman and Akhtor (1976) added a third possible mechanism. They proposed an oxidative reaction involving the β-hydrogen atom and the pyrrole N-H bond to give a conjugated SCHIFF base intermediate. The electron-withdrawing property of the SCHIFF base then facilitates decarboxylation (comp. figure 8).

Only a mechanism not requiring oxygen can be involved in the Coprogen conversion of anaerob organisms. Ehteshamuddin (1968) reported that enzyme preparations of *Pseudomonas* form Proto IX under either aerobic or under anaerobic conditions. Extracts from *Rhodopseudomonas* cells grown in light under reduced oxygen tension converted Coprogen to Proto anaerobically when the extracts were substituted with Mg^{2+}, ATP (NAD(P)) and methionine. Methionine may be replaced by S-adenosyl methionine (Tait 1969). No substitutes were necessary when extracts of cells grown aerobically in the dark were used. The anaerob reaction was sensitive to metal chelating agents. Tait (1969) suggested that the removal of the hydrogen from the propionic acid side chain of Coprogen is linked with an electron transport and energy requiring step. Davis et al. (1973) suggested that the positively charged sulphur atom of S-adenosyl methionine could facilitate the removal of a hydride ion from the β-carbon of the propionic acid side chain.

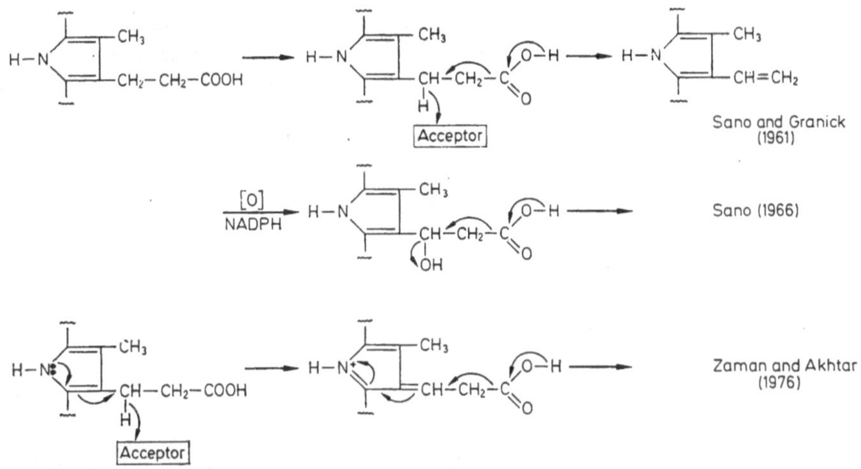

Fig. 8: Possible mechanisms of vinyl group formation from the 2- and 4-propionic acid side chains of uroporphyrinogen

Centrifuging crude extracts of *Chromatium* (Tait 1972a) resulted in a loss of anaerobic activity, which was regained when pellet and supernatant were recombined. The enzyme responsible for aerobic activity could be purified about 60 fold and had an apparent molecular weight of about 44,000.

3.6. The Oxydation of Protoporphyrinogen IX. Protoporphyrinogen Oxydase

Copro, labeled with tritium at the α, β, γ and δ position was converted chemically to Coprogen and subsequently oxidized to Proto IX by a preparation from chicken erythrocytes. Half of the label was lost during the enzymic reaction. If Coprogen was oxidized chemically the label was retained due to the specific isotope effect (Jackson et al. 1974). This finding was the first proof for the supposition that porphyrinogens in the organisms are converted enzymically and stereospecifically to Proto rather than by autoxidation, which is readily observable in Protogen preparations. Corresponding results were reported by Battersby et al. (1976a) using *Euglena* preparations.

Recently enzymes responsible for the conversion of Protogen to Proto have been purified and characterized. Poulson and Polglase (1975) studied the enzyme from yeast mitochondria. This enzyme had a molecular weight about 180,000 and a pH optimum of 7.45. Urogen I and III or Coprogen I and III were not oxidized; mesoporphyrinogen IX was oxidized with less than 20% of the rate observed with Protogen IX. The enzyme was inhibited by Cu^{2+} or Co^{2+}, haem and haemin. A molecular weight of about 35,000 and maximum activity at pH 8.6 to 8.7 was reported for an enzyme partially purified from rat liver mitochondria (Poulson 1976). Molecular oxygen was essential for Protogen oxidation. In a system prepared from *Escherichia coli* electron acceptors other than oxygen were detected. Nitrate and fumarate stimulated anaerobic Protogen oxidation. There was evidence for an involvement of the anaerobic electron transport system (Jacobs and Jacobs 1976, 1977). In preparations from *Staphylococcus epidermidis*, which only form cytochroms when grown aerobically, Protogen oxidase activity was low as compared with those of *E. coli.*

3.7. The Insertion of Magnesium into Protoporphyrin IX. Magnesium Chelatase

After Proto is formed the porphyrin biosynthetic pathway splits into the Fe-porphyrin and the Mg-porphyrin forming branches. Chelatases catalyse the insertion of the metal ions into Proto. Neither protogen nor other porphyrinogens are substrates for this type of enzymes. Porphyrinogens do not form metal chelates. Chelatases inserting Fe^{2+}, Co^{2+} or Zn^{2+} have been investigated and have partly been purified (for a summary see Bogorad 1976). These enzymes do not act on Mg^{2+}. Although Mg-porphyrins accumulate in mutants, plants treated with chelators and in cell free systems (see part 2.3), information about an enzyme inserting Mg^{2+} is spare. Spontaneous insertion of Mg^{2+} into porphyrins does not occur.

Gorchein (1972), however, showed that Mg^{2+} was incorporated into Proto which was added in a lipid sol to cells of semianaerobically grown *Rhodopseudomonas spheroides*. EDTA or related chelators were greatly enhancing. The detectable product was Mg-Proto-ME and not as expected Mg-Proto. This finding was corroborated by earlier observations (Cooper 1963, Jones 1963c, Lascelles 1966) which had shown excretion of Mg-Proto-ME into the culture medium by photosynthetic bacteria. Gorchein (1972) suggested that Mg^{2+} insertion was obligatory coupled with methylation of Proto. This view is supported by the observation that dialysed cells required methionine and ATP or S-adenosyl methionine to form Mg-Proto-ME and that the enhancing effect of the latter substances was inhibited by inhibitors of methylation. The formation of Mg-Proto-ME was assumed to take place at a multienzyme complex catalysing both reactions. Experiments to demonstrate Mg-chelatase activity in cell free extracts failed (Gorchein 1973).

In vitro synthesis of Mg-Proto at a low and non-reproducible synthesis rate was observed by Ellsworth and Lawrence (1973). ^{28}Mg Proto formed when extracts of etiolated wheat seedlings were incubated with $^{28}Mg^{2+}$ and ALA or Protogen, but not when Proto was used.

Very recently *in vitro* insertion of Mg^{2+} into Proto was achieved by Smith and Rebeiz (1977). These investigators used preparations of etioplasts and developing chloroplasts from cucumber cotyledons. The preparations were poised with various additions and cofactors. Earlier Mattheis and Rebeiz (1975) had shown that similar cell free preparations converted exogenous Proto into PChlid. Smith and Rebeiz (1977) detected the formation of Mg Proto (ME) by means of fluorescence emission spectra. Besides Mg-Proto-(ME) other metalloporphyrins were formed. Mg-Proto-(ME) amounted to about 20% of the total metalloporphyrins produced. α,α'-Dipyridyl enhanced the formation of Mg-Proto-(ME). This effect of chelators is also known from *in vivo* experiments (Granick 1961, Gassman and Duggan 1974, Rebeiz et al. 1975b). Mg-Proto and Mg-Proto-ME were not separated in the experiments of Smith and Rebeiz (1977) but by cell-free preparations and cotyledons ALA was converted into Mg-Proto-ME instead of Mg-Proto (Rebeiz et al. 1975a).

Castelfranco et al. (1979) reported that glutamate (α-ketoglutarate) played a role in Mg chelation and that the reaction had a marked requirement for ATP. It seems that glutarate does not only exert influence *via* the synthesis of Proto as is suggested by the biosynthetic pathway of Proto (see also Castelfranco and Schwarcz 1978).

3.8. The Methylisation of Magnesium Protoporphyrin IX. (-)S-Adenosyl-L-methionine Magnesium Protoporphyrin Methyltransferase (E.C. 2.1.1.11)

S-adenosyl-methionine Mg-Proto methyltransferase was first detected by Tait and Gibson (1961) and Gibson et al. (1963) in a chromatophore fraction of *Rhodopseudomonas spheroides*. The authors showed that the ring V methyl ester group of BChl arose from the S-methyl group of methionine *via* the formation of S-adenosyl methionine. By the action of the methyltransferase the S-methyl group of S-adenosyl methionine is transferred to the 6-propionic acid side chain of Mg-Proto. The molar radioactivity of BChl was 66% of that of the methyl donor S-adenosyl -(Me-^{14}C)-methionine. Besides Mg-Proto, Zn and Ca-porphyrins were substrates, in contrast to Fe-and Mn-porphyrins. From other experiments (Green et al. 1957, Rebeiz and Castelfranco 1971a) it can be concluded that the methyl group can come from exogenous formate and methanol.

Ellsworth et al. (1974), working with preparations of wheat leaves, suggested that the methyltransferase has a ping-pong reaction mechanism. First the enzyme forms an intermediate with S-adenosyl-methionine, which separates into S-adenosyl homocysteine and an enzyme bound methyl group. Subsequently this methyl is transferred to Mg-Proto *via* an enzyme-methyl-Mg-Proto intermediate. The enzyme was significantly inhibited by Chlid *a*, PChlid, Mg-Proto-ME and the respective metal-free substances (Ellsworth and St. Pierre 1976). Despite of this finding other properties of the wheat methyl transferase – it is neither regulated by an allosteric site mechanism nor was it found to be inducible by light – led to the conclusion that this enzyme is probably not a key regulatory enzyme in Chl biosynthesis.

If crude homogenates of etiolated wheat seedlings were poised with S-adenosyl-L-methionine, NAD, phosphate, CoA, and ATP ^{3}H-labelled Mg-Proto was converted into labelled PChlid (Ellsworth and Hervish 1975).

The methyl transferase from *Euglena* chloroplasts was purified about 70 fold by Ebbon and Tait (1969). Purification of the enzyme from *Rhodopseudomonas spheroides* was unsuccessfull as it was not possible to solubilize the chromophore enzyme (Gibson et al. 1963). Purification from wheat failed due to the instability of the methyl transferase (Ellsworth et al. 1974). Radmer and Bogorad (1967) used a chloroplast preparation of maize leaves. A pH optimum of 8.4 was found for the methyl transferase of *Rhodopseudomonas*, optima of 7.8 and about 8 were reported for wheat and *Euglena* enzymes, respectively.

Proto was a poor substrate for methyl transferase from *Rhodopseudomonas*, *Euglena*, maize and wheat. The reaction rate amounted to about 10% or less of that observed with Mg-Proto. Radmer and Bogorad (1967) suggested that the Proto which was methylated was in fact Zn-Proto. Zn-Proto may be formed enzymically and non-enzymically and is methylated by the methyl transferase of maize with about one third of the velocity found for Mg-Proto. Consequently it appears that the sequence of reaction steps is Proto → Mg-Proto → Mg-Proto-ME.

3.9. Steps from Magnesium Protoporphyrin IX Methylester to Protochlorophyllide

The steps ensuing the formation of Mg-Proto-ME include the formation of the 4-ethyl group by reduction of the 4-vinyl group and the formation of the ring V.

The conversion of Mg-Proto-ME to PChlid can be followed *in vitro* (Mattheis and Rebeiz 1976). A NADH-dependent enzyme was found to be involved in the reduction of the 4-vinylgroup (Ellsworth and Hsing 1973) and a β-type oxidation mechanism in the formation of the ring V (Ellsworth and Aronoff 1968, 1969). A β-type oxidation had been suggested by Granick (see Granick 1967). Mg-Proto-ME derivatives having the proposed structures, i.e. Mg-Proto-ME with a 6-acrylate, a 6-hydroxy propionate and a 6-ketopropionate side chain, respectively, were identified by Ellsworth and Aronoff (1968, 1969) by means of mass and visible spectroscopic techniques in cultures of *Chlorella* mutants. In addition, Ellsworth and Aronoff (1968, 1969) found derivatives with 2 to none vinyl groups in the tetrapyrrolic moiety, likewise besides PChlid 4-vinyl(4-desethyl)PChlid. These findings suggested that the vinyl group reducing enzyme was not very specific and that reduction could take place at any intermediate step of ring V formation. Two interlinked pathways, one using 2,4-divinyl the other 2-vinyl-4-ethyl derivatives have to be assumed. From their results with etiolated wheat homogenates Ellsworth and Hsing (1973), however, concluded that the 4-vinyl side chain reductive reaction likely occurs before isocyclic ring formation.

PChl still bearing a 4-vinyl group (Mg 2,4-divinylphaeoporphyrin a_5 ME) has also been detected in *Rhodopseudomonas spheroides* where Chl synthesis was blocked by inhibitors (Jones 1963b) or by mutation (Griffiths 1962, Jones 1966). In addition inner seed coats of *Curcurbitaceae* contain this compound (Jones 1966, Houssier and Sauer 1969a, b, Ellsworth and Nowak 1974) (see also part 3.12).

The results so far do not permit a decision in favour of a certain sequence of reactions. On the basis of their results concerning in vitro conversion of 4-vinyl(4-desethyl)PChlid to Chlid Griffiths and Jones (1975) postulated the reduction of the 4-vinyl group to be the last reaction in the biosynthesis of PChlide.

From a study of Belanger and Rebeiz (1979) it appears that both pathways may be used parallel to give two distinct species of Chl *a* and subsequently of Chl *b*.

3.10. The Reduction of Protochlorophyllide, Formation of Chlorophyllide *a*. Accompanying Spectral Changes

The last but one step in the biosynthesis of Chl is the reduction of PChlid. Two hydrogens are introduced on the ring D on the carbons 7 and 8. The hydrogens have *trans* configuration (Fleming 1967, Brockman 1968). In higher plants this reduction of PChlid and formation of Chlid is a light-requiring step. Only photosynthetic bacteria, algae, mosses, and some ferns and gymnosperms are able to execute this step in the dark (see Kirk and Tilney-Bassett 1967).

It is not known which physiological peculiarity or substance distinguishes plants greening in the dark from the other ones, however, the difference does not seem too big. Chl formation in *Pinus* cotyledons was found to be dependent on the contact with the megagametophyte (Schmidt 1924, Bogorad 1950) and developing embryos of wheat produce Chl in the dark, when transplanted to these gametophytes (Bogdanovič 1973a, b). The photoreceptor involved in the photoreduction of PChlid is PChlid itself (Koski et al. 1951, Shibata 1957, Egneus and Sundqvist 1970).

Nevertheless the photoconversion is an enzymic process. Light and the diffusable substance from megagametophytes may both serve to bring about energetic conditions suitable for the conversion. Although proposed by several investigators (see below) the enzymic nature of PChlid photoconversion has unequivocally only been demonstrated recently when Griffiths (1974, 1978) showed *in vitro* photoconversion of added PChlid on etioplast membranes under flash light and continous light.

Since the investigations of Shibata (1957) on PChlid photoconversion in etiolated bean and maize leaves it is known that photoconversion *in vivo* is accompanied by a series of spectral changes besides the spectral shift due to the disappearing PChlid. In all objects subsequently investigated the spectral changes have been found to occur in the same chronological sequence and with only minor deviations at the same wavelengths. (Therefore data only differing within a few nm may be assumed to characterize similar pigment protein complexes.) In those studies also a few additional short-living changes were observed. Step by step it has become tentatively possible to correlate absorption maxima with distinct forms and aggregates of pigments and absorbance changes with biochemical events.

In the red region of the spectrum (where those measurements are done because other plant pigments do not absorb there) 3 forms of PChlid have been detected by their absorption maxima. Young etiolated leaves showed maximum absorption at about 650 nm (P_{650}). During further development in darkness and increasing amounts of PChlid the 650 nm maximum was gradually replaced by an absorption peak at about 636 nm (P_{636}) showing a 650 nm shoulder (Shibata 1957, Sironval et al. 1965). The 636 nm peak was especially prominent in leaves fed with ALA (Sundqvist 1970, Egneus and Sundqvist 1970, Murray and Klein 1971, Gassman 1973). A third form of PChlid has maximum absorption at about 628 nm (P_{628}) (Butler and Briggs 1966, Sironval et al. 1968, Dujardin and Sironval 1970, Kahn et al. 1970, Akoyunoglou and Michalopoulos 1971). The 636 nm absorbing species of PChlid is the only one in maize roots (Björn 1976). As shown and reviewed by several investigators P_{650} is rapidly converted to Chlid (Smith and French 1963, Boardman 1966, Gassman and Bogorad 1967b, Granick and Gassman 1970, Sundqvist 1969, 1970, Thorne 1971a, Klein and Schiff 1972, Gassman 1973). Sironval et al. (1965) extracting freshly illuminated leaves found this pigment to be the first produce of photoconversion. The absorption maximum of the leaves after photoconversion shifts to 684 nm (C_{684}).

In the dark period following illumination, P_{650} appears again concomitantly with a decrease of P_{636}. By a second flash the newly formed P_{650} is converted to Chlid. Thus, consequently P_{636} is converted to Chlid *via* P_{650} (Gassman and Bogorad 1967 b, Granick and Gassman 1970, Sundquist 1969, 1970, 1973, Thorne 1971 a, Gassman 1973). This reaction pattern led to the suggestion that the protein moiety of P_{650} acts as an enzyme. Further evidence for the enzymic nature of the protein moiety of the holochrome came from experiments showing that no new protein synthesis was required for PChlid conversion (Nadler and Granick 1970, Sundquist 1970).

Enzyme substrate complexes are dissociable. An attempt to demonstrate such a property for P_{650} was made by Manetas and Akoyunoglou (1976). They found, feeding labeled ALA to etiolated barley and bean leaves, that there was a free exchange between newly formed labeled and endogenous unlabeled PChlid molecules on the active site of the protein. P_{628} was assumed to be involved in the free exchange. As to the conversion of long wavelength and short wavelength species of PChlide, Cohen and Rebeiz (1978) have drawn similar conclusions. By means of a sensitive spectrofluorometric assay and by an assay using exogenous labeled PChlide Matheis and Rebeiz (1978) succeeded in demonstrating the formation of phototransformable PChlide from nonphototransformable PChlide. The conversion of PChlide to Chlide occurred in continuous light. In addition Griffiths (1976) had pointed out that PChlide-630 (P_{628}) can be activated by NADPH to give PChlide-638/650 (P_{636}/P_{650}). Hitherto P_{628} had always been assumed not to be photoconvertible.
Kahn et al. (1970) demonstrated by means of fluorescence measurements that P_{628} was spatial far from P_{650} as compared with P_{636}. On excitation energy was transferred from P_{636} to P_{650} but not from P_{628}. There was only fluorescence at 655 nm coming from P_{650} besides fluorescence at 630 nm ascribed to P_{628}. However, Kahn and Nielsen (1974) thought that the single band of fluorescence at 655 nm is also consistent with the presence of a single species of photoconvertible PChl(id), although they do not rule out the possible existence of two species P_{636} and P_{650} in dynamic equilibrium. It was also assumed that P_{650} is in an aggregated and the lower wavelength absorbing species are disaggregated forms of PChlid (Butler and Briggs 1966, Lang and Sarvari 1974, Brouers and Sironval 1975; for more literature about this matter see e.g. Brouers 1975, El-Sharkawi and Michel 1975, Sundquist and Klockare 1975, Axelsson 1976a, b). A critical discussion of differences within the P_{650} and P_{636} forms is given by Virgin (1975). In a theoretical approach energy transfer and reduction of PChlide have been correlated by Brouers and Sironval (1978).

During the formation of C_{684} from P_{650} a short living intermediate was detected (Gassman et al. 1968 a, Bonner 1969, Sironval and Brouers 1970, Sironval and Kuyper 1972, Thorne 1971 a, b) showing maximum absorption at 678 nm (C_{678}). In addition Thorne (1971 a, b) found an intermediate preceding the formation of C_{678}. This intermediate had a main absorption band at 668 nm (PC_{668}). He evidenced that two photosteps, with a single photon each, were required to convert P_{650} into C_{684}. The two photons requirement as well as results of Mathis and Sauer (1972, 1973) and (Schultz and Sauer 1972) obtained by means of dichroism analysis suggested that the 668 nm intermediate (PC_{668}) was dimeric containing both PChlid and Chlid. C_{684} was found to be reverted to C_{678} by red light (Bauer and Siegelmann 1972).

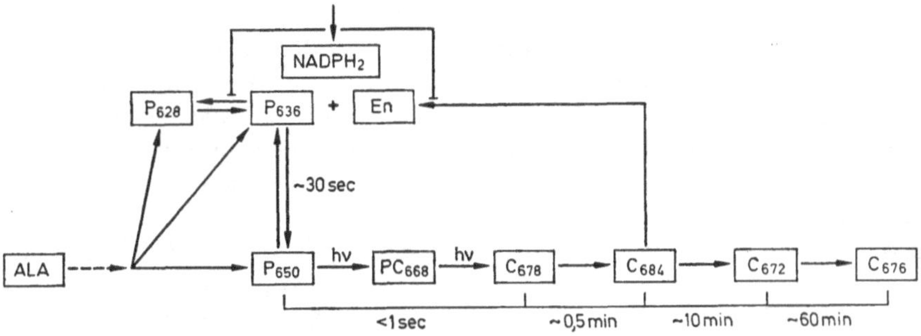

Fig. 9: Synoptic scheme of the sequence of optically detectable forms of protochlorophyllide and chlorophyllide during the formation of chlorophyll *a*

After a light flash the formation of C_{684} from P_{650} takes approximately half a minute. In the course of the next 10 to 20 minutes a shift (named Shibata shift) to 672 nm (C_{672}) is observed. Therafter the maximum of absorption very slowly, during about two hours, moves to 676 nm (C_{676}) (Shibata 1957). A series of additional short living intermediates absorbing in the range between 680 and 750 nm have been detected by Dujardin and Sironval (1977) by trapping them at 178 °K. The intermediates appear to ensue the 680 nm Chlide complex (C_{684}).

C_{672} was reported to be formed from P_{650} without a 684 nm intermediate after a light flash of minor intensity, if PChlid was only partly photoconverted (Litvin and Belyaeva 1971, Virgin and French 1973, Mathis and Sauer 1973). Mathis and Sauer (1973) deduced this P_{672} form from the dimeric PC_{668}. It was also observed that C_{672} could directly arise from P_{636} (Butler and Briggs 1966, Klein and Schiff 1972, v. Wettstein et al. 1971, Kahn et al. 1970, Kahn and Nilsen 1974). From the work of Walter (1974) where additional literature is summarized, however, it appears that C_{684} is also an intermediate in these transitions. Walter (1974) found that leaves rich in P_{636} pass through C_{684} much quicker than do leaves rich in P_{650}.

After a series of proposals (Granick and Gassman 1970, Thorne 1971 a, b, Mathis and Sauer 1973, Virgin and French 1973, Henningsen et al. 1974) and new experimental evidence (Griffiths 1974, 1975, Brodersen 1976) it seems to be a widely subscribed view that P_{650} is a PChlid protein complex activated by NADPH. Comparing the features outlined by Griffith (1976) and Brodersen (1976), however, it seems not quite clear whether P_{636} is PChlid already linked to the enzyme protein. From the scheme of Griffith (1976) (see also Kahn and Nielsen 1974) it appears that P_{636} and P_{650} are both regenerated from P_{628} (PChlide 630) which forms a complex with enzyme and NADPH. After photoconversion of P_{650} to C_{678} Chlide is still linked to the enzyme and there is evidence that it remains on the protein until the shift from 684 to 672 nm occurs. Conditions preventing this spectral shift, i.e. high concentrations of glycerol, also prevent the regeneration of P_{650} (Brodersen 1976). During the transition of C_{684} to C_{672} a rearrangement of etioplast structures takes place (for a summary see Henningsen and Boynton 1974).

Stummann (1979) succeeded in correlating the 680 to 672 nm shift to the halving of the molecular weight of Chlide holochrome. Shortly after photoconversion he was able to separate two species of Chlide holochrome having apparent molecular weights of 74,000 and 29,000 daltons. After illumination Canaani and Sauer (1977) had observed dissociation of a 600,000 dalton holochrome from bean leaves composed of subunits of about 45,000 daltons. Following the 678 to 674 nm shift Chlide was associated with protein of 16,000 daltons.

Griffith (1976) guesses that the enzyme might separate from Chlid before C_{683}; C_{672} could then be a disaggregated form of C_{683}. In any case, the enzyme can recombine with PChlid and NADPH to form new P_{650}. Another prerequisite for P_{650} formation and photoconversion seems to be ATP. Horton and Leech (1975) propose that ATP gives the protein the ability to bind to the PChlide and enables formation of dimeric PChlid protein complexes. Lichtenthaler and Sundqvist (1979) found that PChlide holochrome from wheat leaves was associated with plastoquinone and phylloquinone. PChlide, plastoquinone and phylloquinone were present in a constant molar ratio of 2:1:1. The authors consider the possibility that prenylquinones might serve as electron or hydrogen carriers between NADPH and PChlide.

Sometimes phytylation of Chlid was related to the spectral shift from 684 to 672 nm (Klein 1962, Sironval et al. 1965), but the bulk of evidence favours the view that phytylation takes place during a longer time interval than this shift (Wolff and Price 1957, Boardman 1967, Akoyunoglou and Michalopoulos 1971, Henningsen and Thorne 1974). On the basis of results obtained with chloroplasts mutants of barley Henningsen and Thorne (1974), however, concluded that there is a causal connection between the absorption shift from 682 nm to 672 nm of Chlid and esterification. Mutants not showing this shift fail to esterify Chlid. Since phytylation *per se* does not cause a dramatic absorption change, the spectral change appears to be dependent on coordinated changes of components in the prolamellar body membrane.

If phytylation is inhibited by low temperature, no change in the structure of the prolamellar body occurs, although PChlid is converted to Chlid (Treffry 1970).

Which enzymes and intermediates are involved in phytylation is still discussed (see part 3.11). Chls from bacteria may have alcohols other than phytol (Rapoport and Hamlow 1961). Katz et al. (1972) and Brockmann et al. (1973) and Gloe and Pfennig (1974) found BChl esterified with geranylgeraniol. Traces of alcohols other than phytol, especially geranylgeraniol, have also been detected in Chls of higher plants (Ogawa et al. 1975, Wellburn 1976, Rüdiger et al. 1976). These compounds are considered to be intermediates on the way to phytylation.

3.11. The Phytylation of Chlorophyllide *a*. Chlorophyllase (E.C. 3.1.1.14); Methyl Chlorophyllide *a*; Geranylgeranyl Chlorophyllide *a*

In innumerable experiments Chlase has been shown to catalyse the hydrolysis of Chls to Chlids and phytol. Although under *in vitro* conditions scarcely any formation of Chls from Chlids and phytol has been demonstrated, Chlase is assumed to be involved in the formation of Chls *in vivo*.

Crude Chlase preparations are known since the investigations of Willstätter and Stoll (1913) who used air dried powder of *Heracleum* leaves. Purification of Chlase is rendered difficult by its tight association with plastid lipoproteins and its tendency to aggregate with itself or with other substances.

Ardao and Vennesland (1960) found large amounts of Chlase in a chloroplast lipoprotein complex from spinach. Twenty years before Krossing (1940) had reported that Chlase was in a chloroplast fraction. More recently it was shown (Terpstra 1976) that Chlase from *Phaeodactylum* was associated with pigmented membranes and that it was only active after desintegration of the lamellae. Following the purification procedure of McFeeters (1975) Chlase was associated with material floating after centrifugation. In addition Bacon and Holden (1970) reported that the enzyme was strongly adsorbed to negatively charged material. Corresponding reports can also be found in the publications of other investigators (see below).

For better solubilization of Chlase, it proved to be advantageous to make preparations with detergents or to start preparation with acetone powder of leaves. Desoxycholate (Klein and Vishniak 1961, Ogura 1972) and Triton X-100 (McFeeters et al. 1971, McFeeters 1975) were suitable detergents. Bacon and Holden (1970), however, successfully purified Chlase from sugar beet leaves only using buffer. Sugar beet Chlase had a molecular weight between 30,000 and 38,000 daltons as shown by gel filtration. Chlase from tea leaves was only partly soluble without detergent (Ogura 1972).

Chlase from *Phaeodactylum* seems to be part of the PS 1 system (Terpstra and Goedheer 1979). The enzyme becomes soluble after all lipids are extracted from the membranes. Its activity depends on $MgCl_2$ and dithiothreitol; its molecular weight according to gel electrophoresis is 38,000 daltons (Terpstra 1978). Using gel electrophoresis the molecular weight of Chlase from *Phaseolus* leaves was found to be 30,000 daltons (Moll and Stegwee 1978). Moll and Stegwee (1978) also found synthetic activity of Chlase if the enzyme was incubated with substrates included in liposome membranes.

All purification steps can be performed at room temperature (Ellsworth 1971, McFeeters 1975). Because of the insolubility of the substrates of Chlase in aqueous solution test mixtures have to contain 30–60% acetone or 0,2% Triton X-100. Triton was reported to have significant advantages over assays using high acetone concentrations. (McFeeters et al. 1971). Ellsworth et al. (1976), however, found a decrease in activity using Triton. These differences may be due to different preparation procedures. But in either case, the assay conditions are quite different from the conditions in the plastid lamellae. A pH-optimum between pH 6 and 8 was measured.

Chlase hydrolyses Chl *a* and Chl *b* and the corresponding pheophytins, BChl, ChChl and purpurins (Fischer and Lamprecht 1938, Klein and Vishniak 1961, McFeeters et al. 1971). PChl and 4-vinyl-PChl, different from Chls by the nonreduced ring D, are not converted. Chls lacking the carboxymethyl group on position 10 reduce the reaction velocity. As demonstrated with iso-amylbenzoate (Seiler and Grob 1969) and p-nitrophenyl-acetate (McFeeters et al. 1971) Chlase has no general hydrolysing properties.

Measuring the apparent K_Ms of various substrates of Chlase McFeeters (1975) showed that a decreasing length of the alcohol group (phytyl-, ethyl-, methyl-) has only minor effects on substrate binding. However, when the Mg is removed from the tetrapyrrole ring a decreasing size of the alcohol improves substrate binding. An additional, essential substrate binding site is the 9-ketogroup. Although the 10-carboxymethyl group and the 7,8-hydrogens are essential for Chlase reaction, they have relatively small effects on substrate binding. They are assumed to be important for the proper orientation of the substrate.

Chl forming Chlase activity *in vitro* had been found by Willstätter and Stoll (1913) and Shimizu and Tamaki (1963). These results, however, were not reconfirmed by other investigators. Therefore, until recently the finding that Chlase activity often parallels Chl accumulation (Holden 1963, Shimizu and Tamaki 1968, Böger 1965, Chiba et al. 1967; compare with Ganoza an McFeeters 1976) and the fact that the test mixtures for Chlase are not at all physiological, were the only indications for synthesizing properties of Chlase.

A role of Chlase in the degradation of Chl has been assumed by Sabater and Rodriguez (1978). Procedures which protect Chl from being degraded lower Chlase activity.

New aspects were added by Ellsworth (1971, 1972a, b) and Ellsworth et al. (1976) who found that Chlase preparations contained three activities: 1. Chl *a* hydrolysis, 2. Chlid *a* phytylation and 3. methyl Chlid *a* phytyl transesterification. Methyl Chlid *a* phytyl transesterifying activities had also been observed by Chiba et al. (1967), Hines and Ellsworth (1969), Wellburn (1970), and Ichinose and Sasa (1973). Ellsworth et al. (1976) suggested that these activities were due to two different enzymes: Chlase and a Chlid *a* phytylating enzyme. They propose the hypothesis that the phytylating enzyme may act by binding Chlid *a* and forming a firmly bound methyl Chlid *a* intermediate which is then phytyl transesterified to yield Chl *a*. Since synthesis of [^{14}C]-methyl labeled methyl Chlid could not be detected, it was assumed that the methyl group might be temporarily «loaned» to the bound Chlid by a methionine residue at the active site of the enzyme.

Granick (1967) had suggested that phytol might be added *via* phytyl pyrophosphate rather than by the Chlase reaction. This suggestion, however, was refuted by the results of Watts and Kekwick (1974), who demonstrated that [^{14}C]-phytyl monophosphate was incorporated into Chl as well as [^{14}C]-phytyl pyrophosphate. Addition of unlabeled phytol decreased the incorporation rate of the labeled substances. Watts and Kekwick (1974) assumed that the phosphates are removed before the phytylation.

On the other hand Rüdiger et al. (1977) described a cell-free system from maize leaves which incorporated [^{14}C]-geranylgeranyl pyrophosphate, but not free [^{14}C]-geranylgeraniol into Chl *a* (esterified with geranylgeraniol).

An esterification of Chlid (or methyl Chlid) with preformed phytol is also questioned by observations of Ogawa et al. (1975). By means of thin layer chromatography (Schneider 1968) these investigators identified a chlorophyllous pigment with a R_f value different from Chl *a* to be an intermediate in the conversion of Chlid to Chl *a*. The chromatographic behaviour was similar to geranylgeranyl Chlid *a*. Small amounts of geranylgeraniol as part of Chl were also detected by Wellburn (1970, 1976), Liljenberg (1974), and Rüdiger et al. (1976). The latter also detected dihydrogeranylgeranyl Chlid *a*. Rüdiger et al. (1976) suggested that *in vivo* geranylgeranyl Chlid *a* intermediates were hydrogenated when bound to the tetrapyrrole ring to give Chl *a*. Such a mechanism would explain the finding that the mole-per-mole ratio of Chl and phytol is always close to one (Steffens et al. 1976). In additional investigations Schoch et al. (1977) found tetrahydrogeranylgeraniol (THGG) besides geranylgeraniol (GG), dihydrogeranylgeraniol (DHGG) and phytol (P) to be esterified with Chlid *a*. Kinetic studies on the increase and the decrease of these compounds during Chl *a* synthesis led the authors to propose the following sequential steps for Chl *a* biosynthesis: Chlid *a* → Chl *a* GG → Chl *a* DHGG → Chl *a* THGG → Chl *a* P.

Ensuing studies with bean leaves (Schoch 1978) confirmed these findings. The double bounds of THGG were found to be located at the $\Delta^{2,14}$ positions (Schoch and Schäfer 1978).

These results on esterification and the nature of the primary alcohol attached to Chlide demand new reflections on the role of Chlase.

3.12. Protochlorophyllide and Protochlorophyll

If etiolated leaves are extracted, PChl is found besides PChlid. PChl amounts to about 20% or less of the total quantity of the chlorophyllous pigments in young etiolated leaves (Wolff and Price 1957, Smith et al. 1959, Rudolph 1965, Sironval et al. 1965. Schneider 1968). Higher percentages of PChl, up to about 40% are found in aged leaves and as recently was shown in very young 2 days old bean leaves (Lancer et al. 1976). PChlid and PChl present in an etiolated leaf do not exceed 1/100 of the maximum amount of Chl in a light grown leaf. However, their amounts may artificially be increased by feeding ALA (Granick 1959) (see also parts 3.10; 5).

When the first analytical data on PChl were available, it was scarcely questioned that the esterifying alcohol of PChl was phytol (e.g. see Smith and Benitez 1955, Boardman 1966). But soon some results obtained with etiolated plants gave rise to doubts (Fischer and Bohn 1957, Fischer and Rüdiger 1959). In 1974 Liljenberg showed PChl from barley to be esterified with geranylgeraniol. No phytyl ester was found.

PChls were also isolated from the seed coats of curcurbits (Seybold 1949, Fischer and Bohn 1958, Jones 1966). In a gaschromatographic-mass-spectrometric analysis on seed coat PChls Ellsworth and Nowak (1974) detected all possible C_{20} isoprenoid alcohols between and including geranylgeraniol and phytol.

In addition these PChls could be separated into two species with respect to the tetrapyrrole moiety (Jones 1963b, 1966). One of these species is identical with leaf PChl, the other is a 4-vinyl(4-desethyl)PChl. The latter compound was found to be esterified about 50% by phytol, the normal PChl about 90% (Ellsworth and Nowak 1974). Ellsworth and Nowak (1974) assumed that the more oxidized chlorophyllous compounds—more oxidized on the alcohol moiety as well as on the tetrapyrrole ring than Chls – could be the result of a deficiency in biological reducing agents during ageing.

Probably this variety of PChls in seed coats has no biological meaning. Nevertheless the detection of different PChls shows that the ring V can be formed before and after the formation of the 4-ethyl group and in addition that Chls with alcohols other than phytol may be precursors of phytyl Chlid (see parts 3.11; 3.13).

There has been a general agreement, and thus it was outlined above, that Chl is formed from PChlid *via* Chlid (see also Wolff and Price 1957, Virgin 1960, Godnev et al. 1963, Sudyina 1963, Sironval et al. 1965, Bovey et al. 1974). The recent findings on PChlid conversion in isolated etioplasts and etioplast membranes (Horton and Leech 1975, Griffith 1974) strongly support this view. Lancer et al. (1976) working with very young bean leaves and Cohen and Schiff (1976) studying *Euglena*, however, clearly showed that both PChlide and PChl are photoconverted. Observations indicating PChl photoconversion were communicated already earlier (e.g. Virgin et al. 1963, Rudolph and Bukatsch 1966). According to Rebeiz and Castelfranco (1971a) PChl is not formed from PChlid but is formed from a common precursor by a parallel series of reactions.

Euglena and very young bean leaves are different from the other objects investigated. They contain proplastids rather than etioplasts and form C_{672} from P_{636} without an intervening Shibata shift (Cohen and Schiff 1976, Lancer et al. 1976) (see also 3.10). Under natural conditions greening on the basis of proplastids can be assumed to be the normal mode of leaf development rather than greening on the basis of etioplasts, which only form after prolonged dark periods. It should be demonstrated that all the effects analysed are not effects merely valid for a narrow range of developmental conditions. P_{636} in etiolated leaves is only converted under flash light (e.g. see Sundqvist 1969, Granick and Gassman 1970), a condition certainly not reflecting natural conditions. However, more recent papers (Griffith 1978, Matheis and Rebeiz 1978) show success in overcoming this limitation.

3.13. Isolation and Purification of Protochlorophyll(ide) and Chlorophyll(ide) *a* Holochromes

PChl(id) protein complexes can be isolated and purified in an active state, i.e. under illumination PChl(id) attached to the protein is converted to Chl(id). Attempts, however, to convert added PChlid by the action of isolated holochromes failed (Manetas and Akoyunoglou 1976) so far, although repeated action of holochrome protein within the lamellae of isolated prolamellar bodies was demonstrated recently (Griffith 1974).

Since Smith (1952) and Krasnovski and Kosobutskaya (1952) had first observed that the supernatants of glycerol and buffer extracts, respectively, contained the photoconvertible complex, this complex was intensively studied and characterized (Smith and Kupke 1956, Boardman 1962, Schopfer and Siegelman 1968, Henningsen and Kahn 1971, Henningsen et al. 1974). An improved purification procedure was elaborated by Schopfer and Siegelman (1968). Concerning isolation and stability of the holochrome and its photoconvertibility it proved advantageous to use buffers with a pH higher than 8 and glycerol (Schopfer and Siegelman 1968, Henningsen and

Kahn 1971). Fraction ammonium sulphate precipitation and column chromatography was included in the purification procedures. Mercaptoethanol inactivated the complex.

Comparing the results of various laboratories conspicuous differences in the molecular weights of holochromes exist. These differences certainly reflect differences in the purification procedures and the objects under investigation. Since in general the use of detergents results in minor particle size, it can be assumed that the complexes with high molecular weights are oligomeres. They may contain several PChlid molecules per particle (Schopfer and Siegelman 1968, Henningsen et al. 1974). Recently a variety of detergents was tested (Stummann 1978a), but none had the favourable properties of saponine.

In contrast to the absorption of P_{650} *in vivo*, the main absorption of the isolated holochromes is found about 640 nm and about 644 nm when the complex is isolated in a medium containing saponine. After illumination the absorption peak shifts to 678 nm and during the next hour to 672 nm. Non-detergent bean PChlid holochrome with an absorption at 637 nm immediately was converted to the 673 nm absorbing form (Henningsen et al. 1974). Henningsen et al. (1974) found that the shift from 678 nm to 672 nm was linked to a change in the molecular weight of the holochrome from about 63,000 to 29,000. Similar observations are communicated by Stummann (1979). Bogorad et al. (1968) had already reported that the Chlid which was formed when PChlid was photoconverted was bound to a protein of lower molecular weight than that one of PChlid holochrome. No change in molecular weights was connected with the 644 to 678 nm shift (Henningsen et al. 1974). These observations are in accordance with the idea, outlined in the preceding part, that the PChlid holochrome protein moiety acts as an enzyme. Henningsen et al. (1974) assumed Chlid holochrome to dissociate into a Chlid carrier protein complex and the photoenzyme. – Chl holochrome extracted after a 15 minutes illumination of barley leaves had a molecular weight greater than 100,000 daltons and contained photosystem I activity, leaves extracted after illumination followed by dark period contained the 63,000 dalton and the $>$100,000 dalton form. From the paper of Alscher et al. (1976) it appears that light-stimulated protein synthesis occurs concomitantly with the regeneration of phototransformable PChlid.

Five new classes of Chl protein complexes (molecular weights 100,000, 63,000, 41,000, 39,000, 34,000) were described by Nielsen (1975) to emerge during greening. As shown by simultaneous electron microscopic studies, at early stages of greening the newly formed membrane proteins and Chl were inserted into existing membranes. Thereafter all membrane components appeared to be synthesized.

Extracting etiolated and greening leaves of maize with sodium dodecylsulphate Guignery et al. (1974) found by means of disc electrophoresis that PChlid was associated with two peptides of the molecular weight of 21,000 and 29,000, respectively. After photoconversion *in vivo* the (^{3}H labeled) chromophor remained at the proteins with the molecular weights of 21,000 and 29,000. After 12 hours of illumination Chl was linked to proteins with molecular weights of 25,000 and 70,000. These Chl protein complexes were assumed not to arise from translocation or changes in the proteins but from new synthesis. Acker and Duranton (1975) concluded that these

Table 2: Properties of holochrome proteins

Source	Buffer	Detergent	mol. Weight Pchlide-Hch.	mol. Weight Chlide-Hch.	spectral shift	authors
Phaseolus	Glycin pH 9.6	–	400,000	–	640–675	Smith and Kupke (1956)
Phaseolus	Glycin pH 9.6	–	700,000	–	640–675	Smith (1960)
Phaseolus	Glycin pH 9.6	–	600,000 * ± 50,000	–	640–675	Boardman (1962)
Phaseolus	Tricine pH 8.0	Triton X-100	550,000 ** 300,000	–	639–678–672	Schopfer and Siegelman (1968)
Phaseolus	Tricine pH 8.5	Saponine	100,000 **	–	644–678	Henningsen and Kahn (1971)
Hordeum	Tricine pH 8.5	Saponine	63,000	–	644–678	Henningsen and Kahn (1971)
Hordeum	Tricine pH 8.5	Saponine	63,000 **	29,000 100,000	(642–646)–(677–679)–672	Henningsen et al. (1974)
Phaseolus	Tricine pH 8.5	Saponine	170,000 (500,000)	–	(677–679)–672	Henningsen et al. (1974)
Pisum	Tricine pH 8.5	Saponine	170,000 (500,000)	–	(677–679)–672	Henningsen et al. (1974)
Zea mays	Tris pH 6.8	SDS	21,000 *** 29,000	21,000 29,000 25,000 70,000	–	Guignery et al. (1974)

Determination of molecular weights: * by centrifugation
** by gel filtration
*** by electrophoresis

four Chl protein complexes cannot be related to the four main spectroscopic detectable Chl forms (French et al. 1972) because the relative proportion of the spectroscopic Chl forms remained constant while the relative proportions of the isolated complexes changed.

In the meantime the characterization of pigment protein complexes after the stage of Pchlid and Chlid has made further progress (e.g. see Thornbeer et al. 1979, Krishnan and Guanam 1979). Even pigments being precursors of PChlid like Proto and Mg-Proto have been found to be associated with plastid proteins (Stummann 1978c).

3.14. The Biosynthesis of Chlorophyll *b*

The final step in the synthesis of Chls in green plants is the synthesis of Chl *b*. The 3-methyl group of Chl *a* is oxidized to a formyl group in Chl *b*. The conversion of Chl *a* to Chl *b* is suggested by most of the data available; Chl *a* is the precursor of Chl *b*. However, it is uncertain from which form of Chl *a* Chl *b* derives.

Suggesting evidence for the sequence Chl *a* to Chl *b* came from the finding that there are plants only possessing Chl *a* but no examples for a plant only containing Chl *b*. The argument that Chl *b* plants would be unable to exist because of the lacking reaction center of photosynthesis is refuted by a series of other observations.

Analysing the time course of pigment synthesis, it has always been found that from PChlid first Chl *a* was formed (see part 3.10) and subsequently Chl *b*. First traces of Chl *b* in illuminated etiolated seedlings can be found after 10 to 30 minutes after onset of light (Augustinussen 1964, Thorne and Boardman 1971). The final value for the ratio Chl *a*/Chl *b* which centers about 3 is reached after about 4 hours of illumination (Raven 1973).

Further evidence for the sequence Chl *a* to Chl *b*, although not always quite conclusive, came from experiments with green plants which were fed [^{14}C]-substances. The specific activities of Chl *a* were found higher than those of Chl *b* (Godnev et al. 1963, Shlyk et al. 1963, Duranton 1966), as it can be expected if Chl *b* is derived from Chl *a*. After a while the specific activities of both Chls should be equal, however, in most cases equality in the specific activities was not reported. Chl turnover rates and an observation of Michel-Wolwertz (1963) may be pertinent in explaining these findings. Michel-Wolwertz found that the specific activity of Chl *b* increased progressively with the light intensity. Light intensity was also shown to influence the Chl *a*/Chl *b* ratio (Virgin 1977).

The synthesis of Chl *b via* Chl *a* more cogently was proven by *in vitro* experiments than by the experiments done *in vivo*. *In vitro* conversions of Chl *a* to Chl *b* was first reported by Shlyk and Prudnikova (1967) in homogenates of green barley shoots. Before grinding the shoots were allowed to form labeled Chls from $^{14}CO_2$ in the light. The particles of the homogenate were stirred into a saturated solution of sucrose and incubated in the dark. During dark incubation the specific activity of Chl *b* increased 1.5 times. A summary of the voluminous work performed about Chl *b* synthesis is given by Shlyk (1971).

Ellsworth et al. 1970 used unlabeled homogenates of soybean leaves and added pure [^{14}C]-Chl *a* in combination with acetone, tween and $NADP^+$. Unequivocally the results showed appreciable label in Chl *b* which was only partly due to transphytylation. Labeling of Chl *b* without $NADP^+$ was poor. The reaction proceded in the dark.

In homogenates Chl *b* is not only formed from Chl *a* but also from its early precursor ALA. This was demonstrated by Rebeiz and Castelfranco (1971 a) in homogenates of cucumber cotyledons substituted with ^{14}C ALA. Chl *b* was only formed in homogenates from illuminated cotyledons. A system from greening potato peels synthesizing Chl *b* was described by Ramaswamy and Nair (1974).

Spectral data also were thought to be useful in the discussion about Chl *b* synthesis. By reasons of identical action spectra for the formation of Chl *a* and Chl *b* Ogawa et al. (1973) assumed that Chl *b* is derived from Chl *a*. In contrast to Ogawa et al. (1973) who found a maximum of Chl *b* synthesis in light of about 650 nm, the red absorption peak of PChl(id), Virgin (1977) observed maximum Chl *b* synthesis at about 670 nm. He guessed that the formation of Chl *b* could require an photoreceptor besides PChl(id) (or phytochrome), possibly C_{672} not going through the Shibata shift. This Chl *a* form also might be the immediate precursor of Chl *b*.

Chl *b* synthesis is phytochrome-dependent in so far as a red pulse given 12 hours prior to the onset of white light eliminates the lagging of Chl *b* synthesis behind Chl *a* synthesis (Oelze-Karow and Mohr 1978). It has been shown that Chl *b* is formed at the expense of Chl *a* (Oelze-Karow and Mohr 1978).

Rüdiger et al. (1976) considered the possibility that the transition from Chl *a* to Chl *b* could take place on the stage of Chl *a* geranylgeranyl esters.

Virgin's assumption that the formation of Chl *b* requires an additional light step, however, contradicts the finding that in vitro Chl *b* synthesis takes place in the dark.

Although by isolation never Chlid *b* (except due to the action of chlorophyllase), or a PChlid *b* was found, Kranz (1973) interpreted *in vivo* spectral data as if there were such substances.

4. Location and Site of Synthesis of Chlorophyll Forming Enzymes

Carell and Kahn (1964) had demonstrated that preparations of isolated chloroplasts from *Euglena* and spinach synthesized Uro, Copro, and Proto from ALA. PBG was a substrate, if the plastids were broken; PBG was assumed not to penetrate the plastid envelope. Mg-porphyrins, PChlid and Chl were formed from ALA by isolated greening etioplasts of cucumber cotyledons (Rebeiz and Castelfranco 1971 a, b, Rebeiz et al. 1973) or leaves (Wellburn and Wellburn 1971). According to these findings the series of enzymes converting ALA into Chl is located in the plastids. When the thylacoid fragments were removed from lysed plastid preparations only porphyrins were formed (Rebeiz et al. 1972). Obviously enzymes catalysing later steps in Chl biosynthesis are associated with the lamellar system of the plastids, e. g. Chlase is known to be in a complex with such structures (see part 3.7) and S-adenosyl

Mg-Proto methyltransferase was detected in a chromatophore fraction of *Rhodopseudomonas* (Gibson et al. 1963) and mainly associated with a particulate plastid fraction of *Euglena* (Ebbon and Tait 1969).

Similar conclusions can be drawn from the work of Radmer and Bogorad (1967). From recent experiments of Griffith (1978) and Rüdiger et al. (1977) it is obvious that also NADPH-PChlideoxidoreductase and the CHlid esterifying system are membrane-bound. A study of Smith and Rebeiz (1979) on the intrachloroplastic distribution of intermediates on the pathway to Chl corroborates these earlier findings.

There are indications that in etioplasts PChlid synthesis and PChlid reduction are rather located on prothylakoids than within the prolammellar body itself (Lütz and Klein 1979).

On the other hand, enzymes at the beginning of the Chl biosynthetic chain, e.g. ALAD (Schneider 1970b, Shibata and Ochiai 1975) are readily soluble and care must be taken that these enzymes do not leak out from the plastids during plastid isolation. ALAD and PBGase are present in plastids on very early stages of plastid development, in proplastids and amyloplasts (Schneider 1971b). In addition, Chlase has been detected in cells only containing this type of plastids (Beisenherz and Schneider 1974, see also Holden 1963). Due to lacking *in vitro* assay methods for the enzyme(s) synthesizing ALA, very little is known about the location of ALAE in the plant cell.

Until recently there were only a few indications that ALAE might also be located in the plastids. Wellburn and Wellburn (1971) have shown that labeled glycine and succinate were incorporated into Chl of isolated greening etioplasts. (With regard to the enzymes synthesizing ALA, see 3.1.). Evidence for the location of ALAE within the plastid also comes from time courses for the ability of developing *Heliantus* cotyledons to synthesize ALA which parallel time courses for ALAD and PBGase activity (Schneider 1976). In contrast, the time course of the activity of the mitochondrial enzyme succinyl-CoA synthetase was not correlated with the time courses of plastid enzymes (Schneider 1973a). Fanica-Gaignier and Clément-Métral (1975) found one of the two forms of ALAS of *Rhodopseudomonas spheroides* associated with the chromophores.

Very recently, however, Kannangara and Gough (1977a) and Weinstein and Castelfranco (1978) have shown ALA to be formed in isolated plastids. Even chloroplast stroma preparations were able to form ALA from glutamate (Gough and Kannangara 1977b).

In animal cells, ALAS is located in the mitochondria, the following enzymes through Urogen decarboxylase in the cytoplasm (see Granick 1967). In yeast the distribution of ALAS between mitochondria and cytoplasm depends on the physiological stage of the cells (Porra et al. 1972).

Although the enzymes for Chl synthesis are located in the plastids, they are synthesized in the cytoplasm. This was shown for the enzymes ALAE, ALAD, PBGase, and Chlase in greening tissue cultures of tobacco (Beisenherz and Schneider 1974, Schneider and Beisenherz 1974). Greening tobacco tissue proved to be an excellent system for inhibitor experiments. By chloramphenicol, cycloheximide, and rifamycin SV the activities of ribulose diphosphate carboxylase (RuDPC) and NADP dependent glyceraldehyde dehydrogenase (NADP-GDH) (both located in the plastids) were influenced as expected according to the known sites of their synthesis. RuDPCO, built of subunits synthesized in the cytoplasm and the chloroplasts, was inhibited by all the inhibitors, but NADP-GDH, synthesized by the cytoplasm only, was inhibited by cycloheximide. The chlorophyll forming enzymes ALAE, ALAD, PBGase, and Chlase behaved like NADP-GDH. Consequently these enzymes are synthesized at 80 S ribosomes in the cytoplasm. Slight inhibition of overall Chl biosynthesis by rifamycin SV, despite of lacking enzyme inhibition, points to the fact that proper development of chloroplast structures is also a requirement in Chl biosynthesis besides enzyme synthesis.

The pattern of effects of streptomycin on Chl and RuDPC snythesis in *Euglena* (Diamond 1976) allows a similar interpretation, although this conclusion is not drawn by the author. Findings corresponding to those from tobacco tissue cultures have also been reported from a variety of other plant systems.

E. g. ALAD was tested in bean leaves (Sluiters-Scholten et al. 1973), radish (Shibata and Ochiai 1976), and *Euglena* (Hovenkamp-Obbema and Stegwee 1974), Chlase in *Chlorella* (Ganoza and McFeeters 1976), ALAE in bean and maize leaves (Klein et al. 1975), and S-adenosyl Mg-Proto methyl transferase in *Euglena* (Ebbon and Tait 1969) and wheat (Ellsworth and St. Pierre 1976). Even the light-harvesting Chl *a*/*b* protein is synthesized on 80 s ribosomes (Machold and Aurich 1972). Recently it has been shown (Sundqvist 1978) that the red light-enhanced PChlide accumulation depended on protein synthesis on cytoplasmic ribosomes. Results obtained with PBGase of bean leaves were also in agreement with the general pattern outlined (Bianchi and Stegwee 1979). As for pigment formation in plastids, the state of affairs in *Acetabularia* seems to be the same as in other plants (compare Moore and Shepard 1978).

It has also been reported (Schwartzbach and Schiff 1979) that the light-driven incorporation of leucine into plastid proteins does not depend on plastid DNA. Thus, it seems improbable that the photoreceptor machinery is coded by the plastid DNA. It may be assumed that the sites of primary light response are most likely located in the cytoplasm (Schneider and Beisenherz 1974), provided there is no complicated messenger system between plastid and nucleus.

In contrast to the latter findings a light-dependent increase of ALAD activity in isolated *Avena* etioplasts was ascribed to a *de novo* synthesis at 70 S ribosomes (Hampp and Ziegler 1975).

There was also a report that Chlase from bean leaves was synthesized on chloroplastic ribosomes (Moll et al. 1978).

An increase in the activities of ALAD and PBGase under the influence of CAP has been reported by Averina et al. (1977) from experiments with barley leaves. These authors assume that enzyme synthesis occurs in both the plastid and the cytoplasm, although synthesis in the cytoplasm is thought to dominate.

In another type of experiment Feierabend (1977), however, clearly showed that the enzymes of Chl synthesis leading from ALA to PChlid are synthesized at 80 S ribosomes. Rye seedlings deficient of 70 S ribosomes, due to growth at 32 °C, still formed PChlid when fed with ALA.

Nothing is known about whether 80 S ribosomes, forming enzymes for Chl synthesis, are functionally associated with the chloroplast envelope (see Laulhere and Dorne 1977). Possibilities of how a transfer of enzymes could take place, have been reviewed by Chua and Schmidt (1979) and Gibbs (1979). A key to the understanding of the interdependences of syntheses occurring in the cytoplasm and the plastid may be the finding that chloroplast DNA codes for tRNA from cytoplasmic polyribosomes (McCrea and Hershberger 1978).

5. The Regulation of Chlorophyll Biosynthesis

Chlorophyll accumulation has always been found to be paralleled by plastid growth, changes in plastid structure, and synthesis of nucleic acids, proteins, and lipids in the plastids (for summaries, e.g. Bogorad 1967; Rosinski and Rosen 1972; Henningsen and Boynton 1974; Siddell and Ellis 1977; Kung 1977; Tevini and Lichtenthaler eds. 1977; Mohr 1977; Kasemir 1979).

The synchrony of this variety of processes has made it difficult to separate specific controlls of the greening process from other morphogenetic processes driven by light. However, due to intensive work in recent years, the general pattern of interactions of regulative mechanisms can be outlined. Knowing the general pattern, problems concerning specific controlls of Chl biosynthesis can be formulated more precisely than before. Schemes for the regulation of Chl biosynthesis have been published (e.g. see Schneider 1973c, 1975, Mohr et al. 1974, Bogorad 1976). The role of ALA in plastid development has recently been reviewed by Beale (1978).

Most of the information about regulative processes in Chl biosynthesis comes from experiments with etiolated plants. This developmental stage is characteristic for a short phase during the germination of seeds but not for the greening of leaves and other parts of plants under natural conditions. Etiolated plants have experimental advantages with respect to pigment analysis and rapidly detectable physiological changes. Investigations on Chl biosynthesis in non-etiolated plants, however, would contribute to more comprehensive information.

First reports corroborated by later work (e.g. Withrow et al. 1956, Gassman and Bogorad 1967a, b, Süzer and Sauer 1971) stated that Chl accumulation on illumination of etiolated leaves can roughly be divided into three phases. During the first phase, i.e. seconds after the onset of illumination a small amount of Chl is formed. The second phase is characterized by a lag phase in Chl accumulation lasting from seconds to hours and this phase is finally followed by the third one, the phase of continuous Chl accumulation.

It was known, that during etiolement no Chl but traces of PChl(id) accumulated. PChl(id) amounted to 1/1000 to 1/200 of the Chl content found in illuminated green leaves. By light this PChl(id) was immediately converted to Chl in a mole-for-mole ratio (Smith 1948, Koski 1950). Furthermore (Koski et al. 1951) the light conversion of PChl(id) was found to be most effective at 445 and 650 nm. When it became possible to measure absorption spectra *in vivo* (Shibata 1957), it turned out that the action spectrum for the conversion of PChl(id) was almost identical with the *in vivo* absorption spectrum of PChl(id) (see also part 3.10). It became obvious that PChl(id) catalyzed its own conversion to Chl(id). The first phase in the greening process of etiolated leaves ends when all the PCh(id) present in a convertible form is converted.

The duration of the second phase in light mediated Chl biosynthesis (see e.g. Sisler and Klein 1963, Akoyunoglou 1970) and also the time needed for regeneration of PChl(id) in the dark after light treatment converting it to Chl (Shibata 1957, Butler 1961, Akoyunoglou and Siegelmann 1968) were found to be influenced by the age of the plants; e.g. leaves of five days old bean plants scarcely had a lag phase, experiments with leaves from older plants showed lag phases of about two hours (Sisler and Klein 1963). But even longer lag phases, never followed by a high rate of Chl accumulation were reported and as easily can be observed with etiolated cereal plants, the tips of older etiolated leaves never turn green. The effects of age on PChlide regeneration, the regeneration rate, the amount of PChlide accumulated in the dark and the duration of the Shibata shift were reinvestigated by Ogawa et al. (1978).

To a certain amount the lag phase was influenced by nutritional conditions (Wolff and Price 1961, Eilam and Klein 1962, Klein et al. 1975), although sucrose did not show throughout enhancing effects (Nadler and Granick 1970, Schneider 1973b, Klein et al. 1975). These observations indicate that the effect of light on the greening process is modified by endogenous processes, i.e. by the actual developmental and physiological stage of the cells. Therefore, not unexpected, growth substances (Kirk 1970), water stress (Virgin 1965, Alberte et al. 1975), or ions like iron (Machold and Scholz 1969, Rosen et al. 1977) influencing the physiology of the cell exert strong effects on Chl metabolism. The role of phytohormones (cytokinins) in chloroplast development has been reviewed by Parthier (1979). In addition there are some recent papers on the influence of water stress and hormones on Chl biosynthesis (e.g. Bengtson et al. 1978, Duysen and Freeman 1978).

However, these agents are not at all specific. In most cases the site of action is not known and the observed effects are merely effects of higher order.

Experiments indicating more acurately the cause and origin of the lag phase were experiments analysing the effect of preillumination, i.e. illumination before a more or less extended dark period which is followed by light treatment preceding Chl accumulation measurements.

It was shown that especially red light was effective in eliminating or at least shortening the lag phase (Withrow et al. 1956, Virgin 1961, Augustinussen and Madsen 1965, Rudolph 1965). Several times it was observed that the red light effect could be cancelled by far red light applied immediately after the red light treatment (e.g. see Price and Klein 1961, Mitrakos 1961, Kasemir et al. 1973, Raven 1973, and authors quoted above). These findings were a strong evidence that phytochrome was involved in the process leading to a shorter lag phase.

Since phytochrome is known or at least postulated to influence protein synthesis *via* gene activation, it can be assumed that the lag phase is in some way connected with lacking enzyme or protein synthesis. This assumption was corroborated by numerous reports about the influence of inhibitors of transcription and translation on Chl biosynthesis. In most cases the applications of such substances caused inhibition of Chl accumulation (Gassman and Bogorad 1967a, b, Kirk 1968, Nadler and Granick 1970, Goodenough and Levine 1971, Galling et al. 1973, Klein et al. 1975). Syntheses essential for Chl accumulation could no longer be induced by light. Knowing the multible connections of Chl accumulation with protein synthesis, these results are not surprising. Therefore only those experiments were pertinent which allowed to discriminate between very general effects and effects specifically due to enzymes of the Chl biosynthetic chain. In first attempts ALA, the early precursor of Chl, was applied to inhibited plants. The metabolization of ALA and the accumulating products can be assumed to be indicative for the site of lesion.

Feeding ALA to illuminated plants which were inhibited by inhibitors like chloramphenicol or cycloheximide (Nadler and Granick 1970) resulted in a resumption of Chl synthesis. The experiments indicated that all enzymes except the ALA synthesizing ones (ALAE) were present and that the light requiring synthesis of ALAE was arrested by the inhibitors.

Earlier experiments showing that in etiolated leaves PChl(id) levels considerably increased after ALA application (Granick 1959) had already inferred a regulation of Chl biosynthesis *via* the availability of ALA, *via* enzymes synthesizing ALA.

Since actinomycin D, known as an inhibitor of transcription, had scarcely any effect on Chl biosynthesis, Nadler and Granick (1970) postulated stable mRNA and light-activation of this RNA. Findings of Klein et al. (1975) support the suggestion of Nadler and Granick (1970) that RNA synthesis is not required for the synthesis of ALA during early hours of greening. However, these findings must not be generalized. Apel and Kloppstech (1978) have clearly demonstrated that mRNA coding for the light-harvesting Chl *a/b* protein is massively induced by light. Generally, the interpretation of inhibitor studies should be measured against the knowledge about the location and the site of synthesis of proteins essential in Chl biosynthesis (see part 4).

Testing the effect of chloramphenicol, puromycin and actinomycin in inhibitor experiments (Gassman and Bogorad 1967a, b) no influence on PChl(id) photoconversion was found, however, inhibition of PChl(id) regeneration. ALA restored the synthesis to a certain extent. Early steps in the Chl biosynthetic chain seemed to be missing.

In experiments with *Euglena* (Kirk 1968) ALA could scarcely overcome the blockage set by inhibitors. Kirk concluded that proteins at the end of the biosynthetic pathway rather than at the beginning were limiting Chl synthesis. Experiments with other algae led to similar interpretations (Richards and Nigon 1973, Oh-Hama and Hase 1975). PChl(id) holochrome or thylacoid protein were discussed by Kirk (1968).

Evaluating his results with bean leaves, Margulies (1967) came to the conclusion that it could not be decided whether protein synthesis was required for Chl synthesis or for synthesis in stoichiometric amounts of a protein needed for incorporation and stabilization of chlorophyll in lamelae. And as inferred by more recent studies, it seems that indeed both processes are essential for Chl accumulation (see below). However, comparing different algae and etiolated leaves, one should keep in mind the different modes of growth, cell division and plastid growth, on the one hand; and mere cell elongation if any and rearrangement of plastid structures, on the other hand.

Finally when data on activities of Chl forming enzymes during the greening process were available, it turned out, as already indicated by some of the ALA feeding experiments, that ALA synthesis in various higher plants only took place under illumination (Harel and Klein 1972, Schneider 1973b, 1974, Beale and Castelfranco 1974a, Masoner and Kasemir 1975).

Ensuing syntheses depended on light to a certain extent, but were performed also in darkness. A comparative study on the kinetics of ALAE, ALAD, PBGase and succinyl CoA synthetase in greening tissue cultures of tobacco showed that the role of ALAE in regulating Chl biosynthesis was only exceptional in greening etiolated leaves, but not in greening tissue cultures (Schneider 1971a, 1973a, b; 1976).

Further analysis of Chl biosynthesis in greening etiolated leaves revealed two light effects acting together. Although ALA formation was only detectable as long as plants were illuminated, the actual amounts of ALA synthesized during an assay depended on light treatment before the assay (Schneider 1976, Gehring et al. 1977). These results point toward an influence of light on the formation and on the activity of ALAE.

Blocking Chl accumulation by LA during a developmental phase which would have allowed continuous Chl accumulation resulted in a linear increase of ALA accumulation (Schneider 1973b, Masoner and Kasemir 1975). This finding means that a certain, constant amount of enzyme (ALAE) is present during the whole time of assay. Since in darkness ALA accumulation is almost immediately arrested (Schneider 1973b, Masoner and Kasemir 1976) – despite the enzyme present – it can be concluded that lacking ALA synthesis in the dark is caused by lacking enzyme activation by light.

Enzyme degradation as a cause for the breakdown of ALA biosynthesis in the dark could not be detected to a considerable extent in *Helianthus* cotyledons (Schneider 1976). However, Fluhr et al. (1975) reported ALAE activity of 10 days old, green maize leaves decaying with a half life of about 80 minutes after a 40 minutes lag period. These authors infer that at least in part Chl biosynthesis could be regulated *via* degradation and light-induced synthesis of ALAE, as it was indicated by experiments of several other investigators (e. g. Gassmann and Bogorad 1967a, b, Nadler and Granick 1970, Beale and Castelfranco 1974a).

Even if enzyme degradation occurs in the dark, all observations make clear that in addition to a light-mediated ALAE activation light regulates ALA and Chl biosynthesis *via* the amounts of enzyme protein. This conclusion is corroborated by other, more direct findings. E. g. ALAD and PBGase of *Helianthus* cotyledons which were cut from seedlings grown in D_2O contained more deuterium label when the plants had been cultivated under illumination and not in darkness. The incorporation of label can only be the result of a *de novo* synthesis of these enzyme proteins. Since the time course of ALAE activity during development and illumination paralleled the time courses of ALAD and PBGase activities *de novo* synthesis of ALAE protein may also be assumed to be the reason for a higher ALA accumulation rate.

Numerous experiments (for example see Mohr et al. 1974) have shown that the photoreceptor for this second light effect is phytochrome.

The increases in the activities of ALAD and PBGase in leaves during illumination have mostly been considerably less than the increase in the power to synthesize ALA (Schneider 1973b, 1976). This finding demonstrates the main role of ALAE in the regulation of Chl biosynthesis. Nevertheless increasing activities of the enzymes ensueing ALAE ensure that increasing amounts of ALA are drawn into the Chl biosynthetic pathway. As experiments with algae have shown, ALA accumulation may also be linked to photosynthetic processes (Richards and Nigon 1973, Oh-hama and Senger 1975, Salvador et al. 1976). One may assume that energy or products built up in this pathway enhance ALA and/or ALAE synthesis. Experiments with *Euglena* (Michel 1978) show that PChlide reformation and the synthesis of some unknown metabolite necessary for the photoreduction depend on respiration.

On the basis of our knowledge the lag phase in Chl accumulation can be explained mainly by lacking ALAE protein which is not formed or is degraded faster than it is formed. Formation and decay of enzyme protein are functions of the developmental state of the leaf or the cotyledon which are modified by light or even by circadian rhythmicity (Gehring et al. 1977). The phase of continous Chl accumulation begins when development and light maintain a constant level of ALAE protein. This

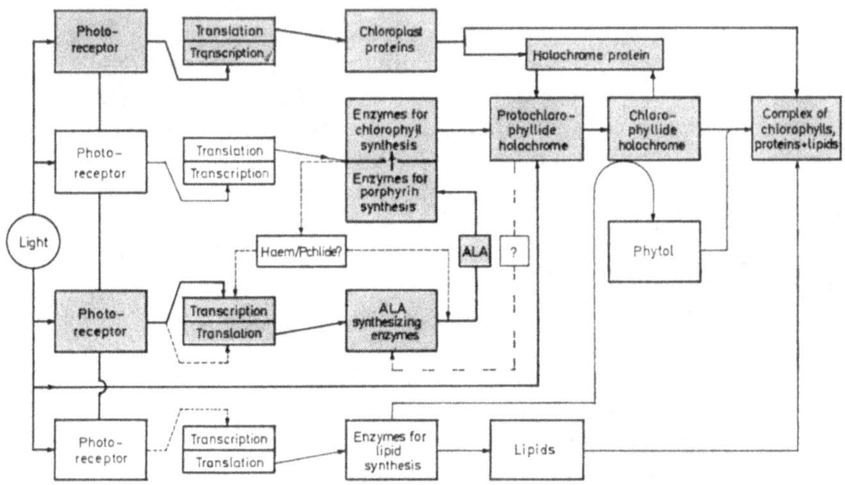

Fig. 10: Scheme on the regulation of chlorophyll biosynthesis. Biosynthetic pathways which are mainly controlled by light are indicated by heavy lines and dark boxes

statement does not exclude a synchronous induction of proteins which are essential for the formation of the Chl complexes in the thylacoids.

How dominant the developmental stage of a cell is, demonstrate experiments with greening cell cultures of tobacco (Schneider 1970a, 1971a, 1973a). These cells do not contain etioplasts but proplastids and amyloplasts (Bergmann and Berger 1966) which become green within a few days of illumination. Activities of ALAE, ALAD, PBGase, succinyl-CoA-synthetase, and chlorophyllase (see also Beisenherz and Schneider 1974) were detectable in dark grown cultures but no PChlid. The presence of ALAE activity in the dark was striking. However, ALA forming enzymes active in the dark were also found in plant organs like roots, tubers, or endosperm (Schneider 1973b). It seems that the greening process of these cells resembles that of potato tubers (Berger and Bergmann 1967) or roots (Björn 1967). Similarities may also be found with algae or bacteria. Results obtained with *Helianthus* cotyledons (Schneider 1976) point toward the possibility that dark activity of ALAE is arrested as soon as etioplasts are developping. Investigations on the pathways leading to ALA in the dark and in the light and on the compartmentation of these pathways would be pertinent to understanding this aspect of regulation (compare 3.1.).

When the tissue cells of tobacco were greening (Schneider 1970a, 1973a) activities (except succinyl-CoA-synthetase a mitochondrial enzyme) increased without a lag phase. The amounts of Chl formed were correlated with the enzyme activities. It can be assumed that plastid structures, enzymes, and Chl form synchronously. Therefore application of ALA did not enhance Chl biosynthesis; in contrast, such a treatment resulted in an inhibition of growth and Chl accumulation (Schneider, 1973a).

Discussing results obtained with maize and bean leaves, Klein et al. (1975) also point out that there are considerable differences in respect to requirements for ALA accumulation in different plants. LA also seems to inhibit processes other than PBG formation (Stobart and Hendry 1978).

Although light perception by PChl and phytochrome is certainly the most important one, other photoreceptors may be or are involved in the regulation of Chl biosynthesis. Blue light was found to be more effective than red light in promoting Chl accumulation in tobacco cells (Bergmann and Berger 1966), potato tuber cells (Berger and Bergmann 1967), and in excised roots (Björn 1967, Richter 1969). Egan et al. (1975) reported that in *Euglena* a pigment absorbing in the blue and less in the yellow light controlled Chl synthesis and in *Cyanidium caldarium*, a pigment system with action maxima at 440, 575 and 645 nm regulated chlorophyll biosynthesis as well as phycobiliprotein and ALA formation (Schneider and Bogorad 1978). This system seems to be composed at least of two photoreceptors as can be seen from the action maxima of different mutants of *Cyanidium*. In *Chlorella protothecoides* maxima of ALA and Chl formation were at 370, 440 and 480 nm (Oh-hama and Senger 1978).

The spectral response curve given by Miller et al. (1979) for ALA synthesis in barley seedlings has maxima at about 650 and 440 nm. However, it is not quite clear whether it is a spectrum for ALAE induction or a spectrum for both induction and activation. The same may be true for the data communicated by Möldgen (1976).

Although new insights in Chl regulating processes were made possible analysing enzyme activities two main problems are yet unsolved: The reaction chain connecting photoreceptor and enzyme formation, a problem beyond the scope of the present discussion, and the mechanism of rapid activation and inhibition of ALAE. In plants without a lag phase synthesis of ALA starts as soon as PChl(id) is reduced by light. If plants are returned to darkness they begin again to accumulate PChl(id) and stop synthesis at about the same amount as before. Anyhow there must be a tight connection between these two reactions. Even the spectrum for light induced ALAE activation seems to resemble the spectrum for photoconversion of PChl(id) (Möldgen 1976). An analysis of the connecting mechanism has not been conclusive so far because of the lack of purified samples of ALAE. However, possibilities concerning connecting mechanisms have been discussed by several authors. Generally it is assumed that ALAE is blocked by precursors of Chl biosynthesis and their derivatives if PChl(id) is not converted to Chl(id). However, there is no information how those substances really act on ALAE in higher plants. There is only the less pertinent report that ALAS was inhibited by Chl(id) (Ramaswamy and Nair 1973) which cannot be a natural inhibitor by reason of its order in the sequence of biosynthetic steps, if we do not assume that ALAE activity is regulated by a biosynthetic step after PChlid reduction. An additional difficulty in interpreting this result is the strong evidence that ALAS plays only a minor role in providing ALA in higher plants (see 3.1). From *Rhodopseudomonas spheroides* it is known that heam and Proto inhibit ALAS (see Lascelles 1968).

Providing dark grown barley leaves with iron chelators resulted in their forming of large amounts of Mg-Proto-ME, Mg-Proto, Proto, and PChl(id) (Duggan and Gassman 1974, Ryberg and Sundquist 1976). Because the chelators arrest the formation of Fe-Proto, the haem content drops after application of the chelators, Duggan and Gassman (1974) guessed that haem or haem protein could be the inhibitors. Some difficulties may exist in interpreting the results because the formation of Proto from Coprogen (Hsu and Miller 1965, Machold and Scholz 1969) and the formation of Chl from Mg-Proto-ME (Jones 1963) are also assumed to be dependent on iron. Gaugh (1972) using mutants of barley which on feeding ALA accumulated various intermediates suggested that any tetrapyrrolic intermediate could directly or indirectly inhibit Chl biosynthesis.

PChl(id) holochrome and haem were considered candidates inhibiting ALAE (Nadler and Granick 1970) and Mohr et al. (1974) assumed allosteric connections between PChl(id) and ALA synthesizing enzymes. Recently Castelfranco and Jones (1975) published a scheme showing protohaem as inhibitor and repressor of ALAE. Rapid turnover of protohaem during Chl accumulation is included as additional regulating process. However, since the concentrations of protohaem in the cell were about the same in darkness and under illumination when Chl was formed, Chl synthesis cannot be regulated by changing protohaem concentrations.

Since in experiments with greening maize leaves (Fluhr et al. 1975) ALA synthesis was not arrested when the PChl(ide) pool was saturated, in this case PChl(id) can also be excluded as direct inhibitor of ALAE. The accumulation curves for ALA

in barley seedlings (Miller et al. 1979) allow a similar interpretation, provided that PChlid accumulates despite the blockage by LA.
Hendry and Stobart (1978 a) found that haemin added to barley leaves inhibited the regeneration of PChlid, the conversion of exogenousely applied ALA to PChlid, and decreased the incorporation of label from glycollate, glycine and glutamate into ALA. Another outcome of this study was that protohaem breakdown was under light control while protohaem synthesis was not.

It is easily seen through the examples that despite all attempts the problem of light-activation and dark-inhibition of ALAE is not solved.

As Fluhr et al. (1975), Bogorad (1976) pointed toward the possibility of an enzyme system mediating between PChl(id) and ALAE. He discussed a regulation of Chl biosynthesis *via* a coupled oxido-reduction system between PChl(id) reductase and a step in ALA synthesis. An oxidative step maintaining ALAE activity were the specific source of electrons for the reduction of PChlid.

Besides inhibition by accumulating intermediates or by a tight functional or spatial connection between PChlid holochrome and ALAE, repression of porphyrin synthesizing enzymes may be taken into account (Granick and Sassa 1971; see also Castelfranco and Jones 1975) although there are no direct experimental data from higher plants supporting this view. Data mainly come from animal systems. The inhibitory effect of ALA observed in tissue cultures of tobacco (Schneider 1973a) or in *Euglena* (Richards and Nigon 1973) might be connected with repression. Experiments with *Euglena* allow Schwartzbach and Schiff (1979) to suggest that ALA is a transcriptional rather than a translational inhibitor. Discussing repression we might think of intermediate products of Chl biosynthesis acting as corepressor but also of an aporepressor derived from plastid proteins not located because of lacking Chl. A model for a mechanism of ALAE repression had to solve problems connected with distance, solubility and diffusion. Chl synthesizing enzymes are coded in the nucleus, synthesized in the cytoplasm (Beisenherz and Schneider 1974, see part 4) and located in the plastid.

Besides in green plants, regulation of Chl biosynthesis was intensely studied in the photosynthetic bacterium *Rhodopseudomonas spheroides* (*Athiorhodaceae*). BChl is formed when this bacterium is growing photosynthetically, i.e. anaerobically in light, but is scarcely detectable in aerated cultures grown in darkness. Small amounts of pigments are formed by cells kept in darkness under reduced oxygen tension. Activities of porphyrin and BChl synthesizing enzymes are found much higher in extracts of anaerobic cells (Lascelles 1959, Davis et al. 1973). As in tobacco tissue (Schneider 1973c) the amount of Chl seems to parallel the activity of these enzymes. However, the involvement of two main factors regulating Chl biosynthesis, light and oxygen, makes the bacterial system even more complex than the system in higher plants, not quite understood at present.

If we assume that light induces the synthesis of Chl forming enzymes in a similar manner as it does in higher plants, only the action of oxygen has to be discussed.

The inhibitory effect of haem on ALAS of *Rhodopseudomonas* is well established (Lascelles 1968). The enzyme has been purified several times (see 3.1). Some of ALAS properties may have regulative significance. Two different forms of ALAS were found, one associated with the chromophores the other located in the cytoplasm. Only the chromatophore enzyme is reported to be light inducible. Both enzymes are inhibited by iron and form inactive complexes with ATP. (Fanica-Gaigner and Clément-Métral 1973a–c, Clément-Métral and Fanica-Gaigner 1975).

If extracts of anaerobically grown *Rhodopseudomonas spheroides* and similar extracts aerated shortly before harvest were freshly prepared the activity in both extracts was low (Marriott et al. 1969). During an hour storage of the extracts ALAS activity in the extracts from the cells grown wholly anaerobically increased several fold. Activity in extracts of oxygenated cells could be increased transferring material from unoxygenated cells (Neuberger et al. 1973a, b). This finding indicated that ALAS was present but inactive in oxygenated cells. The activating material was found to consist of cystintrisulphide, gluthationtrisulphide, and polysulphondisulphonate (Sandy et al. 1975). These substances disappeared quickly when the cells were oxygenated. Two enzymes are responsible for the synthesis and degradation of cystin trisulphide: cystathionase and rhodanese. The amount of cystine trisulphide, however, is assumed to be controlled on the level of substrates for these enzymes rather than by changes in the amounts of these enzymes (de Xifra et al. 1976). At present the mechanisms of oxygen action on tri- and polysulphides is under investigation.

Since other enzymes and proteins of *Rhodopseudomonas* are also quickly influenced by aeration and deaeration (Takemoto and Lascelles 1973) the activating system may play a central role in regulative processes in *Rh. spheroides*. This being so, the role of light action has to be rechecked.

In addition to the regulative mechanisms discussed, the scheme for the regulation of Chl synthesis (Fig. 7) also comprises biosynthetic pathways essential for Chl accumulation but not directly involved in Chl biosynthesis. Like Chl biosynthesis the formation of the pathways for plastid protein and lipid synthesis depend on development and illumination (see quotations at the beginning of this part). Since Chl *b* derives from Chl *a* the formation of Chl *b* is dependent on the same photoreceptors as the formation of Chl *a*. However, it was reported that the step from a Chl *a* form to Chl *b* might be mediated by an additional photoreceptor absorbing at about 670 nm (Virgin 1977 see 3.14).

By what mechanisms phytol synthesis is regulated is not known. However, it is very probable that phytol is only synthesized from its precursors if the latter are bound to Chlid. Phytol synthesis is arrested in darkness and increases with increasing light intensities; the mole-for-mole ratio in respect to Chlid *a* remains close to one (Steffens et al. 1976).

By the balanced interaction of all regulative systems finally the Chl-protein-lipid-complex of the thylacoids, the photosynthetic unit is formed. There are hypotheses how the balance may be achieved, the thorough demonstration of their validity, however, requires further experimental efforts.

References

ACKER, S., and J. DURANTON, 1975: Indépendence des formes spectrales de chlorophylle *a* et des holochromes chlorophylliens. Biochim. Biophys. Acta **387**, 279–287.

AKOGUNOGLOU, G., 1970: The Effect of Age on Phytochrome-Mediated Chlorophyll Formation in Dark grown Bean leaves. Physiol. Plant. **23**, 29–37.

AKOGUNOGLOU, G., and G. MICHALOPOULOS, 1971: The relation between the phytylation and the 682 to 672 nm shift in vivo of chlorophyll *a*. Physiol. Plant. **25**, 324–329.

AKOGUNOGLOU, G., and H. W. SIEGELMANN, 1968: Protochlorophyllide resynthesis in dark grown bean leaves. Plant Physiol. **43**, 66–68.

ALBERTE, S. R., E. L. FICUS and A. W. NAYLOR, 1975: The effects of water stress on the development of the photosynthetic apparatus in greening leaves. Plant Physiol. **55**, 317–321.

AL-HAZIMI, H. M. G., A. H. JACKSON, and D. J. RYDER, 1976: Specificity of coproporphyrinogenoxidase: Conversion of coproporphyrinogen IV into protoporphyrin XIII. Chem. Comm. 188–189.

ALSCHER, R. G., S. P. HAWKES and K. SAUER, 1976: The association of protein synthesis with protochlorophyllide holochrome regeneration in dark grown barley leaves. Biochem. Biophys. Res. Comm. **73**, 240–247.

AOKI, Y., O. WADA, G. URATA, F. TAKAKU and K. NAKAO, 1971: Purification and some properties of δ-aminolevulinate (ALA) synthetase in rabbit reticuloctes. Biochem. Biophys. Res. Comm. **42**, 568–575.

APEL, K., and K. KLOPPSTECH, 1978: The plastid membranes of barley (*Hordeum vulgare*). Light-induced appearance of mRNA coding for the apoprotein of the lightharvesting chlorophyll *a*/*b* protein. Eur. J. Biochem. **85**, 581–588.

ARDAO, C., and B. VENNESTAND, 1960: Chlorophyllase activity of spinach chloroplastin. Plant Physiol. **35**, 368–371.

AUGUSTINUSSEN, E., 1964: On the formation of chlorophyll *a* and chlorophyll *b* in etiolated leaves. Physiol. Plant. **17**, 403–406.

AUGUSTINUSSEN, E., and A. MADSEN, 1965: Regeneration of Protochlorophyll in Etioletad Barley Seedlings following different light treatments. Physiol. Plant. **18**, 828–837.

AVERINA, N. G., N. N. POLIKARPOVA, and A. S. SHLYK, 1977: Influence of chloramphenicol and cycloheximide on the synthesis of δ-aminolevulinate dehydratase in green and greening barley seedlings. Biochemistry (USSR) **42**, 1627–1633.

AXELSSON, L., 1976a: The photostability of different chlorophyll forms in dark grown leaves of wheat: I. Stability to high intensity red light forms appearing after photoreduction of protochlorophyllide. Physiol. Plant. **38**, 327–332.

AXELSSON, L., 1976b: The photostability of different chlorophyll forms in dark grown leaves of wheat. II. Reaction kinetics for the photodecomposition of the 684-form and 673-form. Physiol. Plant. **38**, 333–336.

BACON, M. F., and M. HOLDEN, 1970: Chlorophyllase of sugar-beet leaves. Phytochemistry **9**, 115–125.

BARNARD, G. F., R. ITOH, L. H. HOHBERGER, and D. SHEMIN, 1977: Mechanism of porphobilinogen synthase. Possible role of essential thiol groups. J. Biol. Chem. **252**, 8964–8974.

BARREIRO, O. C. L. DE, 1967: δ-Aminolaevulinate hydro-lyase from yeast. Isolation and purification. Biochim. Biophys. Acta **139**, 479–486.

BARREIRO, O. L. C. DE, 1975: Effect of pyridoxal phosphate cysteine on δ-aminolaevulinate synthetase and dehydratase in Phytochemistry **14**, 2165–2168.

BATLLE, A. M. DEL C., A. BENSON, and C. RIMINGTON, 1965: Purification and properties of coproporphyrinogenase. Biochem. **97**, 731.

BATLLE, A. M., DEL C., and A. M. STELLA, 1978: Delta-aminolaevulinate dehydratase: its mechanism of action. Int. J. Biochem. **9**, 865–870.

BATLLE, A. M. DEL C., A. M. STELLA, A. M. FERRAMOLA, Y. SOPENA, E. A. W. DE XIFRA, and H. A. SANCOVICH, 1978a: Porphyrin biosynthesis. Immobilized enzymes and ligands. X. A novel approach to the study of the relationship between the quarternary structure of aminolaevulinate dehydratase and its activity. Int. J. Biochem. **9**, 401–406.

BATLLE, A. M. DEL C., E. A. W. DE XIFRA, and A. M. STELLA, 1978b: Prophyrin biosynthesis. Immobilized enzymes and ligands XI. A wide variety of applications. Int. J. Biochem. **9**, 407–419.

BATTERSBY, A. R., J. BALDAS, J. COLLINS, D. H. GRAGSON, K. J. JAMES and E. MCDONALD, 1972: Mechanism of biosynthesis of the vinyl groups of protoporphyrin IX. J. chem. Soc. chem. Comm. 1265–1266.

BATTERSBY, A. R., G. L. HODGSON, E. HUNT, E. MCDONALD and J. SAUNDERS, 1976b: Biosynthesis of porphyrin and related macrocycles. Part VI. Nature of the rearrangement process leading to the natural type III porphyrins. J. Chem. Soc. Perkin Trans. I **3**, 273–282.

BATTERSBY, A. R., E. HUNT, E. MCDONALD, J. B. PAINE III, and J. SAUNDERS, 1976c: Biosynthesis of porphyrins and related macrocycles: Enzymic decarboxylation of uroporphyrinogen III.: Structure of an intermediate phyriaporphyrinogen III., and synthesis of the corresponding porphyrin and of two isomeric porphyrins. J. Chem. Soc. Perkin Trans. I. **9**, 1008–1018.

BATTERSBY, A. R., E. MCDONALD, J. R. REDFERN, J. STAUNTON and R. WIGHMAN, 1976a: Biosynthesis of porphyrins and related macrocycles: V. Structural integrity of the type III porphyrinogen macrocycle in an active biolocial system. Studies on the aromatisation of protoporphyrinogen IX. J. Chem. Soc. Perkin Trans. I. **3**, 266–273.

BATTERSBY, A. R., C. J. R. FOOKES, E. MCDONALD, and M. J. MEEGAN, 1978: Biosynthesis of type-III porphyrin: Proof of intact enzymic conversion of the head-to-tail bilane into Urogen-III by intramolecular rearrangement. J. Chem. Soc. Chem. Commun. **5**, 185–186.

BATTERSBY, A. R., C. J. R. FOOKES, G. W. J. MATCHAM, and E. MCDONALD, 1979a: Order of assembly of the four pyrrole rings during biosynthesis of the natural porphyrins. J. Chem. Soc. Chem. Commun. **12**, 539–541.

BATTERSBY, A. R., C. J. R. FOOKES, G. W. J. MATCHAM, E. MCDONALD, and K. E. GUSTAFSON-POTTER, 1979b: Biosynthesis of natural porphyrins. Experiments on the ring-closure steps and with hydroxyanalogue of porphobilinogen. J. Chem. Soc. Chem. Commun. **7**, 316–319.

BAUER, S., and H. W. SIEGELMAN, 1972: Photoconversion of chlorophyllide 684 to chlorophyllide 678. FEBS Lett. **20**, 352–354.

BEALE, S. I., 1970: The biosynthesis of δ-aminolaevulinic acid in Chlorella. Plant. Physiol. **45**, 505–506.

BEALE, S. I., 1971: Studies on the biosynthesis and metabolism of δ-aminolaevulinic acid in *Chlorella*. Plant. Physiol. **48**, 316–319.

BEALE, S. I., 1976: The biosynthesis of δ-aminolaevulinic acid in plants. Phil. Trans. R. Soc. Lond. B **273**, 99–108.

BEALE, S. I., and P. A. CASTELFRANCO, 1973: ^{14}C-incorporation from exogenous compounds into-aminolaevulinic acid by greening cucumber cotyledons. Biochem. Biophys. Res. Comm. **52**, 143–149 (printer's error corrected in vol. 53).

BEALE, S. I., and P. A. CASTELFRANCO, 1974a: The biosynthesis of δ-aminolaevulinic acid in plants. I. Accumulation of δ-aminolaevulinic acid in greening plant tissues. Plant. Physiol. **53**, 291–296.

BEALE, S. I., and C. A. CASTELFRANCO, 1974b: The biosynthesis of δ-aminolaevulinic acid in higher plants. II. Formation of ^{14}C-δ-aminolaevulinic acid from labelled precursors in greening plant tissues. Plant. Physiol. **53**, 297–303.

BEALE, S. I., S. P. GAUGH and S. GRANICK, 1975: The biosynthesis of δ-aminolaevulinic acid from the intact carbon skeleton of glutamic acid in greening barley. Proc. Natu. Acad. Sci. U.S.A. **72**, 2719–2723.

BEALE, S. I., 1978: δ-Aminolevulinic acid in plants: Its biosynthesis, regulation and role in plastid development. Ann. Rev. Plant. Physiol. **29**, 95–120.

BEISENHERZ, W. W., and HJ. A. W. SCHNEIDER, 1974: Syntheseort von Enzymen der Chlorophyllbiosynthese. Ber. Deutsch. Bot. Ges. **87**, 161–166.

BELANGER, F. C., and C. A. REBEIZ, 1979: Chloroplast biogenesis XXVII. Detection of novel chlorophyll and chlorophyll precursors in higher plants. Biochem. Biophys. Res. Commun. **88**, 365–372.

BENGTSON, C., B. KLOCKARE, R. KLOCKARE, S. LARSEN, and C. SUNDQVIST, 1978: The after-effect of waterstress on chlorophyll formation during greening and the levels of abscisic acid and proline in dark grown wheat seedlings. Physiol. Plant **43**, 205–212.

BERGER, C., and L. BERGMANN, 1967: Farblicht und Plastidendifferenzierung in Speichergewebe von *Solanum tuberosum* L. Z. Pflanzenphysiologie **56**, 439–445.

BERGMANN, L., and C. BERGER, 1966: Farblicht und Plastidendifferenzierung in Zellkulturen von *Nicotiana tabacum* var. «Samsun». Planta **69**, 58–69.

BERIDZE, T. G., M. S. ODINTSOVA, N. A. CHEKASHINA and N. M. SISSAKIAN, 1966: The Effect of nucleic acid synthesis inhibitors on the chlorophyll formation of etiolated bean leaves. Biochem. Biophys. Res. Comm. **23**, 683–690.

BIANCHI, A., and D. STEGWEE, 1979: Porphobilinogenase in greening leaves of *Phaseolus vulgaris*. Z. Pflanzenphysiol. **91**, 377–383.

BJÖRN, L. O., 1967: The light requirement for different steps in the development of chloroplasts in excised wheat roots. Physiol. Plant. **29**, 483–499.

BJÖRN, L. O., 1976: The state of protochlorophyll and chlorophyll in corn roots. Physiol. Plant. **37**, 183–184.

BOARDMAN, N. K., 1962: Studies on a protochlorophyll-Protein complex. I. Purification and molecular weight determinations. Biochim. Biophys. Acta **62**, 63–79.

BOARDMAN, N. K., 1966: Protochlorophyll In: L. P. Vernon and R. S. Seely. The chlorophylls. Academic Press, New York pp. 437–479.

BOARDMAN, N. K., 1967: Chloroplast structure and development. In: A. San Pietro, F. A. Greer and T. J. Army (eds.), Harvesting the Sun. Academic press. New York pp. 211–230.

BÖGER, P., 1965: Chlorophyllase of *Chlorella vulgaris* Beijerinck. Phytochem. **4**, 435–443.

BOGDANOVIC, M., 1973a: Chlorophyll formation in the dark. I. Chlorophyll in pine seedlings. Physiol. Plant. **29**, 17–18.

BOGDANOVIČ, M., 1973b: Chlorophyll formation in the dark. II. Chlorphyll in wheat leaves transplanted to pine megagametophytes. Physiol. Plant. **29**, 19–21.

BOGORAD, L., 1950: Factors associated with the synthesis of chlorophyll in the dark in seedlings of *Pinus jeffreyi*. Bot. Gaz. **111**, 221–241.

BOGORAD, L., 1955: Intermediates in the biosynthesis of porphyrins from porphyrinogen. Science **121**, 878–879.

BOGORAD, L., 1958a: The enzymatic synthesis of porphyrins from porphobilinogen. I. Uroporphyrin I. J. Biol. Chem. **233**, 501–505.

BOGORAD, L., 1958b: The enzymatic synthesis of porphyrins from porphobilinogen. II. Uroporphyrin III. J. Biol. Chem. **233**, 510–515.

BOGORAD, L., 1958c: The enzymatic synthesis of porphyrins from porphobilinogen. III. Uroporphyrinogens as intermediates. J. Biol. Chem. **233**, 516–519.

BOGORAD, L., 1967: The role of cytoplasmic units. Control mechanism in plastid development. Develop. Biol. Suppl. **1**, 1–37.

BOGORAD, L., 1976: Ch,orophyll Biosynthesis. In: T. W. Goodwin: Chemistry and Biochemistry of plant pigments. Acad. Press. London, pp. 64–148.

BOGORAD, L., and S. GRANICK, 1953: The enzymatic synthesis of porphyrins from porphobilinogen. Proc. Nat. Acad. Cci. USA **39**, 1176–1188.

BOGORAD, L., and A. JACOBSON, 1964: Inhibition of greening of etiolated leaves by actinomycin D. Biochem. Biophys. Res. Comm. **14**, 113–117.

BOGORAD, L., L. LABER and M. GASSMAN, 1968: Aspects of chloroplast development: Transitory pigment-protein complexes and protochlorophyllide regeneration. In: K. Shibata, A. Takamiga, A. T. Iagendorf and R. C. Fuller, eds., Comparative Biochemistry and Biophysiks of Photosynthesis, Univ. Park Press, State College, Pa. pp. 299–312.

BONNER, B. A., 1969: A short-lived intermediate from in the *in vivo* conversion of protochlorophyllide 650 to chlorophyllide 648. Plant. Physiol. **44**, 739–747.

BOVEY, F., T. OGAWA and K. SHIBATA, 1974: Photoconvertible and non-photoconvertible forms of protochlorophyll(ide) in etiolated bean leaves. Plant. Cell. Physiol. **15**, 1133–1137.

BROCKMANN, H., 1968: Zur absoluten Konfiguration des Chlorophylls. Angew. Chem. **80**, 233–234.

BROCKMANN, H., 1976: Bacteriochlorophyll *e*: structure and stereochemistry of a new type of chlorophyll from *Chlorobiaceae*. Phil. Trans. R. Soc. Lond. B. **273**, 277–285.

BROCKMANN, H., G. KNOBLOCH, I. SCHWEER and W. TROWITSCH, 1973: Die Alkoholkomponente des Bakteriochlorophylls *a* aus *Rhodospirillum rubrum*. Arch. Mikrobiol. **90**, 161–164.

BRODERSEN, P., 1976: Factors affecting the photoconversion of protochloroplihylde to chlorophyllide in etioplast membranes from barley. Photosynthetica **10**, 35–39.

BROUERS, M., 1975: Optical properties of *in vitro* aggregates of protochlorophyllide in nonpolar solvents. II. Fluorescence polarization delayed fluorescence and circular. Photosynthetica **9**, 304–310.

BROUERS, M., and C. SIRONVAL, 1975: Restoration of a $P_{657-647}$ form from $P_{645-638}$ in extracts of etiolated primary bean leaves. Plant. Sci. Lett. **4**, 175–181.

BROUERS, M., and C. SIRONVAL, 1978: The reduction of protochlorophyllide into chlorophyllide. VII. Relations between energy transfer, 690 nm fluorescence emmission, and reduction; a theory. Photosynthetica **12**, 399–405.

BUDZIKIEWICZ, H., and K. TARAZ, 1971: Chlorophyll *c*. Tetrahedron **27**, 1447–1460.

BURNHAM, B. F., 1970: δ-Aminolaevulinic acid synthase. (*Rhodopseudomonas spheroides*.) In: H. Tabor and C. W. Tabor, Methods in enzymology XVII A, Academic Press London, pp. 195–200.

BURNHAM, B. F., and J. LASCELLES, 1963: Control of porphyrin biosynthesis through a negative feedback mechanism. Studies with preparations of δ-aminolaevulate synthetase and δ-aminolaevulate dehydratase from *Rhodopseudomonas spheriodes*. Biochem. J. **87**, 462–472.

BURTON, G., P. E. FAGERNESS, S. HOSOZAWA, P. M. JORDAN, and A. I. SCOTT, 1979: ^{13}C.N.M.R. evidence for a new intermediate, pre-uroporphyrinogen, in the enzymic transformation of porphobilinogen into uroporphorinogens I and III. J. Chem. Soc. Chem. Commun. **5**, 202–204.

Butler, W. L., 1961: Chloroplast Development: Energy transfer and structure. Arch. Biochem. Biophys. **92**, 287–295.

Butler, W. L., and W. R. Briggs, 1966: The relation between structure and pigments during the first stages of proplastid greening. Biochim. Biophys. Acta **112**, 45–53.

Calvaleiro, J. A. S., G. W. Kenner and K. M. Smith, 1973: Biosynthetic intermediates between coproporphyrinogen III and protoporphyrin IX. J. chem. Soc. Comm. 183–184.

Canaani, O. D., and K. Sauer, 1977: Analysis of the subunit of protochlorophyllide holochrome by sodium dodecyl sulfate-poyacrylamide gel electrophoresis. Plant. Physiol. **60**, 422–429.

Caple, M. B., H. Chow, and C. E. Strouse, 1978: Photosynthetic pigments of green sulfur bacteria. The esterifying alcohols of bacterichlorophylls *c* from *Chlorobium limicola*. J. Biol. Chem. **253**, 6730–6737.

Carell, E. C., and J. S. Kahm, 1964: Synthesis of porphyrin by isolated chloroplasts of *Euglena*. Arch. Biochem. Biophys. **108**, 1–6.

Castelfranco, P. A., and O. T. G. Jones, 1975: Protoheme turnover and chlorophyll synthesis in greening barley tissue. Plant Physiol. **55**, 485–490.

Castelfranco, P. A., and S. Schwarcz, 1978: Mg-Protoporphyrin-IX and δ-aminolevulinic acid synthesis from glutamate in isolated greening chloroplasts. Mg-protoporphyrin-IX synthesis. Arch. Biochem. Biophys. **186**, 365–375.

Castelfranco, P. A., J. D. Weinstein, S. Schwarcz, A. D. Pardo, and B. E. Wezelman, 1979: The Mg insertion step in chlorophyll biosynthesis. Arch. Biochem. Biophys. **162**, 592–598.

Chen, T. C., and G. W. Miller, 1974: Purification and characterization of uroporphyrinogen decarboxylase from tobacco leaves. Plant Cell Physiol. **15**, 993–1005.

Chiba, Y., I. Aiga, M. Idemori, Y. Sutoh, K. Matsushita and T. Sasa, 1967: Studies on chlorophyllase of *Chlorella protothecoides*. I. Enzymatic phytylation of methyl chlorophyllide. Plant Cell Physiol. **8**, 623–635.

Chua, N.-H., and G. W. Schmidt, 1979: Transport of proteins into mitochondria and chloroplasts. J. Cell. Biol. **81**, 461–483.

Clément-Métral, J. D., and M. Fanica-Gaignier, 1975: δ-Aminolaevulinic-acid synthetases from *Rhodopseudomonas spheroides* Y. Comparison of the purification and properties of enzymes extracted from bacteria grown in different iron concentrations. Eur. J. Biochem. **59**, 73–77.

Cohen, C. E., and J. A. Schiff, 1976: Events surrounding the early development of *Euglena* chloroplasts. – XI. Protochlorophyll(ide) and its photoconversion. Photochem. Photobiol. **24**, 555–566.

Cohen, C. E., M. B. Bazzaz, S. H. Fullett, and C. A. Rebeiz, 1977: Chloroplat biogenesis. 20. Accumulation of porphyrin and phorbin pigments in cucumber cotyledons during photoperiodic greening. Plant Physiol. **60**, 743–746.

Cohen, C. E., and C. A. Rebeiz, 1978: Chloroplast biogenesis: 22. Contribution of short wavelength and long wavelength protochlorophyll species to greening of higher plants. Plant Physiol. **61**, 824–829.

Cooper, R., 1963: The biosynthesis of coproporphyrinogen, magnesium protoporphyrin monomethyl ester and bacteriochlorophyll by *Rhodopseudomonas capsulata*. Biochem. J. **89**, 100–108.

DAVIES, R. C., A. GARCHEIN, A. NEUBERGER, J. D. SANDY, and G. H. TAIT, 1973: Biosynthesis of Bacteriochlorophyll Nature, Lond. **245**, 15–19.

DAVIES, R. C., and A. NEUBERGER, 1973: Polypyrroles formed from prophyrinogen and amines by uroporphyrinogen synthetase of *Rhodopseudomonas spheroides*. Biochem. J. **133**, 471–492.

DIAMOND, J., 1976: Inhibition of chloroplast development in Euglena by streptomycin: Differential inhibition of the appearance of photosynthesis in the presence of the continued synthesis of chlorophyll. Planta **130**, 145–149.

DOSS, M. (ed.), 1976: Porphyrins in human diseases. Karger, Basel.

DUGGAN, J., and M. GASSMAN, 1974: Induction of Porphyrin Synthesis in etiolated leaves by chelators of iron. Plant Physiol. **53**, 206–215.

DUJARDIN, E., and C. SIRONVAL, 1970: The reduction of protochlorophyllide into chlorophyllide. III. The phototransformability of the forms of the protochlorophyllide lipoprotein complex found in darkness. Photosynthetica **4**, 129–138.

DUJARDIN, E., and C. SIRONVAL, 1977: Transitory pigment-protein complexes similar to photosynthesis active centres during protochlorophyll(ide) photoreduction. Plant Sci. Lett. **10**, 347–355.

DURANTON, J., 1966: Biosynthèse de la chlorophylle *b* au cours du verdissement de plantules de *Zea mays* étiolées. Physiol. Vég. **4**, 75–88.

DUTTON, P. L., 1976: Pheophytins and the photosynthetic reaction center. Photochem. Photobiol. **24**, 655–657.

DUYSEN, M. E., and T. P. FREEMAN, 1978: Thylakoid development and chlorophyll distribution in water-stressed hormone-treated wheat leaves. Can. J. Bot. **56**, 1941–1945.

EBBON, J. G., and G. H. TAIT, 1969: Studies on δ-adenosyl-methionine-magnesium protoporphyrin methyltransferase in *Euglena gracilis* strain z. Biochem. J. **III**, 573–582.

EGAN, J. M., D. DORSKY and J. A. SCHIFF, 1975: Events surrounding the early development of *Euglena* chloroplasts. VI. Action spectra for the formation of chlorophyll, lag elimination in chlorophyll synthesis and appearance of TPN-dependent triose phosphate dehydrogenase and alkaline DNase activition. Plant Physiol. **56**, 318–323.

EGNEUS, H., and C. SUNDQUIST, 1970: An action spectrum for transformation of ALA-chlorophyllide to ALA-chlorophyllide in the wavelength region 605–675 nm. Photosynthetica **4**, 81–83.

EHTESHAMUDDIN, A. F. M., 1968: Anaerobic formation of protoporphyrin IX from coproporphyrinogen III by bacterial preparations. Biochem. J. **107**, 446–447.

EILAM, Y., and S. KLEIN, 1962: The Effect of light intensity sucrose feeding on the fine structure in chloroplasts and on the chlorophyll content of etiolated leaves. J. Cell. Biol. **14**, 169–182.

ELDER, G. H., and J. O. EVANS, 1978: A radiochemical method for the measurement of coproporphyrinogen oxidase and the utilization of substrates other than coproporphyrinogen III by the enzyme from rat liver. Biochem. J. **169**, 205–214.

ELDER, G. H., J. O. EVANS, J. R. JACKSON, and A. H. JACKSON, 1978: Factors determining the sequence of oxidative decarboxylation of the 2- and 4- propionate substituents of coproporphyrinogen oxidase in rat liver. Biochem. J. **169**, 215–223.

ELLSWORTH, R. K., 1971: Studies on Chlorophyllase. I. Hydrolytic and esterification activities of chlorophyllase from wheat seedlings. Photosynthetica **5**, 226–232.

ELLSWORTH, R. K., 1972a: Studies on Chlorophyllase. II. The effects of changes in reaction conditions on hydrolytic and esterification activities observed in preparations from wheat seedlings. Photosynthetica **6**, 32–40.

ELLSWORTH, R. K., 1972b: Studies on chlorophyllase. III. Differences in gel chromatographic behavior of the hydrolytic and esterification activities obtained from wheat seedlings. Photosynthetica **6**, 276–281.

ELLSWORTH, R. K., and H. S. ARONOFF, 1968: Investigations on the biogenesis of chlorophyll *a*. III. Biosynthesis of Mg-Vinylpeoporphyrine a_5 methylester from Mg-Protoporphine IX monomethylester as observed in *Chlorella* mutants. Arch. Biochem. Biophys. **125**, 269.

ELLSWORTH, R. K., and ARONOFF, 1969: Investigations of the biogenesis of chlorophyll a. IV. Isolation and partial characterization of some biosynthetic intermediates between Mg-Protoporphin IX monomethylester and Mg-Vinylpheoporphine a_5, obtained from Chlorella mutants. Arch. Biochem. Biophys. **130**, 374–383.

ELLSWORTH, R. K., J. P. DULLAGHAN and M. E. ST. PIERRE, 1974: The reaction mechanism of -adenosyl-L-methionine: magnesium protoporphyrin IX methyltransferase of wheat. Photosynthetica **8**, 375–383.

ELLSWORTH, R. K., and P. V. HERVISH, 1975: Biosynthesis of protochlorophyllide *a* from Mg-protoporphyrin IX *in vitro*. Photosynthetica **9**, 125–139.

ELLSWORTH, R. K., and A. S. HSING, 1973: The reduction of vinyl side-chains of Mg-protoporphyrin IX monomethyl ester *in vitro*. Biochim. Biophys. Acta **313**, 119–129.

ELLSWORTH, R. K., and G. D. LAWRENCE, 1973: Synthesis of magnesium-protoporphyrin IX *in vitro*. Photosynthetica **7**, 73–86.

ELLSWORTH, R. K., and C. A. NOWAK, 1974: A gaschromatographic-mass spectrometric analysis of the esterifying alcohols of pumkin seed protochlorophylls. Anal. Biochem. **57**, 534–546.

ELLSWORTH, R. K., H. J. PERKINS, J. P. DETWILLER and K. LIU, 1970: On the enzymatic conversion of ^{14}C-labeled chlorophyll *a* to ^{14}C-labeled chlorophyll *b*. Biochim. Biophys. Acta **223**, 275–280.

ELLSWORTH, R. K., and M. E. ST. PIERRE, 1976: Biosynthesis and inhibiton of (–) S-adenosyl-L-methionine: magnesium protoporphyrin methyltransferase of wheat. Photosynthetica, **10**, 291–301.

ELLSWORTH, R. K., R. M. TSUK and L. A. ST. PIERRE, 1976: Studies on Chlorophyllase. IV. Attribution of hydrolytic and esterifying «chlorophyllase» activities observed *in vitro* to two enzymes. Photosynthetica **10**, 312–323.

EL-SHARKAWI, H. M., and B. E. MICHEL, 1975: Energy transfers between the protochlorophyll(ide) forms in lyophilized etiolated bean leaves. Photosynthetica **9**, 277–282.

FANICA-GAIGNIER, M., and J. D. CLÉMENT-MÉTRAL, 1973a: ALA synthetase of *R. spheroides* Y. Purification and some properties. Eur. J. Biochem. **40**, 13–18.

FANICA-GAIGNIER, M., and J. D. CLÉMENT-MÉTRAL, 1973b: Are both fractions of *Rhodopseudomonas spheroides* Y. ALA synthetase isoenzymes? Enzyme **16**, 94–100.

FANICA-GAIGNIER, M., and J. D. CLÉMENT-MÉTRAL, 1973c: Cellular compartimentation of two species of δ-aminolevulinic acid synthetase in a facultative photohetero-trophic bacterium (*Rps. spheroides* Y). Biochem. Biophys. Res. Comm. **55**, 610–615.

FEIERABEND, J., 1977: Capacity for chlorophyll synthesis in heat-Bleached 70 S ribosome deficient rye leaves. Planta **135**, 83–88.

FISCHER, F. G., and H. BOHN, 1957: Über die Bildung und das Vorkommen von Phytol. Ann. Chem. **611**, 224–235.

FISCHER, G., and R. LAMBRECHT, 1938: Verhalten von Chlorophyll-Derivaten gegen Chlorophyllase. Z. physiol. Chem. **253**, 253–260.

FISCHER, F. G., and W. RÜDIGER, 1959: Über den Phytolgehalt der Protochlorophylle und des im Dunkeln gebildeten Chlorophylls. Ann. Chem. **627**, 35–46.

FLEMING, I., 1967: Absolute configuration and structure of chlorophyll. Nature **216**, 151–152.

FLUHR, R., E. HAREL, S. KLEIN and E. MELLER, 1975: Control δ-aminolevulinic acid and chlorophyll accumulation in greening maize leaves upon light-dark transitions. Plant. Physiol. **56**, 497–501.

FRENCH, C. S., J. S. BROWN, and M. C. LAWRENCE, 1972: Four universal forms of chlorophyll *a*. Plant Physiol. **49**, 422–429.

FRENCH, J., D. C. NICHOLSON, and C. REMINGTON, 1970: Identification of the acrylate porphyrin S-411 from *meconium*. Biochem. J. **120**, 393–397.

FRYDMAN, B., R. B. FRYDMAN, A. VALASINAS, E. S. LEVY and G. FEINSTEIN, 1976: Biosynthesis of uroporphyrinogens from porphobilinogen: Mechanism and the nature of the process. Phil. Trans. R. Soc. Lond. B. **273**, 137–160.

FRYDMAN, R. B., and G. FEINSTEIN, 1974: Studies on porphobilinogen deaminase and uroporphyrinogen III cosynthetase from human erythrocytes. Biochim. Biophys. Acta **350**, 358–373.

FRYDMAN, R. B., and B. FRYDMAN, 1970: Purification and properties of porphobilinogen deaminase from wheat germ. Arch. Biochem. Biophys. **136**, 183–202.

FRYDMAN, R. B., A. VALASINAS and B. FRYDMAN, 1973b: Mechanism of uroporphyrinogen biosynthesis from porphobilinogen. Enzyme **16**, 151–159.

FRYDMAN, R. B., A. VALASINAS, S. LEVY and B. FRYDMAN, 1974: The enzymatic incorporation of a tripyrrane into uroporphyrinogen I. FEBS Lett. **38**, 134–138.

GAJDOS, A., and M. GAJDOS-TÖRÖK, 1973: Mechanism of the effect of exogenous glucose on the biosynthesis of porphyrins by *Rhodopseudomonas spheroides*. Enzyme **16**, 101–107.

GALLING, G., C. SALZMANN and E. SPIESS, 1973: Synthese von Chlorophyll und Strukturelementen des Plastiden in *Chlorella* ohne Beteiligung der Chloroplasten-Ribosomes. Planta **114**, 269–284.

GAMES, D. E., A. H. JACKSON and J. R. JACKSON, R. V. BELCHER and S. G. SMITH, 1976: Biosynthesis of protoporphyrin IX from coproporphyrin III. J. Chem. Soc. Chem. Commun. **6**, 187–188.

GANOZA, V. G., and R. F. MCFEETERS, 1976: Chlorophyllase activity during pigmentation changes in *Chlorella protothecoides*. Photosynthetica **10**, 1–6.

GARCIA, R. C., L. C. SAN MARTIN DE VIALE, J. M. TOMIO and M. GRINSTEIN, 1973: Porphyrin Biosynthesis: X. Porphyrinogen carboxylase from avian erythrocytes further properties. Biochim. Biophys. Acta **309**, 203–210.

GASSMAN, M., 1973: The conversion of photoinactive protochlorophyllide$_{633}$ to phototransformabel protochlorophyllide$_{650}$ in etiolated bean leaves treated with δ-aminolevulinic acid. Plant. Physiol. **52**, 550–594.

GASSMAN, M., and L. BOGORAD, 1967: Control of chlorophyll production in rapidly greening bean leaves. Plant. Physiol. **42**, 774–780.

GASSMAN, M., and L. BOGORAD, 1967b: Studies on the regeneration of protochlorophyllide after brief illumination of etiolated bean leaves. Plant. Physiol. **42**, 781–784.

GASSMAN, M., S. GRANICK and D. MAUZERALL, 1968: A rapid spectral change in etiolated red kidney bean leaves following phototransformation of protochlorophyllide. Biochem. Biophys. Res. Comm. **32**, 295–300.

GASSMAN, M., J. PLUSČEC and L. BOGORAD, 1968: δ-aminolevulinic acid transaminase in *Chlorella vulgaris*. Plant. Physiol. **43**, 1411–1414.

GEHRING, H., H. KASEMIR and H. MOHR, 1977: The capacity of chlorophyll a biosynthesis in the mustard seedling cotyledons as modulated by phytochrome and circadian rhythmicity. Planta **133**, 295–302.

GIBBS, S. P., 1979: The route of entry of cytoplasmically synthesized proteins into chloroplasts of algae possessing chloroplast ER. J. Cell Sci. **35**, 253–266.

GIBSON, L. D., W. D. LAVER and A. NEUBERGER, 1958: Initial stages in the biosynthesis of porphyrins. 2. The formation of δ-aminolaevulinic acid from glycine and succinyl-coenzyme A by particles from chicken erythrocytes. Biochem. J. **61**, 618–629.

GIBSON, K. D., A. NEUBERGER and G. H. TAIT, 1963: Studies on the biosynthesis of porphyrin and bacteriochlorophyll by *Rhodopseudomonas spheroides*. 4. S-adenosylmethionine-magnesium protoporphyrin methyltransferase. Biochem. J. **88**, 325–334.

GLOE, A., and N. PFENNIG, 1974: Das Vorkommen von Phytol und Geranylgeraniol in den Bacteriochlorophyllen roter und grüner Schwefelbakterien. Arch. Mikrobiol. **96**, 93–101.

GLOE, A., and N. RISCH, 1978: Bacteriochlorophyll c_s, a new bacteriochlorophyll from *Chloroflexus aurantiacus*. Arch. Microbiol. **118**, 153–156.

GODNEV, T. N., V. L. KALER and R. M. ROTFARB, 1962: On the question of the presence of phytol as a component of the protochlorophyll of etiolated leaves. Biochem. Sect. Transl. Doklady Akad. Nauk. SSSR **140**, 187–189.

GODNEV, T. N., R. M. ROTFARB and N. K. AKULOVICH, 1963: On the nature of certain reaction in the biosynthesis of chlorophyll and factors influencing them. Photochem. Photobiol. **2**, 119–128.

GOODENOUGH, U. W., and R. P. LEVINE, 1971: The effect of inhibitors of RNA and protein synthesis on the recovery of chloroplast ribosomes membrane organization and photosynthetic electron transport in the ac–20 strain of *Chlamydomonas reinhardi*. J. Cell. Biol. **50**, 35–49.

GORCHEIN, A., 1972: Magnesium protoporphyrin chelatase activity in *Rhodopseudomonas spheroides*. Studies with whole cells. Biochem. J. **127**, 97–106.

GORCHEIN, A., 1973: Control of magnesium-protoporphyrin chelatase activity in *Rhodopseudomonas spheroides*. Role of light, oxygen and electron and energy transfer. Biochem. J. **134**, 833–845.

GOUGH, S., 1972: Defective synthesis of porphyrins in barley plastids caused by mutation in nuclear genes. Biochim. Biophys. Acta **286**, 36–54.

GOUGH, S. P., and C. G. KANNANGARA, 1977: Synthesis of δ-aminolevulinate by a chloroplast stroma preparation from greening barley leaves. Carlsberg Res. Commun. **42**, 459–464.

GRANICK, S., 1948a: Protoporphyrin 9 as a precursor of chlorophyll. J. Biol. Chem. **172**, 717–727.

GRANICK, S., 1948b: Magnesium protoporphyrin as a precursor of chlorophyll in *Chlorella*. J. Biol. Chem. **175**, 333–342.

GRANICK, S., 1950: Magnesium vinyl pheoporphyrin a_5 another intermediate in the biological synthesis of chlorophyll. J. Biol. Chem. **183**, 713–730.

GRANICK, S., 1959: Magnesium porphyrins formed by barley seedlings treated with δ-aminolevulinic acid. Plant. Physiol. **34**, XVIII.

GRANICK, S., 1961: Magnesium protoporphyrin monoester and protoporphyrin monomethylester in chlorophyll biosynthesis. J. Biol. Chem. **236**, 1168–1172.

GRANICK, S., 1967: The heme and chlorophyll biosynthesis chain. In: T. W. Goodwin, Biochemistry of chloroplasts. Academic Press. London pp. 373–410.

GRANICK, S., and M. GASSMAN, 1970: Rapid regeneration of protochlorophyllide$_{650}$. Plant. Physiol. **45**, 201–205.

GRANICK, S., and S. SASSA, 1971: δ-Aminolevulinic acid synthetase and the control of heme and chlorophyll synthesis. In: H. J. Vogel (ed.) Metabolic Regulation. Academic Press New York pp. 74–141.

GREEN, M. K., I. ALTMAN, J. E. RICHMOND and K. SALOMON, 1957: Role of formate in the biosynthesis of chlorophyll *a*. Nature **179**, 375.

GRIFFITHS, M., 1962: Further mutational changes in the photosynthetic pigment system of *Rhodopseudomonas spheroides*. J. Gen. Microbiol. **27**, 427–431.

GRIFFITHS, W. T., 1974: Protochlorophyll and protochlorophyllide for chlorophyll synthesis *in vitro*. FEBS Letters **49**, 196–200.

GRIFFITHS, W. T., 1975: Characterization of the terminal stages of chlorophyll(ide) synthesis in etioplast membrane preparations. Biochem. J. **152**, 623–635.

GRIFFITHS, W. T., 1976: Studies *in vitro* on plastid development. Biochem. Soc. Trans. **5**, 88–90.

GRIFFITHS, W. T., and O. T. G. JONES, 1975: Magnesium 2,4 divinylphaeporphyrin a_5 as a substrate for chlorophyll biosynthesis *in vitro*. FEBS Lett. **50**, 355–358.

GRIFFITHS, W. T., 1978: Reconstitution of chlorophyllide formation by isolated etioplast membranes. Biochem. J. **174**, 681–692.

GROB, E. C., and J. P. SEILER, 1967: Beitrag zur Charakterisierung der Chlorophyllase. Chimia **21**, 466–468.

GUIGNERY, G. A., A. LUZZATI and J. DURANTON, 1974: On the specific binding of protochlorophyllide and chlorophyll to different peptide chains. Planta **115**, 227–243.

HAMPP, R., and H. ZIEGLER, 1975: Lichtabhängige Neusynthese von δ-Aminolävulinsäure Dehydratase in isolierten *Avena*-Etioplasten. Planta **124**, 255–260.

HAREL, E., and S. KLEIN, 1972: Light dependent formation of δ-aminolevulinic acid in etiolated leaves of higher plants. Biochem. Biophys. Res. Comm. **49**, 364–370.

HAYASAKA, S., and S. TUBOI, 1974: Control of δ-aminolevulinate synthetase activity in *Rhodopseudomonas spheroides*. III. Partial purification of the fraction I activating enzyme and the occurrence of two forms of fraction II. J. Biochem. (Tokyo) **76**, 157–168.

HAYASKI, N., B. GODA and G. KIKUCHI, 1970: Difference in molecular sizes of δ-aminolevulinate synthetases in the soluble and mitochondrial fractions of rat liver. J. Biochem. Tokyo **67**, 859.

HENDRY, G. A. F., and A. K. STOBART, 1978a: The effect of haem on chlorophyll synthesis in barley leaves. Phytochemistry **17**, 73–77.

HENDRY, G. A. F., and A. K. STOBART, 1978b: Effect of 2,2′-bipyrridyl on porphyrin synthesis in etiolated and light-treated barley leaves. Phytochemistry **17**, 671–674.

HENNINGSEN, K. W., and J. E. BOYNTON, 1969: Macromolecular-physiology of plastids. VII. The effect of a brief illumination on plastids of dark-grown barley leaves. J. Cell. Sci. **5**, 757–793.

HENNINGSEN, K. W., and J. E. BOYNTON, 1974: Macromolecular physiology of plastids. IX. Development of plastid membranes during greening of dark-grown barley seedlings. J. Cell. Sci. **15**, 31–55.

HENNINGSEN, K. W., and A. KAHN, 1971: Photoactive subunits of protochlorophyll(ide) holochrome. Plant. Physiol. **47**, 685–690.

HENNINGSEN, K. W., and S. W. THORNE, 1974: Esterification and spectral shifts of chlorophyll(ide) in wildtype and mutant seedlings developed in darkness. Physiol. Plant. **30**, 82–89.

HENNINGSEN, K. W., S. W. THORNE and N. K. BOARDMAN, 1974: Properties of protochlorophyllide and chlorophyll(ide) holochromes from etiolated and greening leaves. Plant. Physiol. **53**, 419–425.

HEYNINGEN, S. VAN, and D. SHEMIN, 1971: Quaternary structure of δ-aminolevulinate dehydratase from *Rhodopseudomonas spheroides*. Biochemistry **10**, 4676–4682.

HIGUCHI, M., and L. BOGORAD, 1975: The purification and properties of uroporphyrinogen I synthases and uroporphyrinogen III. cosynthetase. Interactions between the enzymes. Ann. N.Y. Acad. Sci. **244**, 401–418.

HINES, G. D., and R. K. ELLSWORTH, 1969: Methyl chlorophyllide a as a probable intermediate in the chlorophyll a pathway. Plant. Physiol. **44**, 1742–1744.

HO, Y. K., and J. LASCELLES, 1971: δ-aminolevulinic acid dehydratase of *Spirillum itersonii* and the regulation of tetrapyrole synthesis. Arch. Biochem. Biophys. **144**, 734–740.

HOLDEN, M., 1961: The breakdown of chlorophyll by chlorophyllase. Biochem. J. **78**, 359–364.

HOLDEN, M., 1963: The purification and properties of chlorophyllase. Photochem. Photobiol. **2**, 175–180.

HOLT, A. S., 1961: Further evidence of the relation between 2-devinyl-2-formyl-chlorophyll *a* and chlorophyll *d*. Canad. J. Bot. **39**, 327–331.

HORTON, D., and R. M. LEECH, 1975: The effect on the photoconversion of protochlorophyllide in isolated etioplasts of *Zea mays*. Plant. Physiol. **56**, 113–120.

HOUSSIER, C., and K. SAUER, 1969a: Optical properties of the protochlorophyll pigments. I. Isolation, characterization and infrared spectra. Biochim. Biophys. Acta **172**, 476–491.

HOUSSIER, C., and K. SAUER, 1969b: Optical properties of the protochlorophyll pigments. II. Electronic absorption, fluorescence and circular dichroism spectra. Biochim. Biophys. Acta **172**, 492–502.

HOVENKAMP-OBBENA, R., and D. STEGWEE, 1974: Effect of chloramphenicol on the development of proplastids in *Euglena gracilis*. I. The synthesis of ribulosediphosphate carboxylase NADP-linked glyceraldehyde-3-phosphate dehydrogenase and aminolaevulinate dehydratase. Z. Pflanzenphys. **73**, 430–438.

HSU, W. P., and G. W. MILLER, 1965: Chlorophyll and porphyrin synthesis in relation to iron in *Nicotiana tabacum* L. Biochim. Biophys. Acta **III**, 393–402.

HSU, W. P., and G. W. MILLER, 1970: Coproporphyrinogerase in tobacco. Biochem. J. **117**, 215–220.

ICHINOSE, N., and T. SASA, 1973: Studies on chlorophyllase of *Chlorella protothecoides*. III. Purification and catalytic properties. Plant Cell Physiol. **14**, 1157–1166.

JACKSON, A. H., D. E. GAMES, P. COUCH, J. R. JACKSON, R. B. BELCHER and S. G. SMITH, 1974: Conversion of coproporphyrinogen III to proporphyrin IX. Enzyme **17**, 81–87.

JACKSON, A. H., D. M. JONES, S. G. SMITH, and G. H. ELDER, 1976: Mesoporphyrinogen VI. A substrate for coproporphyrinogen oxidase. J. Chem. Soc. Chem. Commun. **9**, 322–323.

JACKSON, A. H., H. A. SANCOVICH, A. M. FERRAMOLA, N. EVANS, D. E. GAMES, S. A. MATLIN, G. H. ELDER and S. G. SMITH, 1976: Macrocyclic intermediates in the biosynthesis of porphyrins. Philos. Trans. R. Soc. Lond. B. Biol. Sci. **273**, 191–206.

JACOBS, N. J., and J. M. JACOBS, 1976: Nitrate, fumarate and oxygen as electron acceptors for a late step in microbial heme synthesis. Biochim. Biophys. Acta **449**, 1–9.

JACOBS, N. J., and J. M. JACOBS, 1977: Evidence for involvement of the electron transport system at a late step of anaerobic microbial heme synthesis. Biochim. Biophys. Acta **459**, 141–144.

JONES, O. T. G., 1963a: The inhibition of bacteriochlorophyll biosynthesis in *Rhodopseudomonas spheroides* by 8 hydroxyquinoline. Biochem. J. **88**, 335–343.

JONES, O. T. G., 1963b: Magnesium 2,4 divinylphaeoporphyrin a_5 monomethyl ester, a protochlorophyll-like pigment produced by *Rhodopseudomonas spheroides*. Biochem. J. **89**, 182–189.

JONES, O. T. G., 1963c: The production of magnesium protoporphyrin monomethyl ester by *Rhodopseudomonas spheroides*. Biochem. J. **86**, 429–432.

JONES, O. T. G., 1966: A-protein-protochlorophyll complex obtained from inner seed coats of *Curcurbita pepo*. The resolution of its two pigment groups into true protochlorophyll and a pigment related to bacterial protochlorophyll. Biochem. J. **101**, 153–160.

JORDAN, P. M., and D. SHEMIN, 1973: Purification and properties of uroporphyrinogen I synthetase from *Rhodopseudomonas spheroides*. J. Biol. Chem. **248**, 1019–1024.

JORDAN, P. M., G. BURTON, H. NORDLÖV, M. M. SCHNEIDER, L. PRYDE, and A. I. SCOTT, Pre-uroporphyrinogen: a substrate for uroporphyrinogen III cosynthetase. J. Chem. Soc. Chem. Commin. **5**, 204–205.

KAHN, A., and O. F. NIELSEN, 1974: Photoconvertible protochlorophyll(ide)$_{635/650}$ *in vivo*: A single species or two species in dynamic equilibrium? Biochim. Biophys. Acta **333**, 409–414.

KAHN, A., N. K. BOARDMAN and S. W. THORNE, 1970: Energy transfer between protochlorophyllide molecules: Evidence for multiple chromophores in the photo-active protochlorophyllide-protein complex *in vivo* and *in vitro*. J. Mol. Biol. **48**, 85–101.

KANNANGARA, C. G., and S. P. GOUGH, 1977: Synthesis of δ-aminolevulinic acid and chlorophyll by isolated chloroplasts. Carlsberg Res. Commun. **42**, 441–457.

KANNANGARA, C. G., and S. P. GOUGH, 1978: Biosynthesis of δ-aminolevulinate in greening barley leaves: Glutamate 1-semialdehyde aminotransferase. Carlsberg. Res. Commun. **43**, 185–194.

KASEMIR, H., O. OBERDORFER and H. MOHR, 1973: A twofold action of phytochrom in controlling chlorophyll *a* accumulation. Photochem. Photobiol. **18**, 481–486.

KASEMIR, H., 1979: Control of chloroplast formation by light. Cell Biology, Int. reports **3**, 197–214.

KATZ, J. J., H. H. STRAIN, A. L. HARKNESS, M. H. STUDIER, W. A. SVEC, T. R. JANSON and B. T. COPE, 1972: Esterifying alcohols in the chlorophylls of purple photosynthetic bacteria. A new chlorophyll, bacteriochlorophyll (GG) all-trans-geranylgeranyl bacteriochlorophyllide *a*. J. Amer. Chem. Soc. **94**, 7938–7939.

KAUL, K., and P. S. SABHARWAL, 1974: Kinetin-induced changes in δ-aminolevulinic acid-dehydratase of tobacco callus. Plant Physiol. **54**, 644–648.

KEMMER, T., H. BROCKMANN, and N. RISCH, 1979: The individual components of bacteriopheophorbide methyl esters from bacteriochlorophyll *c* and *d*. Z. Naturforschung **34B**, 633–637.

KENNER, G. W., J. RIMMER, K. M. SMITH and J. F. UNSWORTH, 1976: Studies on the biosynthesis of the *Chlorobium* chlorophylls. Phil. Trans. R. Soc. Lond. B. **273**, 255–276.

KENNER, G. W., J. RIMMER, K. M. SMITH, and J. F. UNSWORTH, 1978: Pyrroles and related compounds. Part 39. Structural and biosynthetic studies of the *Chlorobium* chlorophylls-660 (bacteriochlorophyll *c*). Incorporations of methionine and porphobilinogen. J. Chem. Soc. Perkin Trans. I, 845–852.

KIRK, J. T. O., 1968: Studies on the dependence of chlorophyll synthesis on protein synthesis in *Euglena gracilis*, together with a monogram for determination of chlorophyll concentration. Planta **28**, 200–207.

KIRK, J. T. O., 1970: Biochemical aspects of chloroplast development. Ann. Rev. Plant. Physiol. **21**, 11–42.

KIRK, J. T. O., and R. L. ALLEN, 1965: Dependence of chloroplast pigment synthesis on protein synthesis: effect of actidione. Biochem. Biophys. Res. Commun. **21**, 523–530.

KIRK, J. T. O., and R. A. E. TILNEY-BASSETT, 1967: The plastids. Freeman, London.

KIKUCHI, G., A. KUMAR, P. TALMAGE and D. SHEMIN, 1958: The enzymatic synthesis of δ-aminolevulinic acid. J. Biol. Chem. **223**, 1214–1219.

KLEIN, A. O., 1966: Metabolism during light induced expansion of etiolated leaves. Plant. Physiol. **41**, Lvi.

KLEIN, A. O., and W. VISHNIAC, 1961: Activity and partial purification of chlorophyllase in aqueous systems. J. Biol. Chem. **236**, 2544–2547.

KLEIN, O., and H. SENGER, 1978a: 2 biosynthetic pathways to delta-aminolevulinic acid in a pigment mutant of the green alga, *Scenedesmus obliquus*. Plant Physiol. **62**, 10–13.

KLEIN, O., and H. SENGER, 1978b: Biosynthetic pathways to δ-aminolevulinic acid induced by blue light in the pigment mutant C-2A′ of *Scenedesmus obliquus*. Photochem. Photobiol. **27**, 203–206.

KLEIN, S., 1962: Phytolation of chlorophyllide and formation of lamellae in chloroplasts. Nature **196**, 992–993.

KLEIN, S., E. HAREL, E. NE'EMAN, E. KATZ and E. MELLER, 1975: Accumulation of δ-aminolevulinic acid and its relation to chlorophyll synthesis and development of plastid structure in greening leaves. Plant. Physiol. **56**, 486–497.

KLEIN, S., and J. A. SCHIFF, 1972: The correlated appearance of prolamellar bodies protochlorophyllide species and the Shibata shift during development of bean etioplasts in the dark. Plant. Physiol. **49**, 619–626.

KOMAI, H., and J. B. NEILANDS, 1969: The metalloprotein nature of *Ustilago* δ-aminolevulinate dehydratase. Biochim. Biophys. Acta **171**, 311.

KOSKI, V. M., 1950: Chlorophyll formation in seedlings of *Zea mays* L. Arch. Biochem. Biophys. **29**, 339–443.

KOSKI, V. M., C. S. FRENCH and J. H. C. SMITH, 1951: The action spectrum for the transformation of protochlorophyll to chlorophyll *a* in normal and albino corn seedlings. Arch. Biochem. Biophys. **31**, 1–17.

KOSKI, V. M., and J. H. C. SMITH, 1948: The isolation and spectral absorption properties of protochlorophyll from etiolated barley seedlings. J. Amer. Chem. Soc. **70**, 3558–3562.

KRANZ, A. R., 1973: Monogen kontrollierte Langzeit-Transformationen der Chlorophylle in Blättern von *Arabidopsis thaliana* (L.) Hegnh. Z. Pflanzenphysiol. **70**, 333–349.

KRASNOVSKII, A. A., and L. M. KOSOBUTSKAYA, 1952: Spectral investigations of chlorophyll during its formation in plants and in colloidal solutions of material from etiolated leaves. Dokl. Akad. Nauk. SSSR **85**, 177–180.

KRISHNAN, M., and A. GUANAM, 1979: A basic chlorophyll-protein complex. FEBS Lett. **97**, 322–324.

KROSSING, G., 1940: Versuche zur Lokalisation einiger Fermente in den verschiedenen Zellbestandteilen der Spinatblätter. Biochem. Z. **305**, 359–373.

KUNG, S., 1977: Expression of chloroplast genomes in higher plants. Ann. Rev. Plant. Physiol. **28**, 401–437.

LANCER, H. A., C. E. COHEN and J. A. SCHIFF, 1976: Changing ratios of phototransformable protochlorophyll and protochlorophyllide of bean seedlings developing in the dark. Plant. Physiol. **57**, 369–374.

LÁNG, F., and E. SÁRVÁRI, 1974: Early steps of protochlorophyllide phototransformation in etiolated maize leaves and homogenates. Photosynthetica **8**, 9–17 (1974).

LASCELLES, J., 1959: Adaptation from bacteriochlorophyll in *Rhodopseudomonas spheroides*: Changes in activity of enzymes concerned in pyrrole synthesis. Biochem. J. **72**, 508–518.

LASCELLES, J., 1960: The synthesis of enzymes concerned in bacteriochlorophyll formation in growning cultures of *Rhodopseudomonas spheroides*. J. gen. Microbiol. **23**, 487–498.

LASCELLES, J., 1966: The accumulation of bacteriochlorophyll precursors by mutant and wild-type strains of *Rhodopseudomonas spheroides*. Biochem. J. **100**, 175–183.

LASCELLES, J., 1968: Regulation of porphyrin synthesis. In: Goodwin, T. W., Porphyrins and related compounds, pp. 61–74, Academic Press, London.

LAULHERE, J. P., and A. M. DORNE, 1977: Are cytoplasm ribosomes in chloroplast preparations functionally attached to the chloroplast membrane? Plant Sci. Lett. **8**, 251–256.

LICHTENTHALER, H. K., and C. SUNDQVIST, 1979: Association of plastoquinone-9 and phylloquinone K_1 with photoconvertible protochlorophyllide holochrome. Physiol. Plant. **45**, 381–386.

LILIJENBERG, C., 1974: Characterization and properties of a protochlorophyllide ester in leaves of dark grown barley with geranylgeraniol as esterifying alcohol. Physiol. Plant. **32**, 208–213.

LITVIN, F. F., and O. B. BELYAEVA, 1971: Sequence of photochemical and dark reactions in the terminal stage of chlorophyll biosynthesis. Photosynthetica **5**, 200–209.

LLAMBIAS, E. B. C., and A. M. DEL C. BATLLE, 1971a: Studies on the porphobilinogen deaminase-uroporphyrinogen cosynthetase system of cultured soya-bean cells. Biochem. J. **121**, 327–340.

LLAMBIAS, E. B. C., and A. M. EL C. BATLLE, 1970b: Uroporphyrinogen III cosynthetase. Evidence for the existence of a polypyrrolic substrate in soybean callus tissue. FEBS Lett. **6**, 285–288.

LOHR, J. B., and H. G. FRIEDMANN, 1975: New pathway for δ-aminolevulinic acid biosynthesis from α-ketoglutaric acid by two partially purified plant enzymes. Biochem. Biophys. Res. Commun. **69**, 908–913.

LÜTZ, C., and S. KLEIN, 1979: Biochemical and cytological observations on chloroplast development. VI. Chlorophylls and saponins in prolammellar bodies and prothylakoids separated from etioplasts of etiolated *Avena sativa* L. leaves. Z. Pflanzenphysiol., in the press.

MACHOLD, O., and G. SCHOLZ, 1969: Eisenhaushalt und Chlorophyllbildung bei höheren Pflanzen. Naturwissenschaften **56**, 447–452.

MACHOLD, O., and A. AURICH, 1972: Sites of synthesis of chloroplast lamellar proteins in *Vicia faba*. Biochim. Biophys. Acta **281**, 103–112.

MANETAS, Y., and A. AKOYUNOGLOU, 1976: Enzymic nature of the protein moiety of protochlorophyllide holochrome. Plant Physiol. **58**, 43–46.

MARGULIES, M. M., 1967: Effect of chloramphenicol on chlorophyll synthesis of bean leaves. Plant Physiol. **42**, 218–220.

MARRIOTT, J., A. NEUBERGER and G. H. TAIT, 1969: Control of δ-aminolaevulinate synthetase activity in *Rhodopseudomonas spheroides*. Biochem. J. **111**, 385–394.

MASCIA, P., 1978: Analysis of precursors accumulated by several chlorophyll biosynthetic mutants of maize. Mol. Gen. Genetics **161**, 237–244.

MASONER, M., and H. KASEMIR, 1976: Control of chlorophyll synthesis by phytochrome. I.

The effect of phytochrome on the formation of 5-aminolevulinate in mustard seedlings. Planta **126**, 111–119.
MATHIS, P., and K. SAUER, 1972: Circular dichroism studies on the structure and the photochemistry of protochlorophyllide and chlorophyllide holochrome. Biochim. Biophys. Acta **267**, 498–510.
MATHIS, P., and K. SAUER, 1973: Chlorophyll formation in green bean leaves during the early stages. Plant Physiol. **51**, 115–119.
MATTHEIS, J. R., and C. A. REBEIZ, 1975: The net synthesis of protochlorophyllide from tetrapyrroles by isolated plastids. Plant Physiol. **56**, 164.
MATTHEIS, J. R., and C. A. REBEIZ, 1976: Chloroplast biogenesis: Net synthesis of protochlorophyllide from magnesium-protoporphyrin monoester by developing chloroplasts. J. Biol. Chem. **252**, 4022–4024.
MATTHEIS, J. R., and C. A. REBEIZ, 1978: Chloroplast biogenesis. XXIII. The conversion of exogenous protochlorophyllide into phototransformable protochlorophyllide *in vitro*. Photochem. Photobiol. **28**, 55–60.
MAUZERALL, D., and S. GRANICK, 1958: Porphyrin biosynthesis in erythrocytes. III. Uroporphyrinogen and its decarboxylase. J. Biol. Chem. **232**, 1141–1162.
MCCREA, J. M., and C. L. HERSHBERGER, 1978: Chloroplast DNA codes for tRNA from cytoplasmic polyribosomes. Nature **274**, 717–719.
MCFEETERS, R. F., C. O. CHICHESTER and J. R. WHITAKER, 1971: Purification and properties of chlorophyllase from *Ailanthus altissima* (Tree-of-Heaven). Plant Physiol. **47**, 609–618.
MCFEETERS, R. F., 1975: Substrate specificity of chlorophyllase. Plant Physiol. **55**, 377–381.
MICHEL, J. M., 1978: Dependence of the early steps of chlorophyll(ide) synthesis on respiration in dark-grown *Euglena gracilis*. Planta **141**, 45–50.
MICHEL-WOLWERTZ, M. R., 1963: Photooxidation of chlorophyll pigments in the presence of radioactive precursors. Photochem. Photobiol. **2**, 149–158.
MILLER, G. W., A. DEENNEY, J. K. WOOD, and G. W. WELKIE, 1979: Light-induced delta-aminolevulinic acid in dark-grown barley seedlings. Plant Cell Physiol **20**, 131–143.
MITRAKOS, K., 1961: The participation of the red far red reaction system in chlorophyll-metabolism. Physiol. Plant. **14**, 497–503.
MÖLDGEN, R., 1976: Untersuchungen zur Lichtaktivierung von δ-Aminolävulinat-Synthetase, dem Eingangsenzym der Chlorophyllbiosynthese, mit ausgewählten Wellenlängen. Staatsarbeit, Köln.
MOHR, H., H. DRUMM and H. KASEMIR, 1974: Licht und Farbstoffe. Ber. Deutsch. Bot. Ges. **87**, 49–69.
MOHR, H., 1977: Phytochrome and chloroplast development. Endeavour **1**, 107–114.
MOLL, W. A. W., and D. STEGWEE, 1978: The activity of Triton X-100 soluble chlorophyllase in liposomes. Planta **140**, 75–80.
MOLL, W. A. W., B. DE WIT, and R. LUTTER, 1978: Chlorophyllase activity in developing leaves of *Phaseolus vulgaris*. Planta **139**, 79–83.
MOORE, F. D., and D. C. SHEPHARD, 1978: Chloroplast autonomy in pigment synthesis. Protoplasma **94**, 1–18.
MORI, M., and S. SANO, 1972: Studies on the formation of protoporphyrin IX by anaerobic bacteria. Biochim. Biophys. Acta **264**, 252–262.
MURRAY, H. E., and A. O. KLEIN, 1971: Relationship between photoconvertible and non-convertible protochlorophyllides. Plant Physiol. **48**, 383–388.
NADLER, K., and S. GRANICK, 1970: Controls of chlorophyll synthesis in barley. Plant Physiol. **46**, 240–246.

NANDI, D. L., K. F. BAKER-COHEN and D. SHEMIN, 1968: δ-Aminolevulinic acid dehydratase of *Rhodopseudomonas spheroides*. I. Isolation and properties. J. Biol. Chem. **243**, 1224–1230.

NANDI, D. L., and D. SHEMIN, 1968a: δ-Aminolevulinic acid dehydratase of *Rhodopseudomonas spheroides*. II. Association to polymers and dissociation to subunits. J. Biol. Chem. **243**, 1231–1235.

NANDI, D. L., and D. SHEMIN, 1968b: δ-Aminolevulinic acid dehydratase of *Rhodopseudomonas spheroides*. III. Mechanism of porphobilinogen synthesis. J. Biol. Chem. **243**, 1236–1242.

NANDI, D. L., and D. SHEMIN, 1977: Quaternary structure of δ-aminolevulinic acid synthase from *Rhodopseudomonas spheroides*. J. Biol. Chem. **252**, 2278–2280.

NANDI, D. L., and E. R. WAYGOOD, 1967: Biosynthesis of porphyrins in wheat leaves. II. 5-Aminolaevulinate hydro-lyase. Can. J. Biochem. **45**, 327–336.

NANDI, D. L., 1978a: δ-Aminolevulinic acid synthetase of *Rhodopseudomonas spheroides*. Binding of pyridoxal phosphate to the enzyme. Arch. Biochim. Biophys. **188**, 266–271.

NANDI, D. L., 1978b: Studies on δ-aminolevulinic acid synthase of *Rhodopseudomonas spheroides*. Reversibility of the reaction, kinetic, spectral, and other studies related to the mechanism of action. J. Biol. Chem. **253**, 8872–8877.

NANDI, D. L., 1978c: Lysine as a substrate binding site of porphobilinogen synthase of *Rhodopseudomonas spheroides*. Z. Naturforschung **33 C**, 799–800.

NEUBERGER, A., J. D. SANDY and G. H. TAIT, 1973a: Control of 5-aminolaevulinate synthetase activity in *Rhodopseudomonas spheroides* (The involvement of sulfur metabolism). Biochem. J. **136**, 477–490.

NEUBERGER, A., J. D. SANDY and G. H. TAIT, 1973b: Control of 5-aminolaevulinate synthetase activity in *Rhodopseudomonas spehoroides* (Purification and properties of an endogenous activator of the enzyme). Biochem. J. **136**, 491–499.

NEUBERGER, A., and J. M. TURNER, 1963: γ, δ-Dioxovalerate aminotransferase activity in *Rhodopseudomonas spheroides*. Biochim. Biophys. Acta **67**, 342–345.

NEVE, R. A., R. F. LABBE and R. A. ALDRICH, 1956: Reduced uroporphyrin III in the biosynthesis of heme. J. Amer. Chem. Soc. **78**, 691–692.

NIELSEN, N. C., 1975: Electrophoretic characterization of membrane proteins during chloroplast development in barley. Eur. J. Biochem. **50**, 611–624.

OELZE-KAROW, H., and H. MOHR, 1978: Control of chlorophyll *b* biosynthesis by phytochrome. Photochem. Photobiol. **27**, 189–196.

OGAWA, M., Y. TSUTSUI, and M. KONISHI, 1978: Effect of illumination on absorption peak shifts in spectra of intact etiolated cotyledons of *Pharbitis nil*. II. Effects of leaf age on protochlorophyllide regeneration and the Shibata shift. Plant Cell Physiol. **19**, 127–132.

OGAWA, T., F. BOVEY and K. SHIBATA, 1975: An intermediate in the phytylation of chlorophyllide a *in vivo*. Plant Cell Physiol. **16**, 199–202.

OGAWA, T., Y. INOUE, M. KITAJIMA and K. SHIBATA, 1973: Action spectra for biosynthesis of chlorophylls *a* and *b* and β-carotene. Photochem. Photobiol. **18**, 229–235.

OGURA, N., 1972: Studies on chlorophyllase in tea leaves. III. Properties of soluble and insoluble chlorophyllases. Plant Cell Physiol. **13**, 971–979.

OH-HAMA, T., and E. HASE, 1975: Synthesis of δ-aminolevulinic acid and chlorophyll during chloroplast formation in *Chlorella protothecoides*. Plant Cell Physiol. **16**, 297–303.

OH-HAMA, T., and H. SENGER, 1975: The development of structure and function in the chloroplasts of greening mutants of *Scenedesmus*. III. Biosynthesis of δ-aminolevulinic acid. Plant Cell Physiol. **16**, 395–405.

OH-HAMA, T., and H. SENGER, 1978: Spectral effectiveness in chlorophyll and 5-aminolevulinic acid formation during regreening glucose-bleached cells of *Chlorella protothecoides*. Plant Cell Physiol. **19**, 1295–1299.

PARTHIER, B., 1979: Role of phytohormones (cytokinins) in chloroplast development. Biochem. Physiol. Pflanz. **174**, 173–214.

PLUSCEC, J., and L. BOGORAD, 1970: A dipyrrylmethane intermediate in the enzymatic synthesis of uroporphyrinogen. Biochemistry **9**, 4736–4743.

POLLACK, S. E., and C. S. RUSSELL, 1978: Inhibition of wheat germ porphobilinogen deaminase activity by butanedione. FEBS Lett. **90**, 47–50.

PORRA, R. J., R. BARNES, and O. T. G. JONES, 1972: The level and subcellular distribution of δ-aminolevulinate synthetase activity in semianaerobic and aerobic yeast. Z. Physiol. Chem. **353**, 1365–1368.

PORRA, R. J., and J. F. FALK, 1964: The enzymic conversion of coproporphyrinogen III into protoporphyrin IX. Biochem. J. **90**, 69–75.

PORRA, R. J., and L. H. GRIMME, 1974: Chlorophyll synthesis and intracellular fluctuations of 5-aminolaevulinate formation during the regreening of nitrogen-deficient *Chlorella fusca*. Arch. Biochem. Biophys. **165**, 312–321.

PORRA, R. J., E. A. IRVING and A. M. TENNICK, 1972: The nature of the inhibition of δ-aminolevulinic acid synthetase by hemin. Arch. Biochem. Biophys. **148**, 37–43.

POULSON, R., 1976: The enzymatic conversion of protoporphyrinogen IX to protoporphyrin IX in mamalian mitochondria. J. Biol. Chem. **251**, 3730–3733.

POULSON, R., and W. J. POLGLASE, 1975: The enzymic conversion of protoporphyrinogen IX to protoporphyrin. Protoporphyrinogen oxidase activity in mitochondrial extracts of *Saccharomyces cerevisiae*. J. Biol. Chem. **250**, 1269–1274.

PRICE, L., and W. H. KLEIN, 1961: Red, far-red response and chlorophyll synthesis. Plant Physiol. **36**, 733–755.

RADMER, J. R., and L. BOGORAD, 1967: (–)S-Adenosyl-L-methionine-magnesium protoporphyrin methyltransferase, an enzyme in the biosynthetic pathway of chlorophyll in *Zea mays*. Plant Physiol. **42**, 463–465.

RADMER, R., and L. BOGORAD, 1972: A tetrapyrrylmethane intermediate in the enzymatic synthesis of uroporphyrinogen. Biochemistry **11**, 904–910.

RAMASWAMY, N. K., and P. M. NAIR, 1973: δ-Aminolevulinic acid synthetase from cold-stored potatoes. Biochim. Biophys. Acta **293**, 269–277.

RAMASWAMY, N. K., and P. M. NAIR, 1974: Temperature and light dependency of chlorophyll synthesis in potatoes. Plant. Sci. Lett. **2**, 249–256.

RAPOPORT, H., and H. P. HAMLOW, 1961: Chlorobium chlorophyll-660. The esterifying alcohol. Biochem. Biophys. Res. Comm. **6**, 134–137.

RAVEN, C. W., 1973: Chlorophyll formation and phytochrome. Meded. Landbouwhogesch. Wageningen **73**, 1–108.

REBEIZ, C. A., M. ABOU-HAIDAR, M. YAGHI and P. A. CASTELFRANCO, 1970: Porphyrin biosynthesis in cell-free homogenates from higher plants. Plant Physiol. **46**, 543–549.

REBEIZ, C. A., and P. A. CASTELFRANCO, 1971a: Chlorophyll biosynthesis in a cell-free system from higher plants. Plant Physiol. **47**, 24–32.

REBEIZ, C. A., and P. A. CASTELFRANCO, 1971b: Protochlorophyll biosynthesis in a cell-free system from higher plants. Plant Physiol. **47**, 33–37.

REBEIZ, C. A., and P. A. CASTELFRANCO, 1973: Protochlorophyll and chlorophyll biosynthesis in cell-free systems from higher plants. Ann. Rev. Plant Physiol. **24**, 129–172.

REBEIZ, C. A., J. C. CRANE and C. NISHIJIMA, 1972: The biosynthesis of metal porphyrins by subchloroplastic fractions. Plant Physiol. **50**, 185–186.

REBEIZ, C. A., J. C. CRANE, C. NISHIJIMA and C. C. REBEIZ, 1973: Biosynthesis and accumulation of microgram quantities of chlorophyll by developing chloroplasts *in vitro*. Plant Physiol. **51**, 660–666.

REBEIZ, C. A., J. R. MATHEIS, B. B. SMITH, C. C. REBEIZ and D. F. DAYTON, 1975b: Chloroplast biogenesis. Biosynthesis and accumulation of Mg-protoporphyrin IX monoester and longer wavelength metalloporphyrins by greening cotyledons. Arch. Biochem. Biophys. **166**, 446–465.

REBEIZ, C. A., B. B. SMITH, J. R. MATHEIS, C. C. REBEIZ and D. F. DAYTON, 1975a: Chloroplast biogenesis: Biosynthesis and accumulation of Mg-protoporphyrin IX monoester and other metalloporphyrins by isolated etioplasts and developing chloroplasts. Arch. Biochem. Biophys. **167**, 351–365.

RICHARD, F., and V. NIGON, 1973: La synthèse de l'acide δ-aminolévulinic et de la chlorophylle lors de l'éclairment d'*Euglena gracilis* étiolée. Biochim. Biophys. Acta **313**, 130–449.

RICHTER, G., 1969: Chloroplastendifferenzierung in isolierten Wurzeln. Planta **86**, 299–300.

RISCH, N., H. BROCKMANN, and A. GLOE, 1979: Strukturaufklärung von neuartigen Bakteriochlorophyllen aus *Chloroflexus aurantiacus*. Liebigs Ann. Chem. No. 3, 408–418.

ROSEN, J. A., C. S. PIKE and M. L. GOLDEN, 1977: Zinc, iron, and chlorophyll metabolism in zinc-toxic corn. Plant Physiol. **59**, 1085–1087.

ROSINSKI, J., and W. G. ROSEN, 1972: Chloroplast development: Fine structure and chlorophyll synthesis. Quarterly Rev. Biol. **47**, 160–191.

RUDOLPH, E., 1965: Untersuchungen über den Einfluß der Photoreaktionssysteme auf die Chlorophyllsynthese. Planta **66**, 75–94.

RUDOLPH, E., and F. BUKATSCH, 1966: Die Protochlorophyll(id)-Umwandlung und ihre Beziehung zur Photooxidation der Ascorbinsäure bei etiolierten Keimpflanzen. Planta **69**, 124–134.

RÜDIGER, W., J. BENZ, U. LEMPERT, S. SCHOCH and D. STEFFENS, 1976: Hemmung der Phytol-Akkumulation mit Herbiziden. Geranylgeraniol- und Dihydrogeranylgeraniol-haltiges Chlorophyll aus Weizenkeimlingen. Z. Pflanzenphysiol. **80**, 131–143.

RÜDIGER, W., P. HEDDEN, H.-P. KÖST and D. J. CHAPMAN, 1977: Esterification of chlorophyllide by geranylgeranyl pyrophosphate in a cell-free system from maize shoots. Biochem. Biophys. Res. Comm. **74**, 1268–1272.

RYBERG, M., and C. SUNDQVIST, 1976: The influence of γ-hydroxyquinoline on the accumulation of porphyrins in dark-grown wheat leaves with δ-aminolevulinic acid. Physiol. Plant. **36**, 356–361.

SABATER, B., and M. T. RODRIGUEZ, 1978: Control of chlorophyll degradation in detached leaves of barley and oat through effect of kinetin on chlorophyllase levels. Physiol. Plant. **43**, 274–276.

SALVADOR, G. F., G. BENEY and V. NIGON, 1976: Control of δ-aminolevulinic acid synthesis during greening of dark-grown *Euglena gracilis*, Plant Sci. Lett. **6**, 197–202.

SALVADOR, G. F., 1978: Delta-aminolevulinic acid synthesis from gamma-delta-dioxovaleric acid by a cellular preparations of *Euglena gracilis*. Plant Sci. Lett. **13**, 351–355.

SANCOVICH, H. A., A. M. FERRAMOLA, A. M. D. C. BATLLE, A. KIVILEVICH, and M. GRINSTEIN, 1978: Studies on cow liver porphobilinogen deaminase. Acta Physiol. Lat. Amer. **26**, 379–386.

SANDY, J. D., R. C. DAVIES and A. NEUBERGER, 1975: Control of 5-aminolaevulinate synthetase activity in *Rhodopseudomonas spheroides*. A role for trisulphides. Biochem. J. **150**, 245–257.

SANO, S., 1966: 2,4-Bis-(β-hydroxypropionic acid) deuteroporphyrinogen IX, a possible intermediate between coproporphyrinogen III and protoporphyrin IX. J. Biol. Chem. **241**, 5276–5283.

SANO, S., and S. GRANICK, 1961: Mirochondrial coproporphyrinogen oxidase and protoporphyrin formation. J. Biol. Chem. **236**, 1173–1180.

SCHAER, H., W. A. SVEC, B. T. COPE, M. H. STUDIER, R. G. SCOTT and J. J. KATZ, 1974: Structure of bacteriochlorophyll *b*. J. Amer. Chem. Soc. **96**, 3714–3715.

SCHMIDT, A., 1924: Über die Chlorophyllbildung im Koniferenembryo. Bot. Arch. **5**, 260–282.

SCHNEIDER, HJ. A. W., 1968: Chlorophyllides in green and etiolated leaves. Phytochemistry **7**, 885–886.

SCHNEIDER, HJ. A. W., 1970a: Aktivität und Eigenschaften von δ-Aminolävulinatdehydratase in ergrünenden Zellkulturen von *Nicotiana tabacum*. var. „Samsun". Z. Pflanzenphysiol. **62**, 133–145.

SCHNEIDER, HJ. A. W., 1970b: Anreicherung und Eigenschaften von δ-Aminolävulinatdehydratase aus Spinat (*Spinacea oleracea*). Z. Pflanzenphysiol. **62**, 328–342.

SCHNEIDER, HJ. A. W., 1971a: Light mediated increase in activity of porphobilinogen deaminase/uroporphyrinogen III cosynthetase and δ-aminolevulinate dehydratase in tissue cultures of tobacco. Phytochemistry **10**, 319–321.

SCHNEIDER, HJ. A. W., 1971b: Porphyrinsynthesis in isolated particles of tobacco. Z. Naturforsch. **26b**, 908–912.

SCHNEIDER, HJ. A. W., 1973a: δ-Aminolävulinatsynthetase, δ-Aminolävulinatanreicherung und Chlorophyllsynthese in Zellkulturen von Tabak. Z. Pflanzenphysiol. **69**, 68–76.

SCHNEIDER, HJ. A. W., 1973b: Licht- und entwicklungsbedingte Aktivitätsänderungen von vier aufeinander folgenden Enzymen der Porphyrin- und Chlorophyllbiosynthesekette. Z. Naturforsch. **28c**, 45–58.

SCHNEIDER, HJ. A. W., 1973c: Light mediated induction of δ-aminolevulinate synthetase and related enzymes and a tentative working model for the control of chlorophyll synthesis. Enzyme **16**, 108–114.

SCHNEIDER, HJ. A. W., 1973d: pH-Dependent elution volume of some pyrroles chromatographed on Sephadex G-15. J. Chromatog. **81**, 160–162.

SCHNEIDER, HJ. A. W., 1975: Chlorophylle: Aspekte der Biosynthese und ihrer Regulation. Ber. Deutsch. Bot. Ges. **88**, 83–123.

SCHNEIDER, HJ. A. W., 1976: Enzymic capacities for chlorophyll biosynthesis. Activation and *de novo* synthesis of enzymes. Z. Naturforsch. **31c**, 55–63.

SCHNEIDER, HJ. A. W., and W. W. BEISENHERZ, 1974: Determination of the sites of synthesis of chlorophyll synthesizing enzymes in cell cultures of *Nicotiana tabacum*. Biochem. Biophys. Res. Comm. **60**, 468–473.

SCHNEIDER, HJ. A. W., and L. BOGORAD, 1979: Action spectra for the synthesis of phycobilins, chlorophyll and δ-aminolevulinic acid in *Cyanidium caldarium*. Z. Pflanzenphysiol. **94**, 449–459.

SCHOCH, S., 1978: The esterification of chlorophyllide *a* in greening bean leaves. Z. Naturforsch. **33C**, 712–714.

SCHOCH, S., and W. SCHÄFER, 1978: Tetrahydrogeranylgeraniol, a precursor of phytol in the biosynthesis of chlorophyll *a*. – localization of the double bounds. Z. Naturforsch. **33C**, 408–412.

SCHOCH, S., U. LEMPERT, and W. RÜDIGER, 1977: Über die letzten Stufen der Chlorophyll-Biosynthese. Zwischenprodukte zwischen Chlorophyllid und phytohaltigem Chlorophyll. Z. Pflanzenphysiol. **83**, 427–436.

SCHOPFER, P., and H. W. SIEGELMAN, 1968: Purification of protochlorophyllide holochrome. Plant Physiol. **43**, 990–996.

SCHULTZ, A., and K. SAUER, 1972: Circular dichroism and fluorescence changes accompanying the protochlorophyllide to chlorophyllide transformation in greening leaves and holochrome preparations. Biochim. Biophys. Acta **267**, 320–340.

SCHWARTZBACH, S. D., and J. A. SCHIFF, 1979: Events surrounding the early development of *Euglena* chloroplasts. 13. Photocontrol of protein synthesis. Plant Cell Physiol. **20**, 827–838.

SEYBOLD, A., 1949: Zur Kenntnis des Protochlorophylls. III. Planta **36**, 371–383.

SHEMIN, D., 1970: On the synthesis of heme. Naturwissenschaften **57**, 185–190.

SHEMIN, D., 1975: 5-aminolaevulinic acid dehydratase: Structure, function and mechanism. Philos. Trans. R. Soc. Lond. B. Biol. Sci. **273**, 109–115.

SHETTY, A. S., and G. W. MILLER, 1969: Purification and general properties of δ-aminolaevuliniate dehydratase from *Nicotiana tabacum* L. Biochem. J. **114**, 331–337.

SHIBATA, H., and H. OCHIAI, 1975: Studies on δ-aminolevulinic acid dehydratase during chloroplast development in radish cotyledons. Amino Acid, Nucleic Acid **32**, 16–24.

SHIBATA, H., and OCHIAI, H., 1977: Purification and properties of δ-aminolevulinic acid dehydratase from radish cotyledons. Plant Cell Physiol. **18**, 421–429.

SHIBATA, K., 1957: Spectroscopic studies on chlorophyll formation in intact leaves. J. Biochem., Tokyo **44**, 147–173.

SHIMIZU, S., and E. TAMAKI, 1962: Chlorophyllase of tobacco plants. I. Preparation and properties of water soluble enzyme. Bot. Mag. Tokyo **75**, 462–467.

SHIMIZU, S., and E. TAMAKI, 1963: Chlorophyllase of tobacco plants. II. Enzymic phytylation of chlorophyllide and pheophorbide *in vitro*. Arch. Biochem. Biophys. **102**, 152–158.

SHLYK, A. A., 1971: Biosynthesis of chlorophyll b. Ann. Rev. Plant Physiol. **22**, 169–184.

SHLYK, A. A., V. L. KALER, L. I. VLASENOK and V. I. GAPONENKO, 1963: The final stages of biosynthesis of chlorophylls *a* and *b* in the green leaf. Photochem. Photobiol. **2**, 129–148.

SHLYK, A. A., and I. V. PRUDNIKOVA, 1967: Dark biosynthesis of chlorophyll *b* by fractions of barley chloroplasts. Photosynthetica **1**, 157–170.

SIDDELL, S. G., and R. J. ELLIS, 1977: Protein synthesis in etioplasts. Biochem. Soc. Trans. **5**, 98–102.

SIMONESCU, C. I., B. C. SIMONESCU, R. MORA, and M. LEANCA, 1978: Porphyrin-like compounds genesis under simulated abiotic conditions. Origins of Life **9**, 103–114.

SIRONVAL, C., and M. BROUERS, 1970: The reduction of protochlorophyllide. II. The temperature dependence of the P 657/647–P 688/676 phototransformation. Photosynthetica **4**, 38–47.

SIRONVAL, C., M. BROUERS, J.-M. MICHEL and J. KUIPER, 1968: The reduction of protochlorophyllide into chlorophyllide. Photosynthetica **2**, 268–287.

SIRONVAL, C., and Y. KUYPER, 1972: The reduction of protochlorophyllide into chlorophyllide. IV. The nature of the intermediate P 688–676 species. Photosynthetica **6**, 264–275.

SIRONVAL, C., M. R. MICHEL-WOLWERTZ and A. MADSON, 1965: On the nature and possible functions of the 673 and 684-mμ forms *in vivo* of chlorophyll. Biochim. Biophys. Acta **94**, 344–354.

SISLER, E. C., and W. H. KLEIN, 1963: The effect of age and various chemicals on the lag phase of chlorophyll synthesis in darkgrown bean seedlings. Physiol. Plant **16**, 315–322.

SLUITERS-SCHOLTEN, C. M. T., F. M. VAN DEN BERG and D. STEGWEE, 1973: Aminolaevulinate dehydratase in greening leaves of *Phaseolus vulgaris* L. Z. Pflanzenphysiol. **69**, 217–227.

SMITH, B. B., and C. A. REBEIZ, 1977: Chloroplast biogenesis: Detection of Mg-protoporphyrin chelatase *in vitro*. Arch. Biochem. Biophys. **180**, 178–185.

SMITH, B. B., and C. A. REBEIZ, 1979: Chloroplast biogenesis. 24. Intrachloroplastic localization of the biosynthesis and accumulation of protoporphyrin IX, magnesiumprotoporphyrin monoester, and longer wave length metalloporphyrins during greening. Plant Physiol. **63**, 227–231.

SMITH, J. H. C., 1948: Protochlorophyll, precursor of chlorophyll. Arch. Biochem. Biophys. **19**, 449–454.

SMITH, J. H. C., 1952: Factors affecting the transformation of protochlorophyll to chlorophyll. Carnegie Inst. Wash. Year B. **51**, 151–153.

SMITH, J. H. C., 1960: Protochlorophyll transformation. In: Comparative biochemistry of photoreactive systems (M. B. Allen, ed.) pp. 257–277, Academic Press, New York.

SMITH, J. H. C., and A. BENITEZ, 1955: Chlorophylls: Analysis in plant materials. E. Protochlorophyll. In: Moderne Methoden der Pflanzenanalyse. Vol. 4, pp. 172–177, Springer, Berlin.

SMITH, J. H. C., L. J. DURHAM and C. F. WUSTER, 1959: Formation and bleaching of chlorophyll in albino corn seedlings. Plant Physiol. **34**, 340–345.

SMITH, J. H. C., and C. S. FRENCH, 1963: The major and accessory pigments in photosynthesis. Ann. Rev. Plant Physiol. **14**, 181–224.

SMITH, J. H. C., and D. W. KUPKE, 1956: Some properties of extracted protochlorophyll holochrome. Nature **178**, 751–752.

SMITH, K. M., ed., 1975: Porphyrins and metalloporphyrins. Elsevier, Amsterdam.

STEFFENS, D., I. BLOS, S. SCHOCH and W. RÜDIGER, 1976: Lichtabhängigkeit der Phytolakkumulation. Ein Beitrag zur Frage der Chlorophyll-Biosynthese. Planta **130**, 151–158.

STELLA, A. M., V. E. PARERA, E. B. C. LLAMBIAS and A. M. DEL C. BATLLE, 1971: Pyrrylmethane intermediates in the synthesis of uroporphyrinogen III by soybean callus porphobilinogenase. Biochim. Biophys. Acta **252**, 481–488.

STEVENS, E., and B. FRYDMAN, 1968: Isolation and properties of wheat germ uroporphyrinogen III cosynthetase. Biochim. Biophys. Acta **151**, 429–437.

STEVENS, E., R. B. FRYDMAN and B. FRYDMAN, 1968: Separation of porphobilinogen deaminase and uroporphyrinogen III cosynthetase from human erythrocytes. Biochim. Biophys. Acta **158**, 496–498.

STOBART, A. K., and G. A. F. HENDRY, 1978: Chlorophyll formation and glycine metabolism in laevulinic acid treated barley leaves. Phytochemistry **17**, 993–994.

STUMMANN, B. M., 1978a: The capacity of various detergents to solubilize and stabilize protochlorophyll(ide) holochrome. Physiol. Plant **43**, 173–176.

STUMMANN, B. M., 1978b: Tetrapyrrol protein complexes from wild type barley and barley mutants affecting chlorophyll synthesis. Biochem. Physiol. Pflanz. **173**, 249–269.

STUMMANN, B. M., 1979: Correlation of the 680 to 672 nm spectral shift and the halving of the apparent molecular weight from chlorophyll(ide) holochrome from barley. Physiol. Plant **45**, 122–126.

SUDYINA, E. G., 1963: Chlorophyllase reaction in the last stage of biosynthesis of chlorophyll. Photochem. Photobiol. **2**, 181–190.

SÜZER, S., and K. SAUER, 1971: The sites of photoconversion of protochlorophyllide to chlorophyllide in barley seedlings. Plant Physiol. **48**, 60–63.

SUNDQVIST, C., 1969: Transformation of protochlorophyllide formed from exogenous δ-aminolevulinic acid in continuous light and in flash light. Physiol. Plant. **22**, 147–156.

SUNDQVIST, C., 1970: The conversion of protochlorophyllide$_{636}$ to protochlorophyllide in leaves treated with δ-aminolevulinic acid. Physiol. Plant. **23**, 412–424.

SUNDQVIST, C., 1973: The relationship between chlorophyllide accumulation, the amount of protochlorophyllide$_{638}$ and protochlorophyllide in dark-grown wheat leaves treated with δ-aminolevulinic acid. Physiol. Plant. **28**, 464–470.

SUNDQVIST, C., and B. KLOCKARE, 1975: Fluorescence properties of protochlorophyllide in flash irradiated dark-grown wheat leaves, treated with δ-aminolevulinic acid. Photosynthetica **9**, 62–71.

SUNDQVIST, C., 1978: Red light stimulated accumulation of protochlorophyllide in dark grown leaves treated with delta-aminolevulinic acid. Plant Sci. Lett. **12**, 69–76.

TAIT, G. H., 1969: Coproporphyrinogenase activity in extracts from *Rhodopseudomonas spheroides*. Biochem. Biophys. Res. Comm. **37**, 116–122.

TAIT, G. H., 1972: Coproporphyrinogenase activities in extracts of *Rhodopseudomonas spheroides* and *Chromatium* strain D. Biochem. J. **128**, 1159–1169.

TAIT, G. H., 1972b: 5-Aminolaevulinat synthetase of *Micrococcus denitrificans*. Biochem. J. **128**, 32p.

TAIT, G. H., 1973: Aminolaevulinate synthetase of *Micrococcus denitrificans*. Purification and properties of the enzyme, and the effect of growth conditions on the enzyme activity in cells. Biochem. J. **131**, 389–403.

TAIT, G. H., and K. D. GIBSON, 1961: The enzymic formation of magnesium protoporphyrin monomethyl ester. Biochim. Biophys. Acta **52**, 614–616.

TAKEMOTO, J., and J. LASCELLES, 1973: Coupling between bacteriochlorophyll and membrane protein synthesis in *Rhodopseudomonas spheroides*. Proc. Natn. Acad. Sci. USA **70**, 799–803.

TAMAI, H., Y. SHIOI, and T. SASA, 1979: Purification and characterization of δ-aminolevulinic acid dehydratase from *Chlorella regularis*. Plant Cell Physiol. **20**, 435–443.

TERPSTRA, W., 1976: Chlorophyllase and lamellar structure in *Phaeodactylum tricornutum*. III. Situation of Chlorophyllase in pigmented membranes. Z. Pflanzenphysiol. **80**, 177–188.

TERPSTRA, W., 1978: Chlorophyllase in *Phaeodactylum tricornutum* photosynthetic membranes. Extrability, small-scale purification and molecular weight determination by SDS-gel electrophoresis. Physiol. Plant **44**, 329–334.

TERPSTRA, W., and J. C. GOEDHEER, 1979: Chlorophyllase and photosystem I. Methylviologen photoreduction by DCPIP-ascorbate in the presence of photosynthetic membrane fragments, chlorophyll-Triton X-100 or chlorophyll-chlorophyllase. Physiol. Plant **45**, 367–372.

TEVINI, M., and H. K. LICHTENTHALER (eds.), 1977: Lipids and lipid polymers in higher plants. Springer, Berlin.

THORNBER, I. P., J. P. MARKWELL, and S. REINMAN, 1979: Plant chlorophyll-protein complexes: recent advances. Phytochem. Photobiol. **29**, 1205–1217.

THORNE, S. W., 1971a: The greening of etiolated bean leaves. I. The initial photoconversion process. Biochim. Biophys. Acta **226**, 113–127.

THORNE, S. W., 1971b: The greening of etiolated bean leaves. II. Secondary and further photoconversion processes. Biochim. Biophys. Acta **226**, 128–134.

THORNE, S. W., and N. K. BOARDMAN, 1971: Formation of chlorophyll *b*, and the fluorescence properties and photochemical activities of isolated plastids from greening pea seedlings. Plant Physiol. **47**, 252–261.

TIGIER, H. A., A. M. DEL C. BATLLE and G. A. LOCASCIO, 1970: Porphyrin biosynthesis in the soybean callus tissue system. II. Improved purification and some properties of delta amino-laevulic acid dehydratase. Enzymology **38**, 43–56.

TOMIO, J. M., R. C. GARCIA, SAN MARTIN DE VIALE and M. GRINSTEIN, 1970: Porphyrin biosynthesis. VII. Porphyrinogen carboxylyase from avian erythrocytes. Purification and properties. Biochim. Biophys. Acta **198**, 353–363.

TREFFRY, T., 1970: Phytylation of chlorophyllide and prolamellar body transformation in etiolated peas. Planta **91**, 279–284.

TSUKAMOTO, I., T. YOSHINAGA and S. SANO, 1975: Evidence for histidine as another functional group of δ-aminolevulinic acid dehydratase from beef liver. Biochim. Biophys. Res. Comm. **67**, 294–300.

TUBOI, S., and S. HAYASAKA, 1971: Control of δ-aminolevulinate synthetase activity in *Rhodopseudomonas spheroides*. I. Partial purification and the inactive form of fraction I. Arch. Biochem. Biophys. **146**, 282–290.

TUBOI, S. H., J. KIM, and G. KIKUCHI, 1970a: Occurrence and properties of two types of δ-aminolevulinate synthetase in *Rhodopseudomonas spheroides*. Arch. Biochem. Biophys. **138**, 147–154.

TUBOI, S., H. J. KIM and G. KIKUCHI, 1970b: Differential induction of fraction I and fraction II of δ-aminolevulinate synthetase in *Rhodopseudomonas spheroides* under various incubation conditions. Arch. Biochem. Biophys. **138**, 155–159.

VIRGIN, H. I., 1960: Pigment transformations in leaves of wheat after irradiation. Physiol. Plant. **13**, 155–164.

VIRGIN, H. I., 1961: Action spectrum for the elimination of the lag phase in chlorophyll formation in previously dark-grown leaves of wheat. Physiol. Plant. **14**, 439–452.

VIRGIN, H. I., 1965: Chlorophyll formation and water deficit. Physiol. Plant. **18**, 994–1000.

VIRGIN, H. I., 1975: *In vivo* absorption spectra of protochlorophyll$_{650}$ within the region 530–700 nm. Photosynthetica **9**, 84–92.

VIRGIN, H. I., 1977: The spectral response of light dependent chlorophyll *b* formation. Physiol. Plant. **40**, 45–49.

VIRGIN, H. I., and C. S. FRENCH, 1973: The light induced protochlorophyll-chlorophyll *a*-transformation and the succeeding interconversions of different forms of chlorophyll. Physiol. Plant. **28**, 350–357.

VIRGIN, H. I., A. KAHN and D. v. WETTSTEIN, 1963: The physiology of chlorophyll formation in relation to structural changes in chloroplasts. Photochem. Photobiol. **2**, 83–91.

WALTER, G., 1974: Spektrale Änderungen *in vivo* während der Chlorophyll-Bildung etiolierter Weizenkeimpflanzen nach experimentell variiertem Anteil der Protochlorophyll(id)-Holochrome P_{650} und P_{635} durch Behandlung mit δ-Aminolävulinsäure. Photosynthetica **8**, 40–46.

WARNICK, G. R., and B. F. BURNHAM, 1971: Regulation of porphyrin biosynthesis. Purification and characterization of δ-aminolevulinic acid synthetase. J. Biol. Chem. **246**, 6880–6885.

WATTS, R. B., and R. G. O. KEKWICK, 1974: Factors affecting the formation of phytol and its incorporation into chlorophyll by homogenate of leaves of the french bean *Phaseolus vulgaris*. Arch. Biochem. Biophys. **160**, 469–475.

WELLBURN, A. R., 1970: Studies on the esterification of chlorophyllide. Phytochemistry **9**, 2311–2313.

WEIER, T. E., and D. L. BROWN, 1970: Formation of the prolammellar body in 8-day, dark-grown seedlings. Amer. J. Bot. **57**, 267–275.

WEINSTEIN, J. D., and P. A. CASTELFRANCO, 1978: Mg-protoporphyrin-IX and δ-aminolevulinic acid synthesis from glutamate in isolated greening chloroplasts. δ-Aminolevulinic acid synthesis. Arch. Biochem. Biophys. **186**, 376–382.

WELLBURN, A. R., 1976: Evidence for chlorophyll esterified with geranylgeraniol in newly greened leaves. Biochem. Physiol. Pflanzen **169**, 265–271.

WELLBURN, F. A. M., and A. R. WELLBURN, 1971: Chlorophyll synthesis by isolated intact etioplasts. Biochem. Biophys. Res. Comm. **45**, 747–750.

WETTSTEIN, D. v., K. W. HENNINGSEN, J. E. BOYNTON, G. C. KANNAGARA and O. F. NIELSEN, 1971: The genetic control of chloroplast development in barley. In: Boardman, N. K., A. W. Linnane and R. M. Smillie (eds.), Anatomy and biogenesis of mitochondria and chloroplasts, pp. 205–233, North Holland, Amsterdam.

WEINSTEIN, J. D., and P. A. CASTELFRANCO, 1977: Protoporphyrin IX biosynthesis from glutamate in isolated greening chloroplasts. Arch. Biochem. Biophys. **178**, 671–673.

WILSON, E. L., P. E. BURGER and E. B. DOWDLE, 1972: Beef-liver 5-aminolevulinic acid dehydratase: Purification and Properties. Eur. J. Biochem. **29**, 563–571.

WILLSTÄTTER, R., and A. STOLL, 1913: Untersuchungen über Chlorophyll. Springer, Berlin.

WITHROW, R. B., J. B. WOLFF and L. PRICE, 1956: Elimination of the lag phase of chlorophyll synthesis in dark-grown bean leaves by a pretreatment with low irradiances of monochromatic light. Plant Physiol. **31**, Suppl. XIII.

WOLFF, J. B., and L. PRICE, 1957: Terminal steps of chlorophyll *a* biosynthesis in higher plants. Arch. Biochem. Biophys. **72**, 293–301.

WOLFF, J. B., and L. PRICE, 1961: The effect of sugars on chlorophyll biosynthesis in higher plants. J. Biol. Chem. **235**, 1603–1608.

WONG, K. K., 1978: 4-Vinyl protochlorophyllide excretion by *Rhodopseudomonas gelatinosa* in nicotinamide-enriched medium. Plant Sci. Lett. **13**, 269–273.

WONG, K. K., 1978: Protochlorophyllide excretion by a purple sulphur bacterium. Microbios. Lett. **7**, 15–18.

XIFRA, E. A., WIDER DE A. M. DEL C. BATLLE and H. TIGIER, 1971: δ-Aminolaevulinate synthetase in extracts of cultured soybean cells. Biochim. Biophys. Acta **235**, 511–517.

XIFRA, E. A. WIDER DE, J. D. SANDY, R. C. DAVIS and A. NEUBERGER, 1976: Control of 5-aminolaevulinate synthetase activity in *Rhodopseudomonas spheroides*. Philos. Trans. R. Soc. Lond. B. Biol. Sci. **273**, 79–98.

YAMASAKI, H., and T. MORIYAMA, 1970: Inhibitory effect of α-ketoglutarate: Glyoxylate carboligase activity on porphyrin synthesis in *Mycobacterium phlei*. Biochem. Biophys. Res. Comm. **38**, 638–643.

YAMASAKI, H., and T. MORIYAMA, 1971: δ-Aminolevulinic acid dehydratase of *Mycobacterium phlei*. Biochim. Biophys. Acta **227**, 698–705.

ZAMAN, Z., M. N. ABBOUD and M. AKHTAR, 1972: Mechanism and stereochemistry of vinyl group formation in haem bisynthesis. Chem. Comm. **23**, 1262–1264.

ZAMAN, Z., and M. AKHTAR, 1976: Mechanism and stereochemistry of vinyl-group formation in haem biosynthesis. Eur. J. Biochem. **61**, 215–223.

The Biological Degradation of Chlorophyll in Senescent Tissues

Rudolf B. Kufner

Institut für Botanik und Pharmazeutische Biologie der Universität Würzburg

Green· trees in summer conspicuously show the metabolism of chlorophyll: in the spring and summer thousands upon thousands of tons of chlorophyll are doubtlessly synthesized on the earth. During the summer the degradation and production of chlorophyll are held at approximately equal rates. Finally in the autumn the large quantities of chlorophyll are rapidly and completely broken down (Fig. 1). Whereas the main steps in the biosynthesis of these noticeably green pigments are known (Schneider, 1978), knowledge of their degradation is still quite peicemeal. The gaps become especially clear when one attempts to collect and interpret the few investigations of this problem which have been performed to date. Thus, the following pages are intended to be more an inventory of the « lack » of knowledge than of the « stock » of sound knowledge in the field.

The degradation of chlorophyll in senescent tissues, such as in autumn leaves or in the color transformation of green, unripe fruit to ripe fruit, usually proceeds very

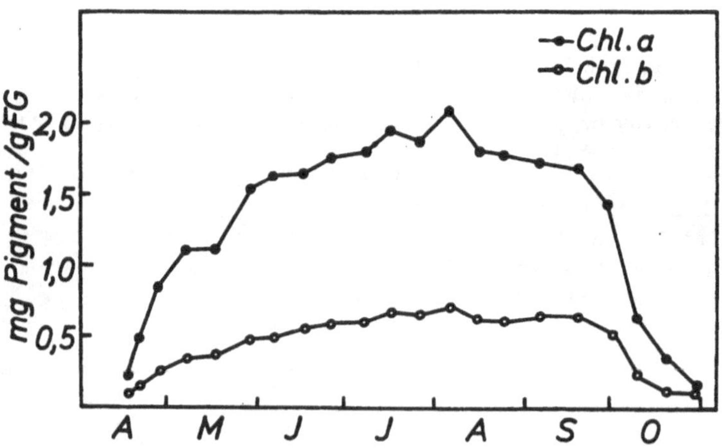

Fig. 1: Change in chlorophyll content of green leaves from *Acer campestre* during their development (pigment in mg/g fresh weight; A, M, J, ... = April, May, June ...). From Kufner, Dissertation, Würzburg, 1977

rapidly without leaving behind higly stable, high-molecular-weight derivatives. An oxidative breakdown to fragments of low molecular weight was therefore initially discussed as a possible degradation pathway (Egle, 1960; Chichester and Nakayama, 1965). These considerations are based on investigations by Noack (1944) who found that in the presence of certain catalysts in vitro chlorophyll derivatives such as chlorophyllides, pheophorbides, etc., are broken down by hydrogen peroxide into the end-products CO_2, NH_3, and H_2O. Egle (1960) applied these results to the biological conditions in the leaf and proposed the following degradation pathway. The chlorophyll is first released from the chlorophyll-protein complex. Then it is split by chlorophyllase into phytol and chlorophyllide. The chlorophyllide is then quickly oxidised with the aid of peroxides and trivalent iron. The first step of this degradation pathway appears to be quite probable. It is supported, among other things, by ultrastructural alterations in the plastids in senescent leaves of the most different plant species (cf., Kufner et al., 1978). The second step, the splitting of chlorophyll into chlorophyllide and phytol has not been generally established to date. The further degradation route, the rapid oxidative breakdown of chlorophyllide, appears to be at least remarkable, since porphyrins happen to be one of the stablest known groups of organic molecules. They can, for example, be detected in stone samples up to 400 million years old (Dunning, 1963).

An indication of the first steps of chlorophyll degradation has also been suspected to be shown by the frequent observation that the degradation of chlorophyll *a* (chla) precedes that of the higher oxidised chlorophyll *b* (chlb) in the breakdown of chlorophyll in senescent tissue (Wolf, 1956).

Thus, the chla/chlb ratio of 3 in green leaves of *Taxus baccata* slowly falls to approximately 2 in senescent leaves (Kufner et al., 1978). Chlb also proves to be generally less sensitive than chla in response to metabolic disturbances, such as in nitrogen-deficient algae cultures (Czygan and Kalb, 1966) and in callus tissue cultures (Scharlemann and Czygan, 1971), and in response to chemical effects in vitro, such as acid treatment (Egle, 1960). That this greater stability of the more highly oxidised chlorophyll components cannot be the key to the understanding of substantial steps in chlorophyll degradation, particularly (photo)oxidative degradation mechanisms, is shown by results from other experimental objects in which chlorophyll degradation proceeds without a shift of this type in the chla/chlb ratio. Thus, Sanger (1971) found a clear shift of the chla/chlb ratio in favour of chlb during the yellowing of leaves from *Quercus ellipsoidalis* and *Populus tremuloides*. In contrast, he did not find such a change during chlorophyll degradation in the leaves of *Corylus americana*. Specht-Jürgensen (1967a and b) also showed a clear shift of the chla/chlb ratio in favour of chlb in the end phase of chlorophyll degradation during yellowing of the leaves of *Ginkgo biloba* while still on the tree; whereas a shift in the ratio hardly occurs at al in isolated leaves senescing under light/dark alternation. In isolated leaves of *Ginkgo biloba* senescing in permanet darkness, chlb was actually degraded much more than chla.

Over the years a series of compounds has become known which could be biological degradation products of chlorophyll in higher plants, e.g.: chlorophyllides, pheophytins, pheophorbides, and chlorines (compilation in Simpson et al., 1976). Pheophytin and other porphyrins have also been found during chlorophyll degradation in one-celled green algae (Czygan, 1968b). These substances are also stored in mutants of algae with atypical chlorophyll metabolism (Czygan, personal communication). Unquestionably, however, many studies will be required in order to be able to discover the catabolic relationships between these compounds and to arrange them in a degradation scheme as attempted by Simpson et al. (1976, Fig. 2). Above all, there is a great deal of unclarity regarding the initial, comparitively simple degradation steps, particularly at the enzymatic level.

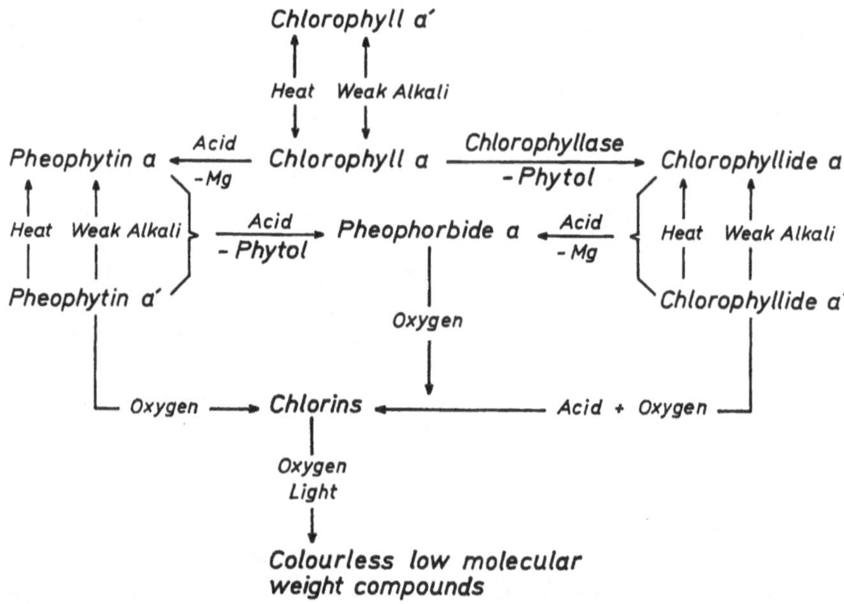

Fig. 2: Hypothetical degradation pathway of chlorophyll (based on: Simpson et al., 1977)

Thus, Csupor (1970) investigated phytol in the leaves and buds of *Acer platanoides* and *Aesculus hippocastanum*. Phytol is esterfied with linoleic acid in the yellow leaves of the *Acer*. Therefore, a new esterfication must occur in degradation of chlorophyll. However, the results of Park et al. (1973) indicate a particularly strong stability of phytol during yellowing. Supposedly a phytol-containing degradation product (or products?) exists which contains the porphyrin ring IV in which nitrogen is present partly in the imid form. The authors of this investigation consider the phytol ester bond to be so strong that they find the phytol to be suitable as a marker for the ring IV during degradation. The stability of the ester bond of phytol also applies to light-induced degradation of chlorophyll in vitro (Jen and Mackinney, 1970a and b; Morris et al., 1973). These few, and to a great extent contradictory, results concerning a relatively simple part of chlorophyll degradation illustrate the insufficient state of knowledge of this process which is so important in view of both its qualitative and quantitative significance in nature.

Even more difficulties will be presented by the explanation of the reactions in the following steps and in the end phase of chlorophyll degradation which Egle (1960) differentiates from biological chlorophyll degradation by calling it abiological pigment breakdown. The overlapping of the low-molecular-weight derivatives of chlorophyll with the low-molecular-weight degradation products of other compounds is the main aspect of this problem. Perhaps improved investigations following the method of McFeeters and Schanderl (1968) could contribute toward a solution. They developed an experimental system based on the incorporation of radioactive chlorophyll into ripening pepper carpels. However, chlorophyll degradation products have not been found to date using this « physiological system ».

Other investigations on objects in which chlorophyll degradation occurs, although not always in conjunction with general senescence and degeneration, show that the overlapping of chlorophyll degradation with a general degeneration in senescent leaves constitutes the main difficulty in explanation of a degradation pathway. In a few of these cases there are concrete indications of a chlorophyll degradation pathway and a reutilization of the degradation products.

Thus, cells of the green algae *Chlorella protothecoides* bleach following addition of glucose or acetate to a nitrogen-deficient medium (Oshio and Hase, 1969a, b). During these processes red pigments whose structure is similar to that of phycobilins are released into the feeding solution. Either they are direct degradation products of chlorophyll or they are synthesized from chlorophyll derivatives which are produced during bleaching.

The studies of Czygan (1967, 1968a, b) also refer to a possibility for reutilization of chlorophyllic degradation products. He investigated the ability of many Chlorococcales species (*Ankistrodesmus braunii*, *Chlorella fusca*, and *Scenedesmus obliquus*, etc.) to produce certain carotenoids, i.e., the so-called secondary carotenoids, during simultaneous chlorophyllic degradation under nitrogen deficiency. All experimental data indicate that these secondary carotenoids are synthesized differently from the primary carotenoids, and to be exact, whenever the formation of chlorophyll is disturbed or when chlorophyll degradation is more intensive than chlorophyll biosynthesis. According to this working hypothesis, the degradation products of chlorophyll are the precursors of the secondary carotenoids.

Finally, Rudolph et al. (1977) and Schmidt-Stohn (1977) found that the synthesis of sphagnorubin, a red membranochromous phenylpropane, in *Sphagnum magellanicum* swamp moss following cold treatment and glucose feeding is accompanied by noticeable changes in the chloroplast pigments. These changes suggest biogenetic relationships between chlorophyll degradation and the formation of sphagnorubin.

At the end of the long path to explanation of the biological degradation of chlorophyll in senescent tissues there will be questions on the further fate of the degradation broducts: which derivatives are lost with the falling leaf? Which are transported to pranches, to the trunk, or possibly even to the roots and stored for long or short periods of time and finally reutilized by the plant? Possibly these questions will open a new point of departure in investigations for explanation of the preceeding degradation steps.

References

CHICHESTER, C. O., and T. O. NAKAYAMA, 1965: Pigment Changes in Senescent and Stored Tissue. In: T. W. Goodwin (ed.): Chemistry and Biochemistry of Plant Pigments. p. 439–458. Academic Press. London-New York.

CSUPOR, L., 1970: Das Phytol in vergilbten Blättern. Planta medica **19**, 37.

CZYGAN, F.-C., 1967: Biosynthese der Carotinoide. Ber. dtsch. Bot. Ges. **80**, 627.

CZYGAN, F.-C., 1968a: Sekundärcarotinoide in Grünalgen, I. Chemie, Vorkommen und Faktoren, welche die Bildung dieser Polyene beeinflussen. Arch. Mikrobiol. **61**, 81.

CZYGAN, F.-C., 1968b: Sekundärcarotinoide in Grünalgen, II. Untersuchungen zur Biogenese. Arch. Mikrobiol. **62**, 209.

CZYGAN, F.-C., und KALB, 1966: Untersuchungen zur Biogenese der Carotinoide in *Trentepohlia aurea*. Z. Pflanzenphysiol. **55**, 59.

DUNNING, H. N., 1963: Geochemistry of organic pigments. In: Berger, I. A. (ed.): Organic Geochemistry. Pergamon Press. Oxford.

EGLE, K., 1960: Biologischer Chlorophyllabbau. In: Ruhland (ed.): Handbuch Pflanzenphysiol. **V/1**, p. 354–386. Springer-Verlag. Berlin-Göttingen-Heidelberg.

JEN, J. J., and MACKINNEY, 1970a: On the photodecomposition of chlorophyll *in vitro*. I. Reaction rates. Photochem. Photobiol. **11**, 297.

JEN, J. J., and G. MACKINNEY, 1970b: On the photodecomposition of chlorophyll *in vitro*. II. Intermediates and breakdown products. Photochem. Photobiol. **11**, 303.

KUFNER, R., F.-C. CZYGAN und L. SCHNEIDER, 1978: Veränderungen des Pigmentgehalts und der Ultrastruktur bei den Plastiden der Nadelblätter von Taxus baccata (L.) während ihrer Entwicklung. Ber. Deutsch. Bot. Ges. **91**, 325.

MCFEETERS, R. F., and S. H. SCHANDERL, 1968: Biological degradation of chlorophyll in a system using bell peppers (*Capsicum frutescens*) J. Food Sci. **33**, 547.

MORRIS, M. M., Y. PARK and G. MACKINNEY, 1973: On the photodecomposition of chlorophyll *in vitro*. J. Agr. Food Chem. **21**, 277.

NOACK, K., 1944: Über den biologischen Abbau des Chlorophylls. Biochem. Z. **316**, 166.

OSHIO, Y., and E. HASE, 1969a: Studies on red pigments excreted by cells of *Chlorella protothecoides* during the process of bleaching induced by glucose or acetate. I. Chemical properties of the red pigments. Plant Cell Physiol. **10**, 41.

OSHIO, Y., and E. HASE, 1969b: Studies on red pigments excreted by cells of *Chlorella protothecoides* during the process of bleaching induced by glucose or acetate. II. Mode of formation of the red pigments. Plant Cell Physiol. **10**, 51.

PARK, Y., M. M. MORRIS and G. MACKINNEY, 1973: On chlorophyll breakdown in senescent leaves. J. Agr. Food Chem. **21**, 279.

RUDOLPH, H., U. KABSCH and G. SCHMIDT-STOHN, 1977: Änderungen des Chloroplastenpigment-Spiegels bei *Sphagnum magellanicum* im Verlauf der Synthese von Sphagnorubin und anderer membranochromer Pigmente. Z. Pflanzenphysiol. **82**, 107.

SANGER, J. E., 1971: Quantitative investigations of leaf pigmets from their inception in buds trough autumn coloration to decomposition in falling leaves. Ecology **52**, 1075.

SCHARLEMANN, W., und F.-C. CZYGAN, 1971: Zusammensetzung und Stoffwechsel der Tetraterpene und Chlorophylle in Kalluskulturen von *Ruta graveolens*. Herba Hungarica **10**, 43.

SCHMIDT-STOHN, G., 1977: Änderung der Plastidenpigmente bei *Sphagnum magellanicum* Brid. in Abhängigkeit von Standort, Verfärbungsgrad und Alter. Z. Pflanzenphysiol. **81**, 289.

SCHNEIDER, H. A. W., Chlorophyll Biosynthesis. Enzymes and Regulation of Enzyme Activities. This Volume.

SPECHT-JÜRGENSEN, J., 1967a: Untersuchungen über Stickstoffverbindungen und Chlorophyll während des Vergilbens der Laubblätter von *Ginkgo biloba*. I. Am Baum alternde Blätter. Flora (Abt. A) **157**, 426.

SPECHT-JÜRGENSEN, J., 1967b: Untersuchungen über Stickstoffverbindungen und Chlorophyll während des Vergilbens der Laubblätter von *Ginkgo biloba*. II. Isolierte Blätter. Flora (Abt. A.) **157**, 471.

SIMPSON, K. L., TUNG-CHING LEE, D. A. RODRIGUEZ und C. O. CHICHESTER, 1976: Metabolism in senescent and stored tissues. In: Goodwin, T. W. (ed.): Chemistry and Biochemistry of Plant Pigments. p. 780–842. Academic Press. London-New York-San Francisco.

WOLF, F. T., 1956: Changes in chlorophyll a und b in autumn leaves. Am. J. Bot. **43**, 714.

Plant Biliproteins

W. Rüdiger

Botanisches Institut der Universität München

The term « biliprotein » has been used for a class of chromoproteins in which the chromophore is a bile pigment (« bilin ») specifically linked to a protein (Rüdiger, 1971). Whereas this specific linkage can be covalent or noncovalent in animal biliproteins (Rüdiger, 1970a), covalent linkages between chromophores and protein have been found in all plant biliproteins. Another universal property of plant biliproteins is the fact that ring A of the chromophores is hydrogenated « in situ » (i.e. when linked to the protein) (see below), in contrast to « normal » bilins (e.g., biliverdin IX α, Fig. 1) and biliproteins of animals.

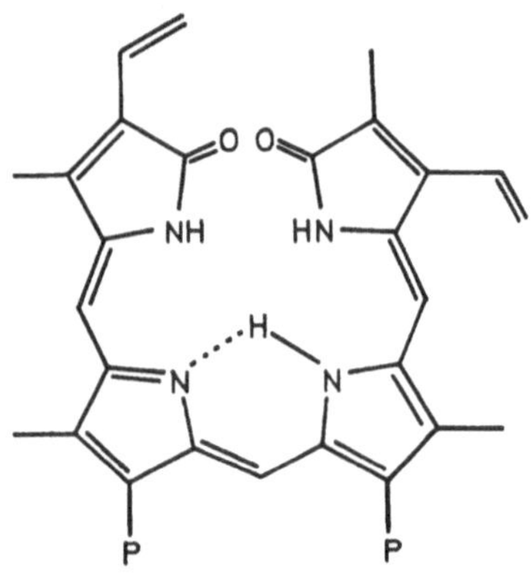

Fig. 1: Biliverdin IXα. The conformation given here (Sheldrick, 1976; Falk and Grubmayr, 1977) is presumbly valid for all bile pigments, but not for biliproteins (Scheer and Kufer, 1977). For convenience, « streched » formula are given for all following bilins independent of the true conformation

These two properties are not shared by « legcholeglobin », a green biliprotein found in old root nodules of Leguminosae as a degradation product of leghemoglobin. The chromophore of legcholeglobin is biliverdin IX α which is noncovalently attached to the globin moiety (Virtanen and Miettinen, 1949; Rüdiger and Lichter, unpublished results). But because leghemoglobin is of bacterial origin, its degradation product cannot be regarded as a true plant biliprotein and will therefore not be discussed further.

Two types of plant biliproteins are known. 1. The first type consists of light-harvesting pigments which act as « antenna » (or « accessory ») pigments in photosynthesis and occur in certain algae; they are called *phycobiliproteins*. 2. *Phytochrome* is a biliprotein of the second type and acts a « reaction center » pigment in a number of photomorphoses in higher plants.

1. Phycobiliproteins

1.1. Properties, Occurrence, Phylogeny, Function

Phycobiliproteins (phycocyanins, phycoerythrins) occur in Cyanophyceae (blue-green algae, which are correctly defined as Cyanobacteria; Stanier and Cohen-Bazire, 1977), Rhodophyceae (red algae) and Cryptophyceae.

The main chromophores of the phycobiliproteins are two phycobilins, namely phycocyanobilin (PCB) and phycoerythrobilin (PEB) (other chromophores occurring occasionally are discussed below). The properties (e.g., electronic spectra) of these two chromophores are modified by the particular proteins such that a number of phycocyanins and phycoerythrins of differing spectral types arise.

The usual nomenclature of phycobiliproteins reflects the history of detection (for a survey see O'hEocha, 1965): C-phycocyanin and C-phycoerythrin were found first in Cyanobacteria; R-phycocyanin and R-phycoerythrin were found in Rhodophyceae. Later B-phycoerythrin (in Bangiophycidae) and allophycocyanin (« another » spectral type) were discovered. These names have been maintained but are now used for biliproteins of a certain spectral type independent of occurrence. Recently a new type of biliprotein, phycoerythrocyanin (Bryant et al., 1976), and a second allophycocyanin (allophycocyanin B) have been described (Glazer and Bryant, 1975). In this terminology, the allophycocyanin detected first should logically be named allophycocyanin A, a name which has not been used so far. The significance of two (Gysi and Zuber, 1976) and three (Zilinskas et al., 1978) different allophycocyanins remains to be solved. Cryptomonad phycobiliproteins are classified according to their absorption maxima into two (Glazer and Cohen-Bazire, 1975) or three (O'Carra and O'hEocha, 1976) phycocyanins and three phycoerythrins (survey: Brooks and Gantt, 1973). Table 1 summarizes the spectral properties and distribution of well-known phycobiliproteins.

Only a range of approximate values for spectral data can be given because these data vary from organism to organism; furthermore, maxima are shifted to the red with higher aggregation. The aggregation state (see below) depends on pH, ionic strength, biliprotein concentration, etc. Spectral data of Table 1 are from dilute biliprotein solutions at near neutral pH (0.5–1.0 absorbance at maximum) and ionic strength equivalent to or near to 0.1 M KCl. The data of Table 1 should be sufficient for the classification of the main biliproteins. Some authors (Hirose et al., 1969) distinguish even two more phycoerythrin types. According to Fujita and Shimura (1974), R-phycoerythrin also occurs in a marine blue-green alga. Reports on the occurrence of double-banded phycoerythrins in Cyanobacteria have to be considered with caution because mono-banded C-phycoerythrin can be transformed to double-

Table 1: Spectral properties and distribution of phycobiliproteins (according to Rüdiger, 1971, complemented[1,2,3])

Biliprotein	Chromophore[a]	λ_{max} of absorption (nm)	λ_{max} of fluorescence (nm)	distribution
allophycocyanin	PCB	(610) 650	655–660	universal in Cyanobacteria and Rhodophyceae
allophycocyanin B[4]	PCB	618, 671	680	universal
C-phycocyanin	PCB	615–620	638–654	universal in Cyanobacteria and many Rhodophyceae
R-phycocyanin	PCB, PEB	555, 618	634–636	most Rhodophyceae
phycoerythrocyanin[5]	PCB, PVL	568	607–610	Anabaena sp.
Cryptomonad				
phycocyanin I[6,7,8]	PCB, PVD? and unknown bilin	581–585 641–645	655–661	various Cryptophyceae
phycocyanin II[5]	PCB and unknown bilin	577, 612	660?	
phycocyanin III[2]	PCB	585, 625–630	not available	Cryptomonas cyanomagna
C-phycoerythrin	PEB	555–562	575	Cyanobacteria, mainly Nostocales
B-phycoerythrin	PEB, PUB	546, 565	575–580	Rhodophyceae: Porphyridiales, Bangiales, Nemalionales
R-phycoerythrin	PEB, PUB	499, 540, 568	578–580	most Rhodophyceae
Cryptomonad				
phycoerythrin I[7,9,10]	PEB	542–545 (560)	584–586	
phycoerythrin II[9]	PEB	555–556		various
phycoerythrin III[9]	PEB	565–568	617	Cryptophyceae

[a] PCB = phycocyanobilin, PEB = phycoerythrobilin, PUB = phycourobilin, PVL = phycobiliviolin, PVD = phycobiliverdin.

[1] Glazer and Hixson, 1975.
[2] O'Carra and O'hEocha, 1976.
[3] Glazer, 1976.
[4] Glazer and Bryant, 1975.
[6] Mörschel and Wehrmeyer, 1975.
[7] MacColl, Berns and Gibbons, 1976.
[8] Glazer and Cohen-Bazire, 1975.
[9] Brooks and Gantt, 1973.

banded pigments by partial proteolysis which easily occurs in crude extracts of many Cyanobacterial cells (Muckle, 1978).

The most widely distributed phycobiliproteins are allophycocyanin and C-phycocyanin. These two seem to be the basic set of biliproteins in blue-green algae and some red algae. C-phycocyanin can be substituted by R-phycocyanin in many (but not all) red algae. The basic set can be supplemented by phycoerythrocyanin or either one of the phycoerythrins. Whereas constant biliprotein ratios are found in red algae (e.g., in *Porphyridium cruentum:* 84% B-phycoerythrin, 11% R-phycocyanin, 5% allophycocyanin, Gantt and Lipschultz, 1974; in *Rhodella violacea:* 58% B-phycoerythrin, 25% C-phycocyanin, 17% allophycocyanin, Koller et al., 1977), the ratios vary according to external conditions in many Cyanobacteria (Bennett and Bogorad, 1973; Tandeau de Marsac, 1977; Wagenmann, 1977).

Immunological studies employing the Ouchterlony double diffusion technique reveal that the phycobiliproteins of Cyanobacteria and red algae belong to three classes which are immunologically unrelated: allophycocyanin, C-phycocyanin and phycoerythrins (Glazer et al. 1971a; Bennett and Bogorad 1971). The immunological relationship within a biliprotein class can be demonstrated, however, even when the sources are very distant organisms, e.g., allophycocyanin from Cyanobacteria and red algae (Glazer, 1976; Cohen-Bazire et al., 1977). The same is true for C-phycocyanin from any source; this compound also cross-reacts with R-phycocyanin (Glazer and Hixson, 1975) and phycoerythrocyanin (Bryant et al., 1976). Cross-reactions have also been reported between C-phycoerythrin, B-phycoerythrin and R-phycoerythrin (Berns, 1967).

However, no immunological relationship was found between C-phycocyanin, allophycocyanin and C-phycoerythrin, on the one hand, and Cryptomonad biliproteins, on the other hand (Berns, 1967; Glazer et al., 1971a, b). Only some immunological relationship between Cryptomonad phycoerythrin and Cryptomonad phycocyanin and between Cryptomonad phycoerythrin and B-phycoerythrin has been reported (MacColl et al. 1976).

Phycobiliproteins of one class show a high degree of conservation of antigenic structure. Subtle differences between phycocyanins from various algae reported in double diffusion experiments (Berns, 1967) could not be reproduced later (Glazer et al., 1971a). However, the very sensitive « phage technique » revealed immunological differences between C-phycoerythrins from various Cyanobacteria (Eder et al., 1978).

The high degree of conservation of immunological properties corresponds to a high degree of homology of amino acid sequences of phycobiliproteins (Brown et al., 1975; Harris and Berns, 1975; Troxler et al., 1975; Glazer et al., 1976; Williams et al., 1978; Freidenreich et al., 1978; Bryant et al., 1978; Zuber, personal communication, Dec. 1977). The sequences are nearly identical in C-phycocyanin from various algae; very similar sequences are found in C-phycoerythrin and B-phycoerythrin. It has been suggested that all biliproteins originate from a common ancestor (Harris and Berns, 1975; Glazer, 1976; Glazer et al., 1976).

Several lines of evidence suggest the phylogenetic development from long-wavelength- to shorter-wavelength-absorbing pigments. Chapman et al. (1968) pointed out that C-phycocyanin is the principal biliprotein in most Cyanobacteria but only a minor component in the more advanced Rhodophyceae. PEB-containing biliproteins are more pronounced in Rhodophyceae than in Cyanobacteria, and PUB-containing biliproteins are restricted to Rhodophyceae. According to Glazer (1976), in terms of the number of chromophores and chromophore-chromophore interactions, allophycocyanin is simpler than either C-phycocyanin or C-phycoerythrin. He pointed out that both the physical location of the biliproteins and the energy transfer order (see below) follow the sequence phycoerythrin → phycocyanin → allophycocyanin. In evolution, the appearance of the acceptor molecule should have preceded that of the donor (for discussion of phylogeny see Brown et al., 1975; Troxler et al., 1975; Bogorad, 1975).

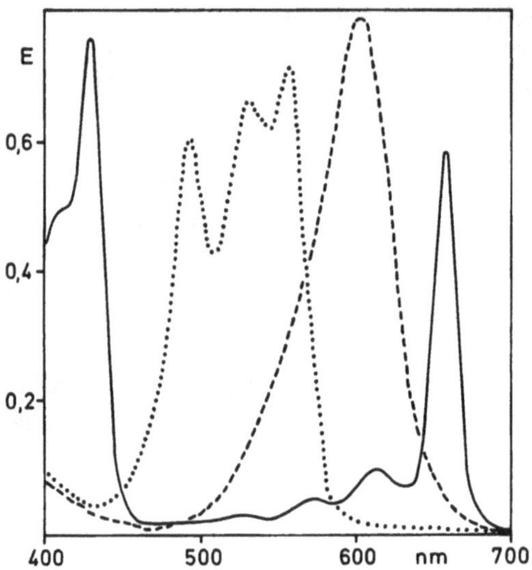

Fig. 2: Electronic spectra of ——— chlorophyll a in diethylether; – – – – C-phycocyanin from *Anacystis nidulans* in phosphate buffer, pH 7; ...... R-phycoerythrin from *Rhodymenia palmata* in phosphate buffer, pH 7

The evolution of shorter-wavelength absorbing phycobiliproteins means a possible ecological advance: chlorophylls absorb only in the red and the blue region of the visible spectrum, phycocyanins and especially phycoerythrins have maximal light absorption between the two chlorophyll bands, i.e., they fill this optical gap (Fig. 2). Biliprotein-containing plants – contrary to green plants – can perform photosynthesis in an environment in which red light has been filtered away by green plants (chlorophyll absorption) or a considerable layer of water (end absorption of OH infrared band). Björn (1975) has compared the solar spectrum at the surface and in 10 m water depths with biliprotein absorption.

The classical investigations on the action spectra of photosynthesis in red algae (Haxo and Blinks, 1950) were followed by determination of the fluorescence *in vivo*

(Arnold and Oppenheimer, 1950; Duysens, 1951; French and Young, 1952) which demonstrated energy migration from the biliproteins to chlorophyll. This energy transfer is very efficient (Tomita and Rabinowitch, 1962). Because phycobiliproteins can account for up to 24% of the dry weight (Myers and Kratz, 1955) or 40% of the total soluble protein (Gantt and Lipschultz, 1974) of algal cells, light harvesting by phycobiliproteins is very effective even at low light intensity. Dale and Teale (1970) developed the principal of «sensitizing» (shorter wavelength-absorbing) and «fluorescing» (longer wavelength-absorbing) chromophores within single biliproteins. This can only be observed if energy coupling occurs between the single chromophores; otherwise separate fluorescence of both types of chromophores would be obtained. The same principle can be applied to different biliproteins in the correct aggregation state. In the sequence phycoerythrin – (phycoerythrocyanin?) – phycocyanin – allophycocyanin – (allophycocyanin B), each former species should act as sensitizing and the subsequent species as fluorescing if energy coupling is efficient. A direct proof for energy transfer from phycoerythrin to allophycocyanin has been obtained by fluorescence determination *in vitro* (Gantt and Lipschultz, 1973) (Fig. 3): intact phycobilisomes (see below) show fluorescence emission only at 675 nm (room temperature) or 685 nm (−196 °C) which is the emission of allophycocyanin in an aggregated state (Gantt et al., 1977). This is remarkable because allophycocyanin is only a minor component. After dissociation of phycobilisomes in *Porphyridium cruentum*, fluorescence of all biliproteins appears according to their relative proportion and that of phycoerythrin predominates (Fig. 3). Thus, effective energy transfer is only possible in intact structures.

Phycobilisomes are granules (diameter 40 nm) on the stroma side of photosynthetic lamellae of Cyanobacteria and red algae as detected by electron microscopy (survey: Gantt, 1975). They contain all phycobiliproteins of the cell, but no chlorophyll. Although phycobiliproteins account for all protein of phycobilisomes in the case of *Porphyridium cruentum* (Gantt and Lipschultz, 1974), colorless proteins are possibly part of phycobilisomes in blue-green algae (Tandeau de Marsac and Cohen-Bazire, 1977). Studies of fluorescence emission, immunoprecipitation and electron microscopy of partially dissociated phycobilisomes led to a phycobilisome model with an allophycocyanin core surrounded first by phycocyanin and then by phycoerythrin in the outer sphere (Gantt et al., 1976). Energy migration occurs with high efficiency in isolated phycobilisomes; measured transfer times correspond to those calculated on the basis of the above-mentioned phycobilisome model (Grabowski and Gantt, 1977). The allophycocyanin part of phycobilisomes seems to be located directly on the photosynthetic lamella (Gantt et al., 1977). In intact cells, light absorbed by phycobilisomes drives photoreaction II but energy transfer from photosystem II to photosystem I is very effective. The yield of this energy transfer in *Porphyridium cruentum* has been determined to range from 0.5 (when the PS II reaction centers are all open) to 0.90–0.95 (when the PS II reaction centers are all closed) (Ley and Butler, 1976). Similary, an effective distribution of light absorbed by biliproteins to photosystems II and I has been found in *Anacystis nidulans* cells (Wang et al., 1977).

Cryptophyceae do not contain phycobilisomes. Cryptomonad biliproteins are

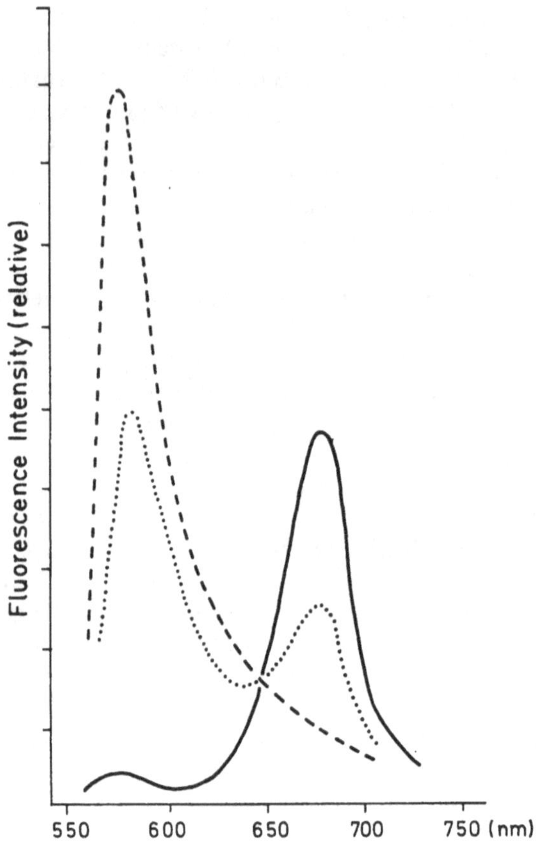

Fig. 3: Fluorescence emission spectra of isolated phycobilisomes from *Porphyridium cruentum*. ——— intact, ... partially, and – – – – totally dissociated phycobilisomes (Gantt and Lipschultz, 1973)

localized within intrathylakoidal spaces (Gantt et al., 1971; Glazer et al., 1971b; Wehrmeyer, 1970). No higher structures can be seen, but merely fine granular material which is believed to be dispersed biliprotein. The mechanism of energy transfer in Cryptophyceae has not yet been investigated in detail.

This difference *in situ* is reflected by aggregation properties of isolated phycobiliproteins *in vitro*. Phycobilisome-forming pigments are highly aggregated in neutral solution (molecular weight 200,000–300,000 daltons), whereas the intrathylakoidally dispersed pigments form only smaller aggregates (30,000–50,000 daltons).

1.2. Structural Organization of Phycobiliproteins

Independent of the aggregation properties, phycobiliproteins are composed of subunits. In most cases, these are light and heavy peptide chains (α- and β-subunits) in the molar ratio 1:1 (Table 2 and 3). Only few exceptions from this 1:1 ratio of subunits have been reported. A 2:3 ratio has been found in one (of three) allophycocyanins from *Nostoc* sp. (Zilinskas et al., 1978). The unusually heavy chain of B-phycoerythrin which makes up 1/6 of the light chain in *Porphyridium cruentum* (Gantt and Lipschultz, 1974) and about the same proportion in *Rhodella violacea* (Koller and Wehrmeyer, 1975) is not the β-chain but a third (γ) subunit (Glazer and Hixson, 1977). The subunits have been separated by electrophoresis or ion exchange chromatography after unfolding of the protein (e.g., with sodium dodecylsulfate or urea). Whereas the size of heavy chains (β-chains) is comparable in less aggregating (Cryptomonad) and aggregating (Cyanobacterial and Rhodophycean) biliproteins, the light chains (α-chains) are smaller in Cryptomonad than in the other phycobiliproteins (Table 3). The size of α- and β-chains is very similar in some biliproteins, especially in allophycocyanin of some algae (Table 2). The subunits cannot be separated by SDS gel electrophoresis in these cases; furthermore, the definition that $\alpha =$ light and $\beta =$ heavy chain does not hold. However, separation was possible, e.g., after unfolding with urea. The correlation with the α and β subunits of other algae was achieved by comparison of amino terminal sequences which are highly conservative and characteristic for α- and β-chains (Glazer et al., 1976), and also by immunochemistry [α- and β-chains cross-react either weakly (Takemoto and Bogorad, 1975) or not at all (Cohen-Bazire et al., 1977)], and by determination of the isoelectric point [α-chains have generally higher isoelectric points than β-chains (Cohen-Bazire et al., 1977)].

Removal of urea from solutions of unfolded subunits (e.g., by dialysis or gel filtration) leads to refolded peptide chains. Native biliproteins containing α- and β-chain in the molar ratio 1:1 have thus been obtained from mixtures of the α- and β-subunits (Glazer and Fang, 1973b; Cohen-Bazire et al., 1977). If isolated α-chains from one organism are mixed with isolated β-chains from another organism in this experiment, hybrid biliproteins are obtained, e.g., hybrid-C-phycocyanins (Glazer and Fang, 1973b) and hybrid allophycocyanins (Cohen-Bazire et al., 1977). Refolding of isolated subunits leads to « native » fluorescing peptide chains. The spectral characteristics of α- and β-subunits of C-phycocyanin (Glazer et al., 1973; Troxler et al., 1975), allophycocyanin (Cohen-Bazire et al., 1977) and C-phycoerythrin (Rüdiger and Zickendraht-Wendelstadt, unpublished results) have been determined. Refolded subunits of allophycocyanins exist in solution as α_2- or β_2-dimers, respectively, (Cohen-Bazire et al., 1977). This fact together with the high degree of homology of amino acid sequences between α- and β-chains of allophycocyanins (Glazer et al., 1976) suggests that contact areas for aggregation are similar in α- and β-chains. Therefore α_2- or β_2-dimers can be formed instead of the native ($\alpha\beta$) units.

Further aggregation stops in Cryptomonad biliproteins either at the ($\alpha\beta$) stage, e.g., phycoerythrin from *Rhodomonas* sp. and *Cryptomonas* sp. (Glazer et al., 1971b; Brooks and Gantt, 1973), or the ($\alpha_2\beta_2$) stage, e.g., phycoerythrin from *Cryptomonas maculata* and *Rhodomonas lens* (Mörschel and Wehrmeyer, 1977; MacColl et al., 1976), phycocyanin from *Chroomonas* sp., and *Hemiselmis virescens* (MacColl et al., 1973; Mörschel and Wehrmeyer, 1975*; Glazer and Cohen-Bazire, 1975).

* With two different subunits ($\alpha^1\alpha^2\beta_2$).

Cyanobacterial and Rhodophytan biliproteins form higher aggregates depending on concentration and pH. In allophycocyanin solutions trimers of the native unit $(\alpha\beta)_3$ predominate [Cohen-Bazire et al., 1977). Trimers $(\alpha\beta)_3$ have also been found in phycoerythrocyanin (Bryant et al., 1976). R- and C-phycocyanin form trimers at pH 7, but form hexamers $(\alpha\beta)_6$ at pH 4.5–5.5 (Glazer et al., 1973; Koller and Wehrmeyer, 1974; Glazer and Hixson, 1975); some monomer $(\alpha\beta)$ has occasionally been

Table 2: Phycobiliproteins of Cyanobacteria: molecular weight of subunits

Organism	mol.-weight · 10^3 α-chain	β-chain	References
C-phycoerythrin			
Aphanocapsa sp.	20.0	22.0	Glazer/Cohen-Bazire, 1971
Phormidium persicinum	19.7	22.0	O'Carra/Killilea, 1971
Fremyella diplosiphon	18.5	20.5	Takemoto/Bogorad, 1975
Nostoc sp.	16.9	20.0	Gray/Gantt, 1975
Pseudanabaena sp.	19.4	22.8	Wagenmann, 1977
Phormidium autumnale	16.0	19.0	Wagenmann, 1977
phycoerythrocyanin			
Anabaena 6411	16.0	20.2	Bryant et al., 1976
Anabaena variabilis	16.0	20.5	Bryant et al., 1976
C-phycocyanin			
Fremyella diplosiphon	16.3	17.6	Bennett/Bogorad, 1971
Aphanocapsa sp.	16.6	20.2	Glazer/Cohen-Bazire, 1971
Synechococcus sp. 6301 *	15.9	19.1	Glazer/Cohen-Bazire, 1971
Nostoc punctiforma Anabaena variabilis Anabaena cylindrica Calothrix scopulorum Anacystis nidulans *	18.5	20.5	O'Carra/Killilea, 1971
Mastigocladus laminosus	ca. 14.0	ca. 14.0	Binder et al., 1972
Oscillatoria agardhii	12.2	14.1	Torjesen/Sletten, 1972
Phormidium luridum	11.9	18.5	Kobayashi et al., 1972
Synechococcus 6301 *	17.7	19.0	Glazer/Fang, 1973 a
Anabaena 6411	17.1	19.0	Glazer/Fang, 1973 b
Nostoc sp.	14.7	16.3	Gray/Gantt, 1975
Phormidium luridum, PC-1	16.9	20.0	Brown et al., 1975
Phormidium luridum, PC-2	16.3	19.1	Brown et al., 1975
Anabaena 6411	16.5	19.4	Bryant et al., 1976
Anabaena variabilis	16.2	18.8	Bryant et al., 1976
allophycocyanin			
Fremyella diplosiphon	16.0	–	Bennett/Bogorad, 1971
Aphanocapsa sp.	16.0	(17.0 + 17.9)	Glazer/Cohen-Bazire, 1971
Synechococcus sp. 6301 *	15.2	(17.2)	Glazer/Cohen-Bazire, 1971
Phormidium luridum	17.3	19.0	Brown et al., 1975
Mastigocladus laminosus	17.2	17.2	Gysi/Zuber, 1976

Table 2 (continued)

Organism	mol.-weight · 10^3 α-chain	β-chain	References
Anabaena 6411	14.9	14.9	Bryant et al., 1976
Anabaena variabilis	14.8	14.8	Bryant et al., 1976
Synechocystis sp.	15.4	17.5	Cohen-Bazire et al., 1977
Synechococcus 6301 *	16.2	17.4	Cohen-Bazire et al., 1977
Synechococcus 6312	17.0	16.0	Cohen-Bazire et al., 1977
Nostoc sp.	16.9	18.4	Zilinskas et al., 1978
allophycocyanin B			
Synechococcus 6301	16.0	17.0	Glazer/Bryant, 1975
Anabaena variabilis	15.3	15.3	Glazer/Bryant, 1975

* Anacystis nidulans is identical with Synechococcus Sp. 6301; only the genus name Synechococcus should now be used (cf., Stanier and Cohen-Bazire, 1977).

Table 3: Phycobiliproteins of Rhodophyceae and Cryptophyceae: molecular weight of subunits

organism	biliprotein		mol.-weight · 10^3 α-chain	β-chain	References
Rhodophyceae					
Acrochaetium virgatulum			19.8	22.5	VAN DER VELDE, 1973
Porphyridium cruentum	B-PE		17.3	(30.0) *	Gantt/Lipschultz, 1974
Porphyridium cruentum	b-PE		17.2	–	Gantt/Lipschultz, 1974
Porphyridium cruentum	B- and b-PE		17.5	17.5	Glazer/Hixson, 1977 **
Rhodella violacea	B-PE I		18.9	(29.2) *	Koller/Wehrmeyer, 1975
Rhodella violacea	B-PE II		18.5	(29.9) *	Koller/Wehrmeyer, 1975
Porphyridium cruentum	R-PC		16.4	18.4	Gantt/Lipschultz, 1974
Porphyridium cruentum	R-PC		18.2	20.5	Glazer/Hixson, 1975
Cyanidium caldarium	C-PC		15.5	18.3	Troxler et al., 1975
Cyanidium caldarium	C-PC		12.5	16.5	Kao et al., 1975
Cryptophyceae					
Cryptomonas sp.	PE III		11.8	19.0	Glazer et al., 1971 b
Cryptomonas ovata	PE III		11.0	17.7	Brooks/Gantt, 1973
Cryptomonas ovata	PE III		9.7	18.2	MacColl et al., 1976
Rhodomonas lens	PE III		9.8	17.7	MacColl et al., 1976
Cryptomonas maculata	PE I		9.9	15.7	Mörschel/Wehrmeyer, 1977
Chroomonas sp.	PC I		10.0	16.0	MacColl et al., 1973
Chroomonas sp.	PC I	α_1	9.2	15.5	Mörschel/Wehrmeyer, 1975
		α_2	10.4		
Hemiselmis virescens Plymouth 157	PC II		10.0	19.0	Glazer/Cohen-Bazire, 1975
Hemiselmis virescens Millport 64	PC I		10.0	19.0	Glazer/Cohen-Bazire, 1975

* presumably γ-chain, cf., Glazer and Hixson, 1977.
** in addition B-PE has a γ-chain of $30.2 \cdot 10^3$ daltons, mol. composition $(\alpha\beta)_6\gamma$.

observed in addition to trimer and hexamer (Brown et al., 1975). Disc-shaped hexames with a central hole seen by electron microscopy of C-phycocyanin solutions (Berns and Edwards 1965) have been interpreted as aggregation of two trimers as part of a beginning crystallization (Bryant et al., 1976). Whereas C-phycoerythrin solutions contain monomers ($\alpha\beta$), dimers $(\alpha\beta)_2$ and hexamers $(\alpha\beta)_6$ (Muckle, 1977), no exact data on the composition of aggregated B-phycoerythrin is available. The heavy peptide chain («γ») makes up only 1/6 of the light chain (α) (Gantt and Lipschultz, 1974; Koller and Wehrmeyer, 1975); only compact aggregates without fine structures are seen under the electron microscope (Gantt 1969).

1.3. Phycobilins and their linkage to the protein

Whereas the variations of phycobiliproteins are large, only a limited number of chromophores exists (Table 1). The chemical structures of the two main chromophores phycocyanobilin (PCB) and phycoerythrobilin (PEB), and of their derivatives phycobiliviolin (PVL) and phycobiliverdin (PVD) are well established. In the case of phycourobilin (PUB) only spectral data which allow a classification of the spectral type are available. The PUB structure proposed by O'Carra and O'hEocha (1976) (see Fig. 4) is based on currently unpublished results.

The first structures to be elucidated were those of PVL and PVD obtained from phycobiliproteins using boiling methanol. Independent techniques, namely, NMR and mass spectroscopy (Crespi et al., 1967; Cole et al., 1967; Chapman et al., 1967; Crespi et al., 1968) and chromic acid oxidation with visible spectroscopy (Rüdiger et al., 1967; Rüdiger and O'Carra, 1969) revealed the structures given in Fig. 4. These structures are characterized by an ethylidene group at ring A and have been confirmed by total synthesis (Gossauer and Hirsch, 1973, 1974; Gossauer, personal communication, 1977).

Splitting of chromophores from the protein using boiling methanol or by other methods (enzymatic methods: Siegelman et al., 1967; Beuhler et al., 1976; HBr/trifluoracetic acid method: Schram and Kroes, 1971) results in a spectral shift, i.e., a structural change of the chromophores. To account for this spectral shift, Rüdiger (1970b) introduced the terms phycobiliviolin and phycobiliverdin for the free pigments (with ethylidene group), as opposed to the terms phycoerythrobilin and phycocyanobilin for the pigments *in situ* (protein-bound) with different spectral characteristics. Some authors use the latter terms for both groups. In such a case it would be useful at least to indicate the spectral characteristics of the particular phycobilin, e.g., by indicating the position of the main absorption band of the cation. Synonyms of this type are given in Fig. 4. The structures of the phycobilins *in situ* – together with the covalent protein linkage at ring A – have been elucidated by 3 sets of data from: 1) spectral characterization, 2) analysis of chromopeptides, and 3) investigation of reactions which break the linkage at ring A.

Characterization of bile pigments by the visible spectrum makes better use of more than of one set of data, e.g., the position of the long-wavelength band of the free base, cation, and zinc complex. In these three sets of data the position of the respective absorption band depends on the number of double bonds in the chromophore system (Fig. 5). Thus, if the spectral data are known, a classification of the spectral type (= number of conjugated double bonds, see Fig. 4) is possible. This empirical

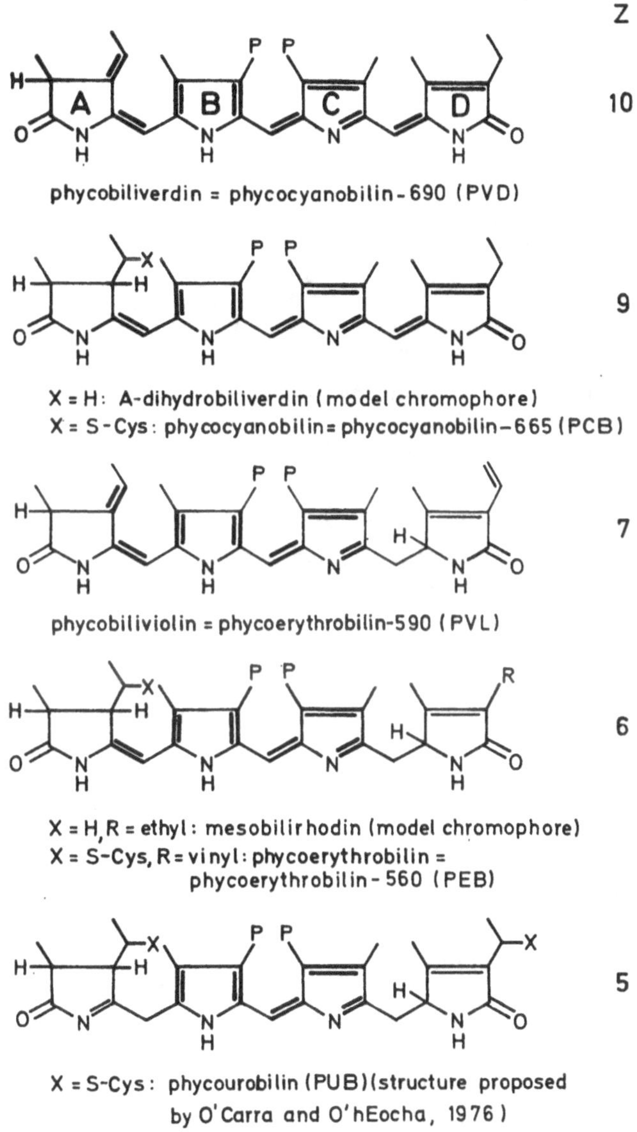

Fig. 4: Structures of phycobilins and related pigments. z = number of double bonds relevant to the chromophore system. Counting of two double bonds in ring C is arbitrary but equally applied to all pigments

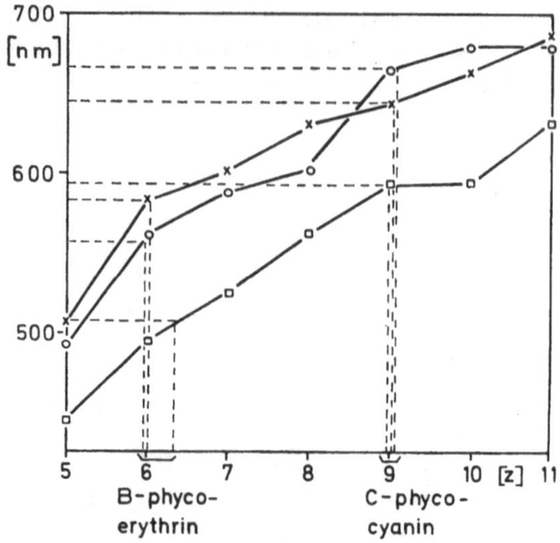

Fig. 5: Absorption maxima of bile pigments in relation to the number of conjugated double bonds (z). □–□–□ free base, ○–○–○ cation, ×–×–× zinc complex (according to data of Köst et al., 1975; Scheer, 1976)

deduction is only valid if all bile pigments have the same conformation. This is presumably true for the most stable isomers, whereas geometrical isomeric bile pigments exhibit different conformation and spectral properties (Falk and Grubmayr, 1977). Phycobiliproteins can be included in this comparison of absorption bands if the specific influence of the protein upon spectral properties of the chromophore is eliminated. This can be achieved by proteolysis or – more simply – by unfolding of the peptide chain (e.g., with urea or guanidinium chloride). Both methods give qualitatively and quantitatively the same results (Köst et al., 1975; Muckle, 1978): proteolyzed or denatured phycoerythrins (B-PE and C-PE) show spectral properties equivalent to 6 double bands, i.e., they have the same chromophore as mesobilirhodin (Rüdiger et al., 1970). Phycocyanin shows 9 double bonds, i.e., the same chromophore as A-dihydrobiliverdin (Scheer 1976). This means that the native chromophores PEB and PCB have hydrogenated A rings (without an ethylidene double bond).

Calculation of the chromophore content of individual subunits (Table 4) is permitted by comparison of extinction coefficients of unfolded biliproteins and their isolated subunits in 8 M urea at pH 2–3 with the extinction coefficients of the cations of phycoerythrobilin, phycocyanobilin and model compounds. Difficulties arise by bleaching of chromophores during unfolding of biliproteins and isolation of subunits (Byfield and Zuber, 1972; Glazer and Hixson, 1977; Muckle, 1978). Extremely low extinction coefficients, such as 25,549 for the α-, and 48,456 for the β-subunit of C-phycoerythrin from *Fremyella diplosiphon* (Takemoto and Bogorad, 1975), must be due to this bleaching, or to some colorless impurities (urea?) in the isolated subunits. Reinvestigation of C-phycoerythrin from the same organism (Muckle

Table 4: Long-wavelength maxima and extinction coefficients of unfolded biliproteins and chromophore content of their subunits. Spectra determined in 8 M urea at pH 2–3 corresponding to cations of chromophores

biliprotein	λ_{max}	$\varepsilon \cdot 10^3$ per $(\alpha\beta)_1$	$\varepsilon \cdot 10^3$ per chromophore	number and type of chromophores α-subunit	β-subunit	Ref.
allophycocyanin	662.5	64.4[a]	32.2	1 PCB	1 PCB	Glazer/Fang 1973a
C-phycocyanin	662.5	106.5	35.5	1 PCB	2 PCB	Glazer/Fang 1973a
R-phycocyanin	662	72.4	36.2	1 PCB	1 PCB	Glazer/Hixson 1975
	560	42.4[b]	42.4		1 PEB	Glazer/Hixson 1975
phycoerythrocyanin	660	~70[c]	~35		2 PCB	Bryant et al. 1976
	590	~43[c]	~43 (?)	1 PVL (?)		Bryant et al. 1976
C-phycoerythrin I	550	261.0	43.5	2 PEB	4 PEB	Muckle/Rüdiger 1977
C-phycoerythrin II	550	203.0	40.5	2 PEB	3 PEB	Muckle 1978
B-phycoerythrin	550	257[d]	42.9	2 PEB	4 PEB	Glazer/Hixson 1977
b-phycoerythrin	550	245.0	40.8	2 PEB	4 PEB	

[a] Glazer and Fang (1973a) calculated only the value 32,200 per subunit = 16,500 daltons.
[b] Measured value: 58,000, of which 15,600 is calculated for 2 PCB at this wavelength.
[c] Only approximate calculation possible because of overlapping bands.
[d] Per $\alpha\beta$ calculated from the original $(\alpha\beta)_6\gamma$, the γ-subunit contains 2 PEB + 2 PUB.

and Rüdiger, 1977) revealed a coefficient of 251,000 corresponding to 6 PEBs with a ratio of 1:2 for the chromophore content of α- and β-subunits. This phycoerythrin belongs to type I with 2 PEBs on the α-, and 4 PEBs on the β-subunit. The same type of C-phycoerythrin is also found in *Aphanocapsa* sp. (Glazer and Hixson, 1975; Muckle and Rüdiger, 1977) and *Phormidium persicinum* (Muckle and Rüdiger, 1977). A second type of C-phycoerythrin with 2 PEBs on the α- and 3 PEBs on the β-subunit is found in *Pseudanabaena* W 1173 (Wagenmann, 1977; Muckle and Rüdiger, 1977) and in *Tolypothrix tenuis* (Muckle and Rüdiger, 1977). If it turns out to be true that evolution proceeds from biliproteins of lower chromophore content to biliproteins with higher chromophore content (Glazer, 1976), then the type II C-phycoerythrin should be the more ancient one.

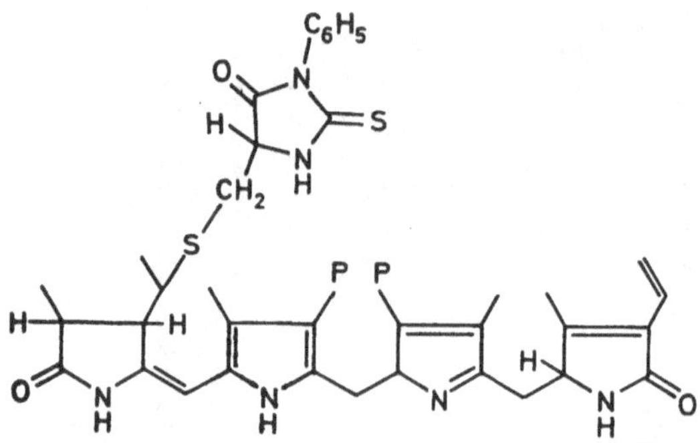

Fig. 6: Structure of PTH-cysteine-chromophore from B-phycoerythrin (Köst-Reyes et al., 1975)

A second type of information on the chromophore structure and its protein linkage *in situ* came from analysis of chromopeptides. The first peptide analyzed with respect to this question (Köst-Reyes et al., 1975) was prepared from B-phycoerythrin (from *Porphyridium cruentum*) by pepsin digestion. It contained the chromophore and the amino acids cysteine, valine and leucine. Cysteine (N-terminal) is split off by Edman degradation in the form of phenylthiohydantoin which still carries the chromophore (Fig. 6). The ethylidene double bond is not present until cleavage of the chromophore from the compound – comparable to the situation in the original phycoerythrin (Köst et al., 1975). The question as to whether cysteine is the linking amino acid of all chromophores was answered in C-phycoerythrin from *Phormidium persicinum* and *Pseudanabaena* W 1173 (Muckle, 1978; Muckle et al., 1978): six and five chromopeptides, respectively, which are present in the trypsin digest in equimolar amounts, were isolated and sequenced (Table 5). In each case, Edman degradation removed the amino acid cysteine together with the chromophore. Because some chromophore is always cleaved during the Edman reaction, chromopeptides lose their colour after a couple of degradation steps, and thus no clear decision can be made on the position of the chromophore. Therefore only small peptides with the linking amino acid close to the N-terminus can be investigated; larger peptides must be further fragmented

Table 5: Sequences of chromopeptides isolated from the trypsin digest of C-phycoerythrin from *Phormidium persicinum* and *Pseudanabaena* W 1173 (Muckle et al., 1978). Cysteine is always removed together with the chromophore during sequencing

	peptide number	*Pseudanabaena* W 1173	*Phormidium persicinum*
α-subunit	1	Gly–Cys–Ser–Pro–Arg	Gly–Cys–Ala–Pro–Arg
	2	Cys–Leu–Arg	Cys–Leu–Arg
β-subunit	3	Gly–Leu–Cys–Arg	Gly–Ala–Cys–Leu–Arg
	4	Phe–Pro–Gly–Val–Gly–Asx–Leu–(X)*–Cys–Ala–Ser–Ile–(Ala, Glx)–Y*	(Phe/Leu)–Leu–(Gly/Val)–(Gly, Leu, Cys, Ala, Ser, Ile, Pro, Thr)
	5	Ile–Leu–Asx–Cys–Leu–(Tyr/Gly)–Asx–Val–(Ser, Glx, Gly_{1-2}, Ala_2, Arg)	Ile–Leu–Glx–Cys–Leu–Gly–Asx–(Val/Leu)–(Pro, Gly, Ile)–(Ser, Ala, Arg) Gly–Ala–Gly–Cys–(Leu/Thr)–Gly–Pro–Leu–Asx–Glx–Gly–Gly–Ile–Val–Ser–Lys

* X and Y together constitute Gly, Glx, Asx, Ser, Thr, and Arg. The position of these amino acids and their distribution between X and Y is unknown.

Table 6: Sequences of the bilin-containing peptides in biliproteins of the phycocyanin class

organism	biliprotein, subunit		Ref.
		75 80 85	
Mastigocladus laminosus	CPC, α	Ala–Ala–Arg–Gly–Lys–Ser–Lys–Cys–Ala–Arg–Asp–	a
Synechococcus 6301	CPC, α	–Thr–Pro–Glu–Gly–Lys–Ala–Lys–Cys–Ala–Arg–Asp	b
		75 80 85	
Mastigocladus laminosus	CPC, β	Arg–Asn–Gly–Thr–Met–Ala–Ala–Cys–Leu–Arg–Asp–	a
		79 85	
Synechococcus 6301	CPC, β	Ala–Ala–Cys–Leu–Arg–Asp–Met	b
Porphyridium aerugineum	CPC, β	Ala–Ala–Cys–Leu–Arg–Asx–Met	c
Porphyridium cruentum	RPC, β	Ala–Ala–Cys–Leu–Arg–Asx–Met	c
		140 145	
Mastigocladus laminosus	CPC, β	Gly–Ile–Thr–Lys–Gly–Asp–Cys–Ser–Ala–Leu–Ile–Ser	a
		145 150 155	
Synechococcus 6301	CPC, β	–Asn–Gly–Ile–Thr–Gln–Gly–Asp(Cys, Ser) Ala–Ile–Ser	b
Porphyridium aerugineum	CPC, β	(Asx, Thr_2–Ser_{1-2}, Glx, Gly_4, Cys, Ala_{1-2}, Ile_2, Leu_3, Tyr, Phe)	c
Porphyridium cruentum	RPC, β	–Gly–Gly–Pro–Ile–Gly–Asp(Cys*, Ser, Glx, Leu, Thr, Glx, Leu, Ala, Ser, Tyr, Phe)	c
Anabaena variabilis	PEC, β	–Lys–Gly–Ile–Thr–Lys–Gly–Asp–Cys–Ser–Gln–Leu–Ile–Ala–	c

CPC = C-phycocyanin; RPC = R-phycocyanin; PEC = phycoerythrocyanin; underlined Cys residues carry the phycocyanobilin chromophore; Cys* carries the phycoerythrobilin chromophore; a = Zuber (personal communication, Dec. 1977); b = Williams/Glazer, 1978; c = Bryant et al., 1978.

first. Another approach employed ^{35}S incorporation onto B-phycoerythrin from *Porphyridium cruentum.* Peptic digestion yielded a large number of chromopeptides, in which the ^{35}S/chromophore ratio was constant throughout. This indicates that a sulphur-containing amino acid (presumably cysteine) is present in all chromopeptides (Köst 1978).

In C-phycocyanins all three chromophores seem to be linked through a thioether bond to cysteine of the peptide chain. Analysis of peptides from C-phycocyanin – although still impure – indicated that cysteine was always present in chromopeptides (Crespi and Smith, 1970). Removal of the chromophore together with cysteine during sequencing was observed in two chromopeptides from *Mastigocladus laminosus* C-phycocyanin (Byfield and Zuber, 1972). This is also true of the third chromopeptide from this organism (Zuber, personal communication, Dec. 1977). In a series of recent papers, Glazer and co-workers investigated the linkages of all chromophores in C-phycocyanins and the β-subunits of R-phycocyanin and phycoerythrocyanin (Williams et al. 1978; Freidenreich et al. 1978; Bryant et al. 1978). These authors found that chemical cleavage of the chromophores either from the subunits or from chromopeptides by Ag^{+}, Hg^{2+}, or HBr in trifluoroacetic acid resulted in the release of previously blocked thiol groups. Because equimolar amounts of chromophore and thiol were released in any case and because each chromophore was singly investigated in the form of the respective chromopeptide, a linkage to cysteine was deduced for all chromophores. Contrary to these findings, Brown and Troxler (1977) reported on the possibility of bilin-apoprotein linkages only in part through cysteine, and in part through another amino acid. According to these authors, 3 cysteine residues and 3 chromophores are present in C-phycocyanin from *Cyanidium caldarium.* Only part of these cysteine residues is blocked (presumably by the chromophores); more than half of the cysteines can be carboxyamidated. Brown and Troxler concluded that those cysteines which can be carboxyamidated cannot carry chromophores; therefore other amino acids have to be discussed for the linkage at ring A of the chromophore. However, the question of whether or not a thioether can be carboxyamidated (to a sulfonium compound) under the applied conditions still has to be checked.

Amino acid sequences around the 3 chromophores are highly conservative in the phycocyanin class, i.e., C-phycocyanin, R-phycocyanin and phycoerythrocyanin (Table 6). Deviating sequences of C-phycocyanin from *Mastigocladus laminosus* have been reported erroneously (Byfield and Zuber, 1972) and are corrected here (Zuber, personal communication, Dec. 1977). With this correction, the earlier assumption of different amino acid sequences around PEB- and PCB-chromophores (Muckle et al., 1978) must be modified. Although only short sequences are available for C-phycoerythrin, homology can be recognized between peptide 2 of PE (Table 5) and the chromophore region of the α-chain of PC (Table 6), furthermore, between peptide 3 of PE and the sequence around Cys81 (or 82) of the PC β-chain, and between peptide 4 of PE and the sequence around Cys 146 (or 152) of PC (cf., Table 5 and 6). Interestingly, homology seems to be more pronounced between peptide 4 of PE and the PEB-carrying region around Cys 152 of RPC than between the same PE peptide and the

PCB-carrying region around the same Cys 152 (or 146) of CPC and PEC. More data are necessary, however, before eventually typical sequences of phycoerythrins and phycocyanins or PEB- and PCB-carrying regions can be recognized.

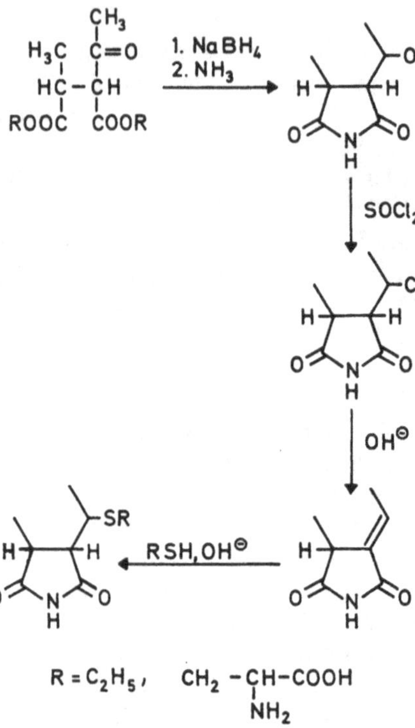

Fig. 7: Synthesis of thioether imides as model compounds for the thioether linkage of biliproteins (Schoch et al., 1974, 1976)

Cysteine as the linking amino acid suggests a thioether linkage similar to that in cytochrome c; which is contrary to earlier suggestions (see Rüdiger, 1971). The direct proof for a thioether linkage came from a study of the cleavage of this linkage. For this purpose, model compounds were synthesized (Fig. 7) and submitted to the same reaction sequence as biliproteins and chromopeptides (Schoch et al., 1974, 1976). The exact analysis of the reaction sequence in the case of model compounds and the good correlation with the reaction in the case of biliproteins and chromopeptides unequivocally proved the thioether linkage in phycoerythrins and phycocyanins. The most conclusive reaction sequence is shown in Fig. 8. Chromic acid oxidation at 20 °C which degrades tetrapyrroles to the respective succinimides (and maleimides) oxidized the thioether to a sulfon. The absolute configuration of one sulfon isomer (with R = ethyl) obtained by this reaction was elucidated by X-ray crystallography (Lotter et al., 1977). The sulfon is easily eliminated in slightly alkaline medium (e.g., dilute ammonia) at room temperature to give the ethylidene imide which is either

E- or Z- configurated depending on the configuration of the sulfon (and the thioether of course) (Klein, 1977; Klein and Rüdiger, 1978). Elimination at 100 °C from biliproteins (Köst and Rüdiger 1974) as well as from model compounds (Schoch et al. 1976) always yields a mixture of the Z- and E-ethylidene compounds. The results of the elimination of biliproteins at room temperature are only compatible with the absolute configuration of the chromophore ring A as indicated in Fig. 9 (Klein, 1977; Klein and Rüdiger, 1978).

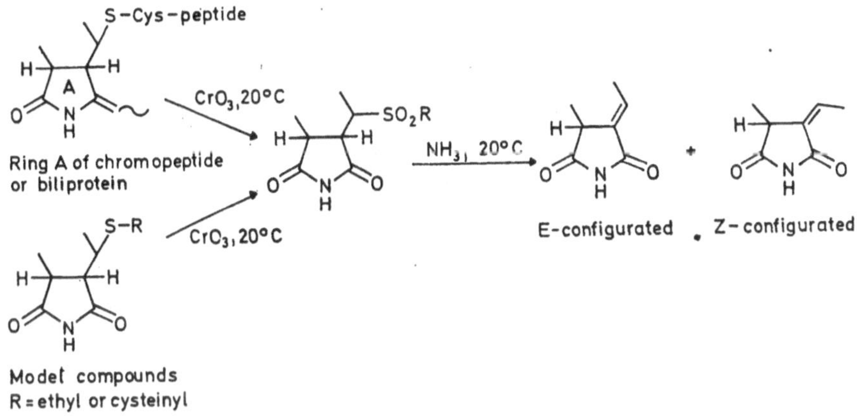

Fig. 8: Elimination reaction («chromic acid-ammonia degradation») of biliproteins and model compounds (Klein, 1977; Klein and Rüdiger, 1978)

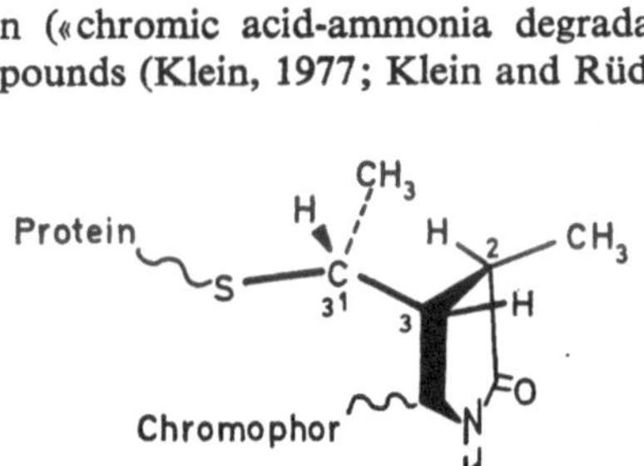

Fig. 9: Absolute configuration at chromophore ring A in biliproteins derived from degradation experiments (Klein, 1977; Klein and Rüdiger, 1978)

1.4. Phycobilin Biosynthesis and Its Regulation

Biosynthesis of bile pigments is better investigated in animals than in plants. Microsomal heme oxygenase oxydizes heme (= protoporphyrin Fe II complex) to the bile pigment biliverdin and carbon monoxide in animals (Tenhunen et al., 1969). When the same reaction takes place in plants, carbon monoxide and phycobilin with specific activities of 1:7 are expected after incorporation of 5-^{14}C-5-aminolevulinic acid (Fig. 10).

Such labelling experiments were first performed by Troxler (1972) with *Cyanidium caldarium*. This alga produces much phycocyanin in the light after a dark period,

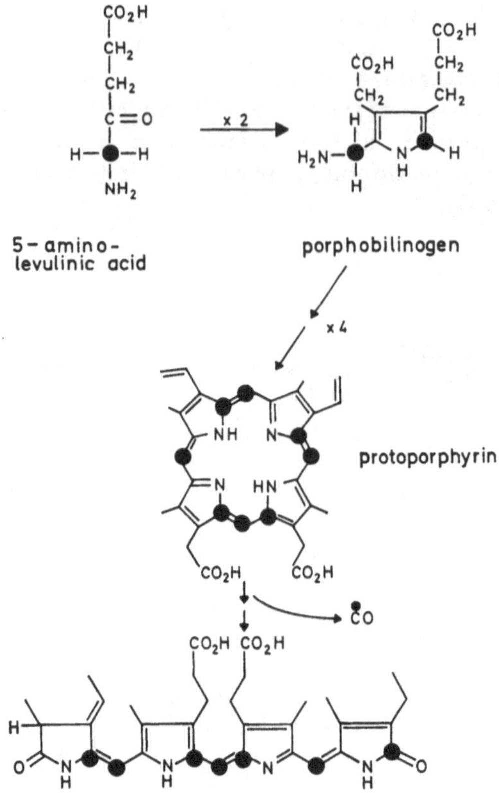

Fig. 10: Biosynthesis of phycobilins. Black dots indicate the expected label after incorporation of 5-^{14}C-aminolevulinic acid

whereby aminolevulinic acid is specifically incorporated into the chromophore (Troxler and Brown, 1970). Indeed the production of carbon monoxide always parallels that of phycobilin; the specific activities are found as expected (Table 7). This is also true for mutants which lack chlorophyll (mutant GGB); mutants II-C and GGB-Y without phycobilin do not produce carbon monoxide. Parallel formation of carbon monoxide and phycobilins has also been proved in the red alga *Porphyridium cruentum* and a number of Cyanobacteria (Troxler and Dokos, 1973). Whereas the specific activities in red algae are 1:7 as expected, aminolevulinic acid is metabolized in Cyanobacteria such that no specific labelling can be observed (Troxler and Brown, 1975).

In conclusion, biosynthesis of phycobilins apparently proceeds via oxidative ring opening of a porphyrin ring analogous the biosynthesis of bile pigments in animals. However, it is not yet clear whether a Fe-porphyrin (or another metal complex) is an intermediate and whether the ethylidene group is introduced before or after the

Table 7: Formation and specific activity of carbon monoxide and phycobiliverdin in *Cyanidium caldarium*. Incorporation of 5-^{14}C-5-aminolevulinic acid in the light (according to Troxler, 1972)

genotype	chlorophyll (μmol)	phycobiliverdin (μmol)	phycobiliverdin (Ipm/μmol)	carbon monoxide (μmol)	carbon monoxide (Ipm/μmol)
wild strain	45	40	7×7600	39	7359
mutant strain III-D-2	65	56	7×8000	54	8100
mutant strain GGB	0	33	7×8800	35	9000
mutant III-C	25	0		0	
mutant GGB-Y	0	0		0	

oxidation of the porphyrin ring. Interestingly, bacteriochlorophyll b has an ethylidene group in the corresponding position (Scheer et al., 1974).

Addition of the thiol group of cysteine to the ethylidene group of phycobilins is presumably the last step of biosynthesis of the biliprotein subunits.

Biosynthesis of phycobiliproteins can be regulated by light. Whereas the red alga *Cyanidium caldarium* produces phycocyanin in the light but not in the dark, some Cyanobacteria (e.g., *Tolypothrix tenuis*) are able to produce biliproteins in the dark as well under some conditions. Whether phycocyanin or phycoerythrin synthesis predominates in the dark depends on the preillumination (red or green light) (for a review and discussion of possible mechanisms see Bogorad, 1975). This is a special case of the long-known chromatic adaptation which has mostly been studied in continuous light. Green light (or fluorescent light) leads to enhanced phycoerythrin formation, red light to (relatively) more phycocyanin and less phycoerythrin. Not all cyanobacteria exhibit chromatic adaptation. Tandeau der Marsac (1977) distinguishes 3 physiological groups among 44 investigated strains of phycoerythrin-containing Cyanobacteria: group I does not adapt chromatically, group II adapts by modification of phycoerythrin synthesis; group III adapts by modification of both phycoerythrin and phycocyanin synthesis. Wagenmann (1977) subdivides groups II and III into those organisms which only reduce phycoerythrin production in red light («partial adaptation») and those which do not produce any phycoerythrin in red light («total adaptation»). Some phycocyanin (and allophycocyanin) is always produced, which is consistent with energy transfer from phycoerythrin via these pigments. Chromatic adaptation is also found in 2 *Phormidium* species which do not produce any phycoerythrin: the pigment observed absorbing at 570–580 nm in green but not in red light (Wagenmann, 1977) is presumably phycoerythrocyanin (Bryant et al., 1976).

Readaptation – i.e., transfer of fluorescent-light-adapted cells to red-light-adaptation and vice versa – has been investigated by Bennett and Bogorad (1973) with *Fremyella diplosiphon*. Fluorescent light induces, and red light stops, phycoerythrin synthesis in this organism; no metabolic breakdown of phycoerythrin is observed under any light in growing cultures. The chromatic adaptation in this organism includes other physiological changes, such as mean filament length reduction in red light (Bennett and Bogorad, 1973) and reduced dark growth rate upon brief green irradiation (Diakoff and Scheibe, 1975). The photoreversibility of these physiological responses led the latter authors to the assumption of a unique phytochrome-like, green-red photoreversible pigment in Cyanophyta, although Bogorad (1975) had pointed out that the action spectra are somewhat different in different organisms, which suggests a variety of photopigments. Preliminary reports on a green-red photoreversible pigment in *Tolypothrix tenuis* (Scheibe, 1972; Diakoff and Scheibe, 1973) have apparently not been followed up. A more recent report on 3 photochromic pigments from Cyanobacteria (Björn and Björn, 1976) also has to be considered as preliminary. The photoreversibility of isolated (?) fractions makes up an absorbance difference of merely about 10^{-3} absorption units whereas the absorption amounts to 1–2 absorption units. The small temperature increase caused by irradiation can easily cause a 0.1 % absorbance change (with difference spectra similar to those published by Björn and Björn (1976) in phycocyanin and phycoerythrins (Rüdiger, unpublished results). The questions of the existence of photoreversible pigments in Cyanobacteria and the mechanism of their physiological action remain to be elucidated.

2. Phytochrome

Many – but not all – plant photomorphoses are induced most effectively by red light (660 nm) and reversed by far-red light (730 nm). The photoreceptor pigment for these photomorphoses has been called « phytochrome ». It exists in two photoreversible forms. The P_r form is biosynthesized in the dark. Red light transforms P_r to the P_{fr} form which causes photomorphoses. The following simplified scheme can be given:

$$P_r \underset{\text{far-red light}}{\overset{\text{red light}}{\rightleftarrows}} P_{fr} \cdots\cdots\rightarrow \text{photomorphoses}$$

Only aspects of phytochrome as a plant pigment will be delt with in the following. For reviews on physiological aspects see Mitrakos and Shropshire (1972) and Smith (1975).

Phytochrome is considered to be a typical pigment of higher plants although some physiological effects due to phytochrome have also been reported in the green alga *Mougeotia* (Haupt, 1959) and the red alga *Porphyra tenera* (Dring, 1967). Phytochrome from the alga *Mesotaenium* and the liverwort *Sphaerocarpos* possibly has spectral properties somewhat different from the pigment of higher plants (a blue shift of 10–20 nm for both photoreversible forms has been observed in partially purified preparations; Taylor and Bonner, 1967) but this question has to be reinvestigated.

Phytochrome has been isolated most often from etiolated oat seedlings (*Avena sativa* L.) which contain much more pigment than other plants (Siegelman and Firer, 1964) or from etiolated annual rye (*Secale cereale*) (for a survey of isolation procedures see Rüdiger, 1972a; Pratt, 1973). Many of the early investigators isolated only « small phytochrome » (peptide chain of 60,000–70,000 daltons) which is a proteolysis product of « large phytochrome » (peptide chain of 120,000 daltons) (for a review see Briggs and Rice, 1972). Kinetics of the formation of «small phytochrome» in extracts of oat seedlings demonstrated that it is not present *in vivo*, but only after extraction, i.e., as a proteolysis artefact of isolation (Grombein and Rüdiger, 1976). Spectral properties of native high molecular phytochrome differ somewhat from those of the proteolyzed pigment (Pratt and Cundiff, 1975; Pratt, 1975).

Immunocytochemical methods show that phytochrome is distributed throughout the cytoplasm in cells of etiolated oat and rice shoots prior to any light treatment (P_r form) but becomes associated with discrete regions of the cell within some minutes after red light treatment (P_{fr} form) (Mackenzie et al., 1975). After far-red irradiation (i.e., = phototransformation back to P_r), phytochrome is slowly (1–2 hours) redistributed throughout the cytoplasm. These properties correspond to the good solubility of P_r, and enhanced pelletability of P_{fr}. The type of binding induced by red irradiation of zucchini squash extracts *in vitro* (Quail et al., 1973) is at least in part artifactual binding of P_{fr} to degraded ribosomal material by electrostatic forces (Quail 1975a, b). Another type of binding is found in many plants after irradiation *in vivo* which can be demonstrated by pelletability in extracts (Grombein et al., 1975b; Pratt and Marmé, 1976). Such phytochrome-containing pellets have been analyzed by density gradient centrifugation (Marmé et al., 1976) and by gel chromatography (Jose, 1977) but the nature of the assumed phytochrome receptor is still under question. It is interesting that redissolved pelleted phytochrome is partially recovered in the exclusion peak of gel chromatography. This fraction could possibly contain material from the assumed receptor (Grombein and Rüdiger, 1976).

The biliprotein nature of phytochrome was proposed as early as 1950 because of the similarities between the action spectra of physiological responses and the absorption spectra of C-phycocyanin and allophycocyanin (for a historical review see Borthwick, 1972). The first indication that this proposal was correct came from experiments of Siegelman et al. (1966). These authors were able to cleave the phytochrome chromophore (in poor yield) from the protein. The cleaved chromophore had spectral and chromatographic properties similar to the chromophore from C-phycocyanin and allophycocyanin, but was not identical to it. Furthermore, it yielded a mono- and a dimethyl ester which were identified chromatographically. Therefore the parent chromophore must have been a dicarboxylic acid. Further investigation was hampered by the poor yield of free chromophore. Poor yields of the chromophore were also found by Kroes (1970) using different cleavage reactions.

To avoid the difficulties in the cleavage reaction, a microdegradation with chromic acid was applied to the chromoprotein phytochrome itself. This reaction splits tetrapyrroles at the methine bridges and yields succinimides or maleimides (depending on the degree of saturation of the pyrrole ring) from the outer rings, and pyrrole dialdehydes from the inner ring. The dialdehydes are further oxidized to maleimides (Rüdiger, 1969). The reaction cannot only be applied to free tetrapyrrole chromophores, but also to biliproteins. The linkages between the protein and the individual rings (after oxidation of pyrroles in the form of maleimide or succinimide rings) have to be cleaved by a separate reaction (for application to phycocyanin and phycoerythrin see p.332).

At first a denatured phytochrome sample was investigated using this method (Rüdiger and Correll, 1969). This sample was no longer photoreversible, but the results proved for the first time that the phytochrome chromophore is a bile pigment yielding the expected degradation products. In particular, the rings with propionic acid side chains are the inner rings because only the pyrrole dialdehyde with this side chain is formed. The results obtained from the outer rings were somewhat different when reexamined using native phytochrome. The vinyl group of ring D was apparently modified in the denatured product of the first investigation, but later proved to be present as such in native phytochrome (Rüdiger, 1972b). The treatment of the denatured sample with acid and alkali which caused a shift of endo- and exocyclic double bonds in ring A (Rüdiger and Correll 1969) is not equivalent to the photochemical transformation: P_r and P_{fr} possess a hydrogenated ring A as shown by spectroscopic properties (Grombein et al., 1975a) and new degradation experiments (Klein et al. 1977). Whereas the previously applied «hydrolytic chromic acid degradation» did not yield any product from ring A of phytochrome (Rüdiger, 1972b), the combined chromic acid-ammonia degradation (see Fig. 8) unequivocally proved the configuration, the hydrogenated ring A, and the thioether linkage, all of which are identical with the phycocyanin and phycoerythrin chromophore (see Fig. 9). The degradation results are qualitatively and quantitatively identical with P_r and P_{fr}. This means that β-pyrrolic positions and protein linkages were not changed during the phototransformation of P_r to P_{fr} (Klein et al., 1977).

Table 8: Absorption maxima (nm) of chromophores of biliproteins and bile pigments in 6 M guanidinium chloride (from Grombein et al., 1975a; Scheer, 1976)

compound	number of conjugated double bonds[b]	Cation ($pH < 2$)	Base (pH 8)	Anion ($pH > 11$)	Zinc complex
mesobiliverdin[a]	11	685	630–655	–	688
phytochrome P_r	9[c]	675–680	620–625	765–770	650 (590)
C-phycocyanin	9	665–670	610 (590)	765–770	640 (590)
A-dihydrobiliverdin[a]	9	665	594	766	638
phytochrome P_{fr}	8[c]	610–615	($\rightarrow P_r$)	($\rightarrow P_r$)	($\rightarrow P_r$)
mesobiliviolin	8	590–600	555	–	–
mesobilipurpurin	8	590–595	540	–	–

[a] In methanol. [b] See Fig. 3. [c] Plus one for the vinyl side chain which is only cross-conjugated.

The spectral properties of unfolded phytochrome (P_r form) are consistent with an A-dihydrobiliverdin chromophore, i.e., with 9 conjugated double bonds, provided that one also considers the cross-conjugated double bond of the vinyl group (Table 8). When the P_{fr} form of phytochrome is unfolded (by guanidiniumchloride) in the dark in acid medium, a spectrum is obtained which is consistent with only 8 conjugated double bonds (Table 8). This form is only stable as a cation. It is transformed into the P_r chromophore if one tries to obtain the free base or anion (Grombein et al., 1975a). Furthermore, it is immediately transformed into the P_r cation in the light (Grombein et al., 1975a). This is illustrated in the following scheme:

P_r (native) 662 nm	$\xrightarrow{(h\nu)660\text{ nm}}$ $\xleftarrow{(h\nu)730\text{ nm}}$	P_{fr} (native) 718 nm
↓ guanidinium chloride pH 1.5		↓ guanidinium chloride pH 1.5
P_r (denatured) 675 nm	$\xleftarrow{(h\nu)600\text{ nm}}$	P_{fr} (denatured) 615 nm

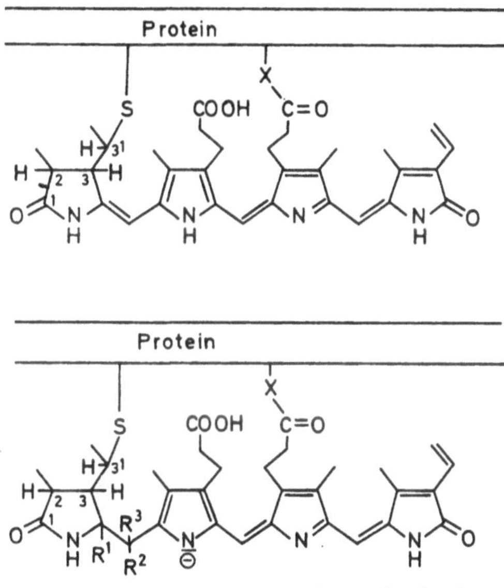

P_r

P_r

Fig. 11: Structures of the phytochrome chromophores in the P_r and P_{fr} form (Klein et al., 1977). R^1 and R^2 are assumed to be functional groups of the protein (thiol ?); R^3 = hydrogen

The chemical structures given in Fig. 11 have been derived from these data for the phytochrome chromophore in the P_r and P_{fr} form (Klein et al., 1977). Whereas the structure of the P_r chromophore is well established, the P_{fr} structure is based on the few available spectral data and on the reactivity of model compounds. The methine bridge between ring A and B of dihydrobiliverdin can easily be eliminated by photooxygenation (yielding a purpurin and a formyltripyrrinone; Scheer et al., 1977) or photoaddition (yielding a photo-dimer under nitrogen with $K_3[Fe(CN)_6]$; Scheer and Krauss, 1977). It is evident from the formulae of the reaction products (Fig. 12) that the methine bridge can undergo oxidation and addition reactions. However, no indication of a hydrogenation has been found in any of the reactions. Whereas the photooxygenation reactions are irreversible, the photodimerization has been reversed by heat (Scheer and Krauss, 1977). But no photoreversion was possible. A photoreversible product could possibly be formed, however, by adding functional groups of the protein (thiol groups?) photochemically at the methine bridge (R^1 and R^2 in Fig. 11).

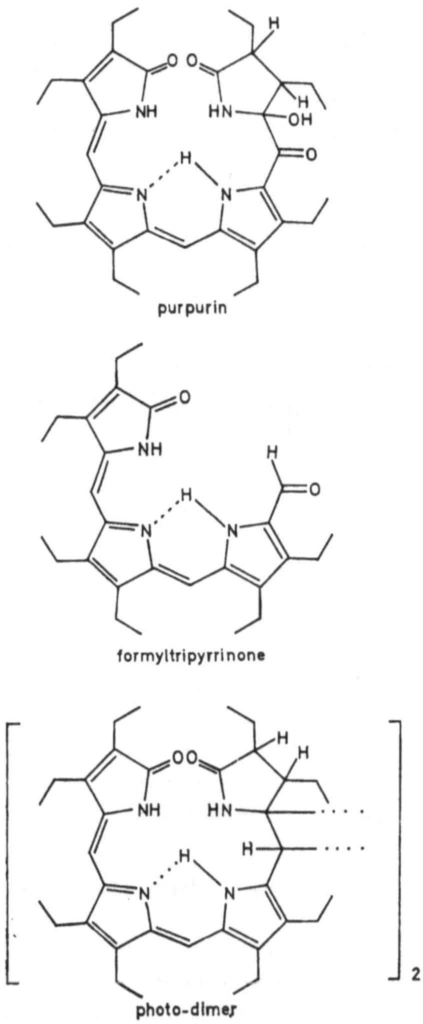

Fig. 12: Photooxygenation and photodimerization products of dihydrobiliverdin (Scheer et al., 1977; Scheer and Krauss, 1977). Stereoisomers of the photodimer are observed; dotted lines are linkages to corresponding methine bridge of the second chromophore forming a cyclobutan ring (Scheer, personal communication, 1977)

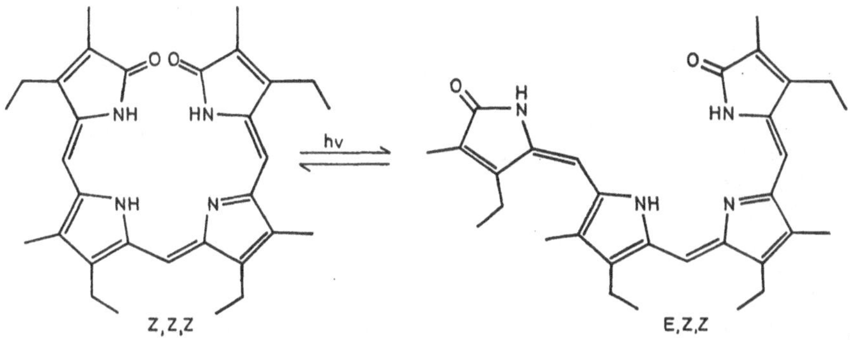

Fig. 13: Geometric isomers of a biliverdin obtained by irradiation. Pigment is adsorbed at Al_2O_3. Photoequilibrium is reached from either isomer (Falk and Grubmayr, 1977)

Another model reaction which is photoreversible has been reported recently (Falk and Grubmayr, 1977). The « normal » Z,Z,Z biliverdin can be photoisomerized to the E,Z,Z compound if it is adsorbed at aluminium oxide (Fig. 13). The hypsochromic shift (38 nm as free base) corresponds to the spectral « loss » of one double bond. It remains to be seen as to whether such a reaction is possible with A-dihydrobiliverdin. In any case, the most probable candidate for the form absorbing at 730 nm is the anion of the respective chromophore (cf., Fig. 5).

The cited investigations of the author were supported by the Deutsche Forschungsgemeinschaft and the Fonds der Chemischen Industrie.

References

ARNOLD, W., and R. R. OPPENHEIMER, 1950: Internal conversion in the photosynthetic mechanism of blue-green algae. J. gen. Physiol. **33**, 423.

BENNETT, A., and L. BOGORAD, 1971: Properties of subunits and aggregates of blue-green algal biliproteins. Biochemistry **10**, 3625–3634.

BENNETT, A., and L. BOGORAD, 1973: Complementary chromatic adaptation in a filamentous blue-green alga. J. Cell. Biol. **58**, 419–435.

BERNS, D. S., and M. R. EDWARDS, 1965: Electron micrographic investigations of C-phycocyanin. Arch. Biochem. Biophys. **110**, 511–516.

BERNS, D. S., 1967: Immunochemistry of Biliproteins. Plant Physiol. **42**, 1569–1586.

BEUHLER, R. J., R. C. PIERCE, L. FRIEDMAN, and H. W. SIEGELMAN, 1976: Cleavage of Phycocyanobilin from C-Phycocyanin. Separation and Mass Spectral Identification of the Products. J. Biol. Chem. **251**, 2405–2411.

BINDER, A., K. WILSON, and H. ZUBER, 1972: C-Phycocyanin from the thermophilic blue-green alga *Mastigocladus laminosus*. Isolation, characterization and subunit composition. FEBS Letters **20**, 111–116.

BJÖRN, L. O., 1975: Photobiologie, Licht und Organismen. G. Fischer Verlag Stuttgart, p. 48.

BJÖRN, G. S., and L. O. BJÖRN, 1976: Photochromic Pigments from Blue-Green Algae: Phycochromes a, b, and c. Physiol. Plant. **36**, 297–304.

BOGORAD, L., 1975: Phycobiliproteins and Complementary Chromatic Adaptation. Ann. Rev. Plant Physiol. **26**, 369–401.

BORTHWICK, H. A., 1972: History of Phytochrome. In: K. Mitrakos and W. Shropshire jr. (Ed.) Phytochrome, Academic Press New York, p. 3–23.

BRIGGS, W. R., and H. V. RICE, 1972: Phytochrome: chemical and physical properties and mechanism of action. Ann. Rev. Plant Physiol. **23**, 293–334.

BROOKS, C., and E. GANTT, 1973: Comparison of phycoerythrins (542, 566 nm) from Cryptophycean algae. Arch. Mikrobiol. **88**, 193–204.

BROWN, A. S., J. A. FOSTER, P. V. VOYNOW, C. FRANZBLAU, and R. F. TROXLER, 1975: Allophycocyanin from the Filamentous Cyanophyte, *Phormidium luridum*. Biochemistry **14**, 3581–3588.

BROWN, A. S., and R. F. TROXLER, 1977: Bilin-Apoprotein Linkages in Rhodophytan Phycobiliproteins: The Role of Cysteine. FEBS Letters **82**, 206–210.

BRYANT, D. A., A. N. GLAZER, and F. A. EISERLING, 1976: Characterization and Structural Properties of the Major Biliproteins of *Anabaena* sp. Arch. Microbiol. **110**, 61–75.

BRYANT, D. A., C. S. HIXSON, and A. N. GLAZER, 1978: Structural Studies on Phycobiliproteins. III. Comparison of Bilin-Containing Peptides from the β Subunits of C-phycocyanin, R-phycocyanin, and Phycoerythrocyanin. J. Biol. Chem. **253**, 220–225.

BYFIELD, P. G. H., and H. ZUBER, 1972: Chromophore-Containing Peptide Sequences in C-Phycocyanin from *Mastigocladus laminosus*. FEBS Letters **28**, 36–40.

CHAPMAN, D. J., W. J. COLE, and H. W. SIEGELMAN, 1967: The Structure of Phycoerythrobilin. J. Amer. chem. Soc. **89**, 5976–5977.

CHAPMAN, D. J., W. J. COLE, and H. W. SIEGELMAN, 1968: Phylogenetic implications of phycocyanobilin and C-phycocyanin. Amer. J. Bot. **55**, 314–316.

COHEN-BAZIRE, G., S. BÉGUIN, S. RIMON, A. N. GLAZER, and D. M. BROWN, 1977: Physico-Chemical and Immunological Properties of Allophycocyanins. Arch. Microbiol. **111**, 225–238.

COLE, W. J., D. J. CHAPMAN, and H. W. SIEGELMAN, 1967: The Structure of Phycocyanobilin. J. Amer. chem. Soc. **89**, 3643–3645.

CRESPI, H. L., L. J. BOUCHER, G. D. NORMAN, J. J. KATZ, and R. C. DOUGHERTY, 1967: Structure of Phycocyanobilin. J. Amer. chem. Soc. **89**, 3642–3643.

CRESPI, H. L., U. SMITH, and J. J. KATZ, 1968: Phycocyanobilin. Structure and Exchange Studies by Nuclear Magnetic Resonance and Its Mode of Attachment in Phycocyanin. A Model for Phytochrome. Biochemistry **7**, 2232–2242.

CRESPI, H. L., and U. H. SMITH, 1970: The chromophore-protein bonds in phycocyanin. Phytochemistry **9**, 205–212.

DALE, R. E., and F. W. J. TEALE, 1970: Number and Distribution of Chromophore Types in Native Phycobiliproteins. Photochem. Photobiol. **12**, 99–117.

DIAKOFF, S., and J. SCHEIBE, 1973: Action spectra for chromatic adaptation in *Tolypothrix tenuis*. Plant Physiol. **51**, 382–385.

DIAKOFF, S., and J. SCHEIBE, 1975: Cultivation in the Dark of the Blue-green Alga *Fremyella diplosiphon*. A Photoreversible Effect of Green and Red Light on Growth Rate. Physiol. Plant. **34**, 125–128.

DRING, M. J., 1967: Phytochrome in red alga, *Porphyra tenera*. Nature **215**, 1411–1412.

DUYSENS, L. M. N., 1951: Transfer of light energy within the pigment systems present in photosynthesizing cells. Nature **168**, 548–550.

EDER, J., R. WAGENMANN, and W. RÜDIGER, 1978: Immunological Relationship Between Phycoerythrins From Various Blue-Green Algae. Immunochemistry **15**, 315–321.

FALK, H., and K. GRUBMAYR, 1977: Ein geometrisch isomeres Bilatrien-abc. Angew. Chem. **89**, 487–488; Angew. Chem. Int. Ed. **16**, 470–471.

FREIDENREICH, P., G. S. APELL, and A. N. GLAZER, 1978: Structural Studies on Phycobiliproteins. II. C-phycocyanin: Amino Acid Sequence of the β Subunit. Specific Cleavage of the α Subunit. J. Biol. Chem. **253**, 212–219.

FRENCH, C. S., and V. K. YOUNG, 1952: The fluorescence spectra of red algae and the transfer of energy from phycoerythrin to phycocyanin and chlorophyll. J. gen. Physiol. **35**, 873–890.

FUJITA, Y., and S. SHIMURA, 1974: Phycoerythrin of the marine blue-green alga *Trichodesmium thiebautii.* Plant Cell Physiol. **15**, 939–942.

GANTT, E., 1969: Properties and ultrastructure of phycoerythrin from *Porphyridium cruentum.* Plant Physiol. **44**, 1629–1638.

GANTT, E., M. R. EDWARDS, and L. PROVASOLI, 1971: Chloroplast structure of the Cryptophyceae. Evidence for phycobiliproteins within intrathylakoidal spaces. J. Cell. Biol. **48**, 280–290.

GANTT, E., and C. A. LIPSCHULTZ, 1973: Energy transfer in phycobilisomes from phycoerythrin to allophycocyanin. Biochim. Biophys. Acta **292**, 858–861.

GANTT, E., and C. A. LIPSCHULZT, 1974: Phycobilisomes of *Porphyridium cruentum*: Pigment analysis. Biochemistry **13**, 2960–2966.

GANTT, E., 1975: Phycobilisomes: Light-Harvesting Pigment Complexes. BioScience **1975**, 781–788.

GANTT, E., C. A. LIPSCHULTZ, and B. A. ZILINSKAS, 1976: Further evidence for a phycobilisome model from selective dissoziation, fluorescence emission, immunoprecipitation, and electron microscopy. Biochim. Biophys. Acta **430**, 375–388.

GANTT, E., C. A. LIPSCHULTZ, and B. A. ZILINSKAS, 1977: Phycobilisomes in Relation to the Thylakoid Membranes. Brookhaven Symp. Biol. **28**, 347–357.

GLAZER, A. N., and G. COHEN-BAZIRE, 1971: Subunit structure of the phycobiliproteins of blue-green algae. Proc. Nat. Acad. Sci. USA **68**, 1398–1401.

GLAZER, A. N., G. COHEN-BAZIRE, and R. Y. STANIER, 1971a: Comparative Immunology of Algal Biliproteins. Proc. Nat. Acad. Sci. USA **68**, 3005–3008.

GLAZER, A. N., G. COHEN-BAZIRE, and R. Y. STANIER, 1971b: Characterisation of phycoerythrin from a *Cryptomonas* sp. Arch. Mikrobiol. **80**, 1–18.

GLAZER, A. N., and S. FANG, 1973a: Chromophore content of blue-green algal phycobiliproteins. J. Biol. Chem. **248**, 659–662.

GLAZER, A. N., and S. FANG, 1973b: Formation of hybrid proteins from the α- and β-subunits of phycocyanins of unicellular and filamentous blue-green algae. J. Biol. Chem. **248**, 663–671.

GLAZER, A. N., S. FANG, and D. M. BROWN, 1973: Spectroscopic Properties of C-Phycocyanin and of Its α and β Subunits. J. Biol. Chem. **248**, 5679–5685.

GLAZER, A. N., and D. A. BRYANT, 1975: Allophycocyanin B (λ_{max} 671, 618 nm). A New Cyanobacterial Phycobiliprotein. Arch. Microbiol. **104**, 15–22.

GLAZER, A. N., and G. COHEN-BAZIRE, 1975: A Comparison of Cryptophytan Phycocyanins. Arch. Microbiol. **104**, 29–32.

GLAZER, A. N., and C. S. HIXSON, 1975: Characterization of R-Phycocyanin. Chromophore content of R-Phycocyanin and C-Phycoerythrin. J. Biol. Chem. **250**, 5487–5495.

GLAZER, A. N., 1976: Phycocyanins: Structure and Function. Photochem. Photobiol. Rev. **1**, 71–115.

GLAZER, A. N., G. S. APELL, C. S. HIXSON, D. A. BRYANT, S. RIMON, and D. M. BROWN,

1976: Biliproteins of Cyanobacteria and Rhodophyta: Homologous family of photosynthetic accessory pigments. Proc. Nat. Acad. Sci. USA **73**, 428–431.

GLAZER, A. N., and C. S. HIXSON, 1977: Subunit Structure and Chromophore Composition of Rhodophytan Phycoerythrins. J. Biol. Chem. **252**, 32–42.

GOSSAUER, A., and W. HIRSCH, 1973: Synthesen von Gallenfarbstoffen II. Totalsynthese des Phycocyanobilins. Tetrahedron Letters **1973**, 1451–1453.

GOSSAUER, A., and W. HIRSCH, 1974: Totalsynthese des racemischen Phycocyanobilins (Phycobiliverdins) sowie eines «Homophycobiliverdins». Liebigs Ann. Chem. **1974**, 1496–1513.

GRABOWSKI, J., and E. GANTT, 1978: Excitation Energy Migration in Phycobilisomes: Comparison of Experimental Results and Theoretical Predictions. Photochem. Photobiol. **28**, 47–54.

GRAY, B. H., and E. GANTT, 1975: Spectral Properties of Phycobilisomes and Phycobiliproteins from the Blue-Green Alga *Nostoc* sp. Photochem. Photobiol. **21**, 121–128.

GROMBEIN, S., W. RÜDIGER, and H. ZIMMERMANN, 1975a: The Structures of the Phytochrome in Both Photoreversible Forms. Hoppe-Seyler's Z. Physiol. Chem. **356**, 1709–1714.

GROMBEIN, S., W. RÜDIGER, L. PRATT, and D. MARMÉ, 1975b: Phytochrome Pelletability in Extracts of Avena Shoots. Plant Science Letters **5**, 275–280.

GROMBEIN, S., and W. RÜDIGER, 1976: On the Molecular Weight of Phytochrome: a New High Molecular Phytochrome Species in Oat Seedlings. Hoppe-Seyler's Z. Physiol. Chem. **357**, 1015–1018.

GYSI, J., and H. ZUBER, 1976: Allophycocyanin I – A Second Cyanobacterial Allophycocyanin? Isolation, Characterization and Comparison with Allophycocyanin II From the Same Alga. FEBS Letters **68**, 49–54.

HARRIS, J. U., and D. S. BERNS, 1975: Sequences of the N-Terminus Portions of Biliproteins. J. Mol. Evol. **5**, 153–163.

HAUPT, W., 1959: Die Chloroplastendrehung bei *Mougeotia*. I. Über den quantitativen und qualitativen Lichtbedarf bei Schwachlichtbewegungen. Planta **53**, 484–501.

HAXO, F. T., and L. R. BLINKS, 1950: Photosynthetic action spectra of marine algae. J. gen. Physiol. **33**, 389–422.

HIROSE, H., S. KUMANO, and K. MADONO, 1969: Spectroscopic studies on phycoerythrins from Cyanophycean and Rhodophycean algae with special reference to their phylogenetical relations. Bot. Mag. Tokyo **82**, 197–203.

JOSE, A. M., 1977: Gel Filtration of Particle-bound Phytochrome. Planta **134**, 287–293.

KAO, O. H. W., M. R. EDWARDS, and D. S. BERNS, 1975: Physical-Chemical Properties of C-Phycocyanin Isolated From an Acido-Thermophilic Eukaryote, *Cyanidium caldarium*. Biochem. J. **147**, 63–70.

KLEIN, G., 1977: Versuche zur Proteinbindung und zur Stereochemie der Chromophore in Biliproteiden und Modellverbindungen. Dissertation Universität München.

KLEIN, G., and W. RÜDIGER, 1978: Über die Bindungen zwischen Chromophor und Protein in Biliproteiden, V. Stereochemie von Modell-Imiden. Liebigs Ann. Chem. **1978**, 267–279.

KLEIN, G., S. GROMBEIN, and W. RÜDIGER, 1977: On the Linkages Between Chromophore and Protein in Biliproteins, VI. Structure and Protein Linkage of the Phytochrome Chromophore. Hoppe-Seyler's Z. Physiol. Chem. **358**, 1077–1079.

KOBAYASHI, Y., H. W. SIEGELMAN, and C. H. W. HIRS, 1972: C-Phycocyanin from *Phormidium luridum*. Isolation of subunits. Arch. Biochem. Biophys. **152**, 187–198.

KÖST, E., 1978: Untersuchungen zur Protein-Chromophor-Bindung unter Verwendung der radioaktiven Markierung mit ^{35}S am Beispiel von B-Phycoerythrin aus *Porphyridium cruentum*. Dissertation Universität München.

KÖST, H. P., and W. RÜDIGER, 1974: Cis-trans-isomere-2-Äthyliden-3-methylsuccinimide aus Phycoerythrin. Tetrahedron Letters **1974**, 3417–3418.

KÖST, H. P., W. RÜDIGER, and D. J. CHAPMAN, 1975: Über die Bindungen zwischen Chromophor und Protein in Biliproteiden. I. Abbauversuche und Spektraluntersuchungen an Biliproteiden. Liebigs Ann. Chem. **1975**, 1582–1593.

KÖST-REYES, E., H. P. KÖST, and W. RÜDIGER, 1975: Über die Bindungen zwischen Chromophor und Protein in Biliproteiden, II. Nachweis von Cystein als bindende Aminosäure in B-Phycoerythrin. Liebigs Ann. Chem. **1975**, 1594–1600.

KOLLER, K. P., and W. WEHRMEYER, 1974: Isolierung und Charakterisierung der Biliproteide von *Rhodella violacea* (Bangiophycidae). Arch. Mirkobiol. **100**, 253–270.

KOLLER, K. P., and W. WEHRMEYER, 1975: B-Phycoerythrin from *Rhodella violacea*. Characterization of Two Isoproteins. Arch. Microbiol. **104**, 255–261.

KOLLER, K. P., W. WEHRMEYER, and H. SCHNEIDER, 1977: Isolation and Characterization of Disc-Shaped Phycobilisomes from the Red Alga *Rhodella violacea*. Arch. Microbiol. **112**, 61–67.

KROES, H. H., 1970: A Study of Phytochrome, Its Isolation, Structure and Photochemical Transformation. Mededelingen Landbouwhogeschool Wageningen, Nederland 70–18, p. 1–112.

LEY, A. C., and W. L. BUTLER, 1976: Efficiency of energy transfer from photosystem II to photosystem I in *Porphyridium cruentum*. Proc. Nat. Acad. Sci. USA **73**, 3957–3960.

LOTTER, H., G. KLEIN, W. RÜDIGER, and H. SCHEER, 1977: Röntgenstrukturanalyse von (Äthylsulfonyl)-Methylsuccinimid. Tetrahedron Letters **1977**, 2317–2320.

MACCOLL, R., W. HABIG, and D. S. BERNS, 1973: Characterization of phycocyanin from *Chroomonas* species. J. Biol. Chem. **248**, 7080–7086.

MACCOLL, R., D. S. BERNS, and O. GIBBONS, 1976: Characterization of Cryptomonad Phycoerythrin and Phycocyanin. Arch. Biochem. Biophys. **177**, 265–275.

MACKENZIE, J. M., Jr. R. A. COLEMAN, W. R. BRIGGS, and L. H. PRATT, 1975: Reversible Redistribution of Phytochrome Within the Cell upon Conversion to Its Physiologically Active Form. Proc. Nat. Acad. Sci. USA **72**, 799–803.

MARMÉ, D., J. BIANCO, and J. GROSS, 1976: Evidence for Phytochrome Binding to Plasma Membrane and Endoplasmic Reticulum. In: Light and Plant Development, H. Smith (Ed.) Butterworths London 1976, p. 95–110.

MITRAKOS, K., and W. SHROPSHIRE jr., 1972: Phytochrome. Academic Press, London and New York.

MÖRSCHEL, E., and W. WEHRMEYER, 1975: Cryptomonad Biliprotein: Phycocyanin-645 from a *Chroomonas* Species. Arch. Microbiol. **105**, 153–158.

MÖRSCHEL, E., and W. WEHRMEYER, 1977: Multiple Forms of Phycoerythrin-545 from *Cryptomonas maculata*. Arch. Microbiol. **113**, 83–89.

MUCKLE, G., 1978: Vergleichende Untersuchungen von C-Phycoerythrinen aus verschiedenen Cyanobakterien unter besonderer Berücksichtigung der Chromophorregionen. Dissertation Universität München.

MUCKLE, G., and W. RÜDIGER, 1977: Chromophore Content of C-Phycoerythrin from Various Cyanobacteria. Z. Naturforsch. **32** C, 957–962.

MUCKLE, G., J. OTTO, and W. RÜDIGER, 1978: On the Linkages Between Chromophore and Protein in Biliproteins, VII. Amino Acid Sequence in the Chromophore Region of

C-Phycoerythrin from *Pseudanabaena* W 1173 and *Phormidium persicinum*. Hoppe-Seyler's Z. Physiol. Chem. **359**, 345–355.

MYERS, J., and W. A. KRATZ, 1955: Relations between pigment content and photosynthetic characteristics in a blue-green alga. J. gen. Physiol. **39**, 11–22.

O'CARRA, P., and S. D. KILLILEA, 1971: Subunit structures of C-phycocyanin and C-phycoerythrin. Biochem. Biophys. Res. Comm. **45**, 1192–1197.

O'CARRA, P., and C. O'HEOCHA, 1976: Algal Biliproteins and Phycobilins. In: Chemistry and Biochemistry of Plant Pigments (T. W. Goodwin, Ed.) Vol. 1, p. 328–376, Academic Press London.

O'HEOCHA, C., 1965: Biliproteins of algae. Ann. Rev. Plant Physiol. **16**, 415–434.

PRATT, L. H., 1973: Comparative Immunochemistry of Phytochrome. Plant Physiol. **51**, 203–209.

PRATT, L. H., 1975: Photochemistry of High Molecular Weight Phytochrome in vitro. Photochem. Photobiol. **22**, 33–36.

PRATT, L. H., and S. C. CUNDIFF, 1975: Spectral Characterization of High-Molecular-Weight Phytochrome. Photochem. Photobiol. **21**, 91–97.

PRATT, L. H., and D. MARMÉ, 1976: Red Light-enhanced Phytochrome Pelletability. Plant Physiol. **58**, 686–692.

QUAIL, P. H., D. MARMÉ, and E. SCHÄFER, 1973: Particle-bound phytochrome from maize and pumpkin. Nature New Biol. **245**, 189–191.

QUAIL, P. H., 1975a: Particle-bound Phytochrome: Association with a Ribonucleoprotein Fraction from *Cucurbita pepo* L. Planta **123**, 223–234.

QUAIL, P. H., 1975b: Particle-bound Phytochrome: The Nature of the Interaction between Pigment and Particulate Fractions. Planta **123**, 235–246.

RÜDIGER, W., P. O'CARRA, and C. O'HEOCHA, 1967: Structure of Phycoerythrobilin and Phycocyanobilin. Nature **215**, 1477–1478.

RÜDIGER, W., 1969: Chromsäure- und Chromatabbau von Gallenfarbstoffen. Hoppe-Seyler's Z. Physiol. Chem. **350**, 1291–1300.

RÜDIGER, W., and D. L. CORRELL, 1969: Über die Struktur des Phytochrom-Chromophores und seine Protein-Bindung. Liebigs Ann. Chem. **723**, 208–212.

RÜDIGER, W., and P. O'CARRA, 1969: Studies on the structures and apoprotein linkages of the phycobilins. Eur. J. Biochem. **7**, 509–516.

RÜDIGER, W., 1970a: Animal Biliproteins. Biochem. J. **119**, 1 p.

RÜDIGER, W., 1970b: Neues aus der Chemie und Biochemie der Gallenfarbstoffe. Angew. Chem. **82**, 527–534; Angew. Chem., Intern. Ed. **9**, 473–480.

RÜDIGER, W., H. P. KÖST, H. BUDZIKIEWICZ, and V. KRÄMER, 1970: Die Struktur von Mesobilirhodin. Liebigs Ann. Chem. **738**, 197–201.

RÜDIGER, W., 1971: Gallenfarbstoffe und Biliproteide. Fortschr. Chem. org. Naturstoffe **29**, 59–139.

RÜDIGER, W., 1972a: Isolation and Purification of Phytochrome. In: K. Mitrakos and W. Shropshire jr., Phytochrome. Academic Press London and New York, p. 107–125.

RÜDIGER, W., 1972b: Chemistry of Phytochrome Chromophore. In: K. Mitrakos and W. Shropshire jr., Phytochrome, Academic Press London and New York, p. 129–141.

SIEGELMAN, H. W., and E. M. FIRER, 1964: Purification of Phytochrome from Oat Seedlings. Biochemistry **3**, 418–423.

SIEGELMAN, H. W., B. C. TURNER, and S. B. HENDRICKS, 1966: The chromophore of phytochrome. Plant Physiol. **41**, 1289–1292.

SIEGELMAN, H. W., D. J. CHAPMAN, and W. J. COLE, 1967: Enzymatic Cleavage of Phycocyanobilin. Arch. Biochem. Biophys. **122**, 261–262.

SMITH, H., 1975: Phytochrome and Photomorphogenesis. An Introduction to the Photocontrol of Plant Development. MacGraw-Hill, London.

SCHEER, H., W. A. SVEC, B. T. COPE, M. H. STUDIER, R. G. SCOTT, and J. J. KATZ, 1974: Structure of Bacteriochlorophyll b. J. Amer. chem. Soc. **96**, 3714–3716.

SCHEER, H., 1976: Studies on Plant Bile Pigments: Characterization of a Model for the Phytochrome P_r Chromophore. Z. Naturforsch. **31**c, 413–417.

SCHEER, H., and C. KRAUSS, 1977: Studies on Bile Pigments – 3. Oxidative Photodimerization of a Phytochrome P_r Model Pigment and Its Thermal Reversion. Photochem. Photobiol. **25**, 311–314.

SCHEER, H., and W. KUFER, 1977: Studies on Plant Bile Pigments, IV: Conformational Studies on C-Phycocyanin from *Spirulina platensis*. Z. Naturforsch. **32**c, 513–519.

SCHEER, H., U. LINSENMEIER, and C. KRAUSS, 1977: Studies on Plant Bile Pigments, II. Chemical and Photochemical Oxygenation of a Phytochrome P_r Chromophore Model Pigment to Purpurins. Hoppe-Seyler's Z. Physiol. Chem. **358**, 185–196.

SCHEIBE, J., 1972: Photoreversible pigment: occurrence in a blue-green alga. Science **176**, 1037–1039.

SHELDRICK, W. S., 1976: Crystal and Molecular Structure of biliverdin dimethyl ester. J. Chem. Soc. Perkin Trans. **2**, 1457–1461.

SCHOCH, S., G. KLEIN, U. LINSENMEIER, and W. RÜDIGER, 1974: Synthese und Reaktionen von Äthyliden-methylsuccinimid. Tetrahedron Letters **1974**, 2465–2468.

SCHOCH, S., G. KLEIN, U. LINSENMEIER, and W. RÜDIGER, 1976: Über die Bindungen zwischen Chromophor und Protein in Biliproteiden. III. Eliminierungsreaktionen an Modellsubstanzen. Liebigs Ann. Chem. **1976**, 549–558.

SCHRAM, B. L., and H. H. KROES, 1971: Structure of Phycocyanobilin. Eur. J. Biochem. **19**, 581–594.

STANIER, R. Y., and G. COHEN-BAZIRE, 1977: Phototropic Prokaryotes: The Cyanobacteria. Ann. Rev. Microbiol. **31**, 225–274.

TAKEMOTO, J., and L. BOGORAD, 1975: Subunits of Phycoerythrin from *Fremyella diplosiphon*: Chemical and Immunochemical Characterization. Biochemistry **14**, 1211–1216.

TANDEAU DE MARSAC, N., 1977: Occurrence and Nature of Chromatic Adaptation in Cyanobacteria. J. Bacteriol. **130**, 82–91.

TANDEAU DE MARSAC, N., and G. COHEN-BAZIRE, 1977: Molecular composition of cyanobacterial phycobilisomes. Proc. Nat. Acad. Sci. USA **74**, 1635–1639.

TAYLOR, A. O., and B. A. BONNER, 1967: Isolation of phytochrome from the alga *Mesotaenium* and liverwort *Sphaerocarpos*. Plant Physiol. **42**, 762–766.

TENHUNEN, R., H. S. MARVER, and R. SCHMID, 1969: Microsomal heme oxygenase. Characterization of the enzyme. J. Biol. Chem. **244**, 6388–6394.

TOMITA, G., and E. RABINOWITCH, 1962: Excitation energy transfer between pigments in photosynthetic cells. Biophys. J. **2**, 483–499.

TORJESEN, P. A., and K. SLETTEN, 1972: C-Phycocyanin from *Oscillatoria agardhii*. I. Some molecular properties. Biochim. Biophys. Acta **263**, 258–271.

TROXLER, R. F., and A. BROWN, 1970: Biosynthesis of phycocyanin in vivo. Biochim. Biophys. Acta **215**, 503–511.

TROXLER, R. F., 1972: Synthesis of bile pigments in plants. Formation of carbon monoxide and phycocyanobilin in wildtype and mutant strains of the alga. *Cyanidium caldarium*. Biochemistry **11**, 4235–4242.

TROXLER, R. F., and J. M. DOKOS, 1973: Formation of carbon monoxide and bile pigment in red and blue-green algae. Plant Physiol. **51**, 72–75.

TROXLER, R. F., and A. S. BROWN, 1975: Metabolism of δ-Aminolevulinic Acid in Red and Blue-Green Algae. Plant Physiol. **55**, 463–467.

TROXLER, R. F., J. A. FOSTER, A. S. BROWN, and C. FRANZBLAU, 1975: The α and β Subunits of *Cyanidium caldarium* Phycocyanin: Properties and Amino Acid Sequences at the Amino Terminus. Biochemistry **14**, 268–274.

VAN DER VELDE, H. H., 1973: The occurrence in red algae of two phycoerythrins with different molecular weights and spectral properties. Biochim. Biophys. Acta **303**, 246–257 (1973).

VIRTANEN, A. I., and J. K. MIETTINEN, 1949: Formation of Biliverdin from Legcholeglobin, the Green Pigment in Leguminous Root Nodules. Acta chem. Scand. **3**, 17–21.

WAGENMANN, R., 1977: Untersuchungen von Phycoerythrin und Phycocyanin aus *Pseudanabaena* W 1173 im Vergleich mit anderen Cyanobakterien (Blaualgen). Dissertation Universität München.

WANG, R. T., C. L. R. STEVENS, and J. MYERS, 1977: Action Spectra for Photoreactions I and II of Photosynthesis in the Blue-Green Alga *Anacystis nidulans*. Photochem. Photobiol. **25**, 103–108.

WEHRMEYER, W., 1970: Zur Feinstruktur der Chloroplasten einiger photoautotropher Cryptophyceen. Arch. Mikrobiol. **71**, 367–383.

WILLIAMS, V. P., and A. N. GLAZER, 1978: Structural Studies on Phycobiliproteins. I. Bilin-Containing Peptides of C-Phycocyanin. J. Biol. Chem. **253**, 202–211.

ZILINSKAS, B. A., B. K. ZIMMERMAN, and E. GANTT, 1978: Allophycocyanin Forms Isolated From *Nostoc* sp. Phycobilisomes. Photochem. Photobiol. **27**, 587–595.

Addendum

Knowledge on plant biliproteins has considerably increased since the completion of this article. Some of the more recent papers in this field are discussed here.

With regard to the function of phycobiliproteins, energy transfer has been investigated in detail in Cryptomonad biliproteins (MacColl and Berns, 1978; Jung et al., 1980), in phycoerythrins (Grabowski and Gantt, 1978; Zickendraht-Wendelstadt et al., 1979), and in intact phycobilisomes (Porter et al., 1978; Searle et al., 1978). Energy transfer is very efficient not only in intact phycobilisomes but also in parts of it as long as native protein conformations and the aggregation status are retained. Investigations on the ultrastructure of phycobilisomes from Cyanobacteria and Rhodophyceae revealed different spatial forms and sizes in different organisms but always the same general architecture (Mörschel et al., 1977; Koller et al., 1978; Yamanaka et al., 1978; Gantt et al., 1979; Wanner and Köst, 1980; Bryant et al., 1979). Disc-shaped biliprotein aggregates combine to double-discs and short stacks which have been isolated in some cases in the form of tripartite units. Disc-shaped aggregates are in accordance with preliminary X-ray data on crystalline C-phycocyanin and B-phycoerythrin (Abad-Zapatero et al., 1977; Hackert et al., 1977; Sweet et al., 1977).

Results from peptide sequence work (including the first complete amino acid sequence of a phycocyanin, Frank et al., 1978) revealed cysteine as that amino acid which is always linked with the chromophores (Glazer, 1977; Zuber, 1978; Brown et al., 1979; Köst and Köst, 1979; Glazer et al., 1979; Rüdiger, 1978). The linkage has been localized at the side chain of ring A by NMR spectroscopy (Lagarias et al., 1979).

Wallin et al. (1978) reported that cysteine is contained in only one chromopeptide out of three chromopeptides obtained from a phycocyanin α-chain. This observation might, however, be artifactual because the α-chain contains only one chromophore. The existence of a second chromophore-peptide linkage besides the thioether has been questioned recently (Troxler et al., 1978a; Glazer, 1977). The view that the covalent attachment of the chromophore occurs as the last step of biliprotein biosynthesis has been supported by biochemical and chemical (model) experiments (Troxler et al., 1978b; Cosner and Troxler, 1978; Klein and Rüdiger, 1979).

Whereas structure and absolute configuration of the phycobilin have been confirmed by total synthesis (Gossauer and Weller, 1978) the conformation of the chromophores in native phycobiliproteins is still being investigated. Arguments which favor extended chromophore conformations in native biliproteins and cyclic chromophore conformations in denatured biliproteins have been summarized by Scheer (1978; 1979). Calculations show that extended conformations in native biliproteins must be highly twisted (Scheer et al., 1979).

Action spectra for chromatic adaptation have again been explained by the existence of a photoreversible pigment in Cyanobacteria (Vogelmann and Scheibe, 1978; Schneider and Bogorad, 1979). The detection of another such pigment, phycochrome d, has been reported (Björn, 1978). But photoreversibility has also been observed in partly unfolded allophycocyanin and phycocyanin (Ohki and Fujita, 1979).

More recent advances in the phytochrome field have been summarized by Rüdiger (1980).

References

Abad-Zapatero, C., J. L. Fox, and M. L. Hackert, 1977: The Quaternary Structure of a Unique Phycobiliprotein: B-Phycoerythrin From Porphyridium cruentum. Biochem. Biophys. Res. Commun. **78**, 266–272.

Björn, G. S., 1978: Phycochrome d, a New Photochromic Pigment from the Blue-Green Alga, Tolypothrix distorta. Physiol. Plant. **42**, 321–323.

Brown, A. S., G. D. Offner, M. M. Ehrhardt, and R. F. Troxler, 1979: Phycobilin-Apoprotein Linkages in the α and β Subunits of Phycocyanin from the Unicellular Rhodophyte, Cyanidium caldarium. J. Biol. Chem. **254**, 7803–7811.

Bryant, D. A., G. Guglielmi, N. Tandeau de Marsac, A. M. Castets, and G. Cohen-Bazire, 1979: The Structure of Cyanobacterial Phycobilisomes: A Model, Arch. Microbiol., **123**, 113-127.

Cosner, J. C., and R. F. Troxler, 1978: Phycobiliprotein Synthesis in Protoplasts of the Unicellular Cyanophyte, Anacystis nidulans. Biochim. Biophys. Acta **519**, 474–488.

Frank, G., W. Sidler, H. Widmer, and H. Zuber, 1978: The Complete Amino Acid Sequence of Both Subunits of C-Phycocyanin from the Cyanobacterium Mastigocladus laminosus. Hoppe-Seyler's Z. Physiol. Chem. **359**, 1491–1507.

GANTT, E., C. A. LIPSCHULTZ, J. GRABOWSKI, and B. K. ZIMMERMANN, 1979: Phycobilisomes from Blue-Green and Red Algae. Isolation Criteria and Dissociation Characteristics. Plant Physiol. **63**, 615–620.

GLAZER, A. N., 1977: Structure and Molecular Organization of the Photosynthetic Accessory Pigments of Cyanobacteria and Red Algae. Mol. Cell. Biochem. **18**, 125–140.

GLAZER, A. N., C. S. HIXSON, and R. J. DELANGE, 1979: Determination of the Number of Thiother-Linked Cysteine Residues in Cytochromes c and Phycobiliproteins. Anal. Biochem. **92**, 489–496.

GOSSAUER, A., and J.-P. WELLER, 1978: Chemical Total Synthesis of (+)-(2R, 16R)- and (+)-(2S, 16R)-Phycoerythrobilin Dimethyl Ester. J. Amer. Chem. Soc. **100**, 5928–5933.

GRABOWSKI, J. and E. GANTT, 1978: Photophysical Properties of Phycobiliproteins from Phycobilisomes: Fluorescence Lifetimes, Quantum Yields, and Polarization Spectra. Photochem. Photobiol. **28**, 39–45.

HACKERT, M. L., C. ABAD-ZAPATERO, S. E. STEVENS jr., and J. L. FOX, 1977: Crystallization of C-Phycocyanin from the Marine Blue-Green Alga Agmenellum quadruplicatum. J. Mol. Biol. **111**, 365–369.

JUNG, J., P.-S. SONG, R. J. PAXTON, M. S. EDELSTEIN, R. SWANSON, and E. E. HAZEN jr., 1980: Molecular Topography of the Phycocanin Photoreceptor from Chroomonas species. Biochemistry, **19**, 24–32.

KLEIN, G., and W. RÜDIGER, 1979: Thioether Formation of Phycocyanobilin: A Model Reaction of Phycocyanin Biosynthesis. Z. Naturforsch. **34**c, 192–195.

KÖST, E., and H.-P. KÖST, 1979: The Protein-Chromophore Bond in B-Phycoerythrin from Porphyridium cruentum. Radiosulfur Labeling Experiments. Eur. J. Biochem., **102**, 83–91.

KOLLER, K.-P., W. WEHRMEYER, and E. MÖRSCHEL, 1978: Biliprotein Assembly in the Disc-Shaped Phycobilisomes of Rhodella violacea. On the Molecular Composition of Energy-Transfer Complexes (Tripartite Units) Forming the Periphery of the Phycobilisome. Eur. J. Biochem. **91**, 57–63.

LAGARIAS, J. C., A. N. GLAZER, and H. RAPOPORT, 1979: Chromopeptides from C-Phycocyanin. The Structure and Linkage of a Phycocyanobilin Bound to the β-Subunit. J. Amer. Chem. Soc., **101**, 5030–5037.

MACCOLL, R., and D. S. BERNS, 1978: Energy Transfer Studies on Cryptomonad Biliproteins. Photochem. Photobiol. **27**, 343–349.

MÖRSCHEL, E., K. P. KOLLER, W. WEHRMEYER, and H. SCHNEIDER, 1977: Biliprotein Assembly in the Disc-Shaped Phycobilisomes of Rhodella violacea. I. Electron Microscopy of Phycobilisomes in Situ and Analysis of their Architecture After Isolation and Negative Staining. Cytobiol. **16**, 118–129.

OHKI, K., and Y. FUJITA, 1979: Photoreversible Absorption Changes of Guanidine-HCl Treated Phycocyanin and Allophycocyanin Isolated from the Blue-Green Alga Tolypothrix tenuis. Plant Cell Physiol. **20**, 483–490.

PORTER, G., C. J. TREDWELL, G. F. W. SEARLE, and J. BARBER, 1978: Picosecond Time-Resolved Energy Transfer in Porphyridium cruentum. Part I. In the Intact Alga. Biochim. Biophys. Acta **501**, 232–245.

RÜDIGER, W., 1978: The Photochemistry and Chemistry of Phytochrome and Phycobiliproteins. Chromophore-Protein Interactions. In: Plant Growth & Light Perception (B. Deutch, B. I. Deutch, A. O. Gyldenholm, eds.). Annual European Symposium on Photomorphogenesis, Aarhus.

RÜDIGER, W., 1980: Phytochrome, a Light Receptor of Plant Photomorphogenesis. Structure and Bonding **40**, 101–140.

SEARLE, G. F. W., J. BARBER, G. PORTER, and C. J. TREDWELL, 1978: Picosecond Time-Resolved Energy Transfer in Porphyridium cruentum. Part II. In the Isolated Light Harvesting Complex (Phycobilisomes). Biochim. Biophys. Acta **501**, 246–256.

SCHEER, H., 1978: Biliproteins: The Chromophore. In: Plant Growth & Light Perception (B. Deutch, B. I. Deutch, A. O. Gyldenholm, eds.), p. 25–52. Annual European Symposium on Photomorphogenesis, Aarhus.

SCHEER, H., 1980: Biliproteine. Angew. Chem., in press.

SCHEER, H., H. FORMANEK, and W. RÜDIGER, 1979: The Conformation of Bilin Chromophores in Biliproteins: Ramachandran-Type Calculations. Z. Naturforsch. **34**c, 1085–1093.

SCHNEIDER, H. A. W., and L. BOGORAD, 1979: Spectral Response Curves For the Formation of Phycobiliproteins, Chlorophyll and Delta-Aminolevulinic Acid in Cyanidium caldarium. Z. Pflanzenphysiol. **94**, 449–459.

SWEET, R. M., H. E. FUCHS, R. G. FISHER, and A. N. GLAZER, 1977: Preliminary Crystallographic Investigations of Two Phycobiliproteins. J. Biol. Chem. **252**, 8258–8260.

TROXLER, R. F., A. S. BROWN, and H.-P. KÖST, 1978a: Quantitative Degradation of Radiolabeled Phycobiliproteins. Chromic Acid Degradation of C-Phycocyanin. Eur. J. Biochem. **87**, 181–189.

TROXLER, R. F., P. KELLY, and S. B. BROWN, 1978b: Phycocyanobilin Synthesis in the Unicellular Rhodophyte Cyanidium caldarium. Biochem. J. **172**, 569–576.

VOGELMANN, T. C., and J. SCHEIBE, 1978: Action Spectra for Chromatic Adaptation in the Blue-Green Alga Fremyella diplosiphon. Planta **143**, 233–239.

WALLIN, R., R. SELSET, and K. SLETTEN, 1978: Characterization of Chromophoric Peptides From C-Phycocyanin. Biochem. Biophys. Res. Commun. **81**, 1319–1328.

WANNER, G., and H.-P. KÖST, 1980: Investigations on the Arrangement and Fine Structure of Porphyridium cruentum Phycobilisomes. Protoplasma, **102**, 97–109.

YAMANAKA, G., A. N. GLAZER, and R. C. WILLIAMS, 1978: Cyanobacterial Phycobilisomes. Characterization of the Phycobilisomes of Synechococcus sp. 6301. J. Biol. Chem. **253**, 8303–8310.

ZICKENDRAHT-WENDELSTADT, B., J. FRIEDRICH, and W. RÜDIGER, 1979: Spectral Characterization of Monomeric C-Phycoerythrin from Pseudanabaena W 1173 and its α and β Subunits: Absorption, Fluorescence, Fluorescence Polarization. Energy Transfer in Isolated Subunits and C-Phycoerythrin. Photochem. Photobiol., **31**, 367–376.

ZUBER, H., 1978: Studies on the Structure of the Light-Harvesting Pigment-Protein-Complexes from Cyanobacteria and Red Algae. Ber. Deutsch. Bot. Ges. **91**, 459–475.

Quinonoid Pigments

E. Leistner

Institut für Pharmazeutische Biologie und Phytochemie
der Westfälischen Wilhelms-Universität Münster

1. Structure and occurrence

Many naturally occurring quinones are either benzo-(I), naphtho- (II) or anthraquinones (III). The remaining quinones can be divided into tetracen- (IV), phenanthrene- (V) and miscellaneous quinones (Fig. 1; Thomson, 1971).

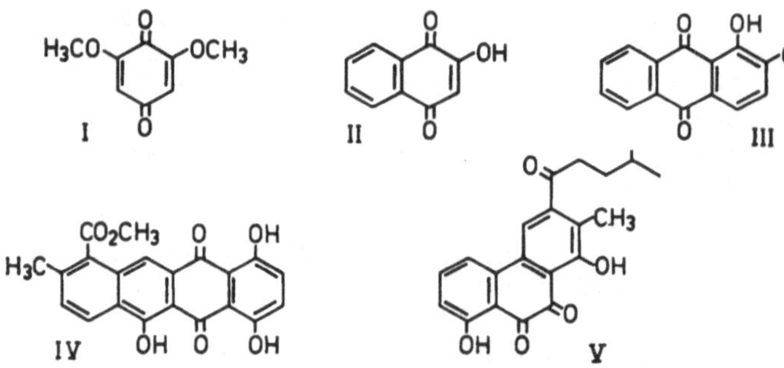

Fig. 1: 2,5-Dimethoxybenzoquinone (I), lawsone (II), alizarin (III), η_1-pyrromycinone (IV) and piloquinone (V) representing different types (benzoquinone, naphthoquinone, anthraquinone, tetracenequinone and phenanthrenequinone) of naturally occurring quinones

Structural variations of quinones are based on the substitution patterns and the fact that quinones may occur as C-glycosides (Hörhammer and Wagner, 1961) or O-glycosides (Miller, 1973). Moreover, dimeric (Fig. 2 VI; Steglich et al., 1973), trimeric (Fig. 2 VII; Cook and Down, 1970), and tetrameric (Thomson, 1971) quinones are encountered in some plants. The intense colour of the fungus *Cortinarius percomis* stems from a quinonoid anthracen dimer (Steglich et al., 1973), whereas in liquid cultures *Penicillium islandicum Sopp* produces dimeric anthraquinones the structures of which have recently been reexamined (Seo et al., 1973; Takeda, et al. 1973).

Quinones are widely distributed because some of them, such as Vitamin K (Newton et al., 1971; Esmon et al., 1975; Fernland and Stenflo, 1975), plastoquinone (Amesz, 1973), and ubiquinone (Swank and Burris, 1969; Koser, 1968) play important physiological rôles (Bentley and Campbell, 1974). These latter quinones, however, occur

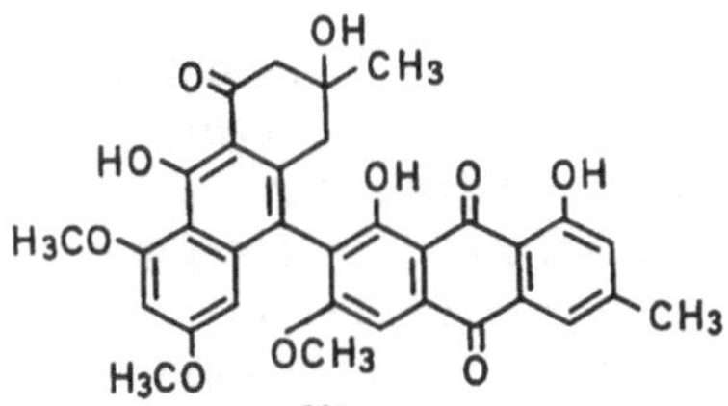

VI

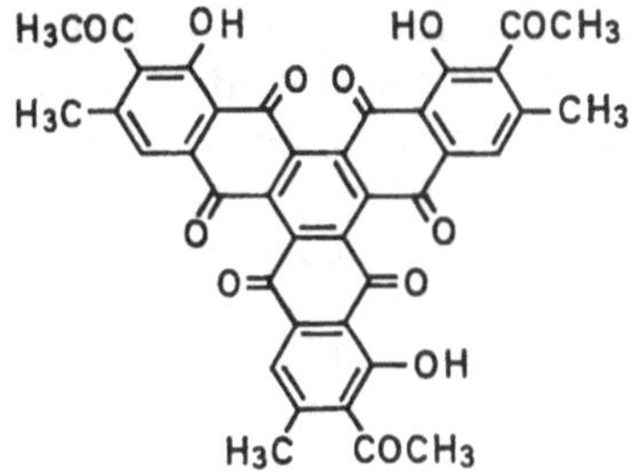

VII

Fig. 2: Anhydro-phlegmacin-9,10-quinone-8′-methyl ether (VI) from *Cortinarius percomis* and trianellinone (VII) from *Dianella revoluta*

in low concentration only and thus do not contribute significantly to the pigmentation of plants.

Quinonoid pigments may occur in all parts of plants, including the flowers. Nodososide (Fig. 3, VIII), for example, has been isolated from flowers of *Cassia nodosa* (Rizvi et al., 1971); and hypericin (Fig. 3, IX) has been found in flowers of *Hypericum perforatum* (Brockmann et al., 1942).

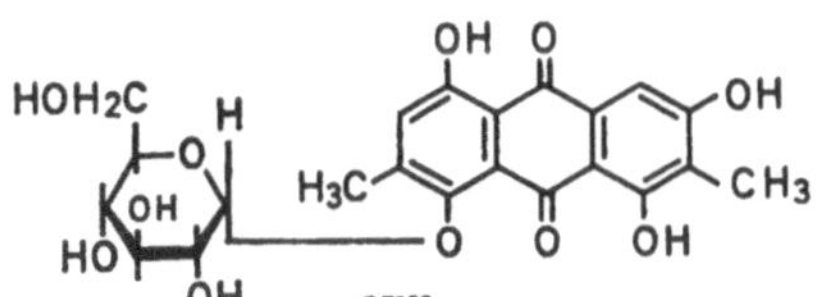

VIII

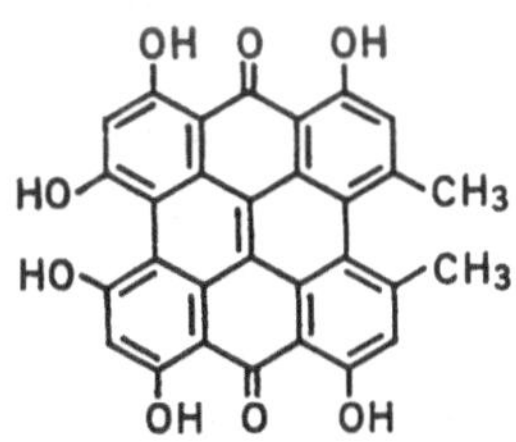

IX

Fig. 3: Quinonoid pigments from flowers of *Cassia nodosa* (nodososide, VIII) and Hypericum *perforatum* (hypericin, IX)

As opposed to many other natural products, quinones are produced in rather large amounts in callus (Baier and Friedrich, 1970; Dhruva et al., 1972; Furuya et al., 1972; Mizukami et al., 1977) and cell suspension (Zenk et al., 1975; Leistner 1975; Bauch and Leistner, 1978a, b) cultures derived from different higher plants. In a detailed study (Zenk et al., 1975) factors influencing the production of quinones in a culture of *Morinda citrifolia* have been investigated. It was found that 2,4-D suppressed anthraquinone formation completely whereas NAA stimulated formation of pigments. The influence of nutritional factors on pigment production was also investigated. It was found that the amount of secondary metabolites produced exceeded the anthraquinone content of the organized plant by a factor of 10 when compared on a dry-weight basis. Similar results were obtained with a cell suspension culture of *Galium mollugo* from which anthraquinones, anthraquinoneglycosides and naphthalenic compounds were isolated (Bauch and Leistner, 1978a).

The composition of metabolites found in the culture may differ from the composition of compounds detected in the organized plant. 5-Hydroxy-digitolutein (Fig. 4, X) has been isolated from callus tissue of *Tectona grandis* although this quinone has not been reported to occur in the *Tectona* plant (Dhruva et al., 1972). Furuya (1972) isolated 3-methyl purpurin (Fig. 4, XI) from callus tissue of *Digitalis lanata*. This compound is an anthraquinone which had not been found in nature previously. Also a hitherto unknown anthraquinone has been isolated from a cell suspension culture of *Morinda citrifolia* (Leistner, 1975).

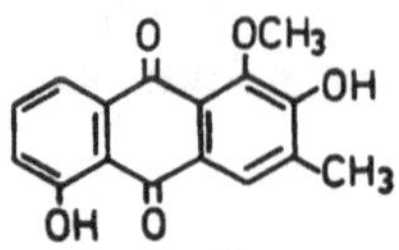

X

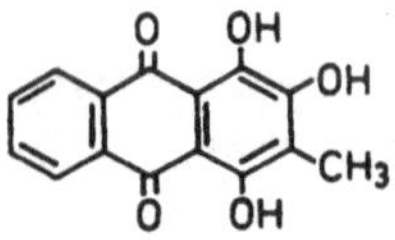

XI

Fig. 4: Anthraquinones from callus cultures of higher plants. 5-Hydroxydigitolutein (X) (*Tectona grandis*) and 3-Methylpurpurin (XI) (*Digitalis lanata*)

2. Biosynthesis

2.1. Biosynthetic pathways leading to benzoquinones

Under physiological conditions benzoquinones, like all other quinones, are in a state of equilibrium with the corresponding hydroquinones which belong to the group of phenolics. The biosynthesis of phenolics and benzoquinones is closely related.

Feeding experiments with specifically labelled precursors have shown that 2-methoxybenzoquinone, 2,6-dimethoxybenzoquinone and the glycoside of benzohydroquinone, viz., arbutin are derived from cinnamic acids. The substitution pattern of the quinones reflects the substitution pattern of the cinnamic acids from which they are derived (Fig. 5; Bolkart and Zenk, 1968a). The first step in the biosynthesis of this type of benzoquinones is the β-oxidation of a cinnamic acid. This reaction has also been demonstrated to occur in cell-free extracts of potato tubers and *Polyporus hispidus* (French et al., 1976). The product of this reaction, viz., p-hydroxybenzoic acid has been shown to be decarboxylated by cell-free preparations obtained from cell suspension cultures of *Glycine max*. Since this reaction is catalyzed by peroxidases, the reaction proceeds oxidatively yielding mono- and oligomeric benzoquinones (Berlin and Barz, 1975).

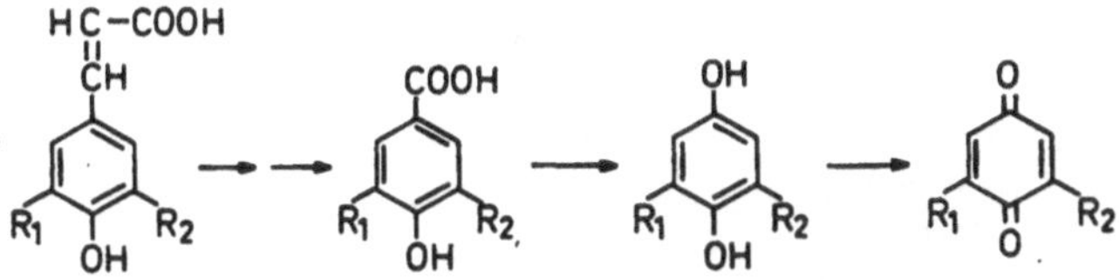

Fig. 5: The biosynthesis of a benzoquinone from a correspondingly substituted cinnamic acid

β-Oxidation, however, is not always the initial reaction involved in the biosynthesis of benzoquinones from phenylpropanoids. Plastoquinone (Whistance and Threlfall, 1970) and the benzoquinones occurring in *Pyrolaceae* (Thomson, 1971) are derived from tyrosine via homogentisic acid and toluhydroquinone.

Benzoquinones can also be derived from mevalonic acid (Fig. 6). Dimethylallylpyrophosphate and isopentenylpyrophosphate are believed to be precursors of thymoquinone (Fig. 6, XII) (Sandermann, 1962) whereas helicobasidin (Fig. 6, XIII), the first fungal benzoquinone which has been demonstrated not to be acetate-derived, is likely to be formed from farnesylpyrophosphate (Fig. 6). Different mechanisms have been considered to be involved in helicobasidin biosynthesis (Bentley and Chen, 1969). Mevalonic acid may not only participate in the biosynthesis of the benzoquinone nucleus, but it may also give rise to the lipophilic side chain of plastoquinone and of benzoquinones occurring in *Pyrolaceae*. A lipophilic side chain is also attached to the acetate-derived nucleus of mavioquinone. This side chain, however, stems from acetate and propionate (Scherrer and Azerad, 1976).

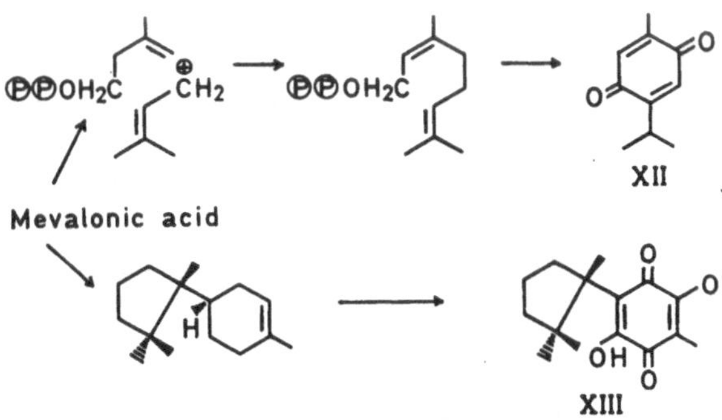

Fig. 6: Biosynthesis of benzoquinones (thymoquinone, XII, and helicobasidin, XIII) from mevalonic acid

It is clear that different biosynthetic pathways lead to similar benzoquinones. The biosynthesis of aurantiogliocladin (XIV) and ubiquinone (XV) in *Gliocladeum roseum* is a unique example: Both benzoquinones are produced by the same organism and have an almost identical substitution pattern, yet aurantiogliocladin (Fig. 7,

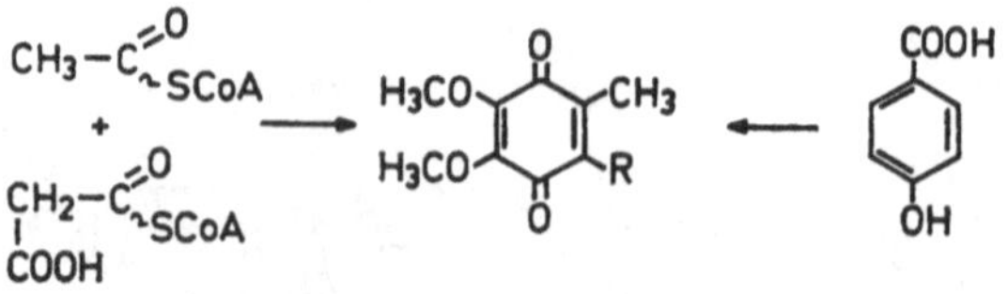

XIV R = CH3
XV R = n isoprene units

Fig. 7: Biosynthesis of two structurally closely related benzoquinones from acetate (aurantiogliocladin, XIV) or p-hydroxybenzoic acid (ubiquinone, XV) in *Gliocladeum roseum*

XIV) is derived from acetate, whereas he nucleus of ubiquinone (Fig. 7, XV) is formed from p-hydroxybenzoic acid (Bentley and Lavate, 1965).

Considering that the biosynthesis of benzoquinones has not yet been investigated exhaustively, it is interesting that there are at least six different biosynthetic pathways which are known to lead to benzoquinones.

2.2. Biosynthetic pathways leading to naphthoquinones

Naphthoquinones, like benzoquinones, may be synthesized within a plant by different pathways. While the origin of a certain type of naphthoquinone (Eugster, 1975) from mevalonic acid seems to be obvious, four other biosynthetic routes leading to napththoquinones have been discovered through a series of feeding experiments. The substitution pattern of naphthoquinones known to date is indicative of the biosynthetic mechanisms by which the compounds are formed.

Alkannin (Fig. 8, XVI) and shikonin are enantiomers of a naphthoquinone with a C_6-side chain. The benzenoid moiety of alkannin is derived from p-hydroxybenzoic acid, whereas the quinonoid ring and the side chain are derived from geranylpyrophosphate (Fig. 8). The sequence of the reactions (substitution, cyclisation, aromatization) involved in the biosynthesis of alkannin is not known (Schmid and Zenk, 1971).

Recent investigations on shikonin biosynthesis shed some light on this problem (Inouye et al. 1979).

The biosynthesis of alkannin and chimaphilin (Fig. 8, XVII) have one thing in common: in both cases a prenyl side chain is attached to an aromatic nucleus. Toluhydroquinone and Dimethylallylpyrophosphate give rise to an intermediate which on cyclization and aromatization yields a dimethylnaphthoquinone (chimaphilin). The methyl group attached to C-atom 7 of chimaphilin arises from C-atom two of mevalonic acid (Bolkart et al. 1968) whereas the methyl group at position two of chimaphilin is derived from the β C-atom of tyrosine (Bolkart and Zenk, 1968b) via homogentisate and toluhydroquinone (Bolkart and Zenk, 1969).

The biosynthesis of plumbagin (Fig. 8, XVIII) seemed to take place in a similar way, because application of labelled tyrosine to *Drosophyllum lusitanicum* resulted in the isolation of labelled plumbagin. Chemical degradation of the naphthoquinone showed, however, that the labelling pattern of plumbagin was inconsistent with a

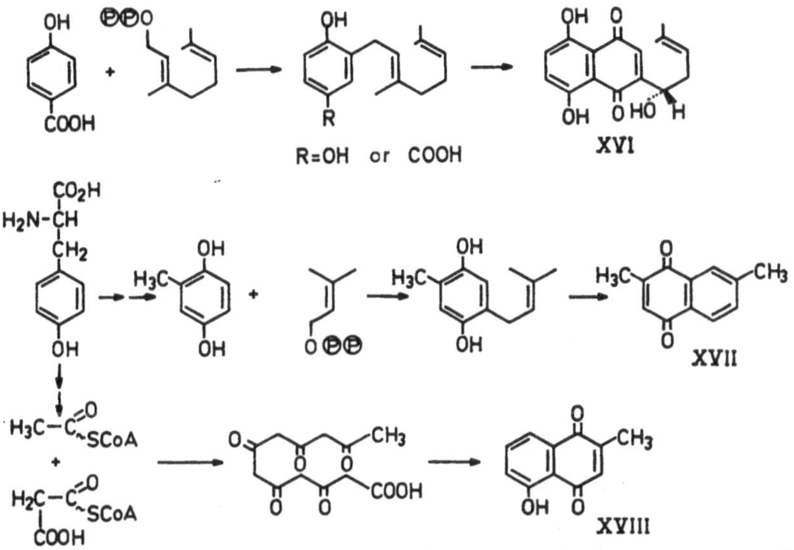

Fig. 8: Three different pathways leading to naphthoquinones in *Boraginaceae* (alkannin, XVI), *Pyrolaceae* (chimaphilin, XVII) and *Droseraceae* (plumbagin, XVIII)

direct incorporation of tyrosine into plumbagin. The labelling pattern was interpretable, however, when it was assumed that tyrosine was degraded via homogentisic acid and acetoacetic acid and then incorporated into plumbagin. Indeed, labelled acetic acid served as a very efficient precursor of plumbagin with an unusually high incorporation rate (27%). Extensive chemical degradation showed that plumbagin is indeed an acetogenin (Durand and Zenk, 1971). Thus it had been shown for the first time that acetate may serve as a precursor of naphthoquinones in higher plants as well, and that higher plants are also able to degrade tyrosine to acetate via homogentisic acid (Durand and Zenk, 1974a and b).

Investigations on the biosynthesis of Vitamin K (XXVIII) (Cox and Gibson, 1966; Leistner et al., 1967; Campbell et al., 1967; Leistner and Zenk, 1971a), lawsone (Fig. 9, II) (Zenk and Leistner, 1967), and juglone (Fig. 9, XXIV) (Leistner and Zenk, 1968b) resulted in the discovery of another biosynthetic pathway leading to naphthoquinones. Shikimic acid had been incorporated into these naphthoquinones with retention of the carboxyl group, suggesting that either shikimic acid, chroismic acid (XIX) or prephenic acid is directly incorporated into these naphthoquinones. Using bacterial mutants, Young (1975) has shown that chorismic acid rather than shikimic or prephenic acid represents the branching point leading to vitamin K_2. Since the pyruvate residue of chorismic acid is split off during the biosynthesis of naphthoquinones, chorismic acid is the source of seven out of ten C-atoms of the naphthoquinone nucleus. The remaining three C-atoms are supplied by α-ketoglutaric acid (Campbell, 1969). In an attempt to rationalize the participation of α-ketoglutaric acid (Fig. 9, XX) and chorismic acid (Fig. 9, XIX) in naphthoquinone biosynthesis (Fig. 9), a hypothetical intermediate, viz., o-succinoylbenzoic acid (Fig. 9, 10, XXI)

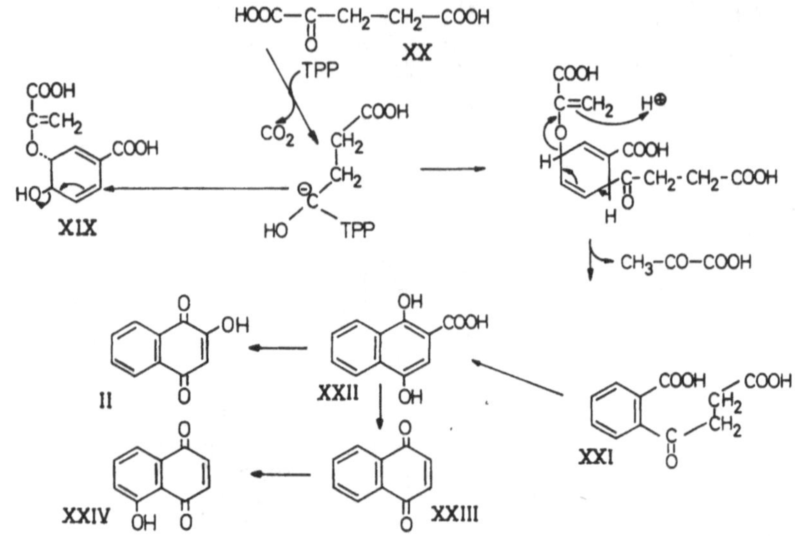

Fig. 9: Hypothetical formation of o-succinoylbenzoic acid (XXI) from chorismic (XIX) acid and α-ketoglutaric acid (XX) and the putative metabolic relation between o-succinylbenzoic acid, 1.4-dihydroxy-2-naphthoic acid (XXII), 1,4-naphthoquinone (XXIII), lawsone (II) and juglone (XXIV)

was proposed to be a precursor of naphthoquinones (Dansette and Azerad, 1970). This precursor stimulated growth of vitamin K-deficient bacteria, and a radioactively labelled sample was incorporated into juglone, lawsone, vitamin K_2 and even anthraquinones (vide infra). o-Succinoylbenzoic acid was shown to be a natural product occurring in *Impatiens balsamina* (Grotzinger and Campbell, 1974) and in mutants of *E. coli* (Young, 1975).

An enzyme system from *E. coli* mediating the conversion of o-succinoylbenzoic acid to 1,4-dihydroxy-2-naphthoic acid (Fig. 9, 10, XXII) has recently been described (Bryant and Bentley, 1976). This enzyme has been found to require ATP and Coenzyme A. 1,4-Dihydroxy-2-naphthoic acid is also a new natural product. It has been shown to be present in the culture broth of an E. coli mutant (Young, 1975). A radioactively labelled sample of 1,4-dihydroxy-2-naphthoic acid has been incorporated into plant naphthoquinones (Müller and Leistner, 1976).

A simultaneous prenylation and decarboxylation reaction are involved in the final steps of the biosynthesis of vitamin K (Fig. 10, XXVIII). This reaction leads to the immediate precursor of vitamin K, viz., demethylmenaquinone (Fig. 10, XXVII) (Shineberg and Young, 1976) which is eventually methylated to give vitamin K (XXVIII) (Samuel and Azerad, 1972).

The biosynthesis of vitamin K differs from the biosynthesis of prenylated naphthalenes (XXV, XXVI) found in *Bignoniaceae* in that as prenylation in these plants takes place at a prearomatic stage. Substrates for the prenylation may be either 2-

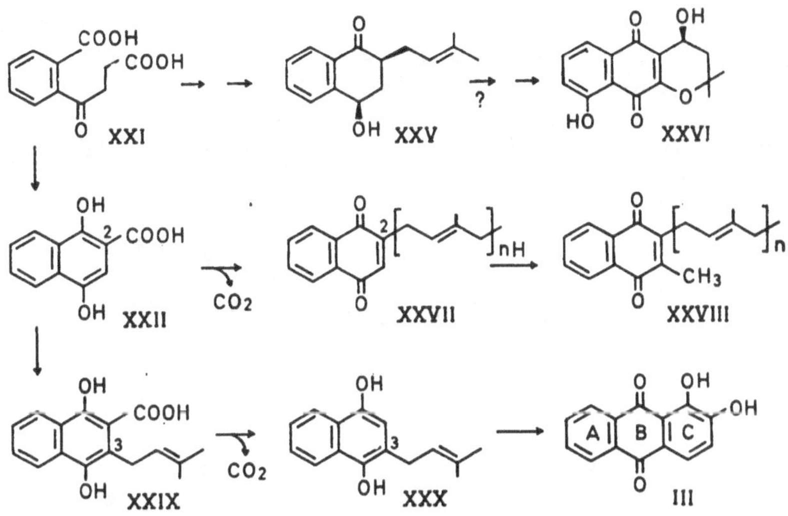

Fig. 10: Different prenylation reactions leading to
a) catalponol (XXV) and 4,9-dihydroxy-α-lapachon (XXVI) in *Catalpa ovata G. Don.* (*Bignoniaceae*),
b) to vitamin K_2 (XXVIII) in *E. coli* (*Enterobacteriaceae*),
c) to alizarin (III) in *Rubia tinctorum* (*Rubiaceae*)

carboxy-4-hydroxy-α-tetralon or 2-carboxy-4-oxo-α-tetralon (Inouye et al., 1976 and 1978). Prearomatic naphthalenic compounds like 4-oxo-α-tetralon and 5-hydroxy-4-oxotetralon are also likely to be involved in the biosynthesis of juglone, the major naphthoquinone occurring in *Juglans* plants (Müller and Leistner, 1978a). The biosynthesis of this naphthoquinone is unusual because, unlike all other naphthoquinones, which are derived from o-succinoylbenzoic acid, juglone is derived from at least one symmetrical intermediate (Leistner and Zenk, 1968b; Leduc et al., 1970; Scharf et al., 1971; Inoue, 1976). These intermediates are likely to be 4-oxo-α-tetralon and/or 1,4-naphthoquinone (Müller and Leistner, 1978a), which have been demonstrated to occur in different Juglans species (Müller and Leistner, 1976 and 1978b).

2.3. Biosynthetic pathways leading to anthraquinones

The structure and substitution patterns of anthraquinones generally reflects their mode of biosynthesis. Birch and Donovan (1955) pointed out that there should be at least two biosynthetic pathways leading to anthraquinones and that acetate is involved in one of these pathways. This hypothesis proved to be fruitful because Gatenbeck was able to show that labelled acetate was specifically incorporated into emodin and islandicin (XXXI) (Gatenbeck, 1960, 1962) (Fig. 11). It is interesting to note that one of the oxygen atoms of the carboxyl group of acetate may be incorporated into the anthraquinones and thus give rise to a specific substitution pattern. As a consequence acetate-derived anthraquinones are usually substituted in both benzenoid rings. Acetate is believed to be incorporated into anthraquinones via a hypothetical polyketide which can be folded in different ways. While the polyketide giving rise to islandicin (XXXI) is folded (Paulick et al., 1975) as indicated in Fig. 11, aloesaponarin (XXXII) from *Aloe saponaria* is likely to be derived by a different folding mechanism (Yagi et al., 1974) (Fig. 11).

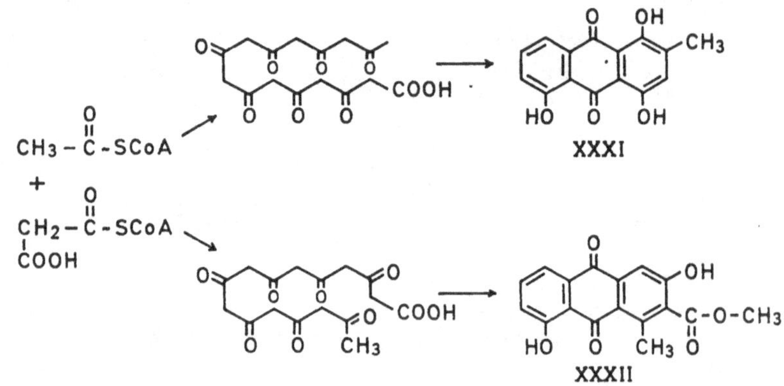

Fig. 11: Different folding mechanism of a hypothetical polyketide involved in the biosynthesis of islandicin (XXXI) and aloesaponarin (XXXII)

Later steps in the biosynthesis of acetate-derived anthraquinones usually involve decarboxylation of the cyclized polyketide; but it has been shown in at least one case that endocrocin is not decarboxylated to give emodin directly (Steglich et al., 1972). Instead, both anthraquinones are formed in a parallel fashion with the nonaromatic precursor to endocrocin possibly being decarboxylated to give rise to the prearomatic precursor of emodin.

Dianthraquinones occurring in *Penicillium islandicum* are derived from emodin and emodinanthrone (Sankawa et al., 1973). It has been proposed that coupling of the monomers proceeds by a Diels-Alderreaction rather than phenol coupling (Seo et al., 1973; Takeda, et al. 1973).

Emodin and chrysophanol (1,8-dihydroxy-3-methylanthraquinone) are consistuents of lower as well as higher plants in which they are also derived from acetate (Leistner and Zenk, 1969; Leistner, 1971; Fairbairn and Muhtadi, 1972).

Anthraquinones occurring in *Rubiaceae*, however, are not derived from acetate but rather by the shikimic-acid-o-succinoylbenzoic acidroute (Leistner and Zenk, 1967a, b; Dansette and Azerad, 1970; Leistner, 1973a). The biosynthesis of these anthraquinones is thus similar to the biosynthesis of naphthoquinones like juglone, lawsone and vitamin K (Fig. 10). The incorporation of o-succinoylbenzoic acid into alizarin (III) accounts for the origin of rings A and B (Fig. 10), while ring C is derived from mevalonic acid via dimethyl-allylpyrophosphate (Leistner and Zenk, 1968c, Burnett and Thomson, 1968b). The metabolic product of o-succinoylbenzoic acid to which the dimethylallylpyrophosphate moiety is attached is not known. 1,4-Naphthoquinone had been considered to play this rôle (Leistner and Zenk 1968a), but this assumption has been rejected because only nonsymmetric intermediates participate in alizarin biosynthesis (Leistner and Zenk, 1971b). As in naphthoquinone biosynthesis, 1,4-dihydroxy-2-naphthoic acid (XXII) has been postulated to be an intermediate in alizarin (III) biosynthesis (Fig. 10) (Leistner and Zenk, 1971b). Dimethylallylpyrophosphate would be attached to the acid in position 3 (Leistner and Zenk, 1971b). This is strongly supported by the finding that mollugin (methyl-2,2-dimethyl-6-hydroxy-2H-naphtho [1,2-b]pyran-5-carboxylate) is a naturally occurring constituent of *Galium mollugo* (Schildknecht et al., 1976), a plant containing anthraquinones which are derived from o-succinoylbenzoic acid (Bauch and Leistner, 1978a and b). This is at variance with the prenylation mechanism observed in vitamin K biosynthesis and the biosynthesis of *Catalpa* metabolites (Fig. 10).

Prenylated 1,4-dihydroxy-2-naphthoic acid (XXIX) may undergo decarboxylation to give desoxylapachol or its corresponding hydroquinone (XXX), a radioactive sample of which has been incorporated into alizarin (III, Fig. 10) (Leistner, 1973a).

Some anthraquinones occurring in the *Rubiaceae* (morindone and soranjidiol) and *Scrophulariaceae* (phomarin) have a substitution pattern akin to those anthraquinones which are derived from acetate. Hence it was reasonable to assume that the *Rubiaceae* and *Scrophulariaceae* produce anthraquinones not only by the o-succinoylbenzoic acid pathway, but also from acetate. It has been shown, however, that in these plants all anthraquinones, regardless of their substitution pattern, are derived from o-succinoylbenzoic acid (Dansette, 1972; Leistner, 1973b). The anthraquinones in the *Gesneriaceae* are also derived from this metabolite (Stöckigt et al., 1973).

3. Biosynthesis and distribution of quinones in the plant kingdom

If quinones are to be used as chemical markers in an attempt to deduce evolutionary relationships between plant taxa one must cope with two major obstacles. Firstly, quinones are not confined to certain plant taxa but are rather widely distributed in the plant kingdom (Thomson, 1971; Zenk and Leistner, 1968). Secondly, the mere occurrence of secondary metabolites within plant species does not suffice to justify their use as chemical markers because their biosynthesis might be different (Merxmüller, 1967; Mentzer, 1966).

As a matter of fact, experiments described in section 2.1 have shown that benzoquinones are derived by six different pathways, in section 2.2 that napthoquinones are produced by plants in five different ways, and in section 2.3 that anthraquinones are formed by two different mechanisms. Only one biosynthetic route is known to give rise to tetracen- and phenanthrenequinones which have been shown to be acetate derived (Thomson, 1971).

Since there are different biosynthetic routes leading to quinones, their employment as chemical characters can be particulary misleading. It has been pointed out, however, that quinones may well serve as chemical markers, provided biosynthetic pathways leading to quinones, rather than quinones themselves are used (Hegnauer,

1971). Indeed, acetate-derived anthraquinones and o-succinoylbenzoic acid-derived anthraquinones are confined to different taxa (Hegnauer, 1971). This observation can be extended to benzo- and naphthoquinonoid pigments.

Only a few quinonoid pigments have been isolated so far from the *Guttiferae-Parietales* branch.* These quinones (benzo- and naphthoquinones) are either derived from acetate or from toluhydroquinone and dimethylallylpyrophosphate. This latter pathway does not seem to occur in the *Rosiflorae-Pinnatae* or within the *Tetracyclicae*. In these taxa, however, quinones are either formed from acetate, from o-succinonylbenzoic acid and mevalonic acid, or from o-hydroxybenzoic acid and geranylpyrophosphate.

The polyacetate route is the most widely distributed. It occurs in microorganism and angiosperms, with the exception of the *Tetracyclicae* in which no polyacetate route leading to quinones has been observed until now (the occurrence of rhein, a typical acetogenin, in a compositae plant (Thomson, 1971) requires reexamination).

Anthraquinones occurring in the taxon *Tetracyclicae* are derived from mevalonic and o-succinoylbenzoic acid.

4. Degradation of quinonoid pigments

Whereas some information on the biosynthesis of quinones is available, little is known about the degradation of this group of compounds. *Claviceps purpurea* splits the ring of emodin (XXXIII) oxidatively (Fig. 12). The product is a benzophenone-carboxylic acid (XXXIV) which is transformed to a xanthon (XXXV). The latter dimerizes yielding ergochromes (XXXVI) (Franck, 1969; Hooper et al., 1971). A similar mechanism has been observed during degradation of questin (XXXVII) to soluchrin (XXXVIII) by *Aspergillus* terreus (Fig. 12) (Curtis et al., 1972).

5. Biological activity of quinonoid pigments

The degradation of fungal anthraquinones to seco-anthraquinones has attracted much attention because this ring opening plays a key rôle in the biosynthesis of aflatoxins in *aspergillus* species (Moss, 1972; Biollaz et al., 1970; Pachler et al., 1976). Difuroanthraquinones of the versicolorin type are subject to a ringopening reaction which leads to a xanthone and eventually to aflatoxins. These materials cause chlorosis (Moss, 1972) or growth inhibition (El-Khadem, 1966) on plants which have been attacked by certain *aspergillus* species. Domestic animals feeding on rice which has been infected by these *aspergillus* species may die or, if a sublethal dose has been ingested, aflatoxins may cause hepatocarcinoma.

Another anthraquinone which also has a difuran ring system and which may be detrimental to plants is dothistromin (XXXIX). This compound is produced by

* The occurrence of quinones in the taxa of the angiosperms has been published elsewhere (Zenk and Leistner, 1968).

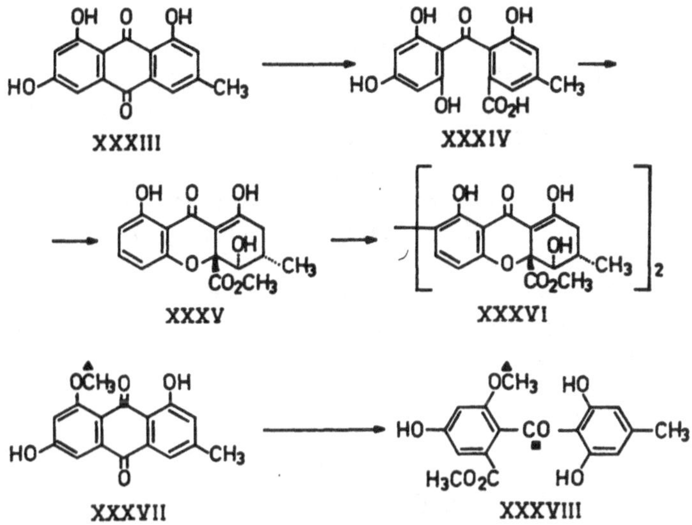

Fig. 12: Degradation of emodin (XXXIII) via a benzophenonecarboxylic acid (XXXIV), and a xanthone (XXXV) to an ergochrome (XXXVI) in *Claviceps purpurea* and degradation of questin (XXXVII) to soluchrin (XXXVIII) in *Aspergillus terreus*

Dothistroma pini (Basset et al., 1970) (Fig. 13). The fungus attacks *Pinus radiata* and other pinus species and causes defoliation, growth inhibition or withering.

Similar observations have been made by Gäumann and his coworkers (Gäumann and Naef-Roth, 1957; Gäumann and Obrist, 1960). They found that the ascomycete *Endothia parasitica* which causes withering of chestnut trees produces two toxins, viz., diaporthin and (+)skyrin (XL), a bisanthraquinone (Fig. 13). Experiments on the plasmolysis of *Spirogyra* and *Rhoeo* suggested that skyrin affects the water permeability of the protoplast.

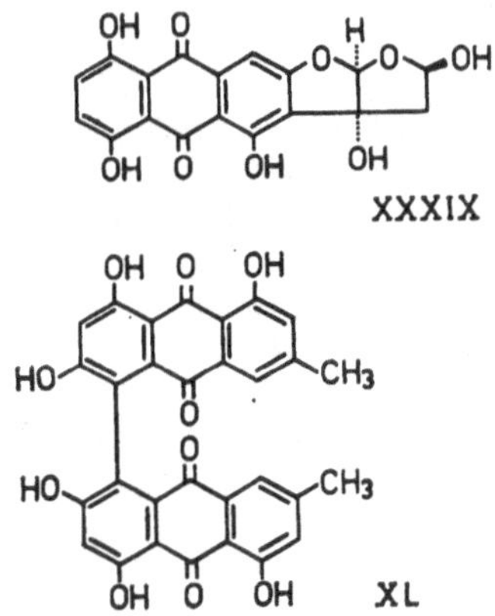

Fig. 13: Dothistromin (XXXIX) and (+) skyrin (XL), two quinones with biological activity

Quinonoid pigments may be advantageous to plants, however. Resistance of teakwood to insects, for example, is attributed to the presence of tectoquinone (Sandermann and Simatupang, 1966). Quinones may also exhibit antifungal or antibactericidal activity. As an example, plumbagin (XVIII) is a fungicide which is produced by *Drosophyllum* (vide supra) and *Ceratostigma willmottianum* (Masuko et al., 1970).

Finally, it should be mentioned that quinones may exhibit pharmacological activities: Anthraquinone glycosides are used as purgatives (Steinegger and Hänsel, 1972) whereas daunomycin and adriamycin are clinically useful chemotherapeutic agents against cancer (Di Marco and F. Arcamone, 1975). Rifamycins which are produced by *Anacardia mediterranei* are used as a cure against tuberculosis (Wehrli and Stehelin, 1971). These naphthoquinones inhibit the DNA-dependent RNA polymerase (Gale et al., 1972).

References

AMESZ, J., 1973: The function of plastoquinone in photosynthetic electron transport. Biochim. Biophys. Acta **301**, 35–51.

BAIER, S., and H. FRIEDRICH, 1970: Anthracen-Derivate in Kalluskulturen aus Senna. Naturwissenschaften **57**, 548.

BASSET, C., M. BUCHANAN, R. T. GALLAGHER, and R. HODGES, 1970: A toxic difurananthraquinone from *Dothistroma pini*. Chem. and Ind. 1659–1660.

BAUCH, H.-J., and E. LEISTNER, 1978a: Aromatic metabolites in a cell suspension culture of *Galium mollugo*. Planta medica **33**, 105–123.

—, and E. LEISTNER, 1978b: Attempts to demonstrate incorporation of labelled precursors into aromatic metabolites in cell suspension cultures of *Galium mollugo*. Planta medica **33**, 124–127.

BENTLEY, R., and D. CHEN, 1969: Helicobasidin: A fungal benzoquinone of isoprenoid origin. Phytochemistry **8**, 2171–2176.

—, and W. LAVATE, 1965: Studies on coenzyme Q. The biosynthesis of aurantiogliocladin and coenzyme Q in moulds. J. Biol. Chem. **240**, 532–540.

—, and I. M. CAMPBELL, 1974: Biological reactions of quinones, in: The chemistry of the quinonoid compounds, S. Patai (ed.) John Wiley, London, New York, Sydney, Toronto.

BERLIN, J., and W. BARZ, 1975: Oxidative decarboxylation of *para*-hydroxybenzoic acids by peroxidases under *in vivo* and *in vitro* conditions. Z. Naturforsch. **30**c, 650–658.

BIOLLAZ, M., G. BUECHI, and G. MILIE, 1970: The biosynthesis of the aflatoxins. J. Amer. Chem. Soc. **92**, 1035–1043.

BIRCH, A. J., and F. W. DONOVAN, 1955: Studies in relation to biosynthesis. V. The structures of some natural quinones. Austral. J. Chem. **8**, 529–533.

BOLKART, K. H., M. KNOBLOCH, and M. H. ZENK, 1968: Mevalonic acid, the precursor of the substituted ring of chimaphilin. Naturwissenschaften **55**, 445.

—, und M. H. ZENK, 1968a: Zur Biosynthese methoxylierter Phenole in höheren Pflanzen. Z. Pflanzenphysiol. **59**, 439–444.

—, and —, 1968b: Tyrosin, a precursor of the quinone ring of 2,7-dimethylnaphthoquinone (chimaphilin). Naturwissenschaften **55**, 444–445.

—, and —, 1969: The homogentisate pathway in the biosynthesis of 2,7-dimethyl-1,4-naphthoquinone (chimaphilin). Z. Pflanzenphysiol. **61**, 356–359.

BROCKMANN, H., F. POHL, K. MAIER und M. N. HASCHAD, 1942: Über das Hypericin, den photodynamischen Farbstoff des Johanniskrautes (*Hypericum perforatum*). Annalen **553**, 1–52.

BRYANT, R. W., and R. BENTLEY, 1976: Menaqinone biosynthesis: Conversion of o-succinylbenzoic acid to 1,4-dihydroxy-2-naphthoic acid and menaquinones by *Escherichia coli* extracts. Biochemistry **15**, 4792–4796.

BURNETT, A. R., and R. H. THOMSON, 1968b: Naturally occurring quinones. Part XV. Biogenesis of the anthraquinones in *Rubia tinctorum* L. (madder). J. Chem. Soc. (C), 2437–2441.

CAMPBELL, I. M., 1969: The roles of alanine, aspartate and glutamate in lawsone biosynthesis in *Impatiens balsamina*. Tetrahedron L., 4777–4780.

—, C. J. COSCIA, M. KELSEY, and R. BENTLEY, 1967: Origin of the aromatic nucleus in bacterial quinones, Biochem. Biophys. Res. Commun. **28**, 25–29.

COOKE, R. G., and J. G. DOWN, 1970: Trianellinone, a novel natural triquinone. Tetrahedron L., 583–584.

COX, G. B., and F. GIBSON, 1966: The role of shikimic acid in the biosynthesis of vitamin K. Biochem. J. **100**, 1–6.

CURTIS, R. F., C. H. HASSALL, and D. R. PARRY, 1972: The biosynthesis of phenols. Part XXIV. The conversion of the anthraquinone questin into the benzophenone, sulochrin, in cultures of *Aspergillus terreus*. J. Chem. Soc. (C), **240**–244.

DANSETTE, P., 1972: Etude sur la biosynthèse des naphtoquinones végétales et bacteriennes. Diss., Univ. Paris-Sud.

—, and R. AZERAD, 1970: A new intermediate in naphthoquinone and menaquinone biosynthesis. Biochem. Biophys. Res. Comm. **40**, 1090–1095.

DHRUVA, B. R., A. V. RAMA RAO, R. SRINIVASAN, and K. VENKATARAMAN, 1972: Structure of a quinone from teak tissue culture. Ind. J. Chem. **10**, 683–685.

DI MARCO, A., and F. ARCAMONE, 1975: DNA Complexing Antibiotics: Daunomycin, Adriamycin and their derivaties. Arzneim. Forsch. **25**, 368–375.

DURAND, R., and M. H. ZENK, 1971: Biosynthesis of plumpagin (5-hydroxy-2-methyl-1,4-naphthoquinone) via the acetate pathway in higher plants. Tetrahedron L., 3009–3012.

—, and —, 1974a: The homogentisate ring-cleavage pathway in the biosynthesis of acetate-derived naphthoqinones of the *Droseraceae*. Phytochemistry **13**, 1483–1492.

—, and —, 1974b: Enzymes of the homogentisate ring-cleavage pathway in cell suspension cultures of higher plants. FEBS Letters **39**, 218–220.

EL-KHADEM, M., G. MENKE and F. GROSSMANN, 1966: Schädigung von Erdnußkeimlingen durch Aflatoxine. Naturwissenschaften **53**, 532.

ESMON, G. T., J. A. SADOWSKI and J. W. SUTTI, 1975: A new carboxylation reaction. The vitamin K-dependent incorporation of $H^{14}CO_3^-$ into prothrombin. J. Biol. Chem. **250**, 4744–4748.

EUGSTER, C. H., 1975: Terpenoide, insbesondere diterpenoide Farbstoffe. Ber. Deutsch. Bot. Ges. **88**, 141–161.

FAIRBAIRN, J. W., and F. J. MUHTADI, 1972: The biosynthesis and metabolism of anthraquinones in *Rumex obtusifolius*. Phytochemistry **11**, 215–219.

FERNLUND, P., and J. STENFLO, 1975: Vitamin K and the biosynthesis of prothrombin. J. Biol. Chem. **280**, 6125–6133.

FRANCK, B., 1969: Struktur und Biosynthese der Mutterkornfarbstoffe. Angew. Chem. **81**, 269–278.

French, C. J., C. P. Vance and G. H. N. Towers, 1976: Conversion of p-coumaric acid to p-hydroxybenzoic acid by cell free extracts of potato tubers and *Polyporus hispidus*.

Furuya, T., H. Kojima, and T. Katenta, 1972: 3-Methylpurpurin and other anthraquinones from callus tissue of *Digitalis lanata*. Phytochemistry **11**, 1073–1076.

Gäumann, E., and S. Naef-Roth, 1957: Über die Toxine der *Endothia parasitica* (Murr.) And. Pflanzenschutzberichte **19**, 9–16.

—, und Obrist, 1960: Über die Schädigung der Wasserpermeabilität pflanzlicher Protoplasten durch einige Welktoxine. Phytopath. Z. **37**, 145–158.

Gale, E. F., E. Cundliffe, P. E. Reynolds, M. H. Richmond, and M. J. Waring, 1972: The molecular basis of antibiotic action. John Wiley and Sons, London, New York, Sydney, Toronto.

Gatenbeck, S., 1960: On the biosynthesis of the pigments from *Penicillium islandicum*. II. Acta Chem. Scand. **14**, 296–302.

—, 1962: The mechanism of the biological formation of anthraquinones. Acta Chem. Scand. **16**, 1053–1054.

Grotzinger, E., and I. M. Campbell, 1974: 4-(2′-caroxyphenyl)-4-oxobutyrate: an obligatory intermediate in lawsone biosynthesis. Phytochemistry **13**, 923–926.

Hegnauer, R., 1971: Pflanzenstoffe und Pflanzensystematik. Naturwissenschaften **58**, 585–598.

Hörhammer, L., and H. Wagner, 1961: New phenolic C-glycosides in plants. In: W. D. Ollis (ed.), Chemistry of natural phenolic compounds. Pergamon Press, Oxford, London, New York, Paris.

Hooper, J. W., W. Marlow, W. B. Whalley, and (in part) A. D. Borthwick and R. Bowden, 1971: The chemistry of fungi. Part LXV. The structures of ergochrysin A, isoergochrysin A, and ergoxanthin, and of secalonic acids A, B, C and D. J. Chem. Soc. (C), 3580–3590.

Inoue, K., S. Ueda, Y. Shiobara, and H. Inouye, 1976: 2-Carboxy-4-hydroxy-α-tetralon, eine Vorstufe bei der Biosynthese des Catalpalactons, des Catalponols und des 4,9-dihydroxy-α-lapachons.

Inouye, H., 1976: personal communication.

Inouye, H., S. Ueda, and K. Inoue, 1978: Biosynthesis of prenylnaphthoquinone congeners in callus cultures of *Catalpa ovata*. Tetrahedron L. 4551–4554.

Inouye, H., S. Ueda, K. Inoue, and H. Matsumara, 1979: Biosynthesis of shikonin in callus cultures of *Lithospermum erythrorhizon*. Phytochemistry **18**, 1301–1308.

Koser, S. A., 1968: Vitamin requirements of bacteria and yeast. Charles C. Thomas Publ., Springfield, Illinois, USA.

Leduc, M. M., P. M. Dansette, and R. G. Azerad, 1970: Incorporation de l'acid shikimique dans le noyau des naphthoquinones d'origine bactérienne et végétale. European J. Biochem. **15**, 428–435.

Leistner, E., 1971: A second pathway leading to anthraquinones in higher plants. Phytochemistry **10**, 3015–3020.

—, 1973a: Mode of incorporation of precursors into alizarin (1,2-dihydroxy-9,10-anthraquinone). Phytochemistry **12**, 337–345.

—, 1973b: Biosynthesis of morindone and alizarin in intact plants and cell suspension cultures of *Morinda citrifolia*. Phytochemistry **12**, 1669–1674.

—, 1975: Isolierung, Identifizierung und Biosynthese von Anthrachinonen in Zellsuspensionskulturen von *Morinda citrifolia* L. Planta medica, Supplement, 214–224.

LEISTNER, E. J. H. SCHMITT, and M. H. ZENK, 1967: α-Naphthol, a precursor of vitamin K_2. Biochem. Biophys. Res. Comm. **28**, 845–850.
—, and M. H. ZENK, 1967a: Incorporation of shikimic acid into 1,2-dihydroxyanthraquinone (alizarin) by *Rubia tinctorum* L. Tetrahedron L., 475–476.
—, und —, 1967b: Ein neuer Biosyntheseweg für Anthrachinone: Der Einbau von Shikimisäure in 1,2-Dihydroxyanthrachinon (Alizarin) und 1,2,4-Trihydroxyanthrachinon (Purpurin) in *Rubia tinctorum* L. Z. Naturforschg. **22** b, 865–868.
—, and —, 1968a: Incorporation of 1,4-naphthoquinone into 1,2-dihydroxyanthraquinone (alizarin) in *Rubia tinctorum* L. Tetrahedron L., 861–863.
—, und —, 1968b: Zur Biogenese von 5-Hydroxy-1,4-naphthochinon (Juglon) in *Juglans regia* L. Z. Naturforsch. **23** b, 259–268.
—, and —, 1968c: Mevalonic acid, a precursor of the substitute benzenoid ring of *Rubiaceae*-anthraquinones. Tetrahedron L., 1395–1396.
—, and —, 1969: Chrysophanol (1,8-dihydroxy-3-methylanthraquinone) biosynthesis in higher plants. Chem. Comm., 210–211.
—, and —, 1971a: Biosynthesis of vitamin K_2. In: B. D. McCormick and L. D. Wright (eds.), Methods in Enzymology, Vol. XVIIIc. Academic Press, New York and London.
—, and —, 1971b: Nonsymmetric incorporation of carboxyl-^{14}C-shikimic acid into alizarin (1,2-dihydroxyanthraquinone) in *Rubia tinctorum* L. Tetrahedron L., 1677–1681.
MASUKO, M., H. EGAWA, A. UEYAMA, K. KOSHIMIZU, K. KOBATA, and J. FUMOTO, 1970: Plumbagin, an antifungal factor in *Ceratostigma willmottianum* (*Plumbaginaceae* plant). Ann. Phytopath. Soc. (Japan) **36**, 286–288.
MENTZER, C., 1966: Biogenetic classification of plant constituents. In: T. A. Swain (ed.), Comparative phytochemistry, 21–31. Academic Press, New York and London.
MERXMÜLLER, H., 1967: Chemotaxonomie? Ber. Deutsch. Bot. Ges. **80**, 608–620.
MILLER, L. P., 1973: Glycosides. In: L. P. Miller (ed.), Phytochemistry, Vol. I. Van Norstrand Reinhold Co., New York, Cincinnati, Toronto, London, Melbourne.
MIZUKAMI, H., M. KONOSHIMA, and M. TABATA, 1977: Effect of nutritional factors on shikonin derivative formation in *Lithospermum* callus cultures. Phytochemistry **16**, 1183–1186.
MOSS, M. O., 1972: Aflatoxin and related mycotoxins. In: J. B. Harborne (ed.), Phytochemical Ecology, 125–144. Academic Press, London and New York.
MÜLLER, W.-U., and E. LEISTNER, 1976: 1,4-Naphthoquinone, an intermediate in juglone (5-hydroxy-1,4-naphthoquinone) biosynthesis. Phytochemistry **15**, 407–410.
—, and —, 1978a: Metabolic relation between naphthalene derivatives in Juglans. Phytochemistry **17**, 1735–1738.
—, and —, 1978b: Aglycones and glycosides of oxygenated naphthalenes and a glycosyltransferase from Juglans. Phytochemistry **17**, 1739–1742.
NEWTON, N. A., G. B. COX, and F. GIBSON, 1971: The function of menaquinone (Vitamin K_2) in Escherichia coli K-12. Biochim. Biophys. Acta **244**, 155–166.
PACHLER, K. G. R., P. S. STEYN, R. VLEGAAR, and P. L. WESSELS, 1976: Carbon-13 nuclear resonance assignments and biosynthesis of aflatoxin B_1 and sterigmatocystin. J. Chem. Soc. P. T. 1, 1182–1189.
PAULICK, R. C., M. L. CASEY, D. F. HILLENBRAND, H. W. WHITLOCK, jr., 1975: A ^{13}C nuclear magnetic resonance study of the biosynthesis of islandicin from $^{13}CH_3{}^{13}CO_2Na$. J. Amer. Chem. Soc. **97**, 5303–5305.
RIZVI, S. A. J., P. C. GUPTA, and R. K. KAUL, 1971: Chemical and spectral studies of

nodososide. A new anthraquinone glycoside from the flowers of *Cassia nodosa*. Planta medica **19**, 222–233.

SAMUEL, O., and R. AZERAD, 1972: C-methylation des desméthylménaquinones: II. Specificité du systeme enzymatique de méthylation de Mycobacterium phlei vis-a-vis du substrat quinonique. Biochimie **54**, 305–317.

SANDERMANN, W., 1962: Metabolism. In: M. Florkin and H. S. Mason (eds.), Comparative biochemistry III: Constituents of life – part A, 591–630.

—, und M. H. SIMATUPANG, 1966: Zur Chemie und Biochemie des Teakholzes (*Tectona grandis* L. *fil.*). Holz als Roh- und Werkstoff **24**, 190–204.

SANKAWA, U., Y. EBIZUKA, and S. SHIBATA, 1973: Biosynthetic incorporation of emodine and emodinanthrone into the anthraquinoids of *Penicillium brunneum* and *P. islandicum*. Tetrahedron L., 2125–2128.

SCHARF, K. H., M. H. ZENK, D. K. ONDERKA, M. CARROLL, and H. G. FLOSS, 1971: Stereochemistry of hydrogen elimination from C-6 of shikimate in naphthoquinone biosynthesis. Chem. Commun., 576–577.

SCHERRER, F., and R. AZERAD, 1976: Biosynthesis of mavioquinone. J. Chem. Soc. Chem. Commun., 128–130.

SCHILDKNECHT, H., F. STRAUB, and V. SCHEIDEL, 1976: Mollugin, ein neuer Farbstoff aus Rhizomen des Wiesen-Labkrautes *Galium mollugo* L. Liebigs Ann. Chem., 1295–1306.

SCHMID, H. V., and M. H. ZENK, 1971: p-Hydroxybenzoic acid and mevalonic acid as precursors of the plant naphthoquinone alkanin. Tetrahedron L., 4151–4155.

SEO, S., U. SANKAWA, Y. OGIHARA, Y. IITAKA, and S. SHIBATA, 1973: Studies on fungal metabolites. XXXII. A renewed investigation on (–)flavoskyrin and its analogues. Tetrahedron **29**, 3721–3726.

SHINEBERG, B., and I. G. YOUNG, 1976: Biosynthesis of bacterial menaquinones: The membrane-associated 1,4-dihydroxy-2-naphthoate octaprenyltransferase of *Escherichia coli*. Biochemistry **15**, 2754–2758.

STEGLICH, W., R. ARNOLD, W. LÖSEL, and W. REININGER, 1972: Biosynthesis of anthraquinone pigments in *Dermocybe*. J. Chem. Soc. (Chem. Comm.), 102–103.

—, E. TÖPFER-PETERSEN, and I. PILS, 1973: Neue Phlegmacin Derivate aus *Cortinarius percomis* (*Agaricales*). Z. Naturforsch. **28**c, 354–355.

STEINEGGER, E., and R. HÄNSEL, 1972: Lehrbuch der Pharmacognosie, Springer Verlag, Berlin, Heidelberg, New York.

STÖCKIGT, J., U. SROCKA, and M. H. ZENK, 1973: Structure and biosynthesis of a new anthraquinone from *Streptocarpus dunnii*. Phytochemistry **12**, 2389–2391.

SWANK, R. T., and R. H. BURRIS, 1969: Restoration by ubiquinone of *Azotobacter vinelandii* reduced micotinamide adenine dinucleotide oxidase activity. J. Bact. **98**, 311–313.

TAKEDA, N., S. SEO, Y. OGIHARA, U. SANKAWA, I. IITAKA, I. KITAGAWA, and S. SHIBATA, 1973: Studies on fungal metabolites. XXXI. Anthraquinoid colouring matters of *Penicillium islandicum* Sopp. and some other fungi (–) Luteoskyrin, (–) Rubroskyrin, (+) Rugulosin and their related compounds. Tetrahedron **29**, 3703–3719.

THOMSON, R. H., 1971: Naturally occurring quinones. Academic Press, London and New York.

WEHRLI, W., and M. STAEHELIN, 1971: Actions of rifamycin. Bact. Rev. **35**, 290–309.

WHISTANCE, G. R., and D. R. THRELFALL, 1970: Biosynthesis of phytoquinones. Homogentisic acid: a precursor of plastoquinones, tocopherols and α-tocopherolquinone in higher plants, green algae and blue-green algae. Biochem. J. **117**, 593–600.

YAKI, A., K. MAKINO, and I. NISHIOKA, 1974: Studies on the constituents of *Aloe saponaria*

Haw. I. The structures of tetrahydroanthracene derivatives and the related anthraquinones. Chem. Pharm. Bull. Tokyo **22**, 1159–1166.

YOUNG, I. G., 1975: Biosynthesis of bacterial menaquinones. Menaquinone mutants of *Escherichia coli.* Biochemistry **14**, 399–406.

ZENK, M. H., and E. LEISTNER, 1967: On the mode of incorporation of shikimic acid into 2-hydroxy-1,4-naphthoquinone (lawsone). Z. Naturforschg. **22** b, 460.

—, and —, 1968: Biosynthesis of quinones. Lloydia **31**, 275–292.

ZENK, M. H., H. EL-SHAGI, and U. SCHULTE, 1975: Planta medica. Supplement, 79–101.

Aknowledgement: The authors work reported herein has been supported by the Deutsche Forschungsgemeinschaft.

Betalains

H. Reznik

Botanisches Institut der Universität Köln

1. Introduction

The term «betalain» was introduced by Marby and Dreiding in 1968 to characterize the red and violet betacyanins and the yellow betaxanthins as derivatives of betalamic acid. Betalains are chromoalkaloid-like, highly water-soluble vacuolar pigments. Mainly flowers, fruits and permanently pigmented leaves are brilliantly coloured by them. The distribution of betalains is restricted to ten families of the Centrospermae. In these families they replace without exception the anthocyanins, which are typical for other angiosperms. Recently, betalains have also been shown to be the pigments of the pilei of mushrooms. They have been detected in the genera *Amanita* and *Hygrocybe* which belong to the Hymenomycetes (Döpp and Musso, 1974; Ardenne et al., 1974). A long-known, easily available betalain which is the prototype of all betalains is the pigment of the red beet (*Beta vulgaris, ssp. vulgaris var. crassa*), called betanin. In 1876, Bischoff had already emphasized the peculiartities of this pigment.

2. The structure of betalains

All betalains may formally be derived from betalamic acid (Fig. 1). They are enaminium derivatives. Betalamic acid, first postulated as a hypothetical intermediate in betalain biosynthesis, is a lemon yellow vacuolar pigment which was later detected as an accompanying compound of betaxanthins and betacyanins (Marby et al., 1972; Reznik, 1975, 1978). The betalain chromophore may be classified as polymethine pigment (Marby and Dreiding, 1968). It represents a pentasubstituted 1,7-diazaheptamethinium resonance system (Fig. 2). This basic structure is responsible for the yellow colour of the betaxanthins (maximum absorption at 470–486 nm). The colour of the betacyanins is due to the integration of the aromatic system of DOPA, which causes a strong bathochromic shift of about 60 nm (maximum absorption 534–554 nm).

The structure of betalains was elucidated in 1960–1966 using the entire arsenal of modern analytical techniques (n.m.r. spectra, mass spectra, etc.). Most of the substantial results have been communicated by Dreiding, Wyler and co-workers (Zurich) and Piattelli and co-workers (Catania). These investigations have been reviewed in detail (Marby and Dreiding, 1968; Marby et al., 1972; Piattelli, 1976).

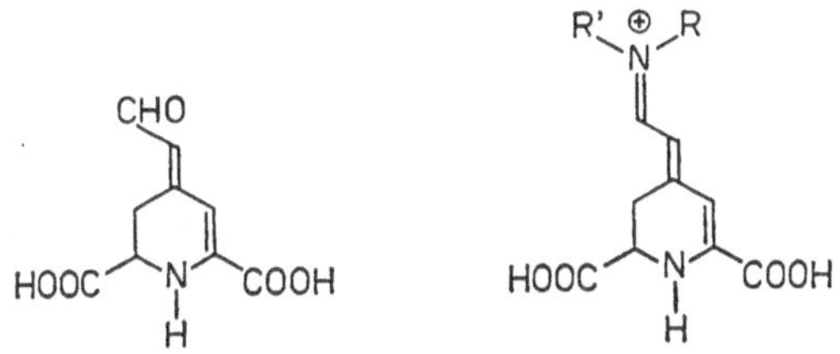

Fig. 1: The chromophore of the betalains. Betalamic acid (left), general formula of the betalains (right) Mabry and Dreiding, 1968)

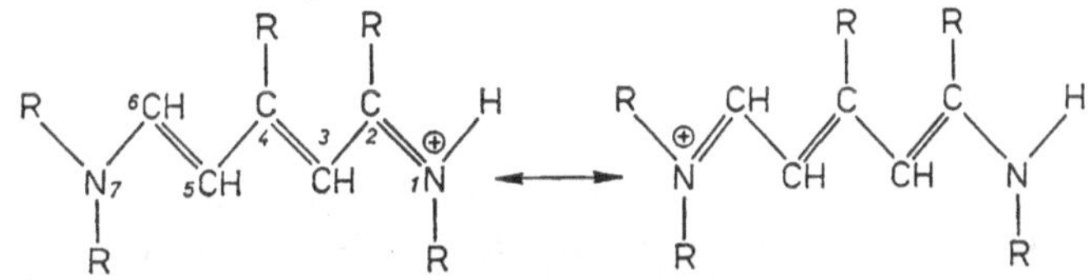

Fig. 2: Resonance structures of the betalain chromophore; 1,7-diazaheptamethin cation (Mabry and Dreiding, 1968)

3. The betacyanins

The natural pigments are mostly O-glycosides. Free aglyca are rather seldom. Irrespective of one exception, the chromophores are betanidin or isobetanidin (Fig. 3). Betanidin differs from isobetanidin by the absolute configuration of the C-15 chiral center; it has an [S] configuration (Wilcox et al., 1965). Enzymatic hydrolysis or mild acidic hydrolysis of the natural glycosides results in formation of betanidin and isobetanidin, respectively. A mixture of both diastereomers is produced by acid hydrolysis or by alcaline media. Flowers of *Carpobrotus acinaciformis* (Aizoaceae) contained non-glycosidated 2-descarboxybetanidin as a minor component (Fig. 4; Piattelli and Impellizzeri, 1970).

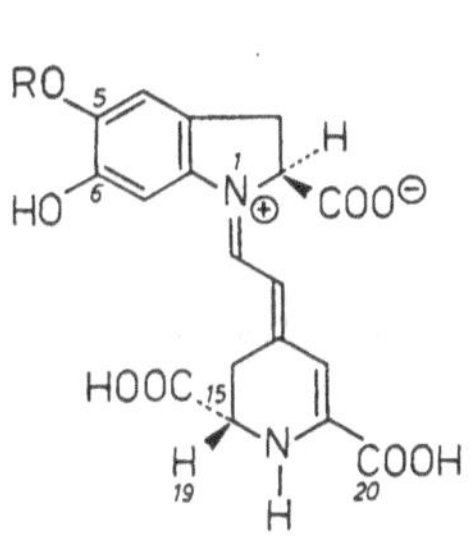

Betanidin

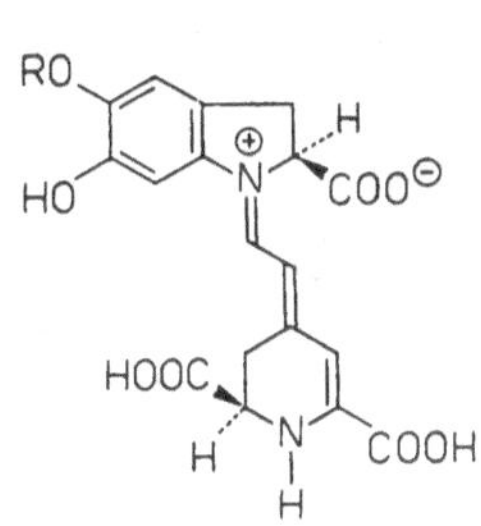

Isobetanidin

Fig. 3

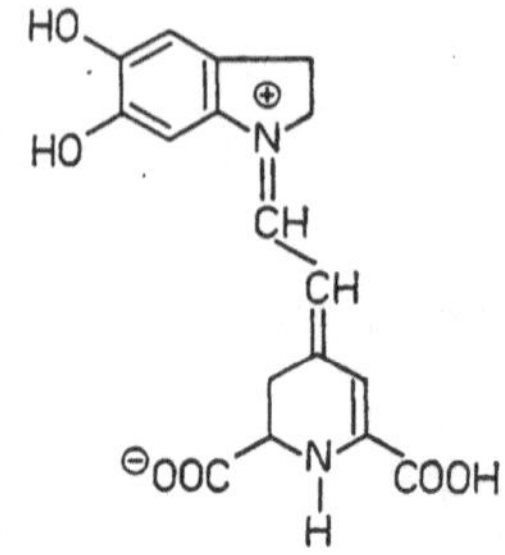

2-Descarboxybetanidin

Fig. 4

All betacyanin chromophores are immonium compounds of betalamic acid and cyclo-DOPA. The formation of glycosides is enabled by hydroxyl groups on positions 5 and 6 of the aromatic part of the molecule. The greater part of the pigments are 5-O-glycosides, but 6-O-glycosides have also been detected (Piattelli, 1976; Table 1, Fig. 5).

Besides the betacyanins which are only combined with sugars, there are also more complex compounds: i.e., various half esters which are formed with sulfuric acid, malonic acid, 3-hydroxy-3-methylglutaric acid. A common compound of this group,

Table 1: Betacyanin glycosides (short selection)
B = betanidin; IB = isobetanidin

Compound	Name	Acids	Deacylated pigment
B 5-O-glucoside	Betanin	–	–
IB 5-O-glucoside	Isobetanin	–	–
B 5-O-sophorosid	Bougainvillein-R I	–	–
B 5-O-(glucuronide)glucoside	Amaranthin	–	–
IB 5-O-(glucuronide)glucoside	Isoamaranthin	–	–
B 6-O-glucoside	Gomphrenin I	–	–
IB 6-O-glucoside	Gomphrenin II	–	–
Sulphuric acid half-ester of betanin	Prebetanin	Sulphuric Acid	B 5-O-glucoside
Malonic acid half-ester of Betanin	Phyllocactin	Malonic Acid	B 5-O-glucoside
Feruloyl-p-coumaroyl amaranthin	Celosianin	p-Coumaric, ferulic Ac.	B 5-O-sophoro-biuronide
Hydroxymethylglutaryl-amaranthin	Iresinin I	Hydroxymethyl-glutaric Acid	B 5-O-sophoro-biuronide

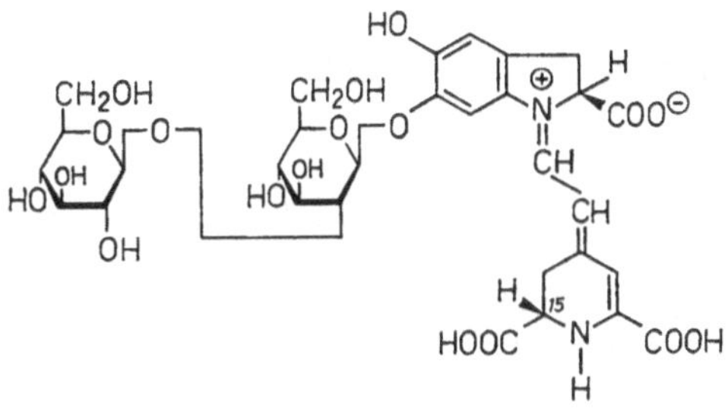

Bougainvillein - v

Fig. 5: Betanidin 6-sophoroside, general structure of bougainvillein-v

Table 2: Betalains of Cactaceae, subfamily Cereoideae (unpublished data)
Flower pigments

	Phyllocactin	Betanin	Indica-xanthin	Mira-xanthin II	Betalamic acid
Southern group					
Cleistocactus	●	●	●		●
Submatucana	●	●	●		
Echinopsis	●	●			
Chamaecereus	●	●	●		●
Pseudolobivia		●	●		●
Lobivia	●	●	●	●	●
Aylostera	●	●	●		
Rebutia	●	●	●		●
Brasilicactus	●	●			
Parodia			●		●
Motocactus	●	●	●		●
Gymnocalycium		●			
Weingartia	●	●	●	●	●
Northern group					
Echinocereus	●	●			
Eriocereus	●	●		●	●
Cereus	●	●			
Astrophytum			●		●
Ferocactus	●	●	●		
Coryphanta	●	●	●		
Mamillaria	●	●			

Number of species investigated: 90

phyllocactin, should be mentioned. It is specific for flowers of Cactaceae (Piattelli and Minale, 1964; Reznik unpublished; Table 2).

A great number of betacyanins are acylated with cinnamic acids. In his latest review, Piattelli (1976) mentions not less than 34 cinnamoyl compounds. p-Coumaric and ferulic acid are the most common acylating compounds. Caffeic and sinapic acid are found less often (Table 1). A compound containing four different cinnamic acids seems to be iresinine III, in which amaranthin is combined with p-coumaric, ferulic, sinapic and caffeic acid (Minale et al., 1966).

4. The betaxanthins

The intensive yellow or orange yellow pigments absorbing between 470 and 486 nm have properties comparable to those of betacyanins. In betaxanthins betalamic acid is combined with an amino acid or an amine (Piattelli, 1976). The structure of only 9 of about 25 betaxanthin pigments detectable on pherograms or in column fractions are known. Substantial knowledge about their constitution is due to the laboratory work of Piattelli. The probably best known yellow betalain is indicaxanthin, which has been detected in fruits of *Opuntia ficus-indica* and other Cactaceae. It contains an L-proline residue (Fig. 6; Piattelli et al., 1964). The chiral C-15 center has an [S] configuration. Our knowledge about other betaxanthins is less complete. Vulgaxanthin I and II from *Beta vulgaris* are characterized by glutamic acid and glutamine, respectively (Fig. 6; Piattelli et al., 1965a). Methionine sulfoxide, aspartic acid, tyramine and dopamine are bound to betalamic acid in miraxanthin I, II, III and V (Figs. 6 and 7; Piattelli et al., 1965b). These pigments have been isolated from yellow flowers of *Mirabilis jalapa*. A hydroxyproline residue is characteristic for portulacaxanthin, which has been detected in flowers of *Portulaca grandiflora* (Piattelli, 1976). Last but not least, the bright yellow dopaxanthin was purified from *Glottiphyllum longum* flowers; it contains dihydroxyphenylalanine (Impellizzeri et al., 1973; Fig. 8). This pigment seems to be common in yellow flowering Aizoaceae, for example, in the highly succulent genera *Glottiphyllum*, *Faucaria*, *Pleiospilos*, *Conophytum* and *Lithops* (Reznik, 1978).

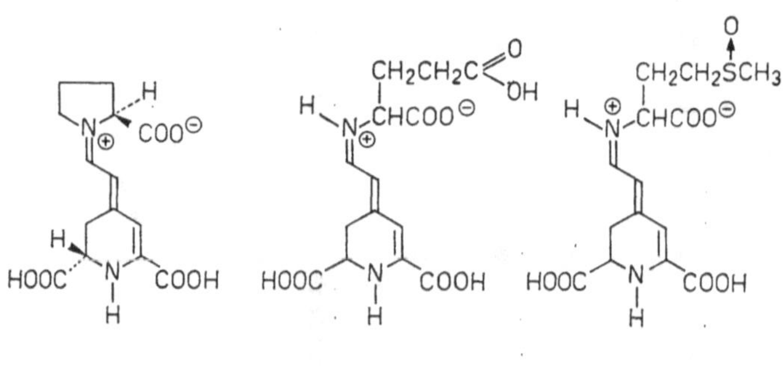

Indicaxanthin Vulgaxanthin-I, Miraxanthin-I

Fig. 6

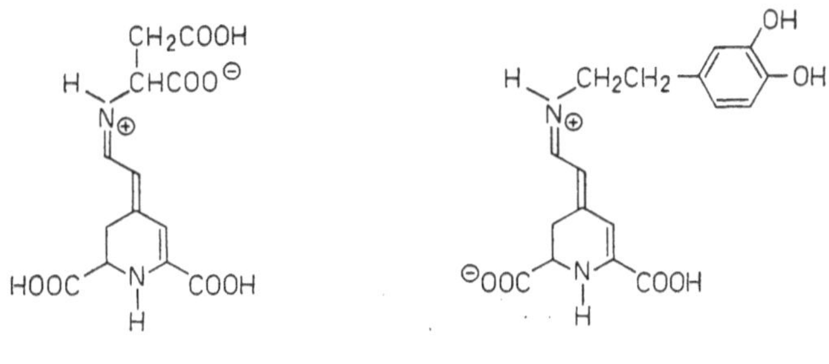

Miraxanthin - II Miraxanthin - V

Fig. 7

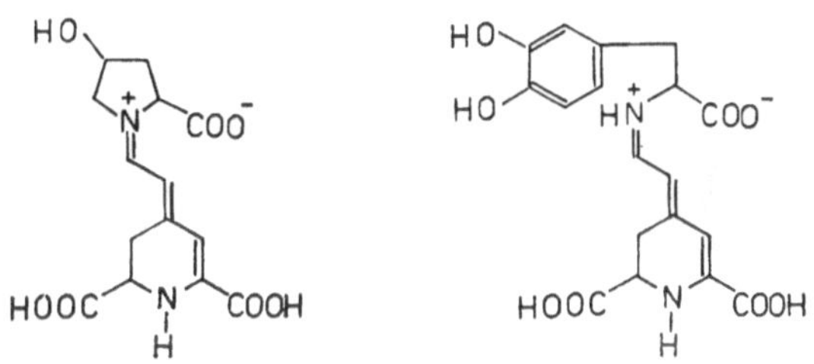

PORTULACAXANTHIN DOPAXANTHIN

Fig. 8

To characterize betaxanthins of yet unknown structure, conversion to neobetaxanthins (maximum absorption 340–360 nm) by means of diazomethane proves to be expedient. This procedure oxidizes the dihydropyridine ring to give a pyridine system (Piattelli et al., 1965; Piattelli, 1976).

5. The betalains of mushrooms

After the taxonomic significance of betalains in families of Centrospermae had been shown many times, it was surprising when the Musso's laboratory reported the occurrence of betaxanthin and other betalain-like pigments in *Amanita* and *Hygrocybe* species. The red cap of the widely distributed toadstool *Amanita muscaria* in particular was thoroughly investigated (Döpp and Musso, 1973, 1974). The methanol extracts were separated by means of electrophoresis, or on DEAE columns which were eluted with a linear gradient of sodium chloride. These procedures revealed up to 9 coloured zones which were partly composed of mixtures of pigments.

A light yellow pigment (max. absorption at about 420 nm) first assumed to be betalamic acid was purified, and after further investigations with *Hygrocybe* it was shown to be a dihydroazepine compound (Ardenne et al., 1974). The latter which is isomeric with betalamic acid was named muscaflavin (Fig. 10, 11). In addition, more than 7

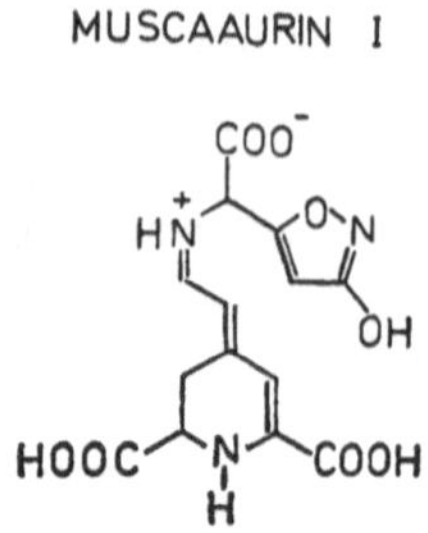

Fig. 9

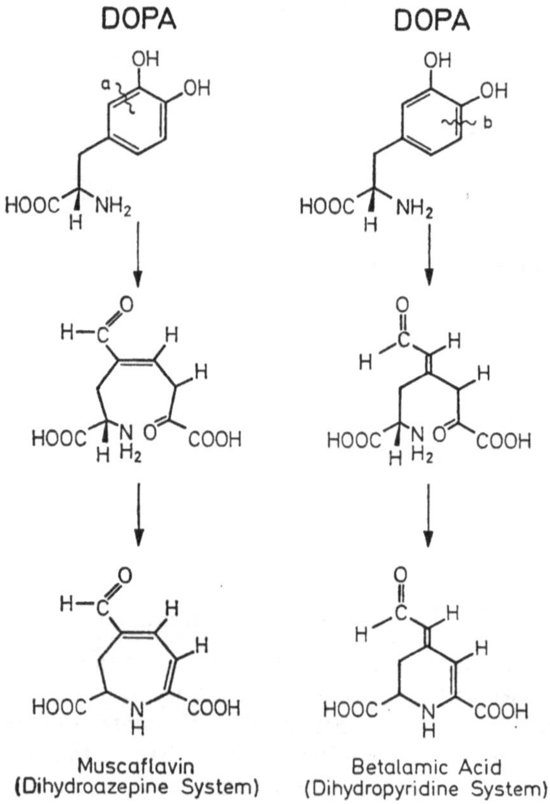

Fig. 10: Formation of betalamic acid and muscaflavin by extradiol cleavage of DOPA

orange yellow «muscaaurines» have been isolated. They are, as also indicated by spectral data (max. absorption about 475 nm) nothing other than betaxanthins. Hydrolysis of chromatographic pure pigments always gives betalamic acid and an amino acid (Döpp and Musso, 1974):

Musca-aurin I	betalamic acid +	ibotenic acid (Fig. 9)
Musca-aurin II	betalamic acid +	unknown amino acid
Musca-aurin III	betalamic acid +	acidic amino acids
Musca-aurin IV		(glu, asp, α-aminoadipinic acid)
Musca-aurin V	betalamic acid +	neutral amino acids (complex mixture, glu-NH_2, asp-NH_2, leu, val, pro
Musca-aurin VII	betalamic acid +	his

Betaxanthins with a muscaflavin chromophore possibly also exist (Ardenne et al., 1974). Three (or more) yellow pigments of *Hygrocybe* have absorption spectra with maxima shifted to shorter wavelengths (max. absorption 465 nm). Besides the already mentioned pale yellow or orange yellow pigments, the red violet muscapurpurine (max. absorption 540 nm) was also found in *Amanita muscaria*. Its structure is assumed to differ from that of the betacyanins.

The unexpected occurrence of betalains in mushrooms may be interpreted as a typical example of biochemical convergence (Piattelli, 1976).

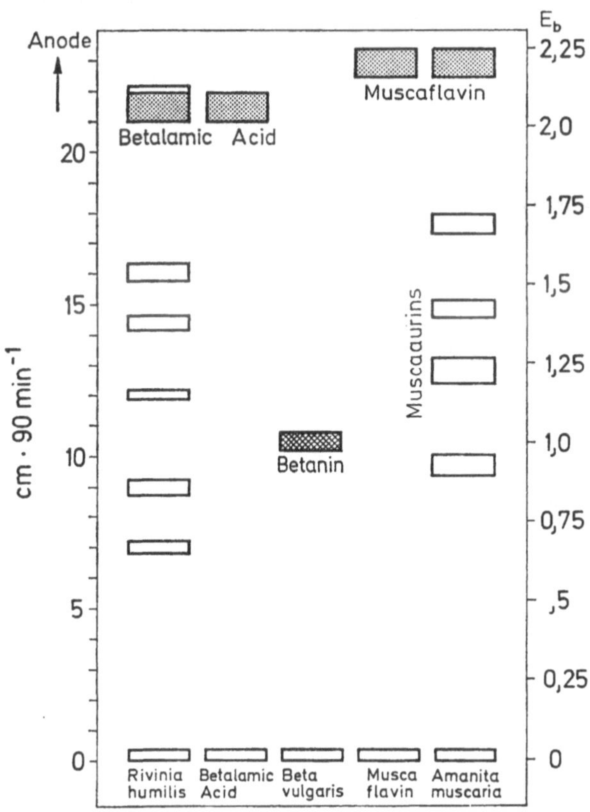

Fig. 11: Separation of betalamic acid, muscaflavin and betaxanthins by high voltage electrophoresis (Reznik, 1978)

6. Betalamic acid and muscaflavin

Stereospecific extradiol cleavages of DOPA result in the two pale yellow pigments betalamic acid and muscaflavin (Fig. 10). The first compound is a dihydropyridine, the latter, a dihydroazepine derivative (Chang et al., 1974; Ardenne et al., 1974).

The modalities of the betalamic acid biogenesis in flower buds of the yellow *Portulaca grandiflora* have been investigated by Chang et al. (1974). The results of feeding experiments with ^{14}C-labelled DOPA corroborate the current hypothesis that oxidative extradiol cleavage of the aromatic ring of DOPA takes place (see biosynthesis section).

Table 3: Betalamic acid in the Aizoaceae and Cactaceae

Number of the species investigated

	Betalamic acid detected	Betalamic acid not detected
Aizoaceae		
Argyroderma	3	1
Bergeranthus	1	
Carpanthea	1	
Conophytum	51	8
Delosperma		7
Dintheranthus	4	
Cephalophyllum	1	
Faucaria	7	1
Glottiphyllum	9	
Hereroa	1	
Lampranthus	1	4
Lithops	33	1
Nananthus	1	
Pleiospilos	16	
Rhombophyllum	3	
Tischleria	1	1
Cactaceae		
Aporocactus	1	
Astrophytum		2
Chamaecereus	2	
Coryphanta		2
Eriocereus	1	
Lobivia	4	1
Mamillaria		26
Mediolobivia	1	1
Notocactus	2	
Opuntia	1	2
Rebutia	3	
Weingartia	2	

The data show that the absolute configuration of L-[S]-DOPA determines the configuration of [S]-betalamic acid as well as the [S] configuration on the C-15 of betanins and the dihydropyridine part of indicaxanthin (Fischer and Dreiding, 1972; Impellizzeri and Piattelli, 1972).

Betalamic acid is not only the common precursor in the biosynthesis of the dihydropyridine moiety of all betalains, it is also a common pale yellow pigment in the vacuoles of betalain-forming plants (Mabry et al., 1972; Reznik, 1978).

A recent investigation on the distribution of free betalamic acid has shown that among 274 species of 10 families of Centrospermae not less than 174 species accumulated this pale yellow compound in the cell sap. Very often betalamic acid is associated with betaxanthins, especially in yellow flowers of Cactaceae and Aizoaceae (Reznik, 1978; Table 3).

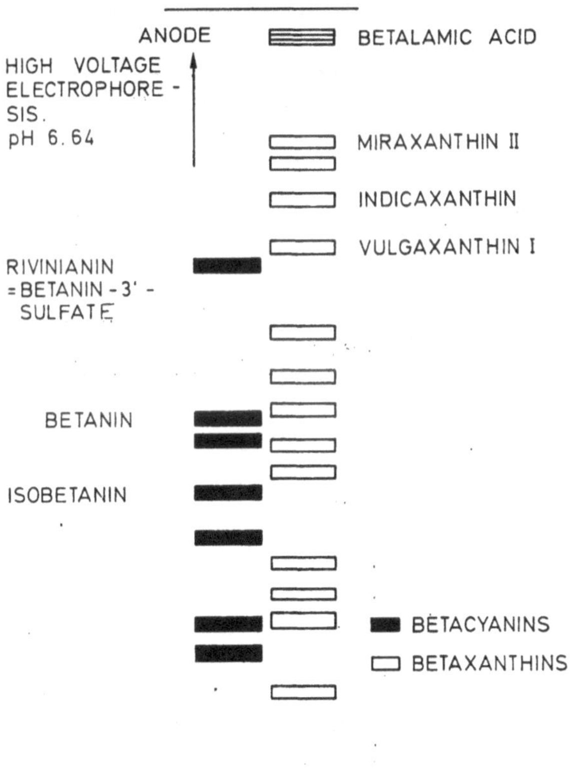

BETALAINS OF PHYTOLACCACEAE

Fig. 12a

Fig. 12a–d: Synoptic representation of the betalain patterns from four systematically related families (unpublished data)

Muscaflavin, the isomer of betalamic acid, is found in the vividly coloured caps of *Amanita muscaria* and several orange yellow species of *Hygrocybe*. These mushrooms, however, are also capable of producing the other alternative of the extradiol cleavage. This is inferred by the finding that the betaxanthins of the musca-aurin group also contain betalamic acid. In contrast, the search for muscaflavins in betalain-containing Centrospermae was futile (Reznik, 1978).

Comment

Purification requires manifold procedures. Chromatography on ion-exchange columns and subsequent chromatography on polyamide by means of methanol-citric acid gradients proved to be expedient (Piattelli, 1976). DEAE sephadex A 25 columns eluted by a linear NaCl gradient have also been very efficient (Döpp and Musso, 1974). For analytical purposes high voltage electrophoresis gives excellent results (Reznik, 1957; Wyler and Dreiding, 1961; Piattelli and Minale, 1964; Reznik, 1978).

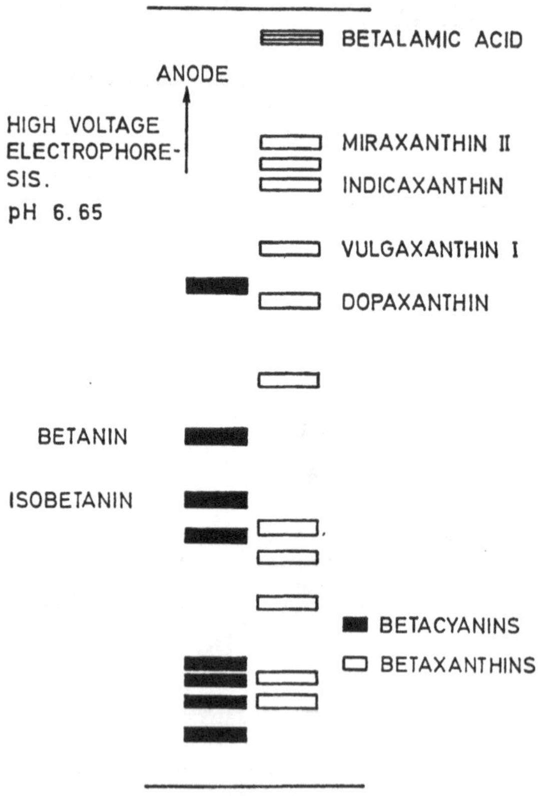

BETALAINS OF AIZOACEAE

Fig. 12b

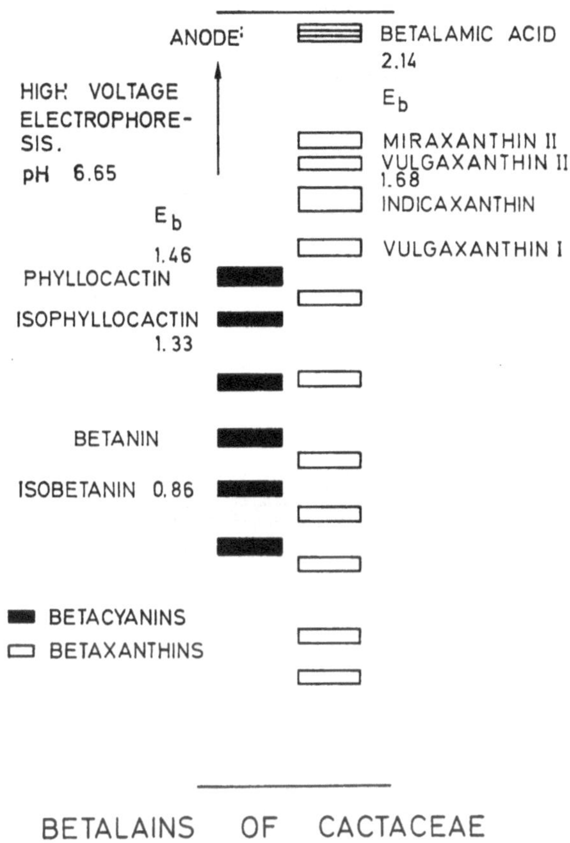

Fig. 12c

This method makes use of the amphoteric character of the betalain molecules. A typical betacyan exists as a cation below pH 2, as an immobile Zwitterion at about pH 2, as a mono-anion between pH 2 and 3.5 and as a bis-anion between pH 3.5 and 7. In the latter case the three carboxyl groups of the betanidin chromophore are dissociated. In addition, the electrophoretic properties of the acylating agents also play a role. The betaxanthins behave similarly. Their electrophoretic mobility is highest in weak acidic buffers. Generally the mobilities of the betacyan ions are related to the motility of betanin (E_b values). With respect to betaxanthins, the mobility may also be related to indicaxanthin (E_i values) (Wyler and Dreiding, 1961; Piattelli et al., 1965).

Electrophoresis often results in highly complicated patterns of coloured zones, the nature of which is not completely known (Figs. 11 and 12a–c). Numerous minor components still await characterization, identification or elucidation of the structure. Degradation procedures have been reviewed by Piattelli (1976).

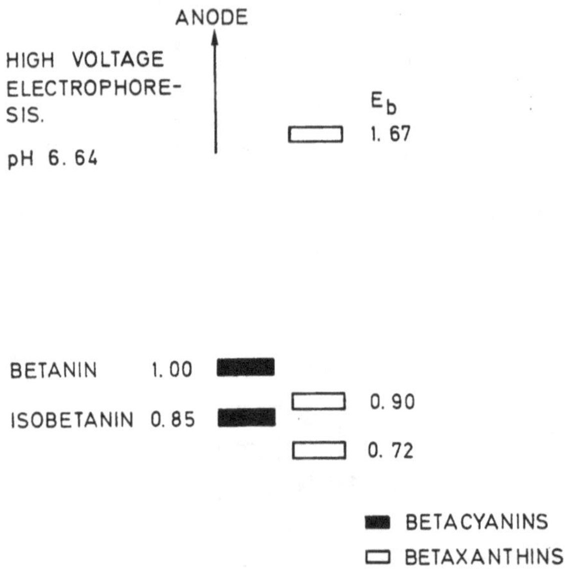

BETALAINS OF DIDIEREACEAE
Fig. 12d

Biosynthesis

At present we are far away from a detailed understanding of the biochemical processes involved in the biosynthesis of betalains. On the whole, however, experiments have demonstrated that the dihydroindol part as well as the dihydropyridine moiety have DOPA as precursor.

Betacyans are built up from two molecules of DOPA. This finding had been postulated in 1963 by Wyler, Mabry and Dreiding. Their supposition has been confirmed by feeding experiments using labelled tyrosine and DOPA (Hörhammer et al., 1964; Garay and Towers, 1966; Minale et al., 1965; Miller et al., 1968). The plants under investigation have mostly been *Beta, Amaranthus* and *Opuntia.*

Minale et al. (1965) fed labelled DOPA to fruits of *Opuntia ficus-indica* and isolated labelled indicaxanthin and betanin. The distribution of radioactivity clearly showed that DOPA was the precursor of the dihydropyridine moiety. Using the same plant, Miller et al. (1968) found that besides the label of the dihydropyridine moiety a minor amount of the label is also present in the dihydroindol moiety.

Liebisch et al. (1969) have confirmed these results by application of 1-^{14}C, ^{16}N-tyrosine or 2-^{14}C, ^{15}N-tyrosine. The labelled nitrogen of the amino group has been localized in both N-heterocycles of betanin.

The mechanism of Cyclo-DOPA formation by an intramolecular amine addition from DOPA, which itself originates from tyrosine, does not create theoretical problems. The transformation of DOPA into a dihydropyridine system, however, has two possibilities with regard to the oxidative opening of the aromatic ring. The first possibility is an intradiol cleavage between the two ortho-situated hydroxyl groups by means of oxygenases of the pyrocatechase type as is known to occur in many microorganisms which degrade ortho-diphenoles. The subsequent ring formation is brought about by linking the nitrogen to the C-5′ of the former aromatic ring. The second possibility is an extradiol cleavage between C-4′ and C-5′ and a bond between the nitrogen and C-3′. This reaction could be catalyzed by an enzyme of the metapyrocatechase type.

The second mechanism was established in 1972 at Zurich and Catania (Fischer and Dreiding, 1972; Impellizzeri and Piattelli, 1972). Impellizzeri and Piattelli fed double labelled tyrosine to fruits of *Opuntia* and isolated the radioactive indicaxanthin (Fig. 13). The ^{14}C-labelled carboxy group of tyrosine has been found as a carboxyl group on the chiral center of the dihydropyridine moiety of indicaxanthin. The tritium label on C-5 of tyrosine has been identified at the postulated position on indicaxanthin. Consequently, the aldehyde group of betalamic acid should have been tritiated. The investigations of Chang et al. (1974) with *Portulaca grandiflora* flowers have confirmed these early findings.

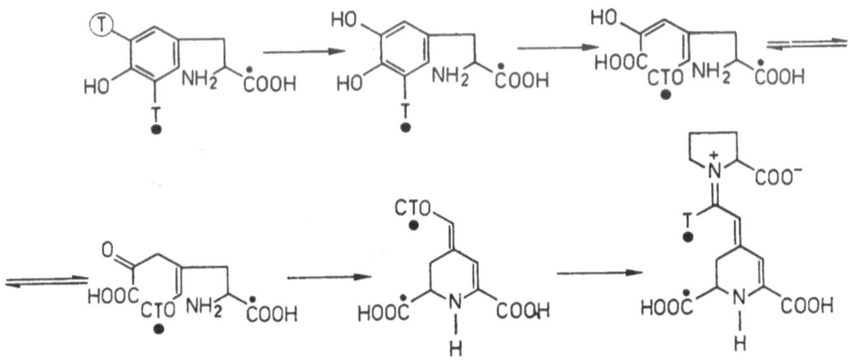

Fig. 13: Proposed pathway for indicaxanthin biosynthesis in Opuntia ficus-indica fruits (Impellizzeri and Piattelli, 1972)

Isolated betacyanins generally contain both C-15 diastereomers, e.g., betanin and isobetanin, phyllocactin and isophyllocactin, etc. However there are also numerous examples for the exclusive formation of betanidin glycosides. It may also be that derivatives of betanidin ([S] configuration) account for the main fraction, whereas derivatives of isobetanidin account for a minor fraction. Seedlings of *Amaranthus tricolor* only contain amaranthin; leaves of older plants, however, contain a mixture of amaranthin and isoamaranthin. These and other observations infer that betanidin pigments are the primary products of the betalain synthesis. Isobetanidin derivatives are the result of a subsequent spontaneous epimerization (Piattelli, 1976). The point in the biosynthetic chain at which the glycosidation of betacyanins takes place is not yet clear. Sciuto et al. (1972) have obtained labelled betanidin-5-O-glycoside after feeding fruits of *Opuntia* with labelled betanidin. This finding has led to the conclusion that the glucosyl-transferase reaction is the last step in the biosynthetic chain. Another differing result has been communicated by Sciuto et al. (1974). Seedlings of *Celosia plumosa* have been able to incorporate cyclo-DOPA (1-^{14}C) as well as cyclo-DOPA (1-^{14}C)-5-glucoside into the dihydroxyindol moiety of amaranthin. In contrast, feeding of betanidin only resulted in minor incorporation rates. This finding may be interpreted by the assumption that intermediates of the betanidin biosynthesis can already be glycosidated. With respect to amaranthin, the following pathways seem to be possible (Piattelli, 1976):

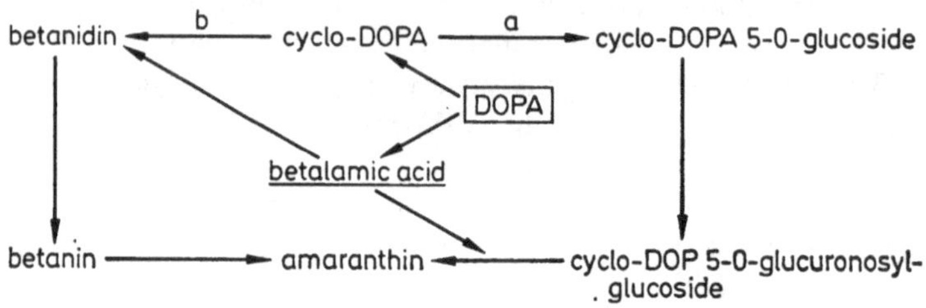

The *Celosia* cultivar (*Celosia plumosa* cv. « Golden Feather ») used by Sciuto et al. (1974) only produces betaxanthins under normal conditions. Amaranthin has only been formed under the specific conditions of the cyclo-DOPA application. It can not excluded that the other amaranthin plants show reactions which accord with the first findings of Sciuto et al. (1972).

A special prospect of forthcoming investigations will be the enzymes of betalain biosynthesis. This field of investigation is still terra icognita. Knowledge of the enzymes responsible for the stereospecific extradiol cleavage of the aromatic DOPA ring will especially contribute to the theoretical understanding of this mechanism.

Physiological aspects of betalain formation

The formation and accumulation of betalain pigments shows many phenomena which corresponds to physiological phenomena during the biogenesis of anthocyanins. Similar to the modalities of pigmentation in plants producing anthocyanins, betalain plants produce betacyanins when wounded, juvenile betacyanins, insolation betacyanins, etc. The occurrence of these responses is mostly restricted to vegetative parts. Light, however, does not seem to play an essential role as in anthocyanin biosynthesis. Comparative investigations have shown that many betalain plants were able to synthesize the pigments in the dark (Wohlpart and Mabry, 1968). The amount of betalains formed in each case was always lower than in the light.

Photoinduction was required for betalain synthesis in *Amaranthus tricolor*, *Amaranthus caudatus* and *Chenopodium rubrum*. These species are intensively studied by several laboratories (Piattelli et al., 1969; Koehler, 1973a; 1975; Guidici de Nicola et al., 1975a, b; Kinsman et al., 1975a, b; Stobart and Kinsman, 1977; and others). It seems that the synthesis of amaranthin is multiply regulated by complex mechanisms including phytochrome and a high energy blue light reaction. Phytohormones also play a role. The effects caused by the interdependence of all factors give rise to various interpretations.

Koehler (1972) has tried to summarize the complex photoregulation of *Amaranthus* in a model (Fig. 14). Kinetine, which maintains high levels of betacyanin in darkness, seems to be highly important (Giudici de Nicola et al., 1972a, b; Koehler, 1972a, b; Piattelli et al., 1971).

Numerous experiments with inhibitors of nucleic acid and protein synthesis (actinomycin, 2-thiouracil, chloramphenicol, puromycin and others) point toward regulation processes on the transcriptional and translational level which exert influence on enzyme activities.

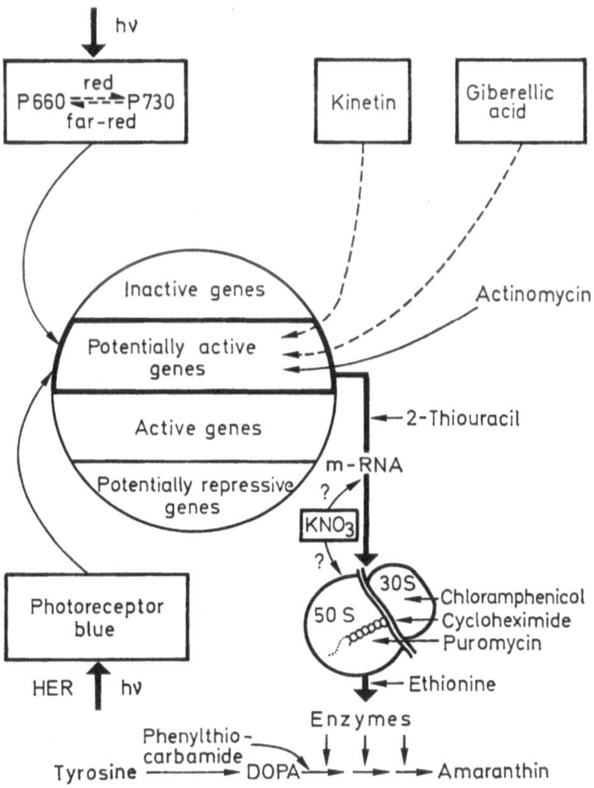

Fig. 14: Regulation of amaranthin biosynthesis, (Working scheme; Koehler, 1972)

Giudici de Nicola et al. (1972a, b) assume that the combined action of light and kinetine may additionally control the availability of high-energy compounds which are necessary for pigment formation. Photoreceptors active in the range of photosynthesis, e.g., photophosphorylation, are considered to play a role in this process, especially under continuous illumination. Furthermore the effect of kinetine on *Amaranthus* has led to an elaborate cytokinin assay (Koehler and Conrad, 1966; Bigot, 1968; Biddington and Thomas, 1973). Conrad (1974) reported that cytokinins can reliably be detected in a standardized *Amaranthus* bioassay $5 \cdot 10^{-10}$ M/l zeatin, $2 \cdot 10^{-9}$ M/l zeatinriboside and $1 \cdot 10^{-8}$ M/l kinetine.

In any case the regulation of the hormonal control of amaranthin formation has been found to be very complex.

Gibberellic acid (GA_3) applied exogenously is an antagonist of cytokinins. The accumulation of betalain pigments ist strongly inhibited by this substance (Colomas et al., 1973; Kinsman et al., 1975; Stobart and Kinsman, 1977).

In discussing experiments with ^{14}C-labelled precursors (tyrosine, DOPA) Stobart and Kinsman (1977) came to the conclusion that GA_3 mainly controls the production and/or availability of tyrosine, whereas kinetine exerts an influence on the biosynthetic pathway between DOPA and the complete pigment:

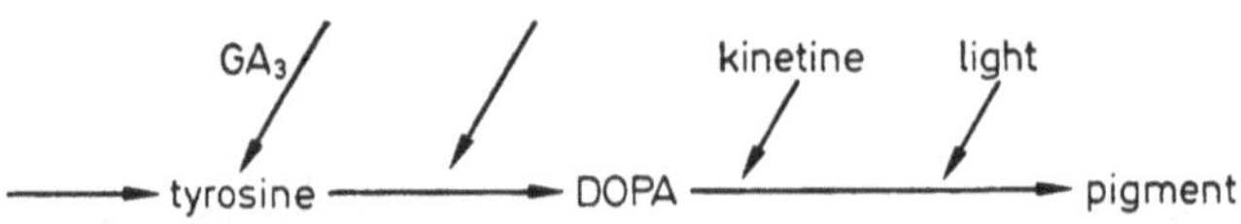

Giudici de Nicola et al. (1975a) worked with isolated cotyledons of *Amaranthus*. After feeding DOPA to illuminated cotyledons the formation of amaranthin, vulgaxanthin II and miraxanthin II was strongly stimulated. According to the interpretation of the authors there is a specific light action on the synthesis of the dihydropyridine moiety (e.g., formation of betalamic acid?).

Comparable to the bioassay based on pigment stimulation by kinetine, GA_3-dependent inhibition of amaranthin formation was used for an elaborate bioassay (Kinsman et al., 1975a). Using seedlings of *Amaranthus caudatus*, GA_1, GA_3 and GA_7 can be detected semiquantitatively.

Tissue cultures:

After beginning failures several investigators finally succeeded in growing tissue cultures with high accumulation rates of betacyanin (Endress, 1976, 1977; Misawa, 1977; Colomas et al., 1978). The highly uniform cells of the cultures have advantages over the complex system of seedlings: the applied substances can more easily penetrate to the zone of action and the substances formed can be isolated easier. Isolating enzymes involved in the synthesis of betalains from tissue cultures may also be advantageous. Colomas et al. (1978) cultivated stem tissue of *Myrtillocactus geometrizans* (Cactaceae) and observed an intensive accumulation of betanin, phyllocactin and indicaxanthin in the callus. The composition of the pigments corresponded with that of the red-violet fruits of *Myrtillocactus*. Endress (1976, 1977) performed numerous series of experiments in order to elucidate the regulation of betacyan formation in the callus of *Portulaca grandiflora*. His interest was expended mainly on the still unsolved cAMP problem.

The formation of betalain and cAMP

Rast et al. (1973) detected a stimulation of amaranthin synthesis by cAMP and dibutyryl-cAMP while working with *Amaranthus caudatus*. They interpreted this finding as a «second messenger» effect of these cyclic nucleotides. Cyclic nucleotides were assumed to be able to replace phytochrome. This interpretation was rejected by Giudici de Nicola et al. (1975b) who investigated the same plant. They held that cAMP acts like an analogon of kinetine. Kinetine as well as cAMP may be conceived of as influencing gene activation by means of the same mechanism. While analysing the cAMP effect in cells of *Portulaca*, Endress (1976, 1977) succeded in identifying native cAMP and cGMP. In addition, several phosphodiesterases and adenylate cyclase were detected. The mode of action of cyclic nucleotides in pigment synthesis is still rather unclear, however, which is also the case in *Portulaca*. On the one hand, a phosphodiesterase which is allosterically regulated by cyclic nucleotides is said to act on the degradation of cAMP and cGMP; and, on the other hand, this enzyme is said to influence the catabolism of RNA.

The involvement of catecholamines, which have been detected in the *Portulaca* callus, is an additional problem. Reference is made in this regard to a publication of Amrhein (1977) which contains a critical discussion of the distribution and function of cAMP in higher plans.

Taxonomic problems

Suggestions made more than 20 years ago are now generally accepted. Betalains (formerly called N-anthocyanins and flavocyanins) are inherent characters of the Centrospermae. The formation of betalains excludes the formation of anthocyanins and vice versa (Reznik, 1955, 1957, 1966; Wyler and Dreiding, 1961; Wohlpart and Mabry, 1968; Mabry et al., 1972b; Piattelli 1976).

Although the ability to form anthocyanins is blocked in the 10 typical betalain families, flavonoids are regularily synthesized; e.g., the typical glycosides of kaempferol, isorhamnetin and quercetin are present in all betalain plants (Mabry et al., 1972b). Flavonols and anthocyanins are products of final ramifications of the differentiation in flavonoid biosynthesis. Therefore, missing anthocyanins seem to be the consequence of a mutation blocking the essential biosynthetic step from dihydroflavonol to anthocyanin. The dehydrogenation step in the biosynthesis of flavonol remains unchanged (Reznik, 1966, 1975). This break in the spectrum of pigments may have favored the evolution of betalains in the ancestral Centrospermae. One may say that the betalains have replaced the anthocyanins by providing all the anthocyanin functions. Investigations on the distribution of betalains have shown that the position of the Caryophyllaceae and Molluginaceae is an exceptional one. They do not possess betalains, and they produce anthocyanins. Therefore, Mabry has proposed division of the Centrospermae into two nearly related orders: The Chenopodiales with the 10 betalain families and the Caryophyllales (Mabry, 1973). Not all taxonomists accept this point of view, however.

Betalains have often been an important criterion in the classification of questionable taxa. Besides the well-known example of the Didiereaceae (Rauh and Reznik, 1961), there are numerous further examples: *Geocarpon* has been perceived to belong to the Caryophyllaceae (Bogle et al., 1971) and the position of *Petiveria* and *Agdestis* as belonging to the Phytolaccaceae has been reconfirmed (Behnke et al., 1974). *Halophytum* is not a Chenopodiaceae, but may belong to the Phytolaccaceae. The betalain pattern (mainly amaranthin and a minor amount of betanin) of the monogeneric group of the Dysphaniaceae of Australia is very similar to that of Chenopodium (Reznik, unpublished; Fig. 15). Morphological criteria of the flowers (Eckardt, 1968) also support the view which places the Dysphaniaceae in the vicinity of *Chenopodium*.

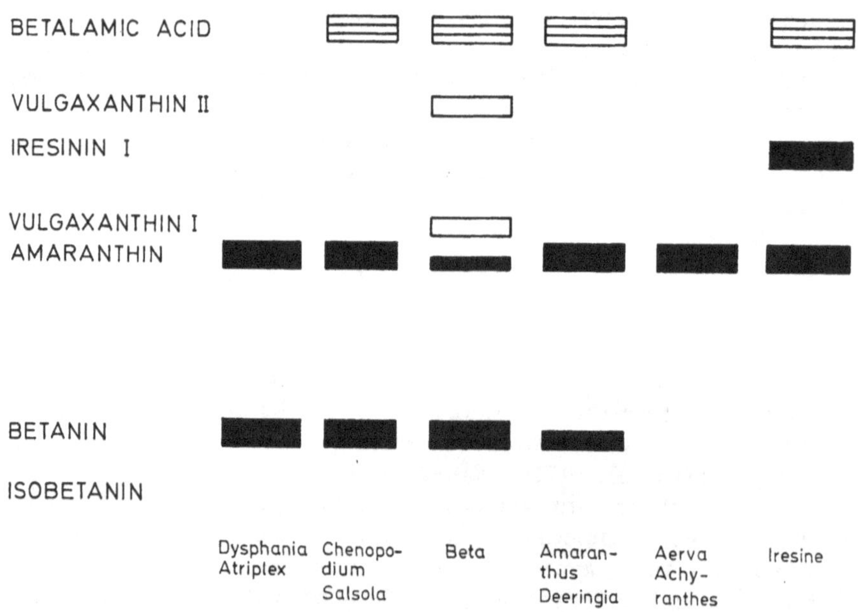

Fig. 15: Betalain patterns in the Chenopodiaceae and Amaranthaceae, including the Australian genus Dysphania (unpublished data)

Finally, what prospect have betalains in systematics? What are the problems? Concerning single families, there is a lack of comparative analyses of the betalain pattern, and concerning known patterns of different genera and species, the interpretation is made difficult by deficiencies in the elucidation of the structures of most of the native betalains.

References

AMRHEIN, N., 1977: The current status of cyclic AMP in higher plants. Ann. Rev. Plant. Physiol. **28**, 123–132.

v. ARDENNE, R., H. DÖPP, H. MUSSO, and W. STEGLICH, 1974: Über das Vorkommen von Muscaflavin bei Hygrocyben (Agaricales) und seine Dihydroazepin-Struktur. Z. Naturforsch. **29** c, 637–639.

BEHNKE, H. D., 1972: Sieve-tube plastids in relation to Angiosperm systematics. – An attempt towards a classification by ultrastructural analysis. Bot. Rev. **38**, 155–197.

–, 1976: Ultrastructure of Sieve-Element Plastids in Caryophyllales (Centrospermae), Evidence for the Delimitation and Classification of the Order. Plant. Syst. Evol. **126**, 31–54.

–, C. CHANG, I. J. EIFERT, and T. MABRY, 1974: Betalains and P-type sieve-tube plastids in Petiveria and Agdestis (Phytolaccaceae). Taxon **23**, 541–542.

–, T. MABRY, I. J. EIFERT, and L. POP, 1975: P-type sieve-element plastids and betalains in Portulacaceae (including Ceraria, Portulacaria, Talinella). Can. J. Bot. **53**, 2103–2109.

BIDDINGTON, N. L., and T. H. THOMAS, 1973: A modified Amaranthus betacyanin bioassay for the rapid determination of cytokinins in plant extracts. Planta **111**, 183–186.

BIGOT, C., 1968: Action d'adénines substitutées sur la synthèse des bétacyanines dans la plantule d'Amaranthus caudatus L. Compt. rend. Acad. Sci., Sér. D, **266**, 349–352.

BISCHOFF, H., 1876: Inaugural-Dissertation der Universität Tübingen.

BOGLE, A. L., T. SWAIN, R. D. THOMAS, and E. D. KOHN, 1971: Geocarpon: Aizoaceae or Caryophyllaceae? Taxon **20**, 473–477.

CHANG, C. P., and T. J. MABRY, 1973: The constitution of the order Centrospermae: rRNA-DNA Hybridization studies among betalain- and anthocyanin producing families. Biochem. Syst. **1**, 185–190.

–, L. KIMLER, and T. J. MABRY, 1974: Biogenesis of betelamic acid. Phytochem. **13**, 2771–2775.

COLOMAS, J., J.-L. LAYTOU, and C. BULARD, 1973: Influence de deux gibbérellines GA_3 et GA_4 sur la biosynthèse de l'amarantine induite par la kinétine chez Amaranthus tricolor L. var. bicolor ruber Hort. C. R. Acad. Sci. Ser. D **277**, 173.

–, 1977: Séparation des pigments bétalaiques par une méthode fondée sur la chromatographie sur gel de Séphadex appliquée à des plantules d'Amaranthus caudatus L. var. pendula. Z. Pflanzenphysiol. **85**, 227–232.

–, P. BARTHE, and C. BULARD, 1978: Séparation et identification des bétalaines synthétisées par les tissus de tige de Myrtillocactus geometrizans cultivés in vitro. Z. Pflanzenphysiol. **87**, 341–346.

CONRAD, K., 1974: Ein sensibilisierter Amaranthus-Cytokinintest (AT 74). Biochem. Physiol. Pflanzen(BPP) **165**, 531–535.

DÖPP, H., and H. MUSSO, 1973: Fliegenpilzfarbstoffe, II. Isolierung und Chromophore der Farbstoffe aus Amanita muscaria. Chem. Ber. **106**, 3473–3482.

–, and –, 1973: Die Konstitution des Muscaflavins aus Amantia muscaria und über Betalaminsäure. Naturwiss. **60**, 477–478.

–, and –, 1974: Eine chromatographische Analysenmethode für Betalainfarbstoffe in Pilzen und höheren Pflanzen. Z. Naturforsch. **29** c, 640–642.

ECKARDT, T., 1968: Zur Blütenmorphologie von Dysphania plantaginella F. v. M. Phytomorphology **17**, 165–172.

ENDRESS, R., 1976: Betacyan-Akkumulation in Kallus von Portulaca grandiflora var. JR unter dem Einfluß von Phytohormonen und Cu^{2+}-Ionen auf unterschiedlichen Grundmedien. BPP **169**, 87–98.

—, 1977: Einfluß möglicher Phosphodiesterase-Inhibitoren und cAMP auf die Betacyan-Akkumulation. Phytochem. **16**, 1549–1554.

FISCHER, N., and A. DREIDING, 1972: Biosynthesis of betalains: On the cleavage of the aromatic ring during the enzymatic transformation of DOPA into betalamic acid. Helv. Chim. Acta **55**, 649–658.

GARAY, A. S., and G. H. N. TOWERS, 1966: Studies on the biosynthesis of amaranthin. Canad. J. Bot. **44**, 231–236.

FRENCH, C. J., R. C. PECKET, and H. SMITH, 1974a: Effect of exogenous DOPA and tyrosine on amaranthin synthesis and pigment type in Amaranthus. Phytochem. **13**, 1505–1511.

GIUDICI DE NICOLA, M., M. PIATELLI, V. CASTROGIOVANNI, and C. MOLINA, 1972a: The effect of short-term irradiation on kinetin-induced amaranthin synthesis in Amaranthus tricolor seedlings. Phytochem. **11**, 1005–1010.

—, —, —, and V. AMICO, 1972b: The effects of light and kinetic on amaranthin synthesis in relation to phytochrome. Phytochem. **11**, 1011–1017.

—, —, and V. AMICO, 1973a: Photocontroll of betaxanthin synthesis in Celosia plumosa seedlings. Phytochem. **12**, 353–357.

—, —, and —, 1973b: Effect of continuous far red on betaxanthin and betacyanin synthesis. Phytochem. **12**, 2163.

—, V. AMICO, S. SCIUTO, and M. PIATTELLI, 1975a: Light controll of amaranthin synthesis in isolated Amaranthus cotyledons. Phytochem. **14**, 479–481.

—, —, and M. PIATTELLI, 1975b: Effects of light and kinetin on amaranthin synthesis induced by cAMP. Phytochem. **14**, 989–991.

HÖRHAMMER, L., H. WAGNER, and W. FRITZSCHE, 1964: Zur Biosynthese der Betacyane. Biochem. Z. **339**, 398–400.

HUNZIKER, J. H., H. D. BEHNKE, I. J. EIFERT, and T. J. MABRY, 1974: Halophytum ameghinoi: A betalain-containing and P-type sieve-tube plastid species. Taxon **23**, 537–539.

IMPELLIZZERI, G., and M. PIATTELLI, 1972: Biosynthesis of Indicaxanthin in Opuntia ficus-indica fruits. Phytochem. **11**, 2499–2502.

—, —, and S. SCIUTO, 1973: Acylated betacyanins from Drosanthemum floribundum. Phytochem. **12**, 2295.

—, —, and —, 1973: A new betaxanthin from Glottiphyllum longum. Phytochem. **12**, 2293.

IMPERATO, F., 1975a: Acylated betacyanins of Portulaca oleracea. Phytochem. **14**, 2091–2092.

—, 1975b: Betanin 3′-sulphate from Rivinia humilis. Phytochem. **14**, 2526–2527.

KIMLER, L., J. MEARS, T. J. MABRY, and H. RÖSLER, 1970: On the question of the mutual exclusiveness of betalains and anthocyanins. Taxon **19** (6), 875–878.

KINSMAN, L. T., N. J. PINFIELD, and A. K. STOBART, 1975a: A gibberellin bioassay based on Betacyanin Production in Amaranthus caudatus seedlings. Plata **127**, 149–152.

—, —, and —, 1975b: The hormonal control of amaranthin synthesis in Amaranthus caudatus seedlings. Planta **127**, 207–212.

KOEHLER, K. H., 1972a: Photocontrol of betacyanin synthesis in Amaranthus caudatus seddlings in the presence of kinetin. Phytochem. **11**, 133–137.

—, 1972b: Action of inhibitors of protein and nucleic acid synthesis on light dependent and kinetin stimulated betacyanin synthesis. Phytochem. **11**, 127–131.

—, 1973a: Die Steuerung der Amaranthinbiosynthese durch das Phytochromsystem (ein-

schließlich Beschreibung der Bestrahlungsanlage und der Lichtenergiemessung). Biol. Zbl. **92**, 307–336.
–, 1973b: Die Betalaine, eine neue Klasse chymochromer Pigmente. Pharmazie **28**, 18–24.
–, 1975: Kinetik der durch Inhibitoren des Protein- und Nukleinsäurestoffwechsels bewirkten Hemmung der Amaranthinbildung bei Amaranthus caudatus-Keimlingen. BPP **168**, 113–122.
–, and K. Conrad, 1966: Ein quantitativer Cytokinintest. Biol. Rundschau **5**, 36–37.

Liebisch, H. W., B. Matschiner, and H. R. Schütte, 1969: Beiträge zur Physiologie und Biosynthese des Betanins. Z. f. Pflanzenphys. **61**, 269–278.

Mabry, T. J., 1973: Is the order Centrospermae monophyletic? Nobel 25, Chemistry in botanical classification, 275–285.
–, A. Taylor, and B. L. Turner, 1963: The betacyanins and their distribution. Phytochem. **2**, 61–64.
–, L. Kimler, and R. A. Larson, 1972a: Hoppe-Seylers Z. phys. Chem. **353**, 127.
–, –, and C. Chang, 1972b: The betalains: Structure, function, and biogenesis, and the plant order Centrospermae. In: V. C. Runeckles, and T. C. Tso (eds.), Recent Advances in Photochem. **5**, 105–134, New York.
–, and A. S. Dreiding, 1968: The Betalains. In: T. J. Mabry, R. E. Alston, and V. C. Runeckles (eds.), Recent Advances in Phytochem. **1**, 145–160. Amsterdam and New York.
–, I. J. Eifert, C. Chang, H. Mabry, C. Kidd, and H. D. Behnke, 1975: Theligonaceae: Pigment and ultrastructural evidence which excludes it from the Order Centrospermae. Biochem. Syst. **3**, 53–55.

Miller, H. E., H. Rösler, A. Wohlpart, H. Wyler, M. E. Wilcox, H. Frohofer, T. J. Mabry, and A. S. Dreiding, 1968: Biogenese der Betalaine. Biotransformation von DOPA und Tyrosin in den Betalaminsäureanteil des Betanins. Helv. Chim. Acta **51**, 1470–1474.

Minale, L., M. Piattelli, and R. A. Nicolaus, 1965: Pigments of Centrospermae. IV. On the biogenesis of indicaxanthin and betanin in Opuntia ficus-indica Mill. Phytochem. **4**, 593–597.
–, –, S. de Stefano, and R. A. Nicolaus, 1966: Pigments of Centrospermae. VI. Acylated betacyanins. Phytochem. **5**, 1037–1052.
–, –, and –, 1967: Pigments of Centrospermae VII. Betacyanins from Gomphrena globosa L. Phytochem. **6**, 703–709.

Misawa, M., 1977: Production of natural substances by plant cell cultures described in Japanese patents. 17–26. In: W. Barz, E. Reinhard, and M. H. Zenk (eds.), Plant tissue culture and its bio-technological application. Springer Verlag, Berlin-Heidelberg-New York.

Piattelli, M., M, Giudici de Nicola, and V. Castrogiovanni, 1969: Photocontrol of Amaranthin Synthesis in Amaranthus tricolor. Phytochem. **8**, 731–736.
–, –, and –, 1972: The effect of kinetin on amaranthin synthesis in Amaranthus tricolor in darkness. Phytochem. **10**, 289–293.
–, and G. Impellizzeri, 1970: 2-Descarboxybetanidin, a minor betacyanin from Carbobrotus acinaciformis. Phytochem. **9**, 2553–2556.
–, and F. Imperato, 1969: Betacyanins of the Family Cactaceae. Phytochem. **8**, 1503–1507.
–, and –, 1970: Betacyanins from Bougainvillea. Phytochem. **9**, 455–458.
–, and –, 1970: Pigments of Bougainvillea glabra. Phytochem. **9**, 2557–2560.
–, and –, 1971: Betacyanins of some Chenopodiaceae. Phytochem. **10**, 3133–3134.

—, and L. Minale, 1964: Pigments of Centrospermae I. Betacyanins from Phyllocactus hybridus hort. and Opuntia ficus-indica Mill. Phytochem. **3**, 307–311.

—, —, and G. Prota, 1964: Isolation, structure and absolute configuration of indicaxanthin. Tetrahedron **20**, 2325–2329.

—, —, and —, 1965: Pigments of Centrospermae – III. Betaxanthins from Beta vulgaris L. Phytochem. **4**, 121–125.

—, and M. Giudici de Nicola, 1970: The inhibition by Actinomycin D and Pyromycin of Light-Stimulated Amaranthin Synthesis. Phytochem. **9**, 785–789.

—, 1976: Betalains. 560–596. In: T. W. Goodwin (Ed.) Chemistry and Biochemistry of Plant Pigments. Academic Press London-New York-San Francisco.

Rast, D., R. Skrivanova, and R. Bachofen, 1973: Replacement of light by dibutyryl-cAMP and c-AMP in betacyanin synthesis. Phytochem. **12**, 2669–2672.

Rauh, W., and H. Reznik, 1961: Zur Frage der systematischen Stellung der Didiereaceen. Bot. Jb. **81**, 94–105.

Reznik, H., 1955: Die Pigmente der Centrospermen als systematisches Element. Zeitschr. f. Bot. **43**, 499–530.

—, 1957: Die Pigmente der Centrospermen als systematisches Element. II. Untersuchungen über das ionophoretische Verhalten. Planta **49**, 406–434.

—, 1966: Zu Chemotaxonomie der Betacyane. Abhandl. d. Deutsch. Akad. d. Wiss. Berlin, Kl. f. Chemie, Geologie u. Biol. Jahrgang 1966, Nr. 3, 57–63.

—, 1975: Betalaine. Ber. Deutsch. Bot. Ges. **88**, 179–190.

—, 1978: Das Vorkommen von Betalaminsäure bei Centrospermen. Z. Pflanzenphysiol. **87**, 95–102.

Sciuto, S., G. Oriente, and M. Piattelli, 1972: Betanidin glucosylation in Opuntia dillenii. Phytochem. **11**, 2259–2262.

—, —, —, G. Impellizzeri, and V. Amico, 1974: Biosynthesis of amaranthin in Celosia plumosa. Phytochem. **13**, 947–951.

Stobart, A. K., and L. T. Kinsman, 1977: The hormonal control of betacyanin synthesis in Amaranthus caudatus. Phytochem. **16**, 1137–1142.

Wilcox, M. E., H. Wyler, and A. S. Dreiding, 1965: Derivate der Betalaminsäure. Helvet. Chim. Acta **48**, 1922–1927.

Wohlpart, A., and S. M. Black, 1973: Accumulation of betanin in disks of Beta vulgaris leaves. Phytochem. **12**, 1325–1329.

—, and T. J. Mabry, 1968: On the light requirement for betalain biogenesis. Plant Physiol. **43**, 457–459.

—, and —, 1968: The distribution and phylogenetic significance of the betalains with respect to the Centrospermae. Taxon **17** (2), 148–152.

Woodhead, S., and T. Swain, 1974: Effect of light on betalain and cinnamic acid biosynthesis in Amaranthus caudatus. Phytochem. **13**, 953–956.

Wyler, H., and A. Dreiding, 1961: Über die Betacyane, die stickstoffhaltigen Farbstoffe der Centrospermen. Experientia **17**, 23–25.

Pigments of Higher Fungi (Macromycetes)

W. Steglich

Institut für Organische Chemie und Biochemie der Universität Bonn

From the beginning of mycology colour and colour reactions of fruiting bodies from higher fungi have been used as valuable taxonomic characters. The elucidation of the chemical background of these phenomena was hampered for a long time by the sensitivity of many pigments and their tendency to form complex mixtures. The advent of modern separation techniques and spectroscopy has led to rapid progress in this field, and during the last decade many pigments of different chemical classes have been elucidated [1, 2, 3, 4]. In several cases these compounds are closely related to each other which makes it appropriate to discuss them with respect to their probable biogenesis. It should be mentioned, however, that rigorous proof of the biosynthesis has been obtained in only a few cases.

1. Pigments of the Shikimate Pathway

Many fungi have developed pathways for the conversion of aromatic amino acids (phenylalanine, tyrosine) into pigments. The first steps in these syntheses are well known reactions of primary metabolism which lead to arylpyruvic acids, L-dopa, hydroxycinnamic acids, hydroxycinnamyl alcohols, and *p*-hydroxybenzoic acid [5, 6].

1.1. Terphenylquinone Group

The most important reaction for the biosynthesis of fungus pigments containing benzene rings is the condensation of two molecules of arylpyruvic acid leading to terphenylquinones [7]:

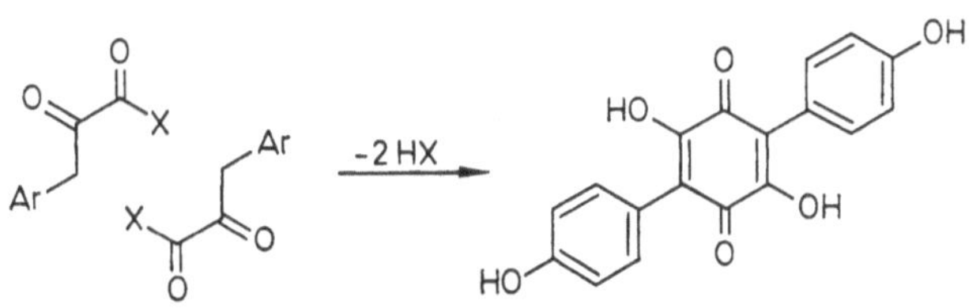

1 R=H
2 R=OH

Compounds of this type, such as polyporic acid (*1*) and atromentin (*2*), have been known for over 100 years [4]. They may be present in fruiting bodies in surprisingly high concentrations; *1* constitutes over 20 percent of the dry weight of *Hapalopilus nidulans*.

The basic system may be modified in numerous ways. *O*-Acylation leads to orange red pigments, such as aurantiacin (*3*) [4] which has been isolated from *Hydnellum* species. Like other terphenylquinones it is present in the fruiting bodies partially as its leuko compound. Phlebiarubrone (*4*) [4] from cultures of *Phlebia strigoso-zonata* contains an *o*-quinone unit fixed by a methylene bridge. By means of oxidative coupling, the terminal 3,4-dihydroxyphenyl residues may react with the hydroxy groups

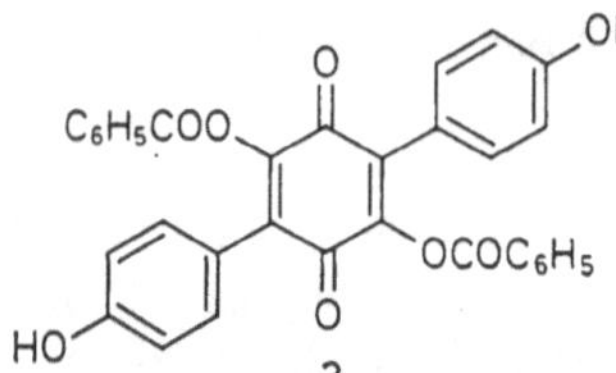

3

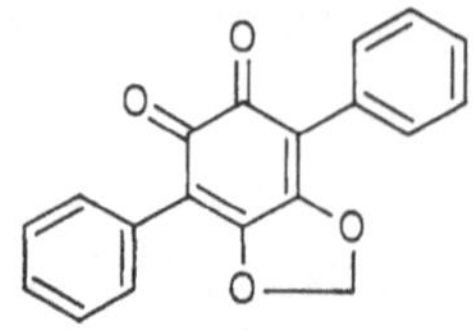

4

at the quinone nucleus to form dibenzofurane derivatives like cycloleucomelone (*5*) [8], cyclovariegatin (*6*) [9], and thelephoric acid (*7*) [4]. Some of these quinones may be recognized by their striking colour changes when treated with aqueous base (*1*: violet, *5*: green; *7*: blue). A more radical change of the basic system appears in hydnuferrugin (*8*) [10] from *Hydnellum* species. The 3,4-dihydroxyphenyl ring of *6* has undergone oxidative ring opening to yield an aldehydo acid, which subsequently cyclizes to *8*.

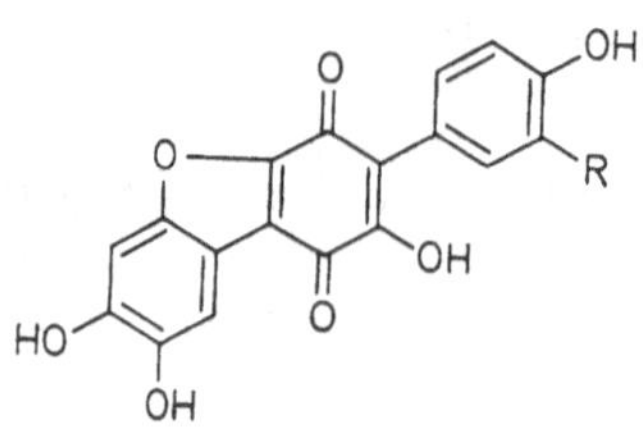

5 R = H
6 R = OH

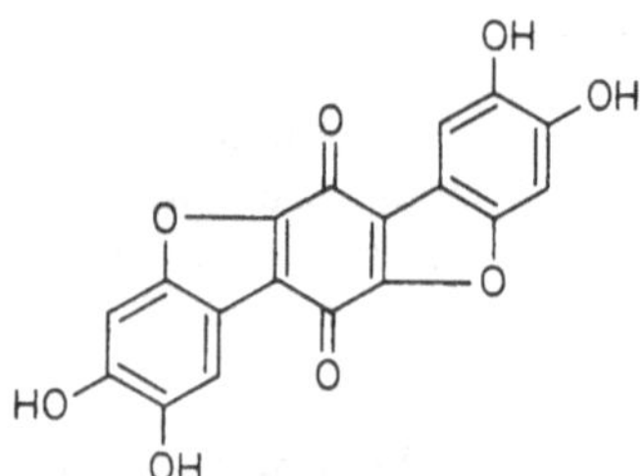

7

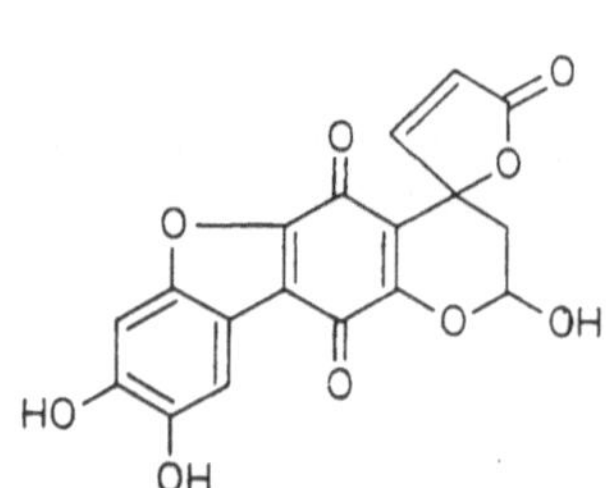

8

Fungi of the order Boletales derive new types of pigments preferably by oxidation of the quinone moiety. Most frequently, terphenylquinones are oxidatively split to yield pulvinic acids [11]:

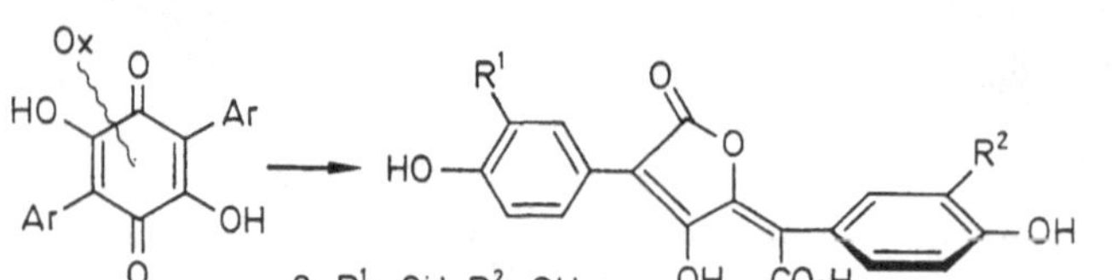

9 R^1 = OH, R^2 = OH
10 R^1 = OH, R^2 = H
11 R^1 = OCH_3, R^2 = OH

Hydroxylated pulvinic acids are responsible for the yellow and red colours of many boletes. Their occurence is restricted to the order Boletales – thus their isolation from

Gomphidius [12], *Rhizopogon* [13] and *Serpula* [14] gives strong chemotaxonomic evidence for the inclusion of these genera in Boletales [14a].

The most common product of this type is variegatic acid (*9*) [15], which causes the typical bluing of several boletes on injury. Compound *9* is thereby oxidized to a quinone methide which is present as its blue anion *12* even at neutral pH, due to its nature as vinylogous carboxylic acid [16]. Even on exposure to air, *9* is slowly transformed into variegatorubin (*13*) [16a], the red pigment of boletes. The remarkable bathochromic shift of the long-wave absorption as a result of this oxidation is caused by flattening of the chromophore on closure of the lactone ring.

9 (λ_{max} = 415 nm)

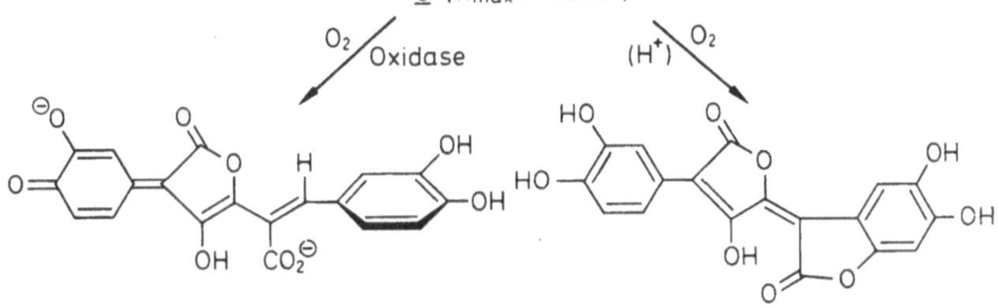

12 (λ_{max} = 595 nm) 13 (λ_{max} = 510 nm)

Xerocomic acid (*10*) shows considerable antibiotic activity against several bacteria [17]. 3-*O*-Methylvariegatic acid (*11*) [18] from cultures of *Hygrophoropsis aurantiaca* is a rare example of 4-hydroxy-3-methoxyphenyl substitution in higher fungi. This substitution pattern could be due to a precursor formed by degradation of pine-wood lignin which is added to the medium.

Some species from Boletales have developed an alternative pathway for oxidation of the dihydroxyquinone ring leading to cyclopentatrione derivatives under ring contraction:

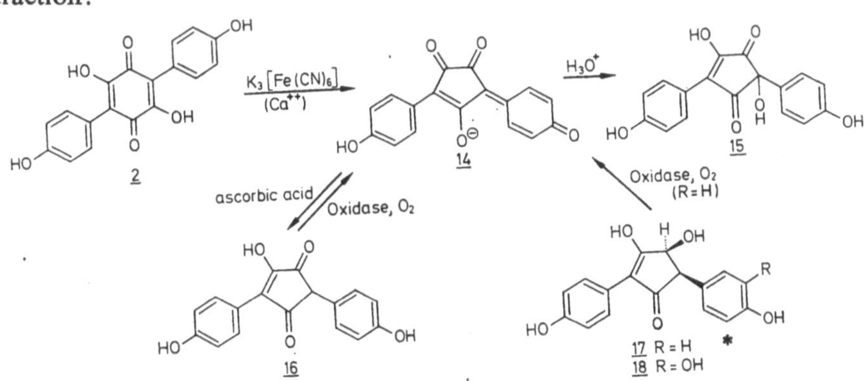

This hypothetical pathway [19] is supported by the oxidation of atromentin (*2*) with K_3 [Fe(CN)$_6$] in the presence of calcium ions yielding the blue anion *14*. Addition of mineral acid or ascorbic acid to the solution leads to the formation of the pale

* Only relative configuration indicated. *17* and *18* are of opposite absolute configuration.

yellow pigments gyroporin (*15*) and gyrocyanin (*16*) respectively [20]. Both occur in fruiting bodies of *Gyroporus cyanescens*, the latter compound being responsible for their intensive bluing on injury. In the gastromycete *Chamonixia caespitosa* the blue anion *14* is produced mainly by oxidation of the colourless chamonixin (*17*) [21]. A related chromogen, involutin (*18*) [22], is present in *Paxillus involutus* where it is responsible for the brown staining of the fruiting bodies on bruising.

Structurally related to the terphenylquinones are several red pigments isolated by Gripenberg from wood infected with *Peniophora sanguinea*. These pigments, e.g., xylerythrin (*19*) [23], are quinone methides which may formally be considered as condensation products of terphenylquinones with *p*-hydroxyphenylacetic acid. Feeding experiments, however, indicate a more complicated biosynthesis [24]. Recently an additional yellow pigment, penioflavin (*20*) [25], has created further problems in considering biogenesis.

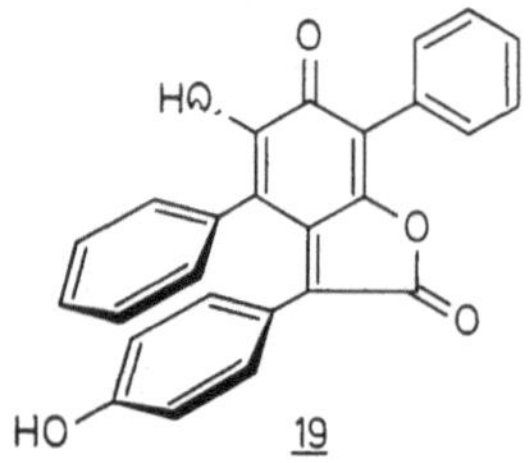

19

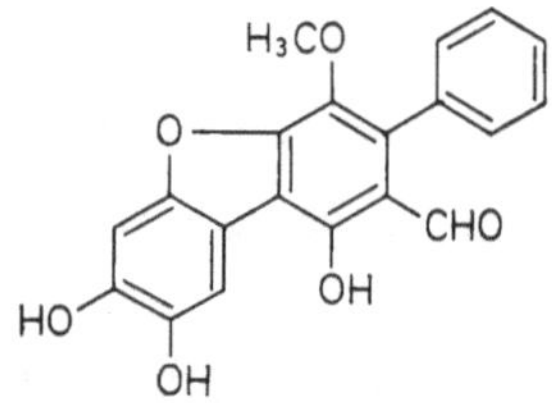

20

Before leaving the terphenylquinone group, attention should be drawn to the grevillins, orange yellow pyrone derivatives from boletes of the genus *Suillus*, which may be considered as terphenylquinones with incomplete biosynthesis. The two arylpyruvic acid units undergo only one C-C bond formation yielding a triketo acid, which subsequently stabilizes under lactone formation:

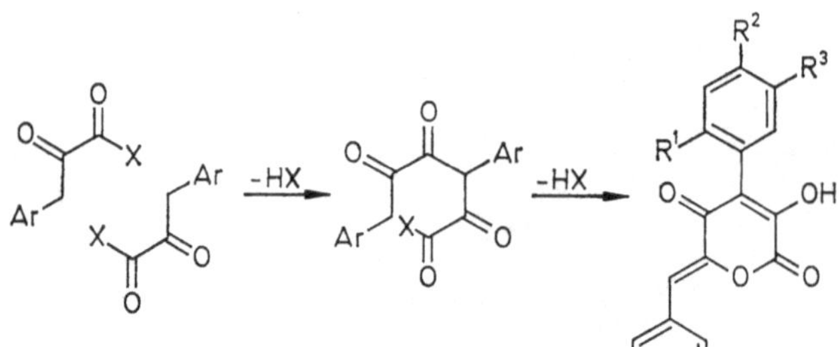

21 R^1=H, R^2=OH, R^3=H, R^4=H
22 R^1=H, R^2=OH, R^3=H, R^4=OH
23 R^1=H, R^2=OH, R^3=OH, R^4=OH
24 R^1=OH, R^2=H, R^3=OH, R^4=OH

So far, grevillins A, B, C and D (*21–24*) [26, 27] have been isolated; the latter possesses an unusual 2,5-dihydroxy pattern in ring A. This may be explained by assuming a biosynthesis *via* oxidative rearrangement of grevillin B [2, 27]. The close connection of the grevillins with terphenylquinones is demonstrated by the quantitative conversion of grevillin B trimethylether (*25*) into leucomelone trimethylether (*26*) on treatment with methanolic methoxide [28]:

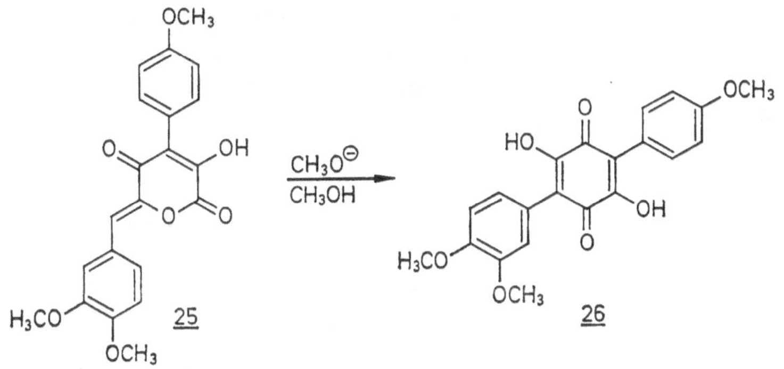

25 26

1.2. Betalains and Muscaflavin

Several toadstools are able to convert tyrosine into L-dopa. This may be stored in the fruiting bodies and is responsible for the remarkable colour reactions (→ red → black) due to melanin formation, which occur on injury (*Strobilomyces floccopus*, *Hygrocybe conica*) [2].

Some *Amanita* species like the well known «fly agaric» (*A. muscaria*) produce yellow, orange, red and purple pigments, which were shown by Döpp and Musso [29] to be betalains – pigments derived from L-dopa by oxidative ring opening – typical for Centrospermae of higher Plants. In *Amanita* some unusual betalains are produced by condensation of betalamic acid (*27*) with rare amino acids present in these fungi. Thus ibotenic acid is used for the formation of musca-aurin-I (*28*), one of the major pigments in *A. muscaria* [29]. Of special interest is the yellow dihydroazepine derivative muscaflavin (*29*) [30], an isomer of *27* which results from an alternative attack of dioxigenase at the catechol unit. Compound *31* has hitherto only been found in *Amanita* and *Hygrocybe* (Agaricales).

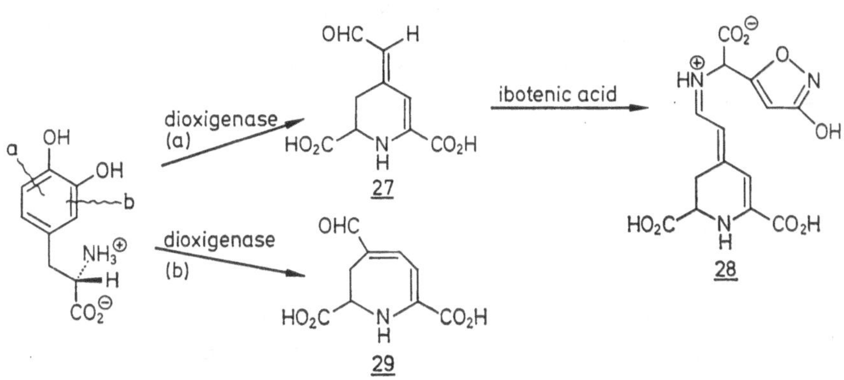

1.3. Styrylpyrone Group

Some wood-inhabiting fungi from Aphyllophorales (*Inonotus, Phaeolus, Phellinus, Hymenochaete*) and Agaricales (*Pholiota, Hypholoma, Gymnopilus*) produce yellow styrylpyrones, such as bisnoryangonin (*31*) [31] and hispidin (*32*) [32]. Biosynthetic investigations with cultures of *Inonotus hispidus* [33] and *Phaeolus schweinitzii* [34] have proven that these pigments are derived from *p*-coumarate and caffeate, respectively, by condensation with two molecules of malonyl-CoA followed by cyclisation of the resulting diketo acid-CoAs.

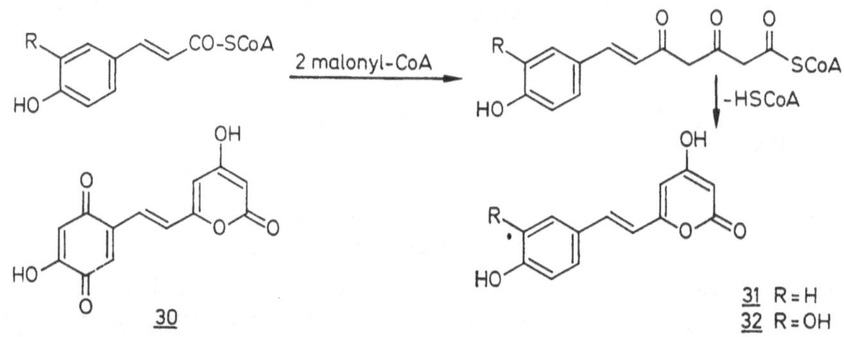

Alternately, *32* can be formed from *31* by direct hydroxylation.

Hispidin (*31*) may be transformed into more elaborate pigments by oxidative processes. Hydroxylation and oxidation give rise to red hymenoquinone (*30*) [34a], responsible for the ox-blood-like colour of the fruiting bodies from *Hymenochaete mougeotii*. Apart from *32 Phellinus pomaceus*, a fungus causing white-rot on plum trees, also produces a minor pigment which was shown to be 3,14'-bihispidinyl (*34*) [*35*]. More complicated dimers are the hypholomins A and B (*36*, *37*) [36], bright-yellow pigments of the common « sulfur tuft » (*Hypholoma fasciculare*) and *Pholiota flammans* [37]. They are accompagnied by the strongly fluorescent fasciculins A and B (*38*, *39*) which lack the CH=CH unit.

The formation of these dimers may easily be explained by a primary oxidation of *32* to dehydrohispidin (*33*) which may attack a second molecule of styrylpyrone at C-3' by means of its electrophilic centers at C-14 (leading to *34*) or C-7 (leading to *35*). The quinone methide *35* subsequently undergoes ring closure to form the dihydrofuran ring of the hypholomins and fasciculins. It should be emphasized, however, that the arylpyrones necessary for the formation of the latter pigments have not yet been isolated from fungi.

Similar oxidation processes may be responsible for the formation of brown-black insoluble polymers (« fungal lignin ») from *32* in aging fruiting bodies of *I. hispidus* [38].

Polyketides whose biosynthesis involves the condensation of an aromatic starter unit with several malonyl-CoAs are probably precursors of the arylpolyene pigments geogenin (*40*) [39] – responsible for the intense yellow colour of *Hydnellum geoge-*

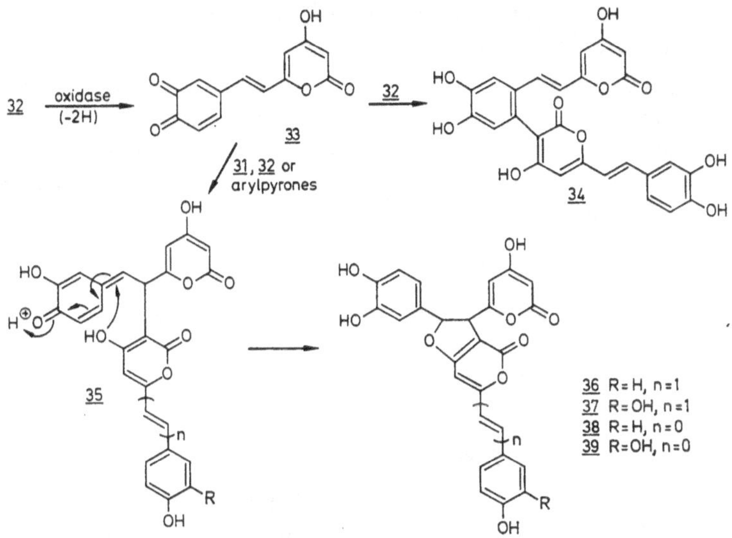

nium – and cortisalin (*41*) [40], a purple red pigment from fruiting bodies of *Cytidia salicina.* The yellow corticocrocin (*42*) [41] from *Corticium croceum* may be formed from a similiar pigment by oxidative degradation.

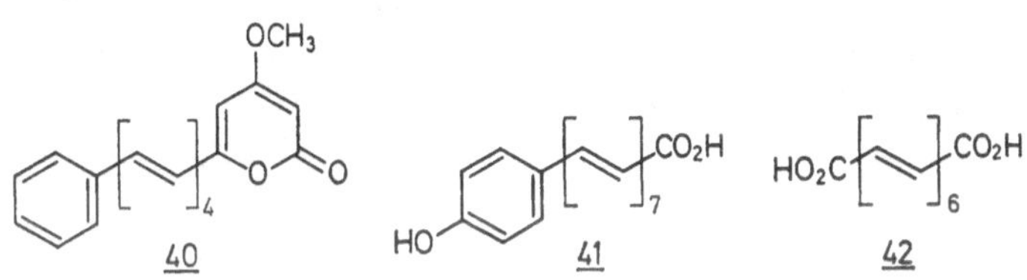

1.4. Purpurogallin Derivatives

Fomes fomentarius, living on beech wood, uses the oxidative dimerization of 2,3,4-trihydroxycinnamyl alcohol (*43*) to produce the brown red purpurogallin derivative fomentariol (*44*) [42].

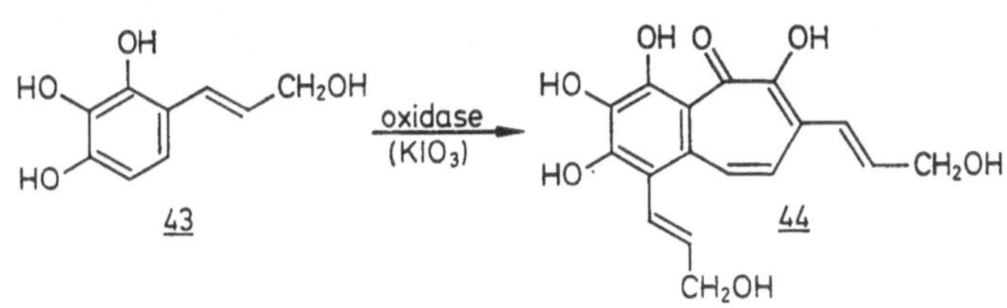

This oxidation may also be accomplished *in vitro* using as an oxidant KIO_3 or an aqueous extract from the fruiting bodies [43]. Compound *44* is accompanied by minor pigments with modified side chains [44].

1.5. Toluquinone Derivatives

The yellow pigment coprinin (*45*) has been isolated from cultures of *Coprinus radians* (= *C. similis*) [45], whereas *Lentinus cyathiformis* (= *L. degener*) produces both *45* and the hydroxy derivative *46* [46] in culture. Packter [47] has found that the quinone ring and the *C*-methyl group of *45* are derived from tyrosine; however, there have been contradictory results by others pointing towards a tetraketide origin [3, 48].

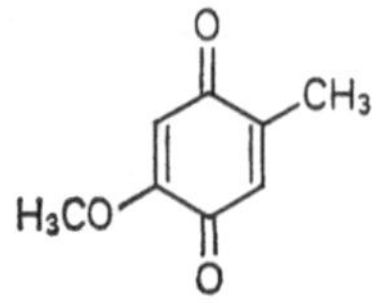

45

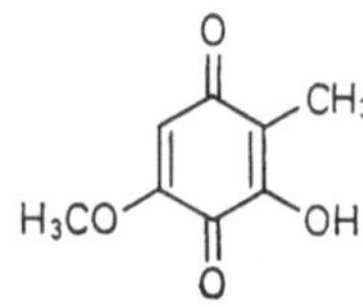

46

1.6. Boviquinone Group

Chroogomphus species and *Suillus bovinus* produce 2,5-dihydroxy-benzoquinones with polyprenyl side chains. The biogenetic origin of these pigments – boviquinone-3 (helveticone) (*47*) [49, 50] and boviquinone-4 (bovinone) (*48*) [51] – is apparently similar to that of the ubiquinones, which are derived from 4-hydroxybenzoic acid [52]. Again oxidative dimers are present as minor components: diboviquinone-4,4 (*49*) in *S. bovinus* and diboviquinone-3,4 (*50*) in *C. rutilus* [50]. Condensation with a C_1 unit may lead to methylenediboviquinone-4,4 (amitenone) (*51*) [4, 51] in *S. bovinus*, and methylenediboviquinone-3,3 and -3,4 (*52, 53*) in *C. rutilus* [50].

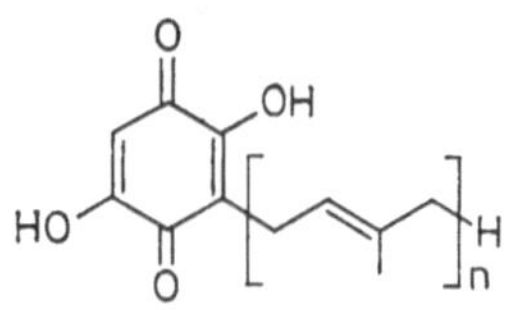

47 n = 3
48 n = 4

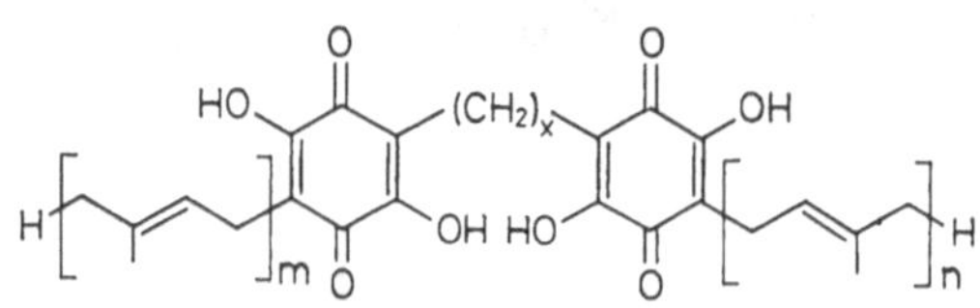

49 m = 4, n = 4, x = 0
50 m = 3, n = 4, x = 0
51 m = 4, n = 4, x = 1
52 m = 3, n = 3, x = 1
53 m = 3, n = 4, x = 1

A unique, naturally occurring ansa compound is tridentoquinone (*54*) [52a], the red pigment of *Suillus tridentinus*, which may be formed biogenetically from *48* via oxidative ring closure linking the terminal double bond to the quinone nucleus.

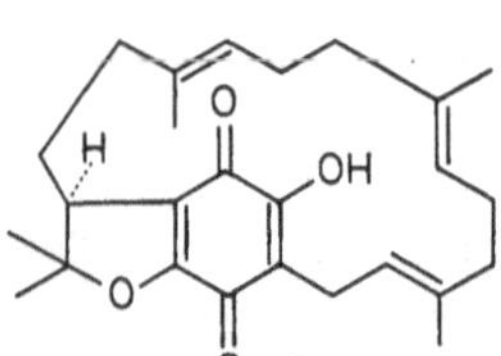

54

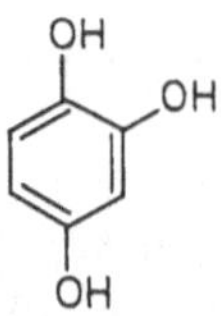

55

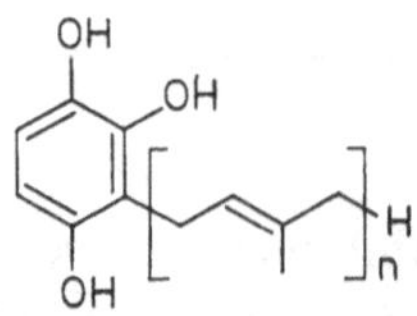

56

The simple chromogen 1,2,4-trihydroxybenzene (*55*) [53] causes the appearance of wine red stains which soon turn black when fruiting bodies of *Gomphidius maculatus* are injured. The isolation of its tetraprenyl derivative *56* – which occurs as a monoacetate in *Suillus granulatus* cap skins [54] – indicates a close biogenetic relationship with the boviquinones.

Whether 2,5-dimethoxybenzoquinone (thermophillin) (*57*), a yellow antibiotic from cultures of *Bjerkandera fumosa* (= *Polyporus fumosus*) [55] and *Lenzites thermophila* [56], is of similar origin can only be decided on the basis of biosynthetic experiments.

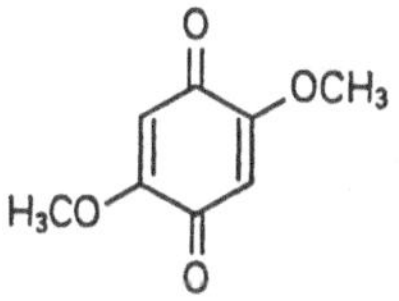

57

2. Pigments of the Acetate/Malonate Pathway

In contrast to its importance for the biosynthesis of metabolites in lower fungi [1, 3], the acetate/malonate pathway is only of limited importance in pigment production in macromycetes, as it is confined to a few basic systems. These systems will be discussed according to Turner's classification [3]. Most of the pigments of this group – with the exception of the anthraquinones – are formed by Ascomycetes (*Daldinia, Bulgaria, Hypoxylon, Chlorosplenium*) or Aphyllophorales (*Gloeophyllum Phlebia, Sarcodontia*) living on wood as substrate.

2.1. Tetraketides

Fruiting bodies of *Albatrellus ovinus* and *A. subrubescens* often show yellow discolourations which are caused by the presence of scutigeral (*58*) [57], a tetraprenylated aromatic trihydroxy aldehyde which shows intense yellow colour on dissolution in alkali. It is closely related to grifolin (*59*) [3], an antibiotic occurring in the same species and in *A. confluens*.

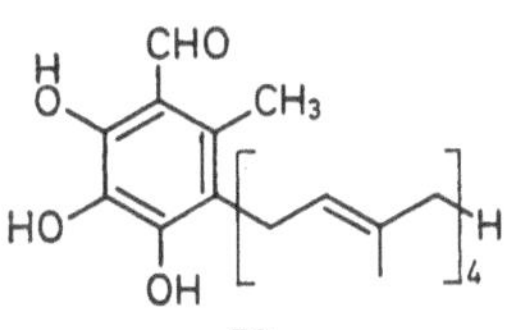

58

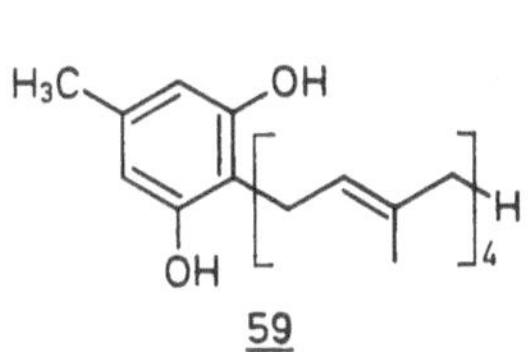

59

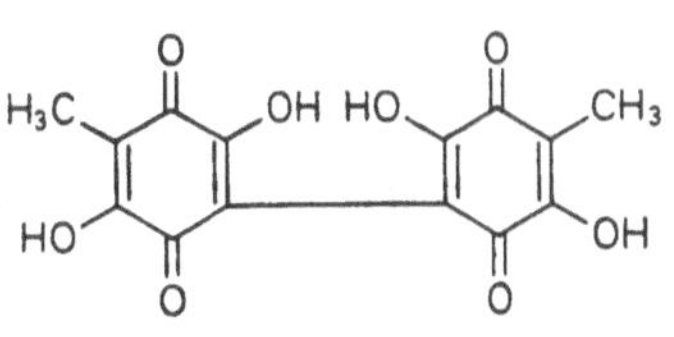

60

Oosporeine (*60*), a biquinone known from many Ascomycetes and Fungi Imperfecti has been isolated together with its leuko derivative from cultures of *Phlebia mellea* and *P. albida* [58].

Probably of tetraketide origin is a brown red pigment, trametin dimethyl ether (*63*), which is obtained from fruiting bodies of the wood-destroying fungi *Gloeophyllum* (*Osmoporus*) *odoratus* and *G. sepiarium* by extraction with methanol [59].

It has been demonstrated that the methoxy groups are introduced during the extraction procedure. From the failure of experiments to extract a similar compound with solvents like DMF it may be concluded that the pigment is present in the sporophore as a polymer, or bound to a polymeric substance. Its structure suggests a biosynthesis from two molecules of 2,4,5-trihydroxy-6-hydroxymethylbenzaldehyde (*62*) analogous to the *in vitro* fluorone synthesis of Hatsuda [60].

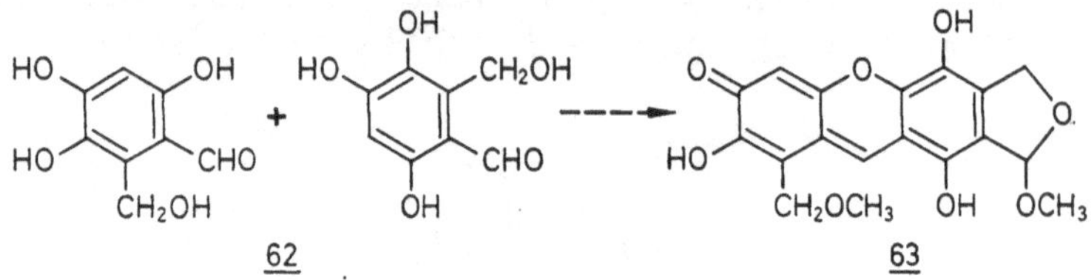

2.2. Pentaketides

Investigation of the brown black sporophores of *Daldinia concentrica* and *Bulgaria inquinans* (Ascomycetes) has led to the interesting discovery that both species produce pigments by oxidative coupling of 1,8-dihydroxynaphthalene (*64*). Bu'Lock [61] isolated mono and dimethyl ethers of *64* from a white mycelium of *D. concentrica*, which provides evidence that the dark red pigment 4,9-dihydroxyperylene-3,10-quinone (*65*) and the black polymers from the sporophores are formed via oxidation of *64* [4].

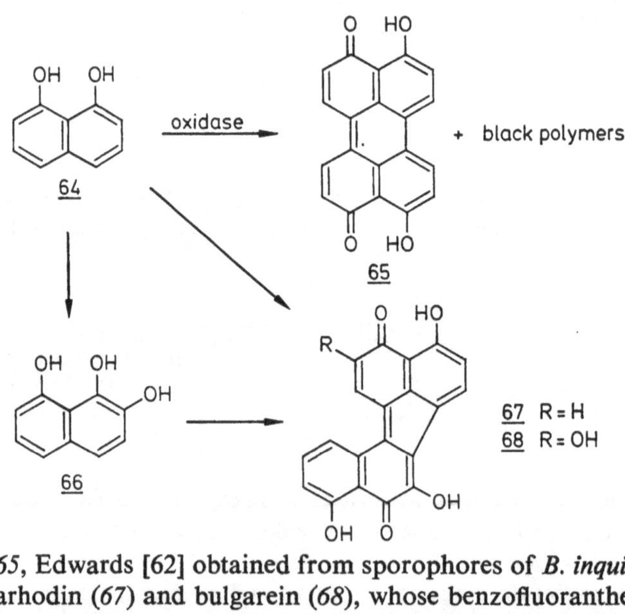

Apart from *65*, Edwards [62] obtained from sporophores of *B. inquinans* the purple pigments bulgarhodin (*67*) and bulgarein (*68*), whose benzofluoranthene system may result from an alternative mode of oxidation probably involving a 1,2,8-trihydroxynaphthalene precursor *66*.

There has been a reported isolation of 6-methyl-1,4-naphthoquinone (*69*) from cultures of *Marasmius graminum* [4]. The biosynthesis of this golden yellow pigment has not yet been elucidated.

2.3. Hexaketides

Azaphilones like (+)-mitorubin (*70*) and the closely related *71–73* are the colouring matters of *Hypoxylon fragiforme* fruiting bodies [63]. (−)-Mitorubrin and the enantiomers of *71* and *73* have been described before as metabolites of *Penicillium* [3].

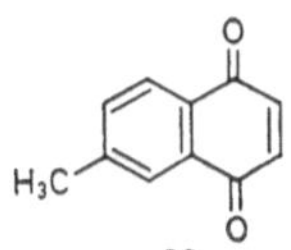

69

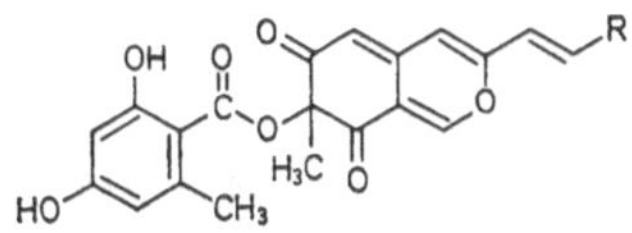

70 R = CH_3
71 R = CH_2OH
72 R = CH_2OAc
73 R = CO_2H

2.4. Octaketides

Anthraquinone pigments are found in great variety in toadstools of the genus *Dermocybe* (Agaricales) [64, 65, 66]. A number of biogenetically closely related pigments which are shown in the following scheme have been isolated from *D. sanguinea:*

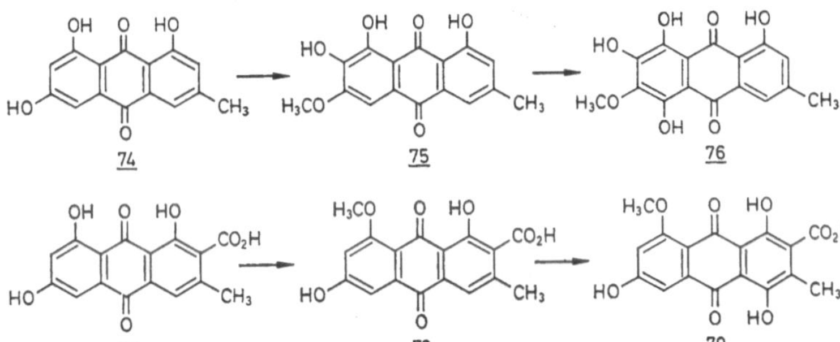

74 75 76
77 78 79

Most of the emodin (*74*) and dermocybin (*76*) are present in the form of the corresponding 1-glucoside, and the carboxylic acids dermolutein (*78*) and dermorubin (*79*) are accompagnied by their 5-chloro derivatives [4]. At first glance endocrocin (*77*) may be considered the biogenetic precursor of *74*. Feeding experiments, however, have provided evidence that ^{14}C-labeled emodin-6-glucoside was used for the biosynthesis of dermoglaucin (*75*) and *76*. This suggests that the common precursor of endocrocin and emodin appears earlier in the biosynthetic pathway [67].

During an investigation of the pigments of *Dermocybe cinnamomeolutea* and *Tricholoma equestre* a dimer, flavomannin-6,6′-dimethyl ether (*80*) has been found which contains the anthraquinone system at a premature stage [68]. Subsequently several yellow pigments of the same type were isolated from fruiting bodies of *Cortinarius vitellinus* [69] and *Tricholoma sulfureum* [70]. The presence of several centers of chirality creates stereochemical problems which will not be further discussed here [69].

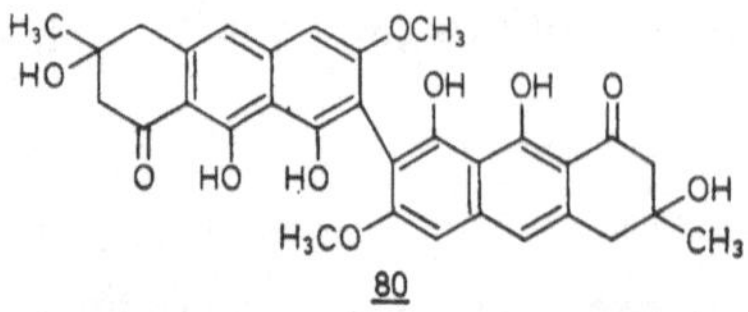

80

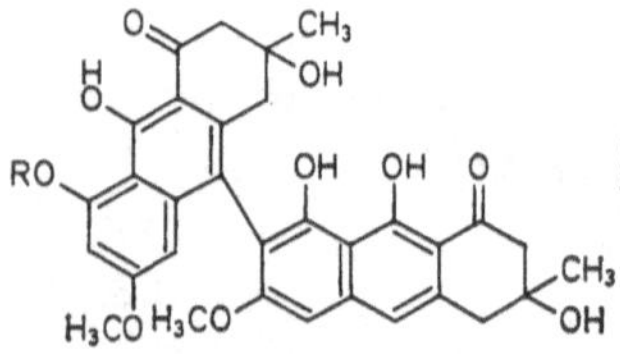

81 R = H
82 R = CH_3

The various shades of yellow and green of *C. odorifer* and *C. percomis* are produced by phlegmacin derivatives which contain the two monomers linked in 7,10′-position. In addition to phlegmacin (*81*) [71] and its 8′-methyl ether (*82*) [72] pigments were found in which the lower half of the molecule has been transformed into an anthrone or anthraquinone unit.

Fruiting bodies of *C. atrovirens* contain the 5,5′-dimers atrovirin (*83*) and skyrin (*84*) [73], the latter being a well known metabolite from fungi and lichens. This metabolite was also recently isolated, in addition to the condensed pigment hypericin (*85*), from dried fruiting bodies of *Dermocybe* collected in New Guinea [74].

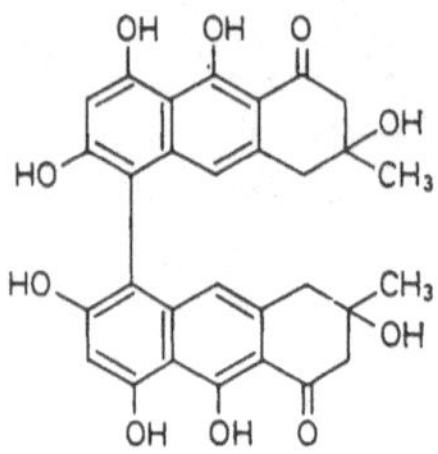

83

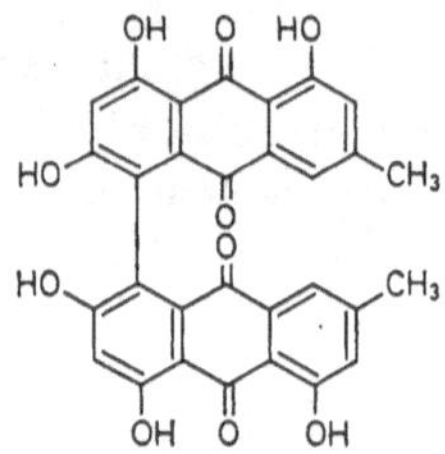

84

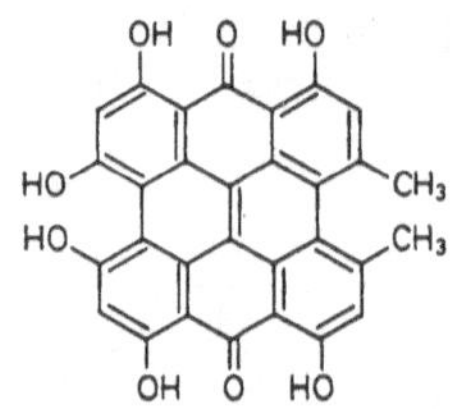

85

C. atrovirens also provided the key substance atrochrysone (*86*) [73] which may be the precursor of all the dimers and of the emodin family. Therefore, the following scheme for the biosynthesis of anthraquinoid compounds in higher fungi may be suggested:

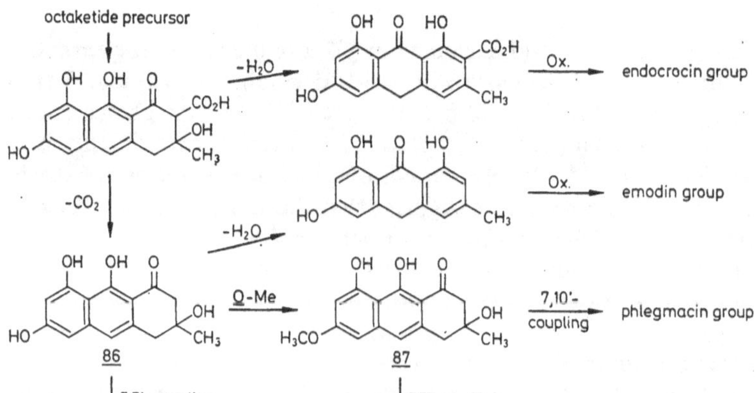

Recently two stereoisomers of phlegmacin (*81*) have been isolated from seedlings of *Cassia torosa* (Leguminosae) [75]. The monomer torosachrysone (*87*) is present in seeds of the same plant [76]; its isomer asperflavine (*88*) occurs in cultures of *Aspergillus flavus* [77]. The configuration of all these compounds from fungi and higher plants still remains to be established.

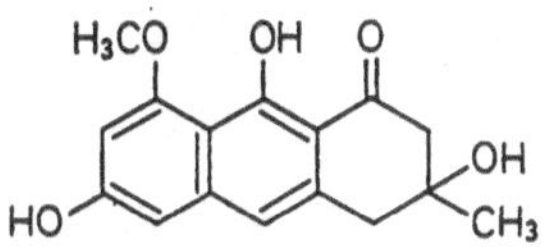

88

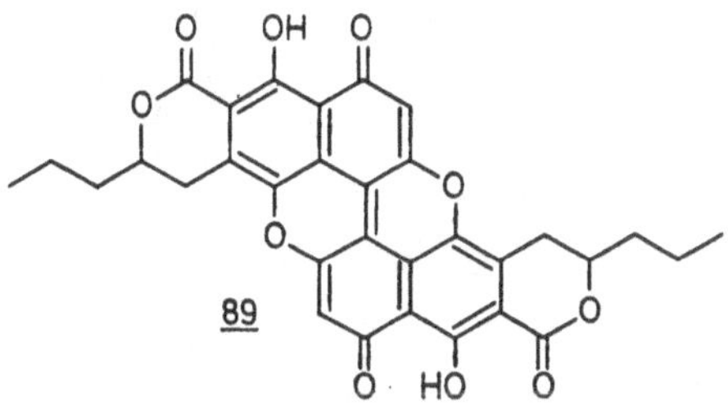

89

A completely different octaketide, obviously formed by oxidative dimerization of a unique precursor, is xylindeine (*89*) [4]. It is produced by *Chlorosplenium aeruginosum* (= *Chlorociboria aeruginosa*) and gives a beautiful green colour to the fruiting bodies and the dead wood on which this fungus lives.

2.4. Sarcodontic Acid

The sulfur yellow colour of fruiting bodies of *Sarcodontia setosa*, a fungus living on old apple trees, is due to the presence of sarcodontic acid (*90*) [78]. This unique compound is probably derived via the acetate/malonate pathway.

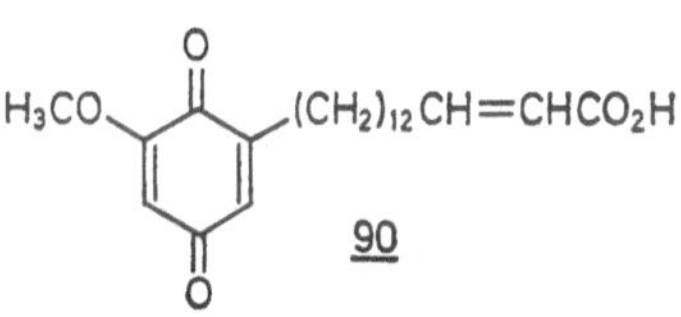

90

3. Pigments of the Mevalonate Pathway

3.1. Sesquiterpenes

Fruiting bodies of *Lactarius* species belonging to the sectio Dapetes [79] produce a latex which may show orange, red or blue colours. The genuine pigment of *L. deliciosus* is the orange yellow 14-hydroxyguai-1,3,5,9,11-pentaene palmitate (*91*) which may be obtained by applying especially careful isolation techniques at −60 °C [80]. On exposure to air and to enzymes present in the latex, *91* undergoes changes responsible for the discolouration of the milk upon cutting the mushrooms. Thus, earlier attempts to isolate the pigments yielded a number of secondary products which are shown in the following scheme [1, 3, 79, 80]. It should be mentioned, however, that some of these compounds may already be present in the intact mushrooms of related species, e.g., lactaroviolin and a still unknown pigment causing the red colour of latex from *L. sangifluus* [69]. Interestingly, Weiss and Harmon [81] recently found that the main pigment responsible for the magnificent blue colour of the latex from the American *L. indigo* is 14-hydroxylactarazulene palmitate, which is closely related to *91*. The lagopodins A, B, C and related benzoquinones have been isolated from cultures of *Coprinus* species, [4]. Their structures are basing on the cuparane skeleton.

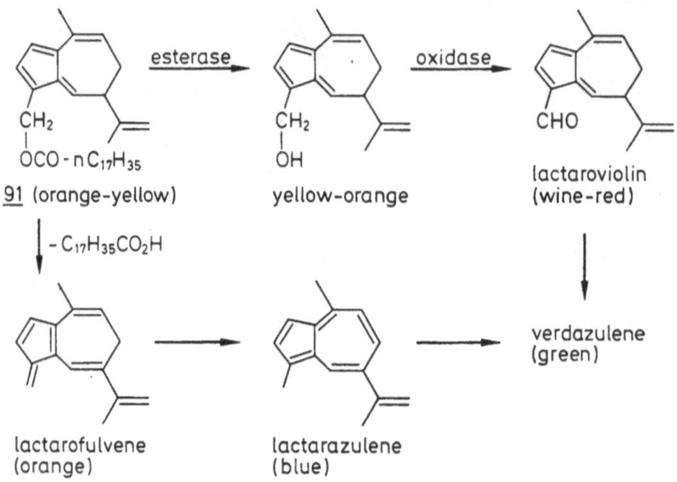

3.2. Carotinoids

Contrary to their wide distribution in the plant and animal kingdom, the occurence of carotinoids in Macromycetes is restricted to a few families and even to single species. Carotinoid pigments are characteristic for Ascomycetes from Pezizales (Pyronemataceae, Sarcoscyphaceae) and Helotiales (Helotiaceae, Hyaloscyphaceae, Bulgariaceae, Geoglossaceae), and for Basidiomycetes from Tremellales (*Tremella mesenterica*), Cantharellales (*Cantharellus*), Dacrymycetales (*Calocera, Dacrymyces*) and Phallales (*Anthurus, Mutinus*) [82, 83]. There are only two reports of carotinoids occuring in Agaricales (β- and γ-carotene as main pigments in fruiting bodies of *Clitocybe venustissima* [84] and *Gerronema* (= *Omphalia*) *chrysophylla* [84a]).

In most cases the predominant pigment is β-carotene, e.g., as the compound responsible for the colour of chantarelles (*Cantharellus cibarius*) [85]. Related species contain lycopene (*C. lutescens*) [85, 86, 87] and canthaxanthin (*92*) (*C. cinnabarinus*, an orange red American species) [88] as their main pigments.

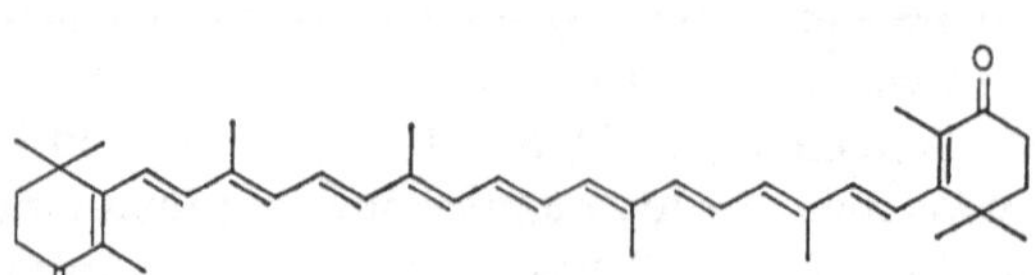

92

The intense red *Anthurus archeri* produces lycopene. In addition the biosynthetic precursors of this acyclic carotene, i. e. phytoene, phytofluene, ζ-carotene, and neurosporene could be isolated from the different phases of development of this Clatraceae [88a].

The beautiful colours of some cup fungi are partially caused by more unusual carotinoids. Thus, plectaniaxanthin (*93*) and 2′-dehydroplectaniaxanthin (*94*) are present in the brilliant red cups of *Sarcoscypha coccinea* where they occur mainly in the form of their di- and monoesters, e.g., with linoleic acid [83, 89]. Aleuriaxanthin (*95*) has been isolated from the orange apothecia of *Aleuria aurantia* [90]. β,γ-Carotene (*96*) with an unusual *exo*-methylene group is present in *Caloscypha fulgens* [91].

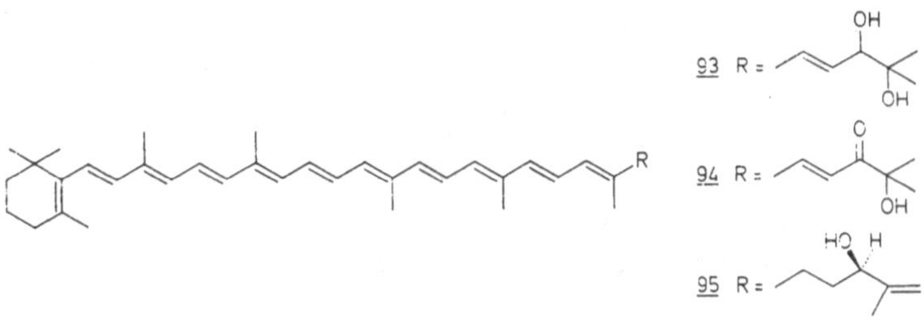

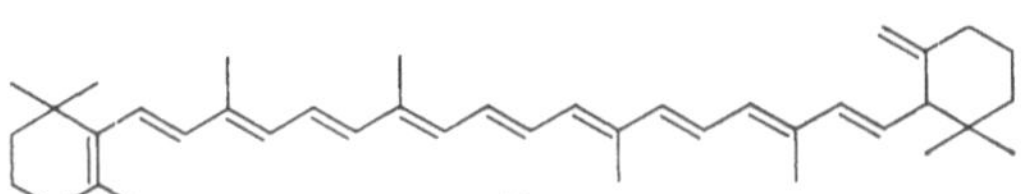

96

4. Nitrogen-containing Pigments

4.1. Phenoxazones

Orange phenoxazone (*97*) and yellow-brown 2-aminophenoxazone (*98*) have recently been isolated from cultures of *Calocybe gambosa* (Agaricales) [92]. More complex red pigments are known from the red sporophores of the wood-rotting *Pycnoporus cinnabarinus* and *P. sanguineus*. The structures of cinnabarin (*99*), cinnabarinic acid (*100*) and tramesanguin (*101*) [3] suggest a biosynthesis via oxidative coupling of 3-hydroxyanthranilic acid, a process for which there is well known laboratory analogy. 3-Hydroxyanthranilic acid is a degradation product of tryptophane.

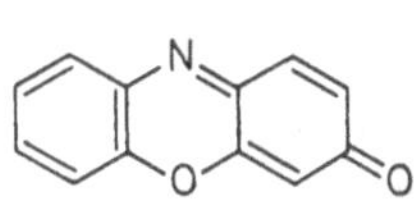

97

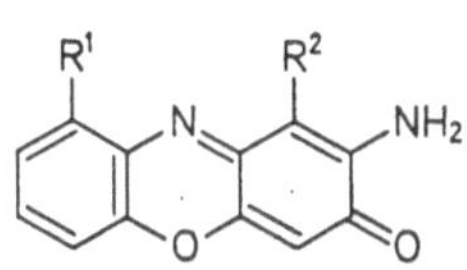

98 $R^1 = H$, $R^2 = H$
99 $R^1 = CH_2OH$, $R^2 = CO_2H$
100 $R^1 = CO_2H$, $R^2 = CO_2H$
101 $R^1 = CO_2H$, $R^2 = CHO$

4.2. Pteridine Group

Several species of *Russula* [93], *Lyophyllum* [94] and *Calocybe* [94] contain riboflavin (*102*) as a yellow, strongly fluorescent pigment. In *Panellus serotinus* only small quantities of *102* are present, the main pigment being its 3-methyl derivative *103* [95].

In careful investigations Eugster and coworkers [1, 93, 96] have shown that the intriguing pigments of *Russula* are closely connected with the riboflavin metabolism.

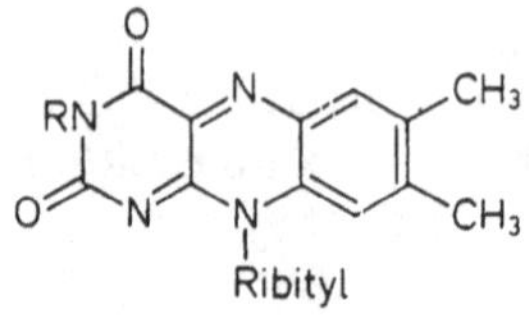

102 R = H
103 R = CH_3

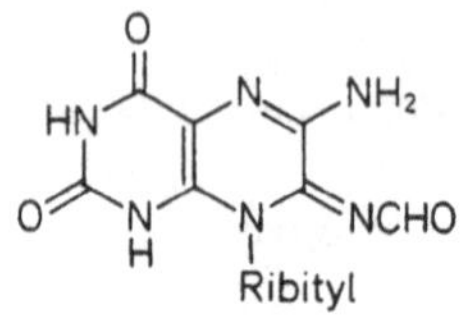

104

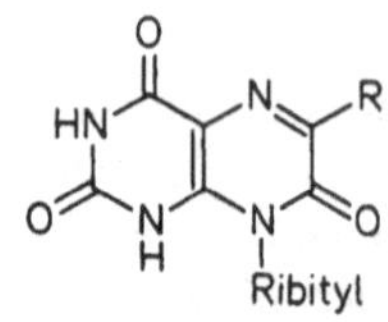

105 R = H
106 R = CH_3

From the complex mixtures of colourless, yellow, red and violet compounds they were able to separate yellow pigments, such as russupteridine-yellow$_{I}$ (*104*), which show intense yellow green fluorescence under UV. The colourless russupteridines *105*, *106* and *107* show blue fluorescence. They may be obtained as degradation products on acid hydrolysis of the more complex red russupteridines, for which dimeric structures presented as partial formula *108* have been suggested. Interestingly, the red dimers contain one ribose residue bound as *N*-glycoside.

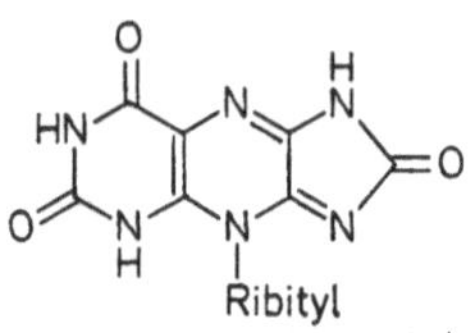

107

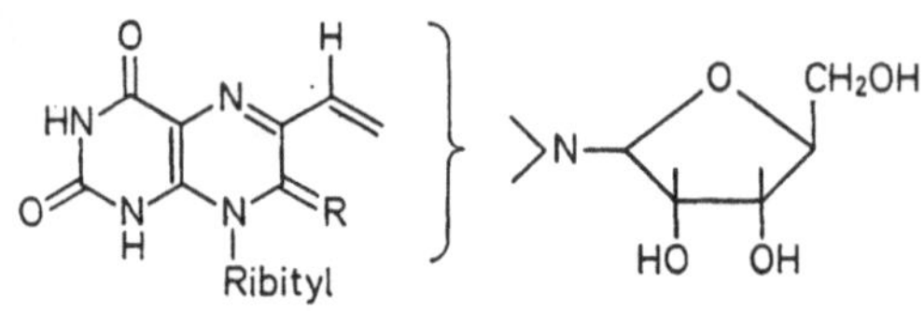

108

4.3. Miscellaneous Pigments

Indigo (*109*) has been reported to occur in cultures of artificial mutants of *Schizophyllum commune* [97] and *Agaricus campester* [98].

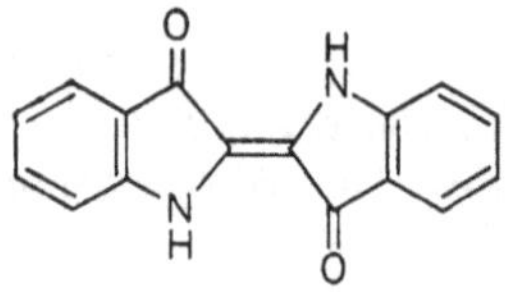

109

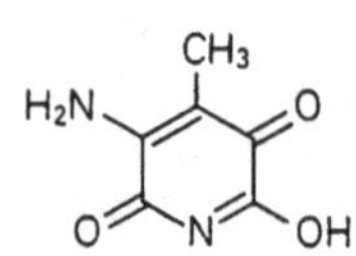

110

The intense green yellow fruiting bodies of *Rhodophyllus incanus* give dark green methanolic extracts from which blue, red and yellow pigments and fluorescent compounds may be separated by sephadex chromatography. The azaquinone formula *110* has been established for the yellow pigment [99].

The flesh of several *Agaricus* species shows a typical reddening on bruising. This is due to the presence of γ-glutaminyl-3,4-benzoquinone (*113*), which may be produced by the action on γ-glutaminyl-4-hydroxybenzene (*111*) of an enzyme with tyrosinase activity which is present in the fruiting bodies of *Agaricus bisporus* [100]. From

A. campester a similar precursor of *113* has been found in the form of the amino acid agaridoxin (*112*) [101].

Reddening due to the action of tyrosinase on tyrosine as the chromogen has been observed with *Russula nigricans* [102] and *Daedaleopsis confragosa* [103].

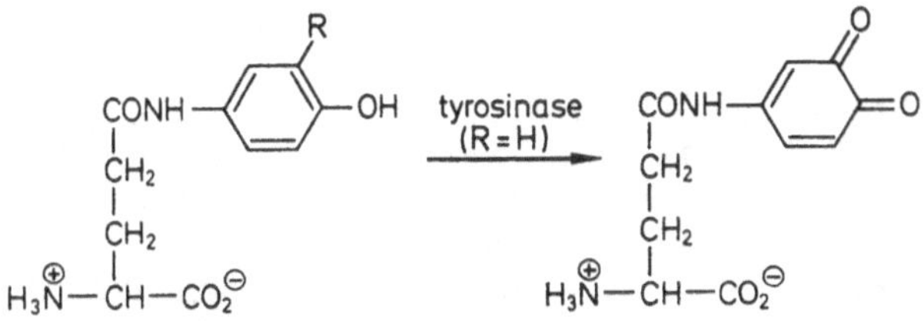

111 R = H
112 R = OH
113

I am greatly indebted to my coworkers who shared with me all the up and downs encountered by an chemist relying on the seasonal appearance of mushrooms. The progress achieved would not have been possible without their enthusiasm, endurance and skill. Of utmost importance was the kind assistance of my colleagues in biology, Professors A. Bresinsky, M. Moser, F. Oberwinkler and Drs. N. Arpin and H. Besl, to whom I would like to express my sincere gratitude. The valuable support of our work by the Deutsche Forschungsgemeinschaft is gratefully acknowledged.

References

[1] EUGSTER, C. H., 1973: Pilzfarbstoffe, ein Überblick aus chemischer Sicht mit besonderer Berücksichtigung der Russulae. Z. f. Pilzk. **30**, 45–96 (includes pigments from micromycetes).

[2] STEGLICH, W., 1975: Pilzfarbstoffe. Chemie in unserer Zeit **9**, 117–123 (discusses pigments derived from shikimate).

[3] TURNER, W. B., 1971: Fungal Metabolites. Academic Press, London, New York (contains numerous pigments arranged according to their biosynthesis).

[4] THOMSON, R. H., 1971: Naturally Occuring Quinones. Academic Press, London, New York (complete treatment of fungal quinones).

[5] HASLAM, E., 1974: The Shikimate Pathway. Butterworths, London.

[6] POWERS, G. H. N., 1969: Metabolism of Cinnamic Acid and its Derivatives in Basidiomycetes, in Perspectives of Phytochemistry, J. B. Harborne and T. Swain Edts., Academic Press, London, New York.

[7] BOSE, A. K., K. S. KHANCHANDANI, P. T. FUNKE, and M. ANCHEL, 1969: Chem. Commun. 1347; CHANDRA, P., G. READ, and L. C. VINING, 1966: Can. J. Biochem. **44**, 403. The incorporation of uniformly ^{14}C-labeled Phe into polyporic acid in fruiting bodies of *Hapalopilus nidulans* has been proved recently: Herrmann, R., and W. Steglich, inpublished.

[8] JÄGERS, E., H.-J. LOHRISCH, und W. STEGLICH: Present in fruiting bodies of *Anthracophyllum* sp. mainly as its diacetate (unpublished).

[9] EDWARDS, R. L., and M. GILL, 1975: J. C. S. Perkin I, 351.

[10] GRIPENBERG, J., 1974: Tetrahedron Letters 619.

[11] MAASS, W. S. G., and A. C. NEISH, 1967: Can. J. Bot. **45**, 59.

[12] Steglich, W., W. Furtner, und A. Prox, 1969: Z. Naturforsch. **24b**, 941.
[13] Steglich, W., I. Pils, und A. Bresinsky, 1971: Z. Naturforsch. **26b**, 37.
[14] Bresinsky, A., 1973: Z. Naturforsch. **28c**, 627.
[14a] Bresinsky, A., und P. Orendi, 1970: Z. f. Pilzk. **36**, 135; Bresinsky, A., 1974: Trav. mycol. dédiés à R. Kühner, Bull. Soc. Linnéenne de Lyon, S. 61; Besl, H., A. Bresinsky, und I. Kronawitter, 1975: Z. f. Pilzk. **41**, 81.
[15] Edwards, R. L., and G. C. Elsworthy, 1967: Chem. Commun. 373; Beaumont, P. C., R. L. Edwards, and G. C. Elsworthy, 1968: J. Chem. Soc. C, 2968.
[16] Steglich, W., W. Furtner, und A. Prox, 1968: Z. Naturforsch. **23b**, 1044.
[16a] Steglich, W., W. Furtner, und A. Prox, 1970: Z. f. Naturforsch. **25b**, 5.
[17] Quack, N., T. Anke, und F. Oberwinkler: unpublished.
[18] Besl, H., A. Bresinsky, L. Kopanski, und W. Steglich, 1978: Z. Naturforsch. **33c**, 820.
[19] Formation of *16* via cyclization of pulvinic acids [22] or the triketo acid intermediate of grevillin biosynthesis may also be considered.
[20] Besl, H., A. Bresinsky, W. Steglich, und K. Zipfel, 1973: Chem. Ber. **106**, 3223.
[21] Steglich, W., A. Thilmann, H. Besl, und A. Bresinsky, 1977: Z. Naturforsch. **32c**, 46.
[22] Edwards, R. L., G. C. Elsworthy, and N. Kale, 1967: J. Chem. Soc. C, 405.
[23] Gripenberg, J., and J. Martikkala, 1969: Acta Chem. Scand. **23**, 2583, and several other publications of J. Gripenberg.
[24] Massow, F. v., 1977: Phytochemistry **16**, 1696.
[25] Gripenberg, J., 1978: Acta Chem. Scand. Ser. B **32**, 75.
[26] Steglich, W., H. Besl, und A. Prox, 1972: Tetrahedron Lett. 4895; Edwards, R. L., and M. Gill, 1973: J. C. S., Perkin I, 1921.
[27] Besl, H., I. Michler, R. Preuss, und W. Steglich, 1974: Z. Naturforsch. **29c**, 784.
[28] Lohrisch, H.-J., und W. Steglich: unpublished; Lohrisch, H.-J., 1978: Thesis, Universität Bonn.
[29] Döpp, H., und H. Musso, 1973: Chem. Ber. **106**, 3473; Döpp, H., und H. Musso, 1974: Z. Naturforsch. **29c**, 640.
[30] Ardenne, R. v., H. Döpp, H. Musso, und W. Steglich, 1974: Z. Naturforsch. **29c**, 637.
[31] Hatfield, G. M., and L. R. Brady, 1968: Lloydia, **31**, 225; a chemotaxonomic survey on the occurence of styrylpyrones is given by Gluchoff-Fiasson, K., and Kühner, R., 1977: C. R. Acad. Sc. Paris, Série D, 1667.
[32] Edwards, R. L., D. G. Lewis, and D. V. Wilson, 1961: J. Chem. Soc. C, 4995.
[33] Perrin, P. W., and G. H. N. Towers, 1973: Phytochemistry **12**, 589; Towers, G. H. N., C. P. Vance, and A. M. D. Nambudiri, 1974: Recent Adv. Phytochem. **8**, 81; Nambudiri, A. M. D., C. P. Vance, and G. H. N. Towers, 1974: Biochim. et Biophys. Acta **343**, 148.
[34] Hatfield, G. M., and L. R. Brady, 1973: Lloydia **36**, 59.
[34a] Klaar, M., und W. Steglich, 1977: Chem. Ber. **110**, 1063.
[35] Klaar, M., und W. Steglich, 1977: Chem. Ber. **110**, 1058.
[36] Fiasson, J.-L., K. Gluchoff-Fiasson, und W. Steglich, 1977: Chem. Ber. **110**, 1047.
[37] Gluchoff-Fiasson, K., and J. Bernillon, 1977: C. R. Acad. Sc. Paris, Série D, **284**, 385.
[38] Bu'Lock, J. D., and H. G. Smith, 1961: Experientia **17**, 552; Bu'Lock, J. D., P. R. Leeming, H. G. Smith, 1962: J. chem. Soc. 2085.
[39] Steglich, W., und B. Jendrny, 1979: Festschrift R. Singer, Sydowia, Beiheft VIII, 378.

[40] GRIPENBERG, J., 1952: Acta Chem. Scand. **6**, 580.
[41] ERDTMAN, H., 1948: Acta Chem. Scand. **2**, 209.
[42] ARPIN, N., J. FAVRE-BONVIN, and W. STEGLICH, 1974: Phytochemistry **13**, 1949.
[43] STEGLICH, W., und L. ZECHLIN, unpublished; ZECHLIN, L., 1978: Thesis Universität Bonn.
[44] GRIPENBERG, J., L. HILTUNEN, and L. NIINISTÖ, 1976: Crystal Structure Comm. **5**, 571; FAVRE-BONVIN, J., K. KAOUADJI, and N. ARPIN, 1977: Phytochemistry **16**, 495; FAVRE-BONVIN, J. N. ARPIN, and J. GRIPENBERG, 1977: ibid. **16**, 1852.
[45] ANCHEL, M., A. HERVEY, F. KAVANAGH, J. POLATNICK, and W. J. ROBBINS, 1948: Proc. Natl. Acad. Sci. U.S. **34**, 498.
[46] PETTERSSON, G., 1966: Acta Chem. Scand. **20**, 45.
[47] PACKTER, N. M., 1969: Biochem. J. **114**, 369; PACKTER, N. M., and M. W. STEWARD, 1967: Biochem. J. **102**, 122.
[48] PETTERSSON, G., 1966: Acta Chem. Scand. **20**, 151.
[49] STEGLICH, W., F. ESSER, und I. PILS, 1971: Z. Naturforsch. **26b**, 336.
[50] BEAUMONT, P. C., and R. L. EDWARDS, 1971: J. Chem. Soc. C, 2582.
[51] BEAUMONT, P. C., and R. L. EDWARDS, 1969: J. Chem. Soc. C, 2398.
[52] BENTLEY, R., 1975: Biosynthesis of Quinones, in Specialist Periodical Reports, Biosynthesis Vol. 2, The Chem. Society, p. 206.
[52a] BESL, H., H.-J. HECHT, P. LUGER, V. PASUPATHY, und W. STEGLICH, 1975: Chem. Ber. **108**, 3675.
[53] ARDENNE, R. v., und W. STEGLICH, 1974: Z. Naturforsch. **29c**, 446.
[54] JÄGERS, E., V. PASUPATHY, and W. STEGLICH,: unpublished.
[55] BU'LOCK, J. D., 1955: J. Chem. Soc. 575.
[56] LEBLANC, G. D., and L.-M. BABINEAU, 1972: Canad. J. Microbiol. **18**, 261.
[57] BESL, H., G. HÖFLE, B. JENDRNY, E. JÄGERS, und W. STEGLICH, 1977: Chem. Ber. **110**, 3770.
[58] TAKESHITA, H., and M. ANCHEL, 1965: Science **147**, 152.
[59] ESSER, F., und W. STEGLICH, unpublished; ESSER, F., 1973: Thesis, Technische Universität Berlin.
[60] HATSUDA, Y., and S. KUYAMA, 1955: Nippon Nogeikagaku Kaishi **29**, 14 (1955); Chem. Abstr. **53**, 16126a (1959).
[61] ALLPORT, D. C., and J. D. BU'LOCK, 1960: J. Chem. Soc. 654; ibid., 1958: 4090.
[62] EDWARDS, R. L., and H. J. LOCKETT, 1976: J. C. S. Perkin I, 2149.
[63] STEGLICH, W., M. KLAAR, und W. FURTNER, 1974: Phytochemistry **13**, 2874.
[64] STEGLICH, W., W. LÖSEL, und V. AUSTEL, 1969: Chem. Ber. **102**, 4104.
[65] STEGLICH, W., und W. REININGER, 1972: Chem. Ber. **105**, 2922.
[66] MOSER, M., 1972: Schweizer Z. f. Pilzkunde **50**, 154 (1972); GRUBER, I., 1969: Thesis Universität Innsbruck; MOSER, M., unpublished.
[67] STEGLICH, W., R. ARNOLD, W. LÖSEL, und W. REININGER, 1972: Chem. Commun. 102; STEGLICH, W., 1972: Hoppe-Seyler's Z. Physiol. Chem. **353**, 124.
[68] STEGLICH, W., E. TÖPFER-PETERSEN, W. REININGER, K. GLUCHOFF, and N. ARPIN, 1972: Phytochemistry **11**, 3299.
[69] STEGLICH, W., und E. TÖPFER-PETERSEN, 1973: Z. Naturforsch. **28c**, 255.
[70] GLUCHOFF, K., and W. STEGLICH, 1974: Traveaux mycologiques dédiés à R. Kühner, numéro spécial de la Bull. Soc. Linnéene de Lyon, S. 163.
[71] STEGLICH, W., und E. TÖPFER-PETERSEN, 1972: Z. Naturforsch. **27b**, 1286.
[72] STEGLICH, W., E. TÖPFER-PETERSEN, und I. PILS, 1973: Z. Naturforsch. **28c**, 354.

[73] JÄGERS, E., und W. STEGLICH: unpublished.
[74] MOSER, M., und G. KELLER: unveröffentlicht.
[75] TAKAHASHI, S., S. KITANAKA, M. TAKIDO, U. SANKAWA, and S. SHIBATA, 1977: Phytochemistry **16**, 999.
[76] TAKIDO, M., S. TAKAHASHI, K. MASUDA, and K. YASUKAWA, 1977: Lloydia **40**, 191.
[77] GROVE, J. F., 1972: J. Chem. Soc. C, 2406.
[78] KŘEPINSKÝ, J., V. HEROUT, F. ŠORM, A. VYSTRČIL, R. PROKEŠ, and G. JOMM, 1965: Coll. Czech. Chem. Comm. **30**, 2626.
[79] SCHMITT, J. A., 1974: Z. f. Pilzkunde **39**, 219.
[80] VOKÁČ, K., Z. SAMEK, V. HEROUT, and F. ŠORM, 1970: Coll. Czech. Chem. Comm. **35**, 1296.
[81] WEISS, U., and HARMON, 1978: personal communication.
[82] FIASSON, J.-L., P. LEBRETON, and N. ARPIN, 1968: Bull. Soc. Nat. et Arch. de l'Ain, 47.
[83] ARPIN, N., 1968: Thèse, Universität Lyon.
[84] ARPIN, N., 1966: C. R. Acad. Sc. Paris **262**, 347.
[84a] FIASSON, J. L., and M. P. BOUCHEZ, 1968: C. R. Acad. Sc. Paris **266**, 1379.
[85] WILLSTAEDT, H., 1937: Svensk. Kem. Tids. **49**, 318.
[86] TURIAN, G., 1960: Arch. Mikrobiol. **36**, 139.
[87] FIASSON, J.-L., and N. ARPIN, 1967: Bull. Soc. Chim. Biol. Fr. **49**, 537.
[88] HAXO, F., 1950: Bot. Gaz. **112**, 228.
[88a] CZYGAN, F.-C., und M. GRÜNSFELDER, 1975: Z. Naturforsch. **30c**, 297.
[89] ARPIN, N., and S. LIAAEN-JENSEN, 1967: Phytochemistry **6**, 995; VACHERON, M. J., G. MICHEL, R. GUILLUY, and N. ARPIN, 1969: Phytochemistry **8**, 897.
[90] ARPIN, N., H. KJØSEN, G. W. FRANCIS, and S. LIAAEN-JENSEN, 1973: Phytochemistry **12**, 2751; BUCHECKER, R., N. ARPIN, and S. LIAAEN-JENSEN, 1976: Phytochemistry **15**, 1013.
[91] ARPIN, N., J.-L. FIASSON, M. P. BOUCHEZ-DANGYE-CAYE, G. W. FRANCIS, and S. LIAAEN-JENSEN, 1971: Phytochemistry **10**, 1595.
[92] SCHLUNEGGER, U. P., A. KUCHEN, und H. CLÉMENCON, 1976: Helv. Chim. Acta **59**, 1383.
[93] ITEN, P. X., S. ARIHARA, und C. H. EUGSTER, 1973: Helv. Chim. Acta **56**, 302.
[94] GLUCHOFF, K., 1974: C. R. Acad. Sci., Ser. D, **279**, 473.
[95] STEGLICH, W., und L. ZECHLIN, 1977: Z. Naturforsch. **32c**, 520.
[96] EUGSTER, C. H., E. F. FRAUENFELDER, und H. KOCH, 1970: Helv. Chim. Acta **53**, 131; EUGSTER, C. H., 1977: Abstracts of Papers, 2nd Internat. Mycol. Congress, Univ. of South Florida, Tampa, Florida, USA, S. 178.
[97] MILES, P. G., H. LUND, and J. R. RAPER, 1956: Arch. Biochem. Biophys. **62**, 1.
[98] FALANGHE, H., and P. A. BOBBIO, 1962: Arch. Biochem. Biophys. **96**, 430.
[99] BACKHAUS, J., I. TOMIDA, and W. STEGLICH: unpublished.
[100] WEAVER, R. F., K. V. RAJAGOPALAN, P. HANDLER, and W. L. BYRNE, 1971: J. Biol. Chem. **246**, 2015.
[101] SZENT-GYORGYI, A., R. H. CHUNG, M. J. BOYAJIAN, M. TISHLER, B. H. ARISON, E. T. SCHOENEWALDT, and J. J. WITTICK, 1976: J. Org. Chem. **41**, 1603.
[102] BOURQUELOT, E., and G. BERTRAND, 1895: C. R. Soc. Biol. **47**, 811.
[103] STEGLICH, W.: unpublished.

Subject Index

Index of Scientific Names

Author Index

FSC
www.fsc.org
MIX
Papier | Fördert gute Waldnutzung
FSC® C083411

Kolibri 360 GmbH
Ferdinand-Jühlke-Straße 7
99095 Erfurt, Deutschland
produktsicherheit@kolibri360.de